ANALYSIS OF
SYSTEMS
IN OPERATIONS
RESEARCH

PRENTICE-HALL INTERNATIONAL SERIES
IN INDUSTRIAL AND SYSTEMS ENGINEERING

W. J. Fabrycky and J. H. Mize, Editors

BLANCHARD *Logistics: Engineering and Management*
FABRYCKY, GHARE, AND TORGERSEN *Industrial Operations Research*
FRANCIS AND WHITE *Facility Layout and Location: An Analytical Approach*
GOTTFRIED AND WEISMAN *Introduction to Optimization Theory*
KIRKPATRICK *Introductory Statistics and Probability*
  for Engineering, Science, and Technology
MIZE, WHITE, AND BROOKS *Operations Planning and Control*
OSTWALD *Cost Estimating for Engineering and Management*
SIVAZLIAN AND STANFEL *Analysis of Systems in Operations Research*
SIVAZLIAN AND STANFEL *Optimization Techniques in Operations Research*
WHITEHOUSE *Systems Analysis and Design Using Network Techniques*

ANALYSIS OF SYSTEMS IN OPERATIONS RESEARCH

B. D. SIVAZLIAN
University of Florida

L. E. STANFEL
University of Texas at Arlington

PRENTICE-HALL, INC., *Englewood Cliffs, New Jersey*

Library of Congress Cataloging in Publication Data

SIVAZLIAN, B. D.
  Analysis of systems in operations research.

  Includes bibliographies.
  1. Operations research. 2. System analysis.
I. Stanfel. L. E., joint author. II. Title.
T57.6.S58   658.4'034   73–22411
ISBN 0–13–033498–7

TO FAYE

© 1975 by Prentice-Hall, Inc.
Englewood Cliffs, New Jersey

10 9 8 7 6 5 4 3 2 1

Printed in the United States of America

PRENTICE-HALL INTERNATIONAL, INC., *London*
PRENTICE-HALL OF AUSTRALIA, PTY. LTD., *Sydney*
PRENTICE-HALL OF CANADA, LTD., *Toronto*
PRENTICE-HALL OF INDIA PRIVATE LIMITED, *New Delhi*
PRENTICE-HALL OF JAPAN, INC., *Tokyo*

CONTENTS

PREFACE   viii

1 Probability Theory   2

1.1 Introduction   2
1.2 Sample Space, Probability, Conditional Probability,
    Independent Events, and Random Variables   3
1.3 Discrete Random Variables   7
1.4 Continuous Random Variables   39
1.5 The Central Limit Theorem   59
    Selected References   60
    Problems   61

2 Stochastic Processes   66

2.1 Introduction   66
2.2 Independent Processes   71
2.3 Independent Processes: Discrete Parameter   72
2.4 Independent Processes: Continuous Parameter   86
2.5 Markov Processes   92
2.6 Markov Chains: Discrete Parameter   94
2.7 Markov Chains: Continuous Parameter   111
2.8 Renewal Theory   129
    Selected References   136
    Problems   136

3 Reliability Theory   140

3.1 Introduction   140
3.2 The Failure Phenomenon; Exogenous- and Endogenous-Type Failures   141
3.3 Statistical Characteristics of a System Subject to Failure   143
3.4 Stochastic Processes Underlying the Failure Phenomenon   152
3.5 Determination of the Failure Characteristics of a System Given
  the Failure Characteristics of its Components—Serial Systems   163
3.6 Methods for Improving the Reliability of a System   165
  Selected References   186
  Problems   187

4 Queuing Theory   190

4.1 Introduction and Historical Background   190
4.2 Characteristics of Queuing Systems   191
4.3 Characteristics of a Queuing Problem   196
4.4 The $M/M/1$ Queuing System   197
4.5 The $M/M/r$ Queuing System   211
4.6 The Modified $M/G/1$ Queuing System   224
4.7 The $M/G/\infty$ Queuing System   242
4.8 Control of Single Server Queuing System   252
  Selected References   258
  Problems   259

5 Inventory Theory for Single-Commodity,
Single-Installation Systems   262

5.1 Introduction   262
5.2 Deterministic Inventory Models   275
5.3 Stochastic Inventory Models   308
  Selected References   342
  Problems   344

6 Inventory Theory for Multicommodity,
Multiinstallation Systems   346

6.1 Introduction   346
6.2 Deterministic Models—The Two-Commodity Problem   353
6.3 Deterministic Models—The Multicommodity Problem   368
6.4 Single-Period Probabilistic Models—
  The Two-Commodity Problem   370
6.5 Single-Period Probabilistic Models—
  The Multicommodity Problem   385
  Selected References   387
  Problems   388

7 Replacement Theory   392

7.1 Introduction   392
7.2 Continuous Review Replacement Systems:
Deterministic Models   398
7.3 Continuous Review Replacement Systems:
Stochastic Models   421
7.4 Periodic Review Replacement Systems:
Deterministic Models   430
7.5 Periodic Review Replacement Systems:
Stochastic Models   432
Selected References   441
Problems   442

8 Information Theory   446

8.1 Introduction   446
8.2 A Measure of Information   447
8.3 Measures of Other Information Quantities   454
8.4 Application 1—Heavy-Coin Problem   462
8.5 Application 2—Odd-Coin Problem   465
8.6 Application 3—Simple Coding Problems   469
Selected References   475
Problems   475

9 Location Theory   480

9.1 Introduction   480
9.2 Unrestricted Single-source Location Problem   482
9.3 Unrestricted Multisource Problem   505
9.4 Restricted Single-source Location Problem   510
9.5 Restricted Multisource Location Problem   513
9.6 Conclusion   516
Selected References   516
Problems   517

APPENDIX: The Laplace Transform   520

A.1 Definition   520
A.2 Some Properties of the Laplace Transforms   523
A.3 The Inverse Laplace Transform   525
A.4 Solving Differential Equations Using Laplace Transforms   527

INDEX   529

PREFACE

The purpose of this book is to provide an introduction to the principal methods in the study of systems encountered in operations research. With its companion volume, *Optimization Techniques in Operations Research* by the same authors, it provides a comprehensive coverage of the subject for junior and senior engineering, operations research, or mathematics students. Prerequisite to the text are the basic courses in differential and integral calculus with elementary differential equations, that are usually covered during the freshman and sophomore years in engineering and science curricula. An introductory course in probability is helpful, but not essential, because the book begins with a chapter on probability theory which is required for the understanding of most of the subsequent chapters. This book is also suitable for graduate students majoring in operations research, management science, economics, or quantitative management with the stated prerequisite, but who have had no preparation in the field.

The book is not a handbook in which the treatment of a given topic consists of a collection of tables and formulas, and care has been taken not to burden the text with unproven relations. The authors feel this would be meaningless from a pedagogical point of view, with respect to both the material under consideration and the preparation and stimulation of the student for further pursuit of the topics. The understanding of the book's scope and content is a first step for the ambitious student who will be delving into the more complex and specialized literature.

Moreover, the book is not an entirely theoretical treatment of the various topics and is not addressed to specialized mathematicians. First, this is not possible in terms of the limited background presumed and the expectation of the audience having an interest in an applied field. Secondly, that sort of theoretical exposition would not serve as the undergraduate treatment we hope to provide. It is believed that many of the basic concepts in the applications of operations research may be presented using mathematical techniques at a moderately low level, without unduly sacrificing some of the rigor. When the presentation becomes, of necessity, less rigorous, we have attempted to provide insight by relying upon illustrations and concrete examples to make a final result intuitively appealing and meaningful.

It is believed that both the set of *optimization techniques* and the *fields of application* constitute operations research. Thus, dynamic programming is an optimization technique, whereas inventory systems is a field in operations research in which such optimization techniques are useful. The two books are separated within this conceptual framework. The present text encompasses the use of quantititave techniques, including the use of probabilistic concepts, in the analysis, formulation, and solution of a number of systems encountered in operations research.

Chapter 1 presents the essentials of probability theory needed in subsequent chapters. Although the content is far from exhaustive, it is hoped that it is adequately self-contained so that the student without preparation in the area will not be at a disadvantage. Students with some previous introductory exposure to probability will feel more at ease in reading the chapter with little or no assistance from an instructor. The chapter presents the theory of discrete and continuous random variables with emphasis on the use of generating functions and Laplace transforms in solving probability problems. It introduces basic notational symbols, definitions, and concepts used in the remaining portion of the book.

Chapter 2, on stochastic processes, exposes the fundamentals of the theory of independent processes, Markov processes, and renewal processes. The theory of denumerable Markov chains is introduced in an elementary and intuitive fashion. General and specialized topics, such as first passage times, classification of states, and others have been purposefully omitted. Only materials which are alluded to in the remaining chapters have been incorporated. The use of flow diagrams is introduced in the section on pure birth and death processes.

Chapter 3 on reliability theory discusses the failure phenomenon and introduces several stochastic models underlying such phenomena. This is followed by the study of serial systems and the development of methods and models for improving system reliability, including the use of redundancy and repair.

Queuing theory is the subject of Chapter 4. Many of the stochastic processes associated with the $M/M/1$, $M/M/r$, $M/G/1$ and $M/G/\infty$ queuing systems are studied in detail. A section on the control of single server queuing system is included. Several numerically solved examples are incorporated illustrating the formulation of economic criteria to arrive at an optimal queue operation.

The theory of inventory management for single-commodity single-installation systems is the content of Chapter 5. Deterministic and stochastic inventory systems operating both under continuous review and periodic review are presented. The chapter includes some new models which have appeared only in article forms. In some sections, a dynamic programming formulation of a model has been used; these sections may be omitted at first reading. In general, a conscious attempt has been made to present the subject in a systematic and coherent way without overextending the framework of the stated prerequisites.

Chapter 6 on the inventory theory for multicommodity multiinstallation systems consists of both original work, and some sections which have already appeared in article forms. It has been included partly to reflect one author's interest in the area, but more importantly, to acquaint the reader with the more complex logistics-type problems encountered in practice. It assumes familiarity with the content of Chapter 5.

It is the intent of Chapter 7 to present replacement theory as a worthy topic in operations research, and to attempt to unify the many deterministic and stochastic elementary models that have or have not appeared in the literature to date. Following the same outline as in Chapter 5, deterministic and stochastic systems operating under continuous and periodic reviews are modeled.

Solving any problem involves the acquisition and utilization of the information available, and therefore, a chapter concerned with the theory of information has been included—Chapter 8. Although it could not be termed a method of operations research, information theory possesses relationships to certain operations research procedures, such as search techniques, and there exist interesting applications to various decision-making problems.

Finally, Chapter 9 dealing with location theory has also been incorporated. Despite its earlier appreciation by mathematicians and economists, and despite its importance in management and military logistics, location theory has only recently begun receiving increased attention by operations researchers. Presently, there exists adequate literature on the subject to justify development of the theory in its own right. The existing theory relies heavily on mathematical programming in formulating and solving the more sophisticated location problems. Except for two brief sections which may be omitted, the elaborate inclusion of such formulations has been avoided to retain consistency with the stated prerequisite.

No attempt has been made to cover any of the topics in the area of statistical decision theory such as Bayesian theory, sequential decision processes, statistical inference, forecasting, etc. . . . Although the authors recognize the importance of the theory, its exposition requires knowledge of the theory of statistics which is not necessary for the understanding of this text. It was also felt that the theory of Markov decision processes can be treated more adequately once the reader is formally acquainted with the content of the present text and its companion volume.

It is imperative for the beginning student to be able to distinguish in operations research the descriptive techniques from the optimization techniques and realize the value of each, and the authors have stressed the importance of both. Realizing the importance of the ability to analyze situations and abstract them into a mathematical model, we have often stated a problem in words, then taken the student through a process of formulating the problem as a deterministic process or a stochastic process, prior to formally solving it.

In this text we have attempted to bring conceptual coherency to many topics which are often presented as separate subjects. For example, the equivalence between replacement theory models and inventory theory models is pointed out. Also, in Chapter 6, the similarity between multicommodity and multiinstallation inventory problems is explained in the presentation. Finally, some of the models in the theory of queue control may be given an equivalent inventory interpretation.

The student is given ample opportunity to develop a capability to formulate and solve mathematically a wide range of operations research problems. A variety of fully worked out examples drawn from such areas as engineering, management, marketing, production, distribution, defense, and others has been included. Many of the numerical examples were solved by computers. A set of carefully selected unsolved problems appears at the end of each chapter, together with a list of selected references. An appendix on the essentiality of Laplace transforms has also been included.

Practicing engineers and scientists whose field of expertise is other than operations research may find the book suitable for an introductory exposition and as a preparation to more specialized topics in operations research.

The accompanying chart shows the logical interdependence between the chapters.

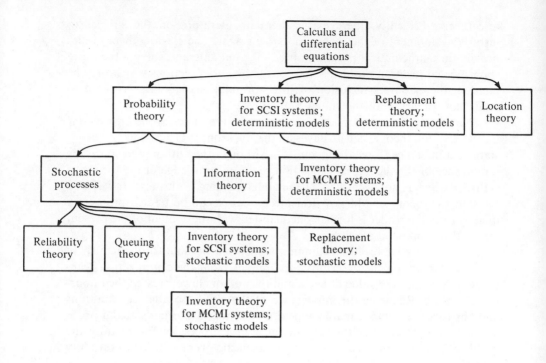

The authors wish to express their gratitude to the many students who have contributed through classroom discussions to the writing of this book. The authors acknowledge in particular the contribution of Dr. J. F. Brown. Dr. T. J. Hodgson made some helpful suggestions on portions of the manuscript. We also thank Edna Larrick and Marcene Brown for their efficient typing services. Last, but not least, the fine cooperation of the editors of Prentice-Hall, Inc., is greatly appreciated.

B. D. SIVAZLIAN
L. E. STANFEL

ANALYSIS OF
SYSTEMS
IN OPERATIONS
RESEARCH

PROBABILITY THEORY

1.1 Introduction

Probability theory, together with the theory of stochastic processes, forms the basis of discussion of several descriptive areas in operations research. For example, the theory of reliability attempts to characterize the failure phenomenon of systems by developing appropriate mathematical models to account for the stochastic nature of the phenomenon. Although the stochastic models encountered in operations research have in general a fairly complex structure, nevertheless, their formulation relies heavily on some basic concepts in probability theory. The present chapter develops some of those concepts which one encounters quite frequently in operations research.

The chapter is essentially divided into two sections: the first section discusses discrete random variables and the use of generating functions; the second section discusses continuous random variables and the use of Laplace transforms. It is hoped the chapter will serve several needs. First, it is self-contained, and it introduces notational symbols and concepts used in subsequent chapters. Second, those students who already have some familiarity in the area may read the chapter to review the essential topics in probability that will be used in the sequel. Finally, an emphasis in the use of

1

transforms should acquaint the student with the more powerful analytic tools used in solving complex probability problems.

Students who are interested in exploring other topical areas in probability theory and in gaining a more detailed knowledge should consult the reference books [1] to [13] appearing at the end of the present chapter.

1.2 Sample Space, Probability, Conditional Probability, Independent Events, and Random Variables

Consider the experiment consisting of the tossing of two coins. The experiment will result in one of four possible outcomes:

Outcome	Coin 1	Coin 2
A_1	Tail	Tail
A_2	Head	Tail
A_3	Tail	Head
A_4	Head	Head

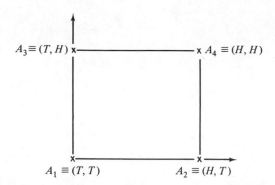

Figure 1.1. Sample space Ω describing the outcomes in tossing two coins.

Letting H denote head and T denote tail, it is possible to describe the outcomes of this experiment as four *sample points* on a *sample space*, as shown in Figure 1.1.

The sample space denoted by Ω contains the descriptions of all possible outcomes of an experiment or of a phenomenon.

The sample space is *finite* if it contains a finite number of description points; it is *countably finite* if there is a one-to-one correspondence between its points and the set of positive integers; otherwise, the sample space is said to be *noncountably infinite*. The sample space described in Figure 1.1 is finite.

An event E is a set of sample points contained in the sample space Ω. For example, in the experiment consisting of the tossing of two coins, let E_1 be the event described as at least one head; then E_1 is the set of points in Ω that includes the sample points (H, H), (H, T), and (T, H). Thus

$$E_1 = A_4 \cup A_2 \cup A_3$$

Clearly,
$$\Omega = A_1 \cup A_2 \cup A_3 \cup A_4$$

and
$$\emptyset = A_i \cap A_j \quad \text{for all } i \neq j$$

where $\emptyset$ is the null set.

An event is said to occur if it can be described by a set of sample points in Ω. Otherwise, we say that the event does not occur. For example, in the previous experiment the event "at least one head" has a proper representation on the sample space; hence it can occur. The event "exactly three tails" has no representation on the sample space; hence it cannot occur and therefore is an *impossible* event.

Two events E_1 and E_2 are said to be *mutually exclusive* in Ω if $E_1 \cap E_2 = \emptyset$. Thus if E_1 denotes the event "at least one head" and E_2 denotes the event "exactly two tails," then E_1 and E_2 are mutually exclusive. A sequence of events $E_1, E_2, \ldots, E_n$ is mutually exclusive if $E_i \cap E_j = \emptyset$ for any $i \neq j$.

A sequence of events $E_1, E_2, \ldots, E_n$ is said to be *exhaustive* in Ω if $E_1 \cup E_2 \cup \ldots \cup E_n = \Omega$. In the previous coin-tossing experiment, it is evident that if one considers each of the outcomes A_1, A_2, A_3, and A_4 as events, then such events are mutually exclusive and exhaustive.

To each event E contained in the sample space Ω, it is possible to associate a real nonnegative number $P\{E\}$, defined as the *probability* of occurrence of the event E and satisfying the following conditions:

(1) $0 \leq P\{E\} \leq 1$.
(2) $P\{\Omega\} = 1$.
(3) If E_1 and E_2 are two mutually exclusive events in Ω, then $P\{E_1 \cup E_2\} = P\{E_1\} + P\{E_2\}$.
(4) If the sequence of mutually exclusive events $E_1, E_2, \ldots, E_n$ refers respectively to each of the possible outcomes represented as sample points in Ω, then

$$P\{E_1 \cup E_2 \cup \ldots \cup E_n\} = P\{\Omega\} = \sum_{i=1}^{i=n} P\{E_i\} = 1$$

On an intuitive basis, probability is often identified with the relative frequency or proportion of occurrences of a given event in a large number of independent experiments. Conditions 1 to 4 imply that the probability of an event can never exceed 1. Furthermore, the probability of a certain event is unity and the probability of an impossible event is zero. In the coin-tossing experiment, assuming that the coins are fair, one may associate to each of the sample points in Ω a probability value of $\frac{1}{4}$, so that, if E_1 refers to the event "at least one head," we can write

$$P\{E_1\} = P\{A_4 \cup A_2 \cup A_3\}$$
$$= P\{A_4\} + P\{A_2\} + P\{A_3\} = \tfrac{3}{4}$$

A concept that plays an important role in stochastic processes is that of conditional events. Let E_1 and E_2 be two events defined over a sample space Ω, with $P\{E_1\} \neq 0$; we define the *conditional probability* of the occurrence of the event E_2, given that the event E_1 has occurred, as

$$P\{E_2 \,|\, E_1\} = \frac{P\{E_1 \cap E_2\}}{P\{E_1\}}$$

The two events E_1 and E_2 are said to be *independent* if

$$P\{E_2 \,|\, E_1\} = P\{E_2\}$$
or
$$P\{E_1 \,|\, E_2\} = P\{E_1\}$$
or
$$P\{E_1 \cap E_2\} = P\{E_1\} \cdot P\{E_2\}$$

So far we have used letters to denote the sample points and events defined on a sample space; alternatively, we could have associated, say, cardinal numbers to each sample point. In applied probability it is often necessary to associate a measurable quantity to the outcomes of an experiment or the events associated with a phenomenon. This leads us to the concept of a random variable.

A random variable or variate X is a real-valued function defined on a sample space Ω, which takes on its possible values according to a probability law or probability function. If we consider the set of aggregate events in a sample space on which a random variable is defined, then we define the probability that the random variable X will have an observed value lying in some set B of the sample space as the probability function $P_X\{B\}$ of the random variable X.

In particular, if we select B to be the set of points in the sample space such that $X \leq x$ for some real number x, then the probability function is called the *distribution function* $\Phi_X(x)$ of the random variable X, where

$$\Phi_X(x) = P\{X \leq x\} \tag{1.1}$$

The distribution function, known sometimes as the *cumulative distribution function* (CDF) completely specifies the probability law of the random variable X.

Going back to the experiment involving the tossing of two coins, suppose that head is assigned a numerical value of 1 and tail is assigned a numerical value of 0. Define now the random variable X to be the arithmetic sum of the outcomes; thus X takes on the values of 2, 1, 1, and 0, respectively, when the outcomes are HH, HT, TH, and TT. The probability of occurrence of each of these outcomes being $\frac{1}{4}$, it follows that the distribution function of X will be defined as

x	$-\infty$	...	0	1	2	...	$+\infty$
$\Phi_X(x)$	0	...	$\frac{1}{4}$	$\frac{3}{4}$	1	...	1

The CDF, Φ_X of a random variable X, is a numerically valued function defined for all x, $-\infty < x < \infty$, with the following properties:

(1) $\Phi_X(x)$ is a nondecreasing function of x.
(2) $\Phi_X(-\infty) = \lim_{x \to -\infty} \Phi_X(x) = 0$.
(3) $\Phi_X(+\infty) = \lim_{x \to \infty} \Phi_X(x) = 1$.

Hence $0 \leq \Phi_X(x) \leq 1$ for $-\infty < x < \infty$. A distribution function can have at most a denumerable number of jumps or discontinuities; furthermore, it is continuous to the right. Conversely, every function $\Phi_X(x)$ that satisfies conditions 1 to 3 can be regarded as the distribution function of some random variable X (Figure 1.2).

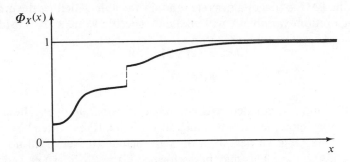

Figure 1.2. A distribution function.

In many practical problems, distribution functions belong to two simple types of random variables: discrete random variables and continuous random variables.

1.3 Discrete Random Variables

i. *Definition: Probability Mass Function*

A random variable X is said to be discrete if its probability law can be expressed by a *probability mass function* (PMF), $\varphi_X(\cdot)$ such that

$$\Phi_X(x) = \sum_{j=-\infty}^{j=x} \varphi_X(j) \tag{1.2}$$

that is, if there exists a function $\varphi_X(\cdot)$ such that $\Phi_X(x)$ can be expressed as a sum. One often identifies the discrete characteristic of a random variable X by saying that X has a *discrete distribution*. In the sequel, we shall restrict our analysis to random variables that take on only nonnegative integer values.

Now from relation (1.1)

$$\Phi_X(x) - \Phi_X(x-1) = P\{X \le x\} - P\{X \le x - 1\}$$
$$= P\{X = x\}$$

Also from relation (1.2)

$$\Phi_X(x) - \Phi_X(x-1) = \sum_{j=0}^{j=x} \varphi_X(j) - \sum_{j=0}^{j=x-1} \varphi_X(j)$$
$$= \varphi_X(x)$$

Thus
$$\varphi_X(x) = P\{X = x\}$$

Hence the PMF $\varphi_X(x)$ of a discrete random variable X defines the probability that the random variable X will take on a specific value x. It is evident that

$$0 \le \varphi_X(x) \le 1, \quad \text{for all } x$$

and

$$\sum_x \varphi_X(x) = 1$$

where the sum is evaluated over the *range of definition* of x. These possible values of x are either finite or denumerably infinite. It may often be convenient to extend the range of definition of X to include the set of all nonnegative integers $0, 1, 2, \ldots$. This may be performed if the PMF, $\varphi_X(x)$, is defined to be equal to zero for those values of X extending outside the proper range of definition of X. With this in mind, we may write, for example, for any non-negative discrete variate X the relation

$$\sum_{x=0}^{\infty} \varphi_X(x) = 1$$

When no ambiguity exists, we shall write $\Phi_X(x)$ and $\varphi_X(x)$ as $\Phi(x)$ and $\varphi(x)$, respectively. We shall often use the symbol r instead of x to emphasize the discreteness of the random variable X $(r = 0, 1, 2, \ldots)$.

ii. *Expectation and Variance*

Two important parameters characterizing the random variable X and which play an important role in probability theory, statistics and stochastic processes, are the expected value and the variance of X.

The *expectation* or *expected value* or *mean* of the discrete random variable X denoted by $E[X]$ or μ_X is defined to be

$$E[X] = \mu_X = \sum_{r=0}^{\infty} r\varphi_X(r) \tag{1.3}$$

More generally, let $g(X)$ be a *function* of the random variable X; we define the expected value of $g(X)$, if it exists, as

$$E[g(X)] = \sum_{r=0}^{\infty} g(r)\varphi_X(r)$$

Thus

$$E[X^2] = \sum_{r=0}^{\infty} r^2 \varphi_X(r) \tag{1.4}$$

The *variance* of the discrete random variable X is the expected value of the squared deviation of the random variable from its mean. It is usually denoted by σ_X^2 or $\text{Var}[X]$

$$\text{Var}[X] = \sigma_X^2 = E[(X - \mu_X)^2] = \sum_{r=0}^{\infty} (r - \mu_X)^2 \varphi_X(r)$$

$$= \sum_{r=0}^{\infty} r^2 \varphi_X(r) - 2\mu_X \sum_{r=0}^{\infty} r\varphi_X(r) + \mu_X^2 \sum_{r=0}^{\infty} \varphi_X(r)$$

$$= \sum_{r=0}^{\infty} r^2 \varphi_X(r) - \mu_X^2 \qquad\qquad (1.5)$$

$$= E[X^2] - \mu_X^2$$

The quantity σ_X taken to be the positive square root of σ_X^2 is known as the *standard deviation* of the random variable X. Sometimes, for simplicity, we shall drop the subscript X in μ_X and σ_X^2.

iii. *Generating Function and Probability Generating Function*

Let $a_0, a_1, a_2, \ldots$ be a sequence of real numbers. We define the *generating function* of this sequence to be the function

$$A(s) = \sum_{r=0}^{\infty} a_r s^r = a_0 + a_1 s + a_2 s^2 + \cdots + a_r s^r + \cdots$$

provided the series converges in some given interval $|s| < s_0$, where s_0 is a real positive number. We note in particular that the element a_r of the sequence can be identified as the coefficient of s^r in the power series expansion of the function $A(s)$.

Let $\varphi_X(r)$, $r = 0, 1, 2, \ldots$, be the PMF of the discrete random variable X; we define the *probability generating function (PGF) of the random variable X* as

$$G_X(s) = \sum_{r=0}^{\infty} s^r \varphi_X(r) = \varphi_X(0) + \varphi_X(1)s + \cdots + \varphi_X(r)s^r + \cdots$$

Since $0 \leq \varphi_X(r) \leq 1$ for all r, the series $G_X(s)$ always converges over the interval $|s| \leq 1$; this convergence can be established by comparing the series expansion of $G_X(s)$ with the geometric series

$$\sum_{r=0}^{\infty} s^r = 1 + s + s^2 + \cdots + s^r + \cdots, \quad |s| < 1$$

and by noting that at $s = 1$, $G_X(1) = 1$, and at $s = -1$, $-1 < G_X(-1) < 1$. When no ambiguity exists, we shall drop the subscript X from $G_X(s)$ and write it as $G(s)$.

The probability generating function is an important tool in probability theory and related areas. Given the PMF of X, it is always possible to form the PGF of X. Conversely, given the PGF of X, one can always recover the PMF of X, the correspondence between the PGF and PMF being unique.

If the PGF of X is given, it is possible to recover the PMF of X in one of the following ways:

(1) If $G_X(s)$ can be expanded as a power series of s, then $\varphi_X(r)$, $r = 0, 1, 2, \ldots$, is identified as the coefficient of s^r. Thus $\varphi_X(0)$ is the coefficient of s^0, $\varphi_X(1)$ is the coefficient of s^1, and so on.

(2) Given $G_X(s)$, it is possible to obtain successively the derivatives of any order of $G_X(s)$ taken with respect to s.

The evaluation of these derivatives at $s = 0$ will then generate the PMF $\varphi_X(r)$. To see this, we write

$$G(s) = \varphi_X(0) + \varphi_X(1)s + \varphi_X(2)s^2 + \cdots + \varphi_X(r)s^r + \cdots$$

$$\frac{dG(s)}{ds} = G'(s) = \varphi_X(1) + 2\varphi_X(2)s + \cdots + r\varphi_X(r)s^{r-1} + \cdots$$

$$\frac{d^2G(s)}{ds^2} = G''(s) = 2\varphi_X(2) + \cdots + r(r-1)\varphi_X(r)s^{r-2} + \cdots$$

In general, for $r = 1, 2, \ldots$, we can write

$$\frac{d^rG(s)}{ds^r} = G^{(r)}(s) = r!\varphi_X(r) + (r+1)r(r-1)\ldots 2\varphi_X(r+1)s + \cdots$$

$$= \sum_{k=r}^{\infty} (k)(k-1)\ldots(k-r+1)\varphi_X(k)s^{k-r}$$

If we set $s = 0$, we obtain, successively,

$$\varphi_X(0) = G(0)$$

$$\frac{dG(s)}{ds}\bigg|_{s=0} = G'(0) = 1\varphi_X(1) \quad \text{or} \quad \varphi_X(1) = \frac{1}{1!}G'(0)$$

$$\frac{d^2G(s)}{ds^2}\bigg|_{s=0} = G''(0) = 2\varphi_X(2) \quad \text{or} \quad \varphi_X(2) = \frac{1}{2!}G''(0)$$

and, in general, we have, for $r = 1, 2, \ldots$,

$$\frac{d^rG(s)}{ds^r}\bigg|_{s=0} = G^{(r)}(0) = r!\varphi_X(r) \quad \text{or} \quad \varphi_X(r) = \frac{1}{r!}G^{(r)}(0)$$

The result, of course, is well known in calculus, since one can identify the power series expansion of a function with the Maclaurin's series expansion. The same techniques may be used to obtain the elements of a sequence from their generating function.

It is possible to determine the mean and variance of a discrete random variable X directly from its PGF $G(s)$ without establishing first its PMF. If we set $s = 1$ in the expression for $dG(s)/ds$, we obtain

$$\left.\frac{dG(s)}{ds}\right|_{s=1} = G'(1) = \varphi_X(1) + 2\varphi_X(2) + \cdots + r\varphi_X(r) + \cdots$$

$$= \sum_{r=1}^{\infty} r\varphi_X(r) = \sum_{r=0}^{\infty} r\varphi_X(r)$$

Hence $\qquad G'(1) = E[X] \qquad\qquad\qquad\qquad\qquad\qquad\qquad (1.6)$

If we now set $s = 1$ in the expression for $d^2G(s)/ds^2$, we obtain

$$\left.\frac{d^2G(s)}{ds^2}\right|_{s=1} = G''(1) = 2\varphi_X(2) + 3\cdot 2\varphi_X(3) + \cdots + r(r-1)\varphi_X(r) + \cdots$$

$$= \sum_{r=2}^{\infty} r(r-1)\varphi_X(r)$$

$$= \sum_{r=2}^{\infty} r^2\varphi_X(r) - \sum_{r=2}^{\infty} r\varphi_X(r)$$

Adding and subtracting the quantities $1\cdot\varphi_X(1) + 0\cdot\varphi_X(0)$ on the right-hand expression, we obtain

$$G''(1) = \sum_{r=2}^{\infty} r^2\varphi_X(r) + 1\cdot\varphi_X(1) + 0\cdot\varphi_X(0)$$

$$- \left[\sum_{r=2}^{\infty} r\varphi_X(r) + 1\cdot\varphi_X(1) + 0\cdot\varphi_X(0)\right]$$

$$= \sum_{r=0}^{\infty} r^2\varphi_X(r) - \sum_{r=0}^{\infty} r\varphi_X(r)$$

$$= E[X^2] - E[X]$$

Therefore, using expression (1.6), we obtain

$$E[X^2] = G''(1) + G'(1) \qquad\qquad\qquad\qquad (1.7)$$

The expression for the variance is

$$\text{Var}[X] = E[X^2] - [E[X]]^2$$
$$= G''(1) + G'(1) - [G'(1)]^2 \qquad\qquad (1.8)$$

As a final remark, we remind the reader that $G_X(1) = 1$; this last relation may be used to check that a derived PGF is a valid expression.

iv. *Complementary Cumulative Function*

We define the *complementary cumulative function* $H_X(r)$ of the discrete random variable X to be the probability that the random variable X will have a value exceeding r. Thus

$$H_X(r) = P\{X > r\} = \sum_{j=r+1}^{\infty} \varphi_X(j)$$

Denote by $Q(s)$ the generating function of $H_X(r)$, $|s| < 1$, then

$$Q(s) = \sum_{r=0}^{\infty} s^r H_X(r) = \sum_{r=0}^{\infty} s^r \sum_{j=r+1}^{\infty} \varphi_X(j)$$

The region over which the double sum is evaluated is represented in Figure 1.3. If we interchange the order of summation, we obtain

$$Q(s) = \sum_{j=1}^{\infty} \sum_{r=0}^{r=j-1} s^r \varphi_X(j)$$

$$= \sum_{j=1}^{\infty} \varphi_X(j) \sum_{r=0}^{r=j-1} s^r$$

Since $\sum_{r=0}^{r=j-1} s^r = (1 - s^j)/(1 - s)$, it follows that

$$Q(s) = \frac{1}{1-s} \sum_{j=1}^{\infty} (1 - s^j)\varphi_X(j)$$

or

$$Q(s) = \frac{1}{1-s}\left[\sum_{j=1}^{\infty} \varphi_X(j) - \sum_{j=1}^{\infty} s^j \varphi_X(j)\right]$$

If we add and subtract the quantity $\varphi_X(0)$ on the right-hand side, we obtain

$$Q(s) = \frac{1}{1-s}\left[\sum_{j=0}^{\infty} \varphi_X(j) - \sum_{j=0}^{\infty} s^j \varphi_X(j)\right]$$

$$= \frac{1}{1-s}[1 - G_X(s)]$$

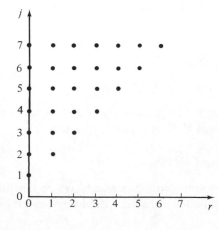

Figure 1.3. Region of summation in $\sum_{r=0}^{\infty} \sum_{j=r+1}^{\infty}$.

The mean of the random variable X can be obtained directly from a knowledge of either $H_X(r)$ or $Q(s)$. Suppose that $H_X(r)$ is given; then we can write

$$\sum_{r=0}^{\infty} H_X(r) = \sum_{r=0}^{\infty} \sum_{j=r+1}^{\infty} \varphi_X(j)$$

The region of summation is the one represented in Figure 1.3. Hence, interchanging the order of summation, we obtain

$$
\begin{aligned}
\sum_{r=0}^{\infty} H_X(r) &= \sum_{j=1}^{\infty} \sum_{r=0}^{r=j-1} \varphi_X(j) \\
&= \sum_{j=1}^{\infty} \varphi_X(j) \sum_{r=0}^{r=j-1} 1 \\
&= \sum_{j=1}^{\infty} j\varphi_X(j) = \sum_{j=0}^{\infty} j\varphi_X(j) \\
&= E[X]
\end{aligned}
\tag{1.9}
$$

Suppose now that the expression for $Q(s)$ is given. We note that

$$\lim_{s \to 1} Q(s) = \lim_{s \to 1} \frac{1 - G_X(s)}{1 - s}$$

The expression $[1 - G_X(s)]/(1 - s)$ is indeterminate at $s = 1$; we may use L'Hospital's rule to obtain

$$\lim_{s \to 1} Q(s) = \lim_{s \to 1} \frac{-G'_X(s)}{-1} = G'_X(1) = E[X]$$

To summarize, the following expressions may be used to evaluate the expectation of the discrete random variable X:

$$E[X] = \sum_{r=0}^{\infty} r\varphi_X(r)$$

$$E[X] = G'_X(1)$$

$$E[X] = \sum_{r=0}^{\infty} H_X(r)$$

$$E[X] = \lim_{s \to 1} Q(s)$$

v. *Examples of Discrete Distributions*

a. Uniform Distribution

A discrete random variable X will be said to have a *uniform distribution* if its PMF is given by

$$\varphi_X(r) = \begin{cases} \dfrac{1}{a}, & r = 0, 1, 2, \ldots, a-1 \\ 0, & \text{otherwise} \end{cases}$$

where $a \geq 1$. To obtain an expression for the PGF of X, we write

$$G_X(s) = \sum_{r=0}^{\infty} s^r \varphi_X(r)$$

$$= \sum_{r=0}^{r=a-1} s^r \frac{1}{a} = \frac{1}{a} \sum_{r=0}^{r=a-1} s^r$$

$$= \frac{1}{a}(1 + s + s^2 + \cdots + s^{a-1})$$

$$= \frac{1}{a} \frac{1 - s^a}{1 - s}$$

By applying L'Hospital's rule once, we note that $\lim_{s \to 1} G_X(s) = 1$. We shall now derive the value of $E[X]$ using the expression for $G_X(s)$:

$$G_X'(s) = \frac{1}{a} \frac{-(1-s)as^{a-1} + (1 - s^a)}{(1-s)^2}$$

Since at $s = 1$ the expression for $G_X'(s)$ is indeterminate, we obtain, using L'Hospital's rule,

$$\lim_{s \to 1} G_X'(s) = \lim_{s \to 1} \frac{1}{a} \frac{-[-as^{a-1} + (1-s)a(a-1)s^{a-2}] - as^{a-1}}{-2(1-s)}$$

$$= \lim_{s \to 1} \frac{1}{a} \frac{as^{a-1} - (1-s)a(a-1)s^{a-2} - as^{a-1}}{-2(1-s)}$$

$$= \frac{a-1}{2}$$

The expression for $\text{Var}[X]$ can similarly be obtained from $G_X(s)$, although the computation is quite tedious.

In general, it is possible to define a uniform distribution whose range lies between two arbitrary integers, say b and c, where $0 \leq b \leq r \leq c$. The PMF of X is then defined as

$$\varphi_X(r) = \begin{cases} 0, & r = 0, 1, \ldots, b-1 \\ \dfrac{1}{c-b}, & r = b, b+1, \ldots, c-1 \\ 0, & r = c, c+1, \ldots \end{cases}$$

The discrete uniform distribution plays an important role in the theory of random numbers and its application to Monte Carlo simulation.

b. BERNOULLI DISTRIBUTION

A discrete random variable X is said to have a *Bernoulli distribution* if its PMF is given by

$$\varphi_X(r) = \begin{cases} p^r q^{1-r}, & r = 0, 1 \\ 0, & r = 2, 3, \ldots \end{cases}$$

where $0 < p < 1$ and $p + q = 1$. The PGF of X is

$$G_X(s) = \sum_{r=0}^{\infty} s^r \varphi_X(r)$$

$$= q + ps$$

Since $G'_X(s) = p$

and $G''_X(s) = 0$

we obtain, using relations (1.6) and (1.8),

$$E[X] = G'_X(1) = p$$

$$\text{Var}[X] = G''_X(1) + G'_X(1) - [G'_X(1)]^2$$

$$= p - p^2 = p(1 - p) = pq$$

The Bernoulli distribution in which the random variable X takes only the values of 0 and 1 with probabilities q and p, respectively, is the simplest type of the discrete distributions. It is encountered in probability models involving a sequence of trials, each trial resulting in either a success outcome or a failure outcome.

c. BINOMIAL DISTRIBUTION

A discrete random variable X is said to have a *binomial distribution* if its PMF is given by

$$\varphi_X(r) = \begin{cases} \binom{n}{r} p^r q^{n-r}, & r = 0, 1, 2, \ldots, n \\ 0, & r = n + 1, \ldots \end{cases}$$

where $0 < p < 1$, $p + q = 1$, and $n \geq 1$. We note that if $n = 1$, we recover the Bernoulli distribution. To obtain the PGF of X, we write

$$G_X(s) = \sum_{r=0}^{\infty} s^r \varphi_X(r)$$

$$= \sum_{r=0}^{r=n} s^r \binom{n}{r} p^r q^{n-r}$$

$$= \sum_{r=0}^{r=n} \binom{n}{r} (ps)^r q^{n-r}$$

$$= (ps + q)^n$$

Now

$$G'_X(s) = np(ps + q)^{n-1}$$

and

$$G''_X(s) = n(n-1)p^2(ps + q)^{n-2}$$

Using relations (1.6) and (1.8), we obtain

$$E[X] = G'_X(1) = np$$

$$\text{Var}[X] = G''_X(1) + G'_X(1) - [G'_X(1)]^2$$

$$= n(n-1)p^2 + np - n^2p^2$$

$$= -np^2 + np = np(1-p)$$

$$= npq$$

The binomial distribution describes the number of successes in an experiment involving n independent trials, each trial resulting in a success or a failure.

d. POISSON DISTRIBUTION

A discrete random variable X is said to have a *Poisson distribution* with parameter λ if its PMF is given by

$$\varphi_X(r) = e^{-\lambda} \frac{\lambda^r}{r!}, \quad r = 0, 1, 2, \ldots$$

where $\lambda > 0$. The expression for the PGF of X is obtained as follows:

$$G_X(s) = \sum_{r=0}^{\infty} s^r \varphi_X(r)$$

$$= \sum_{r=0}^{\infty} s^r e^{-\lambda} \frac{\lambda^r}{r!}$$

$$= e^{-\lambda} \sum_{r=0}^{\infty} \frac{(\lambda s)^r}{r!}$$

$$= e^{-\lambda} e^{\lambda s}$$

$$= e^{-\lambda(1-s)}$$

Computing $G'_X(s)$ and $G''_X(s)$, we obtain

$$G'_X(s) = \lambda e^{-\lambda(1-s)}$$
$$G''_X(s) = \lambda^2 e^{-\lambda(1-s)}$$

Thus
$$E[X] = G'_X(1) = \lambda$$

and
$$Var[X] = G''_X(1) + G'_X(1) - [G'_X(1)]^2$$
$$= \lambda^2 + \lambda - \lambda^2 = \lambda$$

In the Poisson distribution the mean is equal to the variance. The Poisson distribution can be regarded as a limiting form of a binomial distribution in which $n \to \infty$, $p \to 0$, but np remains finite and equal to λ. The Poisson distribution describes the Poisson or birth process studied later in the theory of stochastic processes.

e. GEOMETRIC DISTRIBUTION

A discrete random variable X is said to have a *geometric distribution* if its PMF is given by

$$\varphi_X(r) = pq^r, \quad r = 0, 1, 2, \ldots$$

where $0 < p < 1$ and $p + q = 1$. The expression for the PGF of X is

$$G_X(s) = \sum_{r=0}^{\infty} s^r \varphi_X(r)$$

$$= \sum_{r=0}^{\infty} s^r pq^r$$

$$= p \sum_{r=0}^{\infty} (qs)^r$$

Since $|s| \le 1$ and $0 < q < 1$, $|qs| < 1$; hence the sum expression can be identified as a geometric series with ratio term equal to qs. The value of such a series being $1/(1 - qs)$, it follows that

$$G_X(s) = \frac{p}{1 - qs}$$

Now
$$G'_X(s) = \frac{pq}{(1 - qs)^2}$$

and
$$G''_X(s) = \frac{2pq^2}{(1 - qs)^3}$$

Hence
$$E[X] = G'_X(1) = \frac{pq}{(1 - q)^2} = \frac{pq}{p^2} = \frac{q}{p}$$

and
$$\text{Var}[X] = G''_X(1) + G'_X(1) - [G'_X(1)]^2$$
$$= \frac{2pq^2}{p^3} + \frac{q}{p} - \frac{q^2}{p^2}$$
$$= \frac{2q^2}{p^2} + \frac{q}{p} = \frac{q^2 + pq}{p^2} = \frac{q}{p^2}$$

The geometric distribution is encountered in some simple queuing models. An alternative form of the geometric distribution is

$$\varphi_X(r) = \begin{cases} 0, & r = 0 \\ pq^{r-1}, & r = 1, 2, \ldots \end{cases}$$

In this last form the geometric distribution describes the probability of the first success in a sequence of independent r Bernoulli trials in which the probability of success at each trial is p.

f. NEGATIVE BINOMIAL DISTRIBUTION

A discrete random variable X is said to have a *negative binomial distribution* if its PMF is given by

$$\varphi_X(r) = \binom{n-1+r}{r} p^n q^r, \quad r = 0, 1, 2, \ldots$$

where $0 < p < 1$, $p + q = 1$, and $n = 1, 2, \ldots$. The PGF of X is

$$G_X(s) = \sum_{r=0}^{\infty} s^r \varphi_X(r)$$
$$= \sum_{r=0}^{\infty} s^r \binom{n-1+r}{r} p^n q^r$$
$$= p^n \sum_{r=0}^{\infty} \binom{n-1+r}{r} (qs)^r$$
$$= p^n \sum_{r=0}^{\infty} \frac{(n+r-1)!}{r!(n-1)!} (qs)^r$$
$$= p^n \sum_{r=0}^{\infty} \binom{-n}{r} (-qs)^r, \; |s| \le 1$$
$$= \frac{p^n}{(1-qs)^n} = \left(\frac{p}{1-qs}\right)^n$$

Thus
$$G'_X(s) = \frac{p^n n q (1-qs)^{n-1}}{(1-qs)^{2n}} = \frac{np^n q}{(1-qs)^{n+1}}$$

and
$$G''_X(s) = np^n q \frac{(n+1)q(1-qs)^n}{(1-qs)^{2n+2}} = \frac{n(n+1)p^n q^2}{(1-qs)^{n+2}}$$

from which

$$E[X] = G'_X(1) = \frac{nq}{p}$$

and

$$\text{Var}[X] = G''_X(1) + G'_X(1) - [G'_X(1)]^2$$

$$= n(n+1)\frac{q^2}{p^2} + \frac{nq}{p} - n^2\frac{q^2}{p^2}$$

$$= n\frac{q^2}{p^2} + n\frac{q}{p} = \frac{n(q^2 + pq)}{p^2} = \frac{nq}{p^2}$$

In a sequence of independent Bernoulli trials with probability of success p at each trial, the total number of trials r to achieve exactly n successes, $r \geq n$, is a random variable with a negative binomial distribution.

vi. *Jointly Distributed Discrete Random Variables*

a. DEFINITION AND EXPECTATION

Going back to the experiment involving the tossing of two coins in which the numerical values of 0 and 1 are respectively assigned to the outcomes I and II, it is conceivable to define on each sample point of the space Ω a pair of random variables X_1 and X_2 describing, respectively, the numerical outcomes of the first and second coin. We may then talk about the probability that the random variables X_1 *and* X_2 will take on values defined over some set of points of the sample space Ω. For example, we may like to determine the probability that $X_1 = 0$ and $X_2 = 1$ and write this as $P\{X_1 = 0, X_2 = 1\}$, or we may like to know $P\{X_1 = 1, X_2 \leq 0\}$.

In general, let X_1 and X_2 be two discrete random variables defined on a given sample space. By similarity to the theory of discrete variate, we define the *joint distribution function* of X_1 and X_2, $\Phi_{X_1, X_2}(r_1, r_2)$, as

$$\Phi_{X_1, X_2}(r_1, r_2) = P\{X_1 \leq r_1, X_2 \leq r_2\}$$

The *joint probability mass function* (joint PMF) of X_1 and X_2, $\varphi_{X_1, X_2}(r_1, r_2)$, is the probability that X_1 will take on the value of r_1 and X_2 will take on the value of r_2; hence

$$\varphi_{X_1, X_2}(r_1, r_2) = P\{X_1 = r_1, X_2 = r_2\}$$

where $0 \leq \varphi_{X_1, X_2}(r_1, r_2) \leq 1$ for all values of r_1 and r_2. Restricting our attention to random variables X_1 and X_2 defined only on the set of nonnegative integers, we can write the two following basic relations:

$$\Phi_{X_1, X_2}(r_1, r_2) = \sum_{j=0}^{j=r_2} \sum_{i=0}^{i=r_1} \varphi_{X_1, X_2}(i, j)$$

and

$$\sum_{r_2=0}^{\infty} \sum_{r_1=0}^{\infty} \varphi_{X_1, X_2}(r_1, r_2) = 1$$

When no ambiguity exists, the subscripts X_1, X_2 may be dropped from $\Phi_{X_1,X_2}(r_1, r_2)$ and $\varphi_{X_1,X_2}(r_1, r_2)$ to read, respectively, $\Phi(r_1, r_2)$ and $\varphi(r_1, r_2)$.

Suppose that we extend the range of values of X_1 and X_2 to include all nonnegative integers by setting $\varphi_{X_1,X_2}(r_1, r_2) = 0$ for those values of X_1 and X_2 extending outside the proper range of definition of X_1 and X_2. Then we may, in general, represent the sample space Ω by the set of points in R^2, that is, the set of points defining the two-dimensional space of real nonnegative integers. This is illustrated in Figure 1.4.

Let $g(X_1, X_2)$ be a *function* of the random variables X_1 and X_2; the *expectation* of $g(X_1, X_2)$, $E[g(X_1, X_2)]$, if it exists, is defined as

$$E[g(X_1, X_2)] = \sum_{r_2=0}^{\infty} \sum_{r_1=0}^{\infty} g(r_1, r_2)\varphi_{X_1,X_2}(r_1, r_2)$$

b. Marginal and Conditional Probability Mass Functions

Let $\varphi_{X_1,X_2}(r_1, r_2)$ be the joint PMF of the random variables X_1 and X_2; the *marginal PMF* of the random variable X_1 is defined as

$$\varphi_{X_1}(r_1) = \sum_{r_2=0}^{\infty} \varphi_{X_1,X_2}(r_1, r_2)$$

and the marginal PMF of the random variable X_2 is defined as

$$\varphi_{X_2}(r_2) = \sum_{r_1=0}^{\infty} \varphi_{X_1,X_2}(r_1, r_2)$$

Suppose now that $\varphi_{X_1}(r_1) > 0$; we define the *conditional PMF* of X_2,

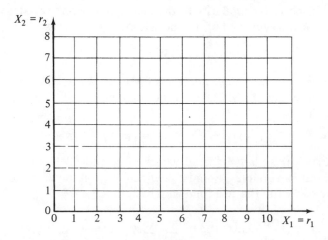

Figure 1.4. Sample space Ω for a pair of discrete random variables X_1 and X_2.

given that $X_1 = k_1$, $\varphi_{X_2|X_1}(r_2|k_1)$, as

$$\varphi_{X_2|X_1}(r_2|k_1) = P\{X_2 = r_2 | X_1 = k_1\}$$
$$= \frac{P\{X_2 = r_2, X_1 = k_1\}}{P\{X_1 = k_1\}}$$
$$= \frac{\varphi_{X_1,X_2}(k_1, r_2)}{\varphi_{X_1}(k_1)}$$

Similarly, if $\varphi_{X_2}(r_2) > 0$, the conditional PMF of X_1, given that $X_2 = k_2$, is defined as

$$\varphi_{X_1|X_2}(r_1|k_2) = P\{X_1 = r_1 | X_2 = k_2\} = \frac{\varphi_{X_1,X_2}(r_1, k_2)}{\varphi_{X_2}(k_2)}$$

The *conditional expectation* of X_2, given that $X_1 = k_1$, written as

$$E[X_2 | X_1 = k_1]$$

is defined by

$$E[X_2 | X_1 = k_1] = \sum_{r_2=0}^{\infty} r_2 \varphi_{X_2|X_1}(r_2|k_1)$$

Similarly, we define the conditional expectation of X_1, given that $X_2 = k_2$, as

$$E[X_1 | X_2 = k_2] = \sum_{r_1=0}^{\infty} r_1 \varphi_{X_1|X_2}(r_1|k_2)$$

The following important relations are left for the reader to prove:

$$E[X_1] = \sum_{r_2=0}^{\infty} E[X_1 | X_2 = r_2]\varphi_{X_2}(r_2)$$

and

$$E[X_2] = \sum_{r_1=0}^{\infty} E[X_2 | X_1 = r_1]\varphi_{X_1}(r_1)$$

c. INDEPENDENT RANDOM VARIABLES

From the definition of independent events, we shall say that two discrete random variables X_1 and X_2 are *independent random variables* if

$$P\{X_1 = r_1, X_2 = r_2\} = P\{X_1 = r_1\} \cdot P\{X_2 = r_2\}$$

for all r_1 and r_2. Written in terms of distribution functions and probability mass functions, the notion of independence of X_1 and X_2 implies that

$$\Phi_{X_1,X_2}(r_1, r_2) = \Phi_{X_1}(r_1)\Phi_{X_2}(r_2)$$

and

$$\varphi_{X_1,X_2}(r_1, r_2) = \varphi_{X_1}(r_1)\varphi_{X_2}(r_2)$$

Using the concept of conditional probability mass function, we see that if

X_1 and X_2 are independent random variables then

$$\varphi_{X_2|X_1}(r_2|k_1) = \varphi_{X_2}(r_2), \quad \text{for all } k_1$$

and
$$\varphi_{X_1|X_2}(r_1|k_2) = \varphi_{X_1}(r_1), \quad \text{for all } k_2$$

It can easily be verified that if X_1 and X_2 are independent random variables then

$$E[X_1X_2] = E[X_1]E[X_2]$$

d. Functions of Random Variables and Sums of Random Variables: Convolutions, Examples

Let $g(X_1, X_2)$ be a *function* of the two random variables X_1 and X_2. In general, this function $g(\cdot, \cdot)$ is itself a random variable, say X. A problem that often arises in probability theory consists in obtaining the distribution function of the random variable $X = g(X_1, X_2)$, knowing the joint distribution function of X_1 and X_2. Of course, the same type of problem would have been encountered if, given the distribution function of the single random variable X_1, we were asked to determine the distribution function of a function of X_1. Define $\varphi_{X_1, X_2}(r_1, r_2)$ to be the joint PMF of the random variables X_1 and X_2, and let $\Phi_X(r)$ be the distribution function of the random variable $X = g(X_1, X_2)$. Then we may obtain $\Phi_X(r)$ as follows:

$$\Phi_X(r) = P\{X \le r\} = P\{g(X_1, X_2) \le r\}$$
$$= \sum\sum_{g(r_1, r_2) \le r} \varphi_{X_1, X_2}(r_1, r_2)$$

where the summation is carried over the set of points in the sample space Ω for which $g(X_1, X_2) \le r$.

An important special case arises when $g(X_1, X_2) = X_1 + X_2$, that is, when the function of the two random variables X_1 and X_2 is their sum; if we set $X = X_1 + X_2$, then

$$\Phi_X(r) = \sum\sum_{r_1+r_2 \le r} \varphi_{X_1, X_2}(r_1, r_2)$$

This represents the probability that the sum of the random variables X_1 and X_2 will have values less than or equal to r. As illustrated in Figure 1.5, the summation encompasses the set of points in the two-dimensional sample space defining X_1 and X_2, bounded by $r_1 = 0$, $r_2 = 0$, and $r_1 + r_2 = r$.

Alternatively, we can write the following expressions for $\Phi_X(r)$:

$$\Phi_X(r) = \sum_{r_2=0}^{r_2=r} \sum_{r_1=0}^{r_1=r-r_2} \varphi_{X_1, X_2}(r_1, r_2) \tag{1.10}$$

$$= \sum_{r_1=0}^{r_1=r} \sum_{r_2=0}^{r_2=r-r_1} \varphi_{X_1, X_2}(r_1, r_2) \tag{1.11}$$

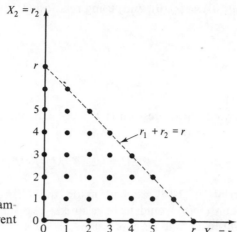

Figure 1.5. Set of points in sample space Ω describing the event $X_1 + X_2 \le r$.

With a knowledge of $\Phi_X(r)$, the PMF of X, $\varphi_X(r)$, can then be obtained from the relation

$$\varphi_X(r) = \Phi_X(r) - \Phi_X(r-1) \tag{1.12}$$

Exercise 1.1

Show that if $X = X_1 + X_2$ then $E[X] = E[X_1] + E[X_2]$. (*Hint:* Form the complementary cumulative function of X.)

A situation of special significance to probability and statistics is when the random variables X_1 and X_2 are independently distributed with respective PMF's $\varphi_{X_1}(r_1)$ and $\varphi_{X_2}(r_2)$. We can then obtain an expression for the PMF of $X = X_1 + X_2$ involving a single summation. We can write, for example, using relation (1.10),

$$\Phi_X(r) = \sum_{r_2=0}^{r_2=r} \sum_{r_1=0}^{r_1=r-r_2} \varphi_{X_1}(r_1)\varphi_{X_2}(r_2)$$

$$= \sum_{r_2=0}^{r_2=r} \varphi_{X_2}(r_2) \sum_{r_1=0}^{r_1=r-r_2} \varphi_{X_1}(r_1)$$

$$= \sum_{r_2=0}^{r_2=r} \varphi_{X_2}(r_2)\Phi_{X_1}(r-r_2) \tag{1.13}$$

Hence
$$\Phi_X(r-1) = \sum_{r_2=0}^{r_2=r-1} \varphi_{X_2}(r_2)\Phi_{X_1}(r-1-r_2) \tag{1.14}$$

Using relation (1.12) and subtracting expression (1.14) from expression (1.13), we obtain

$$\varphi_X(r) = \sum_{r_2=0}^{r_2=r} \varphi_{X_2}(r_2)\Phi_{X_1}(r-r_2) - \sum_{r_2=0}^{r_2=r-1} \varphi_{X_2}(r_2)\Phi_{X_1}(r-1-r_2)$$

$$= \sum_{r_2=0}^{r_2=r-1} \varphi_{X_2}(r_2)[\Phi_{X_1}(r-r_2) - \Phi_{X_1}(r-1-r_2)] + \varphi_{X_2}(r)\Phi_{X_1}(0)$$

Now, $\Phi_{X_1}(0) = \varphi_{X_1}(0)$, and, using relation (1.12), we have

$$
\begin{aligned}
\varphi_X(r) &= \sum_{r_2=0}^{r_2=r-1} \varphi_{X_2}(r_2)\varphi_{X_1}(r-r_2) + \varphi_{X_2}(r)\varphi_{X_1}(0) \\
&= \sum_{r_2=0}^{r_2=r} \varphi_{X_2}(r_2)\varphi_{X_1}(r-r_2)
\end{aligned}
\tag{1.15}
$$

Similarly, using expression (1.11) we can obtain the following alternative relation for $\varphi_X(r)$:

$$
\varphi_X(r) = \sum_{r_1=0}^{r_1=r} \varphi_{X_1}(r_1)\varphi_{X_2}(r-r_1)
\tag{1.16}
$$

If in relation (1.15) we had made the change in variable $r_2 = r - r_1$, expression (1.16) would have resulted. The summation term appearing in the right-hand side of expressions (1.15) and (1.16) are called *convolution* of the functions $\varphi_{X_1}(r)$ and $\varphi_{X_2}(r)$ and are written as $\varphi_{X_1}(r) * \varphi_{X_2}(r)$. It is evident that $\varphi_{X_1}(r) * \varphi_{X_2}(r) = \varphi_{X_2}(r) * \varphi_{X_1}(r)$; we can then state the following:

If X_1 and X_2 are two discrete independent random variables with PMF $\varphi_{X_1}(r_1)$ and $\varphi_{X_2}(r_2)$, the PMF of $X = X_1 + X_2$ is obtained by forming the convolution of the PMF of X_1 and X_2.

Exercise 1.2

If X_1 and X_2 are two independent discrete random variables and if $X = X_1 + X_2$, show that

$$
\text{Var}[X] = \text{Var}[X_1] + \text{Var}[X_2]
$$

Example 1.1

Consider the experiment involving the throw of two symmetrical six-sided dice, each numbered 1, 2, 3, 4, 5, 6. Let X_1 and X_2 be random variables denoting, respectively, the numbers appearing on the first and second die. Assuming that X_1 and X_2 are independent random variables, find the PMF of the following random variables:

(a) $X = X_1 + X_2$.
(b) $X = \max\{X_1, X_2\}$.

If $\varphi_{X_1}(r_1)$ and $\varphi_{X_2}(r_2)$, $r_1, r_2 = 0, 1, 2, \ldots$, denote the respective probability mass functions of X_1 and X_2, then

$$
\varphi_{X_1}(r_1) = \begin{cases} 0, & r_1 = 0 \\ \frac{1}{6}, & r_1 = 1, 2, 3, 4, 5, 6 \\ 0, & r_1 = 7, 8, \ldots \end{cases}
$$

$$
\varphi_{X_2}(r_2) = \begin{cases} 0, & r_2 = 0 \\ \frac{1}{6}, & r_2 = 1, 2, 3, 4, 5, 6 \\ 0, & r_2 = 7, 8, \ldots \end{cases}
$$

Since X_1 and X_2 are independently distributed, their joint PMF is

$$\varphi_{X_1, X_2}(r_1, r_2) = \varphi_{X_1}(r_1)\varphi_{X_2}(r_2)$$

There are several ways of approaching this problem. The simplest procedure is to take advantage of the symmetry property. Each sample point on the two-dimensional sample space Ω (see Figure 1.6) describing the outcome "the number appearing on the first die and the number appearing on the second die" has an associated probability of $\frac{1}{6} \cdot \frac{1}{6} = \frac{1}{36}$. It is then an easy matter to determine by a simple counting process $P\{X = r\}$ for each of the random variables $X = X_1 + X_2$ and $X = \max\{X_1, X_2\}$. In what follows, we use a method which is general in scope and is not limited to any particular problem.

(a) $X = X_1 + X_2$.

Let $\Phi_X(r)$ be the distribution function of the random variable $X = X_1 + X_2$; since X_1 and X_2 are defined for $r_1, r_2 = 0, 1, 2, \ldots,$ X will be defined for $r = 0, 1, 2, \ldots$; then

$$\Phi_X(r) = P\{X_1 + X_2 \leq r\} = \sum\sum_{0 \leq r_1 + r_2 \leq r} \varphi_{X_1}(r_1)\varphi_{X_2}(r_2)$$

where the region of summation consists of the set of points in Ω lying inside and on the region defined by the set of points $r_1 = 0$, $r_2 = 0$, and $r_1 + r_2 = r$ (see Figure 1.5). Hence

$$\Phi_X(r) = \sum_{r_2=0}^{r_2=r} \sum_{r_1=0}^{r_1=r-r_2} \varphi_{X_1}(r_1)\varphi_{X_2}(r_2), \quad r = 0, 1, 2, \ldots$$

We then obtain the following expressions for $\Phi_X(r)$ depending on the values taken by r.

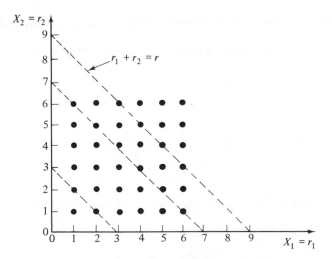

Figure 1.6. Sample space Ω describing the throw of two dice and set of points describing the event $0 \leq X_1 + X_2 \leq r$.

For $0 \leq r \leq 1$,

$$\Phi_X(r) = \sum_{r_2=0}^{r_2=r} \sum_{r_1=0}^{r_1=r-r_2} 0$$

$$= 0$$

For $2 \leq r \leq 6$,

$$\Phi_X(r) = \sum_{r_2=1}^{r_2=r} \sum_{r_1=1}^{r_1=r-r_2} \frac{1}{36}$$

$$= \frac{1}{36} \sum_{r_2=1}^{r_2=r} (r - r_2)$$

$$= \frac{1}{36}[(r-1) + (r-2) + \cdots + 1 + 0]$$

$$= \frac{r(r-1)}{2 \cdot 36}$$

For $7 \leq r \leq 12$,

$$\Phi_X(r) = \sum_{r_2=1}^{r_2=r} \sum_{r_1=1}^{r_1=r-r_2} \frac{1}{36} - 2 \sum_{r_2=7}^{r_2=r} \sum_{r_1=1}^{r_1=r-r_2} \frac{1}{36}$$

$$= \frac{1}{36} \left[\sum_{r_2=1}^{r_2=r} (r - r_2) - 2 \sum_{r_2=7}^{r_2=r} (r - r_2) \right]$$

$$= \frac{1}{36} \{[(r-1) + (r-2) + \cdots + 1 + 0]$$
$$- 2[(r-7) + (r-8) + \cdots + 1 + 0]\}$$

$$= \frac{1}{36} \left[\frac{r(r-1)}{2} - 2\frac{(r-7)(r-6)}{2} \right]$$

For $13 \leq r < \infty$,

$$\Phi_X(r) = \sum_{r_2=1}^{r_2=6} \sum_{r_1=1}^{r_1=6} \frac{1}{36}$$

$$= 1$$

To obtain the PMF of $X, \varphi_X(r)$, we make use of the relation $\varphi_X(r) = \Phi_X(r) - \Phi_X(r-1)$; we obtain

For $0 \leq r \leq 1$,

$$\varphi_X(r) = 0$$

For $r = 2$,

$$\varphi_X(r) = \Phi_X(2) - \Phi_X(1)$$
$$= \frac{2(2-1)}{2 \cdot 36} - 0$$
$$= \frac{1}{36}$$

For $3 \le r \le 6$,

$$\varphi_X(r) = \frac{r(r-1)}{2 \cdot 36} - \frac{(r-1)(r-2)}{2 \cdot 36}$$
$$= \frac{r-1}{36}$$

For $r = 7$,

$$\varphi_X(r) = \Phi_X(7) - \Phi_X(6)$$
$$= \frac{7 \cdot 6}{2 \cdot 36} - \frac{6 \cdot 5}{2 \cdot 36}$$
$$= \frac{6}{36}$$

For $8 \le r \le 12$,

$$\varphi_X(r) = \frac{1}{36}\left[\frac{r(r-1)}{2} - 2\frac{(r-7)(r-6)}{2}\right]$$
$$- \frac{1}{36}\left[\frac{(r-1)(r-2)}{2} - 2\frac{(r-8)(r-7)}{2}\right]$$
$$= \frac{1}{2 \cdot 36}[(r-1)(r-r+2) - 2(r-7)(r-6-r+8)]$$
$$= \frac{13-r}{36}$$

For $13 \le r < \infty$,

$$\varphi_X(r) = 0$$

The results for the PMF $\varphi_X(r)$ may be summarized as

$$\varphi_X(r) = \begin{cases} 0, & 0 \le r \le 1 \\ \dfrac{r-1}{36}, & 2 \le r \le 6 \\ \dfrac{13-r}{36}, & 7 \le r \le 12 \\ 0, & 13 \le r < \infty \end{cases}$$

The plot of the function $\varphi_X(r)$ is shown in Figure 1.7.

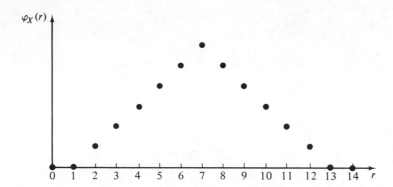

Figure 1.7. Plot of the PMF $\varphi_X(r)$ when $X = X_1 + X_2$ in Example 1.1.

It is also possible to obtain the PMF $\varphi_X(r)$ by using the convolution form of $\varphi_X(r)$. This is left as an exercise to the reader.

(b) $X = \max \{X_1, X_2\}$.

Let $\Phi_X(r)$ be the distribution function of the random variable $X = \max \{X_1, X_2\}$; X is defined for $r = 0, 1, 2, \ldots$, and we have

$$\Phi_X(r) = P\{\max\{X_1, X_2\} \le r\} = \sum_{0 \le \max\{r_1, r_2\} \le r} \varphi_{X_1}(r_1)\varphi_{X_2}(r_2)$$

To determine the region of summation, we note that

$$\max\{r_1, r_2\} = \begin{cases} r_1, & \text{if } r_1 \ge r_2 \\ r_2, & \text{if } r_2 \ge r_1 \end{cases}$$

Thus, if $r_1 \ge r_2$, the inequality $0 \le \max \{r_1, r_2\} \le r$ is equivalent to the inequality $0 \le r_1 \le r$; and if $r_2 \ge r_1$, the inequality $0 \le \max \{r_1, r_2\} \le r$ is equivalent to the inequality $0 \le r_2 \le r$. Therefore, the region of summation is the set of points in Ω lying inside and on the region defined by the set of points $r_1 = 0$, $r_2 = 0$, $r_1 = r$, and $r_2 = r$ (see Figure 1.8); hence

$$\Phi_X(r) = \sum_{r_2=0}^{r_2=r} \sum_{r_1=0}^{r_1=r} \varphi_{X_1}(r_1)\varphi_{X_2}(r_2), \quad r = 0, 1, 2, \ldots$$

We then obtain the following expressions for $\Phi_X(r)$ depending on the values taken by r:

For $r = 0$,

$$\Phi_X(r) = \sum_{r_2=0}^{r_2=0} \sum_{r_1=0}^{r_1=0} 0$$

$$= 0$$

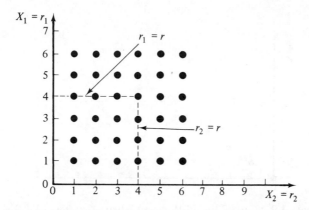

Figure 1.8. Sample space Ω describing the throw of two dice, and set of points describing the event $0 \leq \max\{X_1, X_2\} \leq r$.

For $1 \leq r \leq 6$,

$$\Phi_X(r) = \sum_{r_2=1}^{r_2=r} \sum_{r_1=1}^{r_1=r} \frac{1}{36}$$

$$= \frac{r^2}{36}$$

For $7 \leq r < \infty$,

$$\Phi_X(r) = \sum_{r_2=1}^{r_2=6} \sum_{r_1=1}^{r_1=6} \frac{1}{36}$$

$$= 1$$

And it is easy to verify that the PMF of X, $\varphi_X(r)$, is given by

$$\varphi_X(r) = \begin{cases} 0, & r = 0 \\ \dfrac{2r-1}{36}, & 1 \leq r \leq 6 \\ 0, & 7 \leq r < \infty \end{cases}$$

The reader may verify that

$$\sum_{r=0}^{\infty} \varphi_X(r) = \sum_{r=1}^{r=6} \frac{2r-1}{36} = 1$$

The plot of the function $\varphi_X(r)$ is shown in Figure 1.9.

e. GENERALIZATION

The previous results related to the joint distribution of two random variables X_1 and X_2 can easily be extended to n random variables $X_1, X_2,$

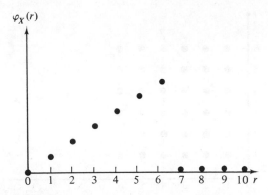

Figure 1.9. Plot of the PMF $\varphi_X(r)$ when $X = \max\{X_1, X_2\}$ in Example 1.1.

$\ldots, X_n$. Thus, if $\{X_i\}$, $i = 1, 2, \ldots, n$, is a sequence of discrete random variables, the joint PMF of $X_1, X_2, \ldots, X_n$, written as $\varphi_{X_1, X_2, \ldots, X_n}(r_1, r_2, \ldots, r_n)$, is defined to be

$$\varphi_{X_1, X_2, \ldots, X_n}(r_1, r_2, \ldots, r_n) = P\{X_1 = r_1, X_2 = r_2, \ldots, X_n = r_n\}$$

and the joint distribution function of $X_1, X_2, \ldots, X_n$ is defined to be

$$\Phi_{X_1, X_2, \ldots, X_n}(r_1, r_2, \ldots, r_n) = P\{X_1 \leq r_1, X_2 \leq r_2, \ldots, X_n \leq r_n\}$$

$$= \sum_{i_n=0}^{i_n=r_n} \cdots \sum_{i_2=0}^{i_2=r_2} \sum_{i_1=0}^{i_1=r_1} \varphi_{X_1, X_2, \ldots, X_n}(i_1, i_2, \ldots, i_n)$$

The sequence of random variables $\{X_i\}$, $i = 1, 2, \ldots, n$, is then defined over the n-dimensional sample space Ω in $\boldsymbol{R}^n$, $\boldsymbol{R}$ being the set of real nonnegative integers.

The marginal PMF of the random variable X_1, for example, is defined to be

$$P\{X_1 = r_1\} = \varphi_{X_1}(r_1) = \sum_{r_n=0}^{\infty} \sum_{r_{n-1}=0}^{\infty} \cdots \sum_{r_2=0}^{\infty} \varphi_{X_1, X_2, \ldots, X_n}(r_1, r_2, \ldots, r_n)$$

and the marginal joint PMF of $X_2, X_3, \ldots, X_n$ is

$$P\{X_2 = r_2, \ldots, X_n = r_n\} = \varphi_{X_2, \ldots, X_n}(r_2, \ldots, r_n)$$

$$= \sum_{r_1=0}^{\infty} \varphi_{X_1, X_2, \ldots, X_n}(r_1, r_2, \ldots, r_n)$$

Similar definitions can be introduced for any of the other combinations of variables.

We may define the conditional PMF of X_n, given that $X_1 = k_1, X_2 = k_2, \ldots, X_{n-1} = k_{n-1}$, to be

$$\varphi_{X_n|X_1,X_2,\ldots,X_{n-1}}(r_n|k_1,k_2,\ldots,k_{n-1})$$

$$= P\{X_n = r_n | X_1 = k_1, X_2 = k_2, \ldots, X_{n-1} = k_{n-1}\}$$

$$= \frac{P\{X_1 = k_1, X_2 = k_2, \ldots, X_{n-1} = k_{n-1}, X_n = r_n\}}{P\{X_1 = k_1, X_2 = k_2, \ldots, X_{n-1} = k_{n-1}\}}$$

$$= \frac{\varphi_{X_1,X_2,\ldots,X_n}(k_1,k_2,\ldots,k_{n-1},r_n)}{\varphi_{X_1,X_2,\ldots,X_{n-1}}(k_1,k_2,\ldots,k_{n-1})}$$

The sequence of discrete random variables $\{X_i\}$, $i = 1, 2, \ldots, n$, is said to form a *sequence of mutually independent random variables*, or the random variables $X_1, X_2, \ldots, X_n$ are said to be *stochastically independent*, if

$$P\{X_1 = r_1, X_2 = r_2, \ldots, X_n = r_n\} = P\{X_1 = r_1\}P\{X_2 = r_2\} \ldots P\{X_n = r_n\}$$

for all $r_1, r_2, \ldots, r_n$. The notion of independence implies that

$$\varphi_{X_1,X_2,\ldots,X_n}(r_1, r_2, \ldots, r_n) = \varphi_{X_1}(r_1)\varphi_{X_2}(r_2) \ldots \varphi_{X_n}(r_n)$$

and
$$\Phi_{X_1,X_2,\ldots,X_n}(r_1, r_2, \ldots, r_n) = \Phi_{X_1}(r_1)\Phi_{X_2}(r_2) \ldots \Phi_{X_n}(r_n)$$

Hence, for example, it follows that

$$\varphi_{X_n|X_1,X_2,\ldots,X_{n-1}}(r_n|k_1,k_2,\ldots,k_{n-1}) = \varphi_{X_n}(r_n)$$

We can also introduce the notion of a *function* of the random variables $X_1, X_2, \ldots, X_n$, say $g(X_1, X_2, \ldots, X_n)$. In general, this function is itself a random variable X, $X = g(X_1, X_2, \ldots, X_n)$. For example, if

$$X = g(X_1, X_2, \ldots, X_n) = X_1 + X_2 + \cdots + X_n$$

the function $g(X_1, X_2, \ldots, X_n)$ defines the random variable X equal to the sum of the random variables $X_1, X_2, \ldots, X_n$. If in particular the random variables $\{X_i\}$, $i = 1, 2, \ldots, n$, are mutually independent, and if $X = X_1 + X_2 + \ldots + X_n$, then

$$\varphi_X(r) = \varphi_{X_1}(r) * \varphi_{X_2}(r) * \cdots * \varphi_{X_n}(r)$$

where the sign $*$ denotes the convolution operation as previously defined. The convolution operation is commutative and associative. For example, the function $\varphi_X(r)$ may be computed as follows: we first form the convolution of the function $\varphi_{X_2}(r)$ and $\varphi_{X_1}(r)$; the resultant function is then convolved with $\varphi_{X_3}(r)$, and so on.

As a special case, if the X_i's have the same PMF, say $\varphi(r)$, we simply write

$$\varphi_X(r) = \varphi *^{(n)} (r)$$

to mean the nth-fold convolution of $\varphi(r)$ with itself.

vii. *Application to the Use of Probability Generating Functions*

a. SUMS OF INDEPENDENT RANDOM VARIABLES, EXAMPLES

We have seen that if X_1 and X_2 are two independently distributed random variables, with respective PMF's $\varphi_{X_1}(r_1)$ and $\varphi_{X_2}(r_2)$, the PMF of the sum $X = X_1 + X_2$ can be obtained by forming the convolution of $\varphi_{X_1}(r)$ with $\varphi_{X_2}(r)$, and is given by

$$\varphi_X(r) = \sum_{r_1=0}^{r_1=r} \varphi_{X_1}(r_1)\varphi_{X_2}(r - r_1)$$

The function $\varphi_X(r)$ is defined for all values of $r, r = 0, 1, 2, \ldots$, if, as previously decided, each of the functions $\varphi_{X_1}(r)$ and $\varphi_{X_2}(r)$ are defined for all values of $r, r = 0, 1, 2, \ldots$ Let now $G_X(s)$ be the PGF of the random variable X:

$$G_X(s) = \sum_{r=0}^{\infty} s^r \varphi_X(r)$$

We shall obtain an important relation between the PGF $G_X(s)$ and the PGF's $G_{X_1}(s)$ and $G_{X_2}(s)$ of the random variables X_1 and X_2. Substituting the convolution form of $\varphi_X(r)$ in the expression for $G_X(s)$, we obtain

$$G_X(s) = \sum_{r=0}^{\infty} s^r \sum_{r_1=0}^{r_1=r} \varphi_{X_1}(r_1)\varphi_{X_2}(r - r_1)$$

The region over which the double-sum expression is to be evaluated is shown in Figure 1.10. Interchanging the order of summation, we obtain

$$G_X(s) = \sum_{r_1=0}^{\infty} \sum_{r=r_1}^{\infty} s^r \varphi_{X_1}(r_1)\varphi_{X_2}(r - r_1)$$

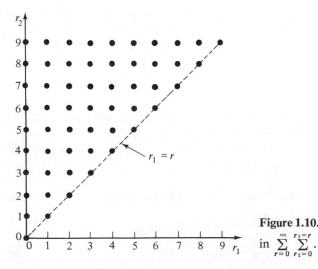

Figure 1.10. Region of summation in $\sum_{r=0}^{\infty} \sum_{r_1=0}^{r_1=r}$.

Let $r = j \overset{.}{+} r_1$; then

$$G_X(s) = \sum_{r_1=0}^{\infty} \sum_{j=0}^{\infty} s^{j+r_1} \varphi_{X_1}(r_1) \varphi_{X_2}(j)$$

$$= \sum_{r_1=0}^{\infty} s^{r_1} \varphi_{X_1}(r_1) \sum_{j=0}^{\infty} s^j \varphi_{X_2}(j)$$

$$= G_{X_1}(s) G_{X_2}(s)$$

We can now state the following important result: *If X_1 and X_2 are independent discrete random variables, then the PGF of $X = X_1 + X_2$ is the product of the PGF's of X_1 and X_2.*

This result can be extended to sums of any number $k, k = 1, 2, \ldots,$ of independent discrete variates $X_1, X_2, \ldots, X_k$. Thus, if $X = X_1 + X_2 + \ldots + X_k$, then

$$G_X(s) = G_{X_1}(s) G_{X_2}(s) \ldots G_{X_k}(s)$$

$$= \prod_{i=1}^{i=k} G_{X_i}(s)$$

In particular, if $X_1, X_2, \ldots, X_k$ are identically and independently distributed with the common PGF $G(s)$, then

$$G_X(s) = [G(s)]^k, \quad k = 1, 2, \ldots$$

In the next set of examples we show how these results can be applied to solve some problems in probability theory and statistics.

Example 1.2

Let $X_1, X_2, \ldots, X_k, k = 1, 2, \ldots,$ be k independently distributed random variables. If

$$X = \sum_{i=1}^{i=k} X_i$$

show that

$$E[X] = \sum_{i=1}^{i=k} E[X_i] \quad \text{and} \quad \text{Var}[X] = \sum_{i=1}^{i=k} \text{Var}[X_i]$$

Suppose that $k = 2$, and let $Z_2 = X_1 + X_2$; then the PGF of Z_2 is

$$G_{Z_2}(s) = G_{X_1}(s) G_{X_2}(s)$$

The first- and second-order derivatives of $G_{Z_2}(s)$ are, respectively,

$$G'_{Z_2}(s) = G'_{X_1}(s) G_{X_2}(s) + G_{X_1}(s) G'_{X_2}(s)$$

and $\qquad G''_{Z_2}(s) = G''_{X_1}(s) G_{X_2}(s) + 2G'_{X_1}(s) G'_{X_2}(s) + G_{X_1}(s) G''_{X_2}(s)$

From expression (1.6),

$$E[Z_2] = G'_{Z_2}(1) = G'_{X_1}(1)G_{X_2}(1) + G_{X_1}(1)G'_{X_2}(1)$$
$$= E[X_1] + E[X_2]$$

From expression (1.8),

$$\begin{aligned}
\text{Var}[Z_2] &= G''_{Z_2}(1) + G'_{Z_2}(1) - [G'_{Z_2}(1)]^2 \\
&= G''_{X_1}(1)G_{X_2}(1) + 2G'_{X_1}(1)G'_{X_2}(1) + G_{X_1}(1)G''_{X_2}(1) \\
&\quad + G'_{X_1}(1)G_{X_2}(1) + G_{X_1}(1)G'_{X_2}(1) \\
&\quad - [G'_{X_1}(1)G_{X_2}(1) + G_{X_1}(1)G'_{X_2}(1)]^2 \\
&= G''_{X_1}(1) + G'_{X_1}(1) - [G'_{X_1}(1)]^2 \\
&\quad + G''_{X_2}(1) + G'_{X_2}(1) - [G'_{X_2}(1)]^2 \\
&= \text{Var}[X_1] + \text{Var}[X_2]
\end{aligned}$$

The final result can be established by induction. Let $Z_{k-1} = \sum_{i=1}^{i=k-1} X_i$; then $X = Z_{k-1} + X_k$, where Z_{k-1} and X_k are independently distributed discrete random variables. Assume now that the results hold true for Z_{k-1}, $k > 3$; then

$$E[Z_{k-1}] = \sum_{i=1}^{i=k-1} E[X_i] \quad \text{and} \quad \text{Var}[Z_{k-1}] = \sum_{i=1}^{i=k-1} \text{Var}[Z_i]$$

Since

$$E[X] = E[Z_{k-1}] + E[X_k]$$

and

$$\text{Var}[X] = \text{Var}[Z_{k-1}] + \text{Var}[X_k]$$

the results follow immediately.

Example 1.3

We consider here the experiment described in Example 1.1, which involved the throw of two symmetrical six-sided dice each numbered 1, 2, ..., 6. The numbers appearing on the first and second dice are random variables denoted, respectively, by X_1 and X_2. Find the PGF of $X = X_1 + X_2$; hence determine the probability that the throw will result in a sum of 10.

The PGF of X_1 is given by

$$\begin{aligned}
G(s) &= \sum_{r_1=0}^{\infty} s^{r_1} \varphi_{X_1}(r_1) \\
&= \sum_{r_1=1}^{r_1=6} s^{r_1} \frac{1}{6} \\
&= \frac{1}{6}(s + s^2 + s^3 + s^4 + s^5 + s^6) \\
&= \frac{s}{6} \frac{1 - s^6}{1 - s}
\end{aligned}$$

Since the PGF of X_2 is the same as that of X_1, it follows that the PGF of $X = X_1 + X_2$ is

$$G_X(s) = \frac{1}{36}(s + s^2 + s^3 + s^4 + s^5 + s^6)^2$$

$$= \frac{s^2}{36}\left(\frac{1 - s^6}{1 - s}\right)^2$$

$$= \sum_{r=0}^{\infty} s^r \varphi_X(r)$$

where $\varphi_X(r)$ is the PMF of X. To determine the probability that the throw will result in a sum of 10, we need to find $P\{X = 10\} = \varphi_X(10)$, which is the coefficient of s^{10} in the power series expansion of $G_X(s)$. This can easily be seen to be $\frac{1}{36}(1 + 2) = \frac{3}{36}$.

Example 1.4

Let $\{X_i\}$, $i = 1, 2, 3, \ldots, n$, be a sequence of n identically distributed independent discrete random variables. Find the PMF of

$$X = \sum_{i=1}^{i=n} X_i$$

when each of the X_i has the following distribution:

(a) Bernoulli.
(b) Poisson.
(c) Geometric.

(a) If each X_i has a Bernoulli distribution with PMF

$$\varphi(r) = \begin{cases} p^r q^{1-r}, & r = 0, 1 \\ 0, & r = 2, 3, \ldots \end{cases}$$

$0 < p < 1$, $p + q = 1$, then the PGF of each X_i is $G(s) = q + ps$. Hence the PGF of X is

$$G_X(s) = (q + ps)^n$$

which can be identified as the PGF of a random variable with the binomial distribution

$$\varphi_X(r) = \begin{cases} \binom{n}{r} p^r q^{n-r}, & r = 0, 1, 2, \ldots, n \\ 0, & r = n + 1, \ldots \end{cases}$$

(b) If each X_i has a Poisson distribution with PMF

$$\varphi(r) = e^{-\lambda}\frac{\lambda^r}{r!}, \quad r = 0, 1, 2, \ldots$$

$\lambda > 0$, then the PGF of each X_i is $G(s) = e^{-\lambda(1-s)}$. Hence the PGF of X is

$$G_X(s) = [e^{-\lambda(1-s)}]^n = e^{-n\lambda(1-s)}$$

Thus X is Poisson distributed with parameter $n\lambda$.

(c) If each X_i has a geometric distribution with PMF

$$\varphi(r) = pq^r, \quad r = 0, 1, 2, \ldots$$

$0 < p < 1$, $p + q = 1$, then, the PGF of each X_i is $G(s) = p/(1 - qs)$. Hence the PGF of X is

$$G_X(s) = \left(\frac{p}{1 - qs}\right)^n$$

which can be identified as the PGF of a random variable having the negative binomial distribution

$$\varphi_X(r) = \binom{n - 1 + r}{r}p^n q^r, \quad r = 0, 1, 2, \ldots$$

b. Compound Distributions

Let $\{X_i\}$, $i = 1, 2, \ldots, N$, be a sequence of mutually independent identically distributed discrete random variables with common PMF $\varphi(r)$, and common PGF $G(s)$. Suppose now that N itself is a discrete random variable independently distributed from all X_i, with PMF $\varphi_N(n)$, $n = 1, 2, \ldots$. The discrete random variable

$$Y = X_1 + X_2 + \cdots + X_N$$

is said to have a *compound distribution*. We shall denote by $\varphi_Y(k)$ the marginal PMF of the random variable Y. Let now $\varphi_{Y,N}(k, n)$ be the joint PMF of the random variables Y and N; then the conditional PMF of Y, given $N = n$, is

$$\varphi_{Y|N}(k \,|\, n) = \frac{\varphi_{Y,N}(k,n)}{\varphi_N(n)}, \quad \text{for all } n = 1, 2, \ldots$$

Hence, for $n = 1, 2, \ldots$, we have

$$\varphi_{Y,N}(k, n) = \varphi_{Y|N}(k \,|\, n) \cdot \varphi_N(n)$$

and the marginal PMF of Y is then given by

$$\varphi_Y(k) = \sum_{n=1}^{\infty} \varphi_{Y,N}(k, n)$$

$$= \sum_{n=1}^{\infty} \varphi_{Y|N}(k \mid n) \cdot \varphi_N(n)$$

Since the X_i, $i = 1, 2, \ldots, N$, are independently and identically distributed random variables, it follows that for a given $N = n$ their sum $Y = X_1 + X_2 + \cdots + X_n$ has a PMF given by

$$\varphi_{Y|N}(k \mid n) = \varphi*^{(n)}(k)$$

Hence
$$\varphi_Y(k) = \sum_{n=1}^{\infty} \varphi*^{(n)}(k)\varphi_N(n)$$

Let now $G_Y(s)$ be the PGF of the random variable Y; then

$$G_Y(s) = \sum_{k=0}^{\infty} s^k \varphi_Y(k)$$

or
$$G_Y(s) = \sum_{k=0}^{\infty} s^k \sum_{n=1}^{\infty} \varphi*^{(n)}(k)\varphi_N(n)$$

Interchanging the order of summation, we obtain

$$G_Y(s) = \sum_{n=1}^{\infty} \varphi_N(n) \sum_{k=0}^{\infty} s^k \varphi*^{(n)}(k)$$

For a given $N = n$, let $G_{Y|N}(s, n)$ denote the PGF of the random variable Y, conditional on N, $n = 1, 2, \ldots$; then

$$G_{Y|N}(s, n) = \sum_{k=0}^{\infty} s^k \varphi_{Y|N}(k \mid n)$$

$$- \sum_{k=0}^{\infty} s^k \varphi*^{(n)}(k)$$

$$= [G(s)]^n$$

Hence
$$G_Y(s) = \sum_{n=1}^{\infty} \varphi_N(n)[G(s)]^n$$

We now attempt to obtain an expression for $G_Y(s)$ and introduce the PGF of the random variable N, $G_N(s)$, to be

$$G_N(s) = \sum_{n=1}^{\infty} s^n \varphi_N(n)$$

The expression for the PGF of Y can be written as

$$G_Y(s) = \sum_{n=1}^{\infty} [G(s)]^n \varphi_N(n)$$

Since $|G(s)| \leq 1$ for all $|s| \leq 1$, then comparing the expression for $G_Y(s)$ with the expression for $G_N(s)$, we obtain

$$G_Y(s) = G_N[G(s)]$$

This defines the PGF of the random variable Y, having marginal or unconditional PMF $\varphi_Y(k)$, in terms of the two known PGF's $G_N(s)$ and $G(s)$.

We note that, since $G(1) = 1$ and $G_N(1) = 1$,

$$G_Y(1) = G_N[G(1)] = G_N(1) = 1$$

a check to the effect that the random variable Y possesses a valid PMF $\varphi_Y(k)$. Theoretically, one could obtain expressions for $\varphi_Y(0)$, $\varphi_Y(1)$, ... by evaluating the function $G_Y(s)$ and its derivatives at $s = 0$. For example,

$$\varphi_Y(0) = G_Y(0) = G_N[G(0)] = G_N[\varphi(0)]$$

However, even if the functional forms of $G_N(s)$ and $G(s)$ are known, the evaluation of the specific form of $\varphi_Y(k)$ is not always an easy matter.

Fortunately, given the respective means and variances of the random variables X_i, $i = 1, 2, \ldots, n$, and N, it becomes possible to obtain an expression for the mean and variance of the random variable Y without a knowledge of the specific form of the function $G_Y(s)$. Differentiating the expression for $G_Y(s)$ once with respect to s, we obtain

$$G_Y'(s) = \frac{dG_Y(s)}{ds} = \frac{dG_Y(s)}{dG(s)} \cdot \frac{dG(s)}{ds}$$

$$= G_N'[G(s)] \cdot G'(s)$$

Thus $\qquad E[Y] = G_Y'(1) = G_N'[G(1)] \cdot G'(1)$

Since $G(1) = 1$, it follows that

$$E[Y] = G_N'(1) \cdot G'(1) = E[N] \cdot E[X].$$

In a similar way, one may show that

$$\text{Var}[Y] = \text{Var}[N] \cdot \{E[X]\}^2 + E[N] \cdot \text{Var}[X]$$

Exercise 1.3

If $\varphi(r) = pq^r$ $(0 < p < 1; p + q = 1; r = 0, 1, 2, \ldots)$, and $\varphi_N(n) = p_1 q_1^{n-1}$ $(0 < p_1 < 1; p_1 + q_1 = 1; n = 1, 2, 3, \ldots)$, show that Y has the geometric distribution

$$\varphi_Y(y) = \frac{pp_1}{1 - pq_1} \left(\frac{q}{1 - pq_1}\right)^y, \quad y = 0, 1, 2, \ldots$$

Remark

Assume now that N is a discrete random variable independently distributed from all X_i with PMF $\varphi_N(n)$, $n = 0, 1, 2, \ldots$. The discrete random variable Y defined as

$$Y = \begin{cases} 0, & N = 0 \\ X_1 + X_2 + \cdots + X_N, & N = 1, 2, \ldots \end{cases}$$

is also said to have a compound distribution. The expression for the PGF of Y may be shown as before to be

$$G_Y(s) = \sum_{n=0}^{\infty} [G(s)]^n \varphi_N(n)$$
$$= G_N[G(s)]$$

The expressions for $E[Y]$ and $\text{Var}[Y]$ remain unchanged.

Exercise 1.4

In the compound Poisson distribution, the random variable N has the Poisson distribution

$$\varphi_N(n) = e^{-\lambda} \frac{\lambda^n}{n!}, \quad n = 0, 1, 2, \ldots$$

Then
$$G_N(s) = e^{-\lambda(1-s)}$$

and
$$G_Y(s) = e^{-\lambda[1-G(s)]}$$

Show that
$$P\{Y = 1\} = \lambda \varphi(1) \cdot P\{Y = 0\}$$

Find $E[Y]$ and $\text{Var}[Y]$.

1.4. Continuous Random Variables

i. *Definition: Probability Density Function*

A random variable is said to be continuous if its probability law can be expressed by a *probability density function* (PDF) $\varphi_X(\cdot)$ such that

$$\Phi_X(x) = \int_{-\infty}^{x} \varphi_X(u) \, du$$

that is, if there exists a function $\varphi_X(\cdot)$ such that the distribution function $\Phi_X(x)$ of the random variable X can be expressed as an integral. The random variable X is then said to have a *continuous distribution*, and the sample space Ω over which X is defined is the set of real numbers. Clearly,

$$\varphi_X(x) = \frac{d}{dx} \Phi_X(x)$$

The probability that the random variable X will take on values between x and $x + dx$ is

$$P\{x < X \le x + dx\} = \varphi_X(x)\, dx$$

The probability that the random variable X will take on values between a and b, $a < b$, is

$$P\{a < X \le b\} = \Phi_X(b) - \Phi_X(a)$$
$$= \int_a^b \varphi_X(x)\, dx$$

The probability density function $\varphi_X(x)$ of a continuous random variable X possesses the properties

$$\varphi_X(x) \ge 0, \quad \text{for all } x$$

and

$$\int_x \varphi_X(x)\, dx = 1$$

The integration is carried over all possible values of x known to be the proper range of definition of the random variable X. The range of definition of X may be either finite or infinite. Here again, as in the discrete case, it is convenient to extend the range of definition of X to include the set of points $\{x : 0 \le x < \infty\}$ or $\{x : -\infty < x < \infty\}$ by defining $\varphi_X(x)$ to be equal to zero whenever X lies outside its proper range of definition. We may then write, in general, for example, the relation $\int_{-\infty}^{\infty} \varphi_X(x)\, dx = 1$. It should be noted, however, that for a continuous random variable the value of $\varphi_X(x)$, at a particular value of x, may be unbounded.

Again, when no ambiguity exists, we shall drop the subscript X from $\Phi_X(x)$ and $\varphi_X(x)$ to read, respectively, $\Phi(x)$ and $\varphi(x)$.

ii. *Expectation and Variance*

Let X be a continuous random variable with PDF $\varphi_X(x)$, $-\infty < x < \infty$. We define the *expectation* or *expected value* or *mean* of X, and write $E[X]$ or μ_X as

$$E[X] = \mu_X = \int_{-\infty}^{\infty} x \varphi_X(x)\, dx$$

Let again $g(X)$ be a *function* of the random variable X; we define the expectation of $g(X)$, if it exists, as

$$E[g(X)] = \int_{-\infty}^{\infty} g(x)\, \varphi_X(x)\, dx$$

The *variance* of the continuous random variable X, denoted by Var$[X]$ or σ_X^2, is defined to be $E[(X - \mu_X)^2]$:

$$\text{Var}[X] = \sigma_X^2 = E[(X - \mu_X)^2] = \int_{-\infty}^{\infty} (x - \mu_X)^2 \varphi_X(x)\, dx$$

$$= \int_{-\infty}^{\infty} x^2 \varphi_X(x)\, dx - 2\mu_X \int_{-\infty}^{\infty} x\varphi_X(x)\, dx + \mu_X^2 \int_{-\infty}^{\infty} \varphi_X(x)\, dx$$

$$= E[X^2] - \mu_X^2$$

The *standard deviation* of X, σ_X, is defined in the same way as for discrete random variables. The kth moment about the origin of the random variable X is defined as $E[X^k]$.

iii. *Moment Generating Functions and Laplace Transforms*

Let X be a continuous random variable with PDF $\varphi_X(x)$, $-\infty < x < \infty$. The *moment generating function* $M_X(\theta)$ of X is defined, for any real number θ, by

$$M_X(\theta) = E[e^{\theta X}] = \int_{-\infty}^{\infty} e^{\theta x} \varphi_X(x)\, dx$$

Because the moment generating function of X may not always exist, it is more convenient to introduce the notion of the *Laplace transform $\bar{\varphi}_X(s)$ of the density function $\varphi_X(x)$ of X,* defined for $s \geq 0$, to be

$$\bar{\varphi}_X(s) = E[e^{-sX}] = \int_{-\infty}^{\infty} e^{-sx} \varphi_X(x)\, dx$$

Clearly, $M_X(\theta) = \bar{\varphi}_X(-\theta)$. The function $\varphi_X(x)$ is obtained as the inverse Laplace transform of $\bar{\varphi}_X(s)$ (see, e.g., Appendix).

Expanding e^{-sx} as a power series in x, we obtain for the Laplace transform of the PDF $\varphi_X(x)$

$$\bar{\varphi}_X(s) = \int_{-\infty}^{\infty} \left[1 - sx + \frac{(sx)^2}{2!} - \cdots + \frac{(-sx)^k}{k!} + \cdots \right] \varphi_X(x)\, dx$$

$$= \int_{-\infty}^{\infty} \sum_{k=0}^{\infty} \frac{(-sx)^k}{k!} \varphi_X(x)\, dx$$

$$= \sum_{k=0}^{\infty} \frac{(-s)^k}{k!} \int_{-\infty}^{\infty} x^k \varphi_X(x)\, dx$$

$$= \sum_{k=0}^{\infty} \frac{(-s)^k}{k!} E[X^k]$$

Thus the coefficient of s^k in the power series expansion of the Laplace transform $\bar{\varphi}_X(s)$ of $\varphi_X(x)$ yields expressions for the kth moment of X about the origin. It is also evident that the quantity $(-)^k E[X^k]$ can be obtained by evaluating the kth derivative of $\bar{\varphi}_X(s)$ with respect to s, at the point $s = 0$. In particular, we have

$$\bar{\varphi}_X(0) = 1$$

$$\frac{d\bar{\varphi}_X(s)}{ds}\bigg|_{s=0} = \bar{\varphi}'_X(0) = -E[X]$$

$$\frac{d^2\bar{\varphi}_X(s)}{ds^2}\bigg|_{s=0} = \bar{\varphi}''_X(0) = E[X^2]$$

Hence

$$\text{Var}[X] = \bar{\varphi}''_X(0) - [\bar{\varphi}'_X(0)]^2$$

Also

$$\bar{\varphi}_X(s) = 1 - \mu_X s + \frac{1}{2!}(\sigma_X^2 + \mu_X^2)s^2 - \cdots$$

Example 1.5

Let X be a continuous random variable with PDF $\varphi_X(x)$, $0 < x < \infty$. Define $H_X(x) = \int_x^\infty \varphi_X(u)\, du$ as the *complementary cumulative function* of X. Show that $E[X] = \int_0^\infty H_X(x)\, dx$.

We have

$$\int_0^\infty H_X(x)\, dx = \int_0^\infty \int_x^\infty \varphi_X(u)\, du\, dx$$

Interchanging the order of integration in the right-hand expression yields

$$\int_0^\infty H_X(x)\, dx = \int_0^\infty \varphi_X(u) \int_0^u dx\, du$$

$$= \int_0^\infty u\varphi_X(u)\, du = E[X]$$

Example 1.6

Let X be a continuous random variable with PDF

$$\varphi_X(x) = \begin{cases} 0, & -\infty < x \le 0 \\ \dfrac{1}{\sqrt{\pi}} \dfrac{e^{-x}}{\sqrt{x}}, & 0 < x < \infty \end{cases}$$

Show that $\varphi_X(x)$ is a proper density function.

We note first that this particular density function has a discontinuity at

$x = 0$, since $\lim_{x \to 0^+} \varphi_X(x) = \infty$. Let now

$$A = \frac{1}{\sqrt{\pi}} \int_0^\infty \frac{e^{-u}}{\sqrt{u}} \, du$$

Making the change in variable $u = x^2$ in the integral term, we obtain

$$A = \frac{1}{\sqrt{\pi}} \int_0^\infty \frac{e^{-x^2}}{x} 2x \, dx = \frac{2}{\sqrt{\pi}} \int_0^\infty e^{-x^2} \, dx$$

Now

$$A^2 = \left(\frac{2}{\sqrt{\pi}} \int_0^\infty e^{-x^2} \, dx \right)\left(\frac{2}{\sqrt{\pi}} \int_0^\infty e^{-y^2} \, dy \right)$$

$$= \frac{4}{\pi} \int_0^\infty \int_0^\infty e^{-(x^2+y^2)} \, dx \, dy$$

If we make a change to polar coordinates (r, θ) by setting $x = r \cos \theta$ and $y = r \sin \theta$, we obtain

$$A^2 = \frac{4}{\pi} \int_0^{\pi/2} \int_0^\infty e^{-r^2} r \, dr \, d\theta$$

$$= \frac{4}{\pi} \left(\int_0^{\pi/2} d\theta \right)\left(\frac{1}{2} \int_0^\infty d(-e^{-r^2}) \right) = 1$$

Hence

$$\int_0^\infty \frac{1}{\sqrt{\pi}} \frac{e^{-u}}{\sqrt{u}} \, du = 1$$

iv. *Examples of Continuous Distributions*

a. RECTANGULAR DISTRIBUTION

A continuous random variable X is said to have a *rectangular* or *uniform distribution* over the interval $(b, c]$ if its PDF is given by

$$\varphi_X(x) = \begin{cases} \dfrac{1}{c - b}, & b < x \le c \\ 0, & \text{otherwise} \end{cases}$$

An important special case is when X is uniformly distributed over the range $0 < x \le a$, in which case

$$\varphi_X(x) = \begin{cases} 0, & -\infty < x \le 0 \\ \dfrac{1}{a}, & 0 < x \le a \\ 0, & a < x < \infty \end{cases}$$

For this special PDF, the expression for the Laplace transform is

$$\bar{\varphi}_X(s) = \int_{-\infty}^{\infty} e^{-sx} \varphi_X(x)\, dx$$

$$= \int_0^a e^{-sx} \frac{1}{a}\, dx$$

$$= \frac{1 - e^{-as}}{as}$$

We next show how to use $\bar{\varphi}_X(s)$ to evaluate $E[X]$. We have

$$\bar{\varphi}'_X(s) = \frac{as \cdot ae^{-as} - a(1 - e^{-as})}{(as)^2}$$

$$= \frac{a^2 s e^{-as} - a + ae^{-as}}{a^2 s^2}$$

Applying L' Hospital's rule once, we obtain

$$\lim_{s \to 0} \bar{\varphi}'_X(s) = \lim_{s \to 0} \frac{a^2 e^{-as} - a^3 s e^{-as} - a^2 e^{-as}}{2a^2 s}$$

$$= \lim_{s \to 0} \left(-\frac{ae^{-as}}{2} \right) = -\frac{a}{2}$$

Hence $E[X] = \dfrac{a}{2}$

By forming $\bar{\varphi}''_X(s)$ and evaluating $\lim_{s \to 0} \bar{\varphi}''_X(s)$, we find in a similar fashion

$$E[X^2] = \frac{a^2}{3}$$

Thus $\text{Var}[X] = \dfrac{a^2}{3} - \dfrac{a^2}{4} = \dfrac{a^2}{12}$

Exercise 1.5

    If X has a general uniform distribution with PDF $\varphi_X(x)$, $b < x \leq c$, zero otherwise, determine $E[X]$, $E[X^2]$ and $\text{Var}[X]$, using the basic definition of these quantities.

    b. NEGATIVE EXPONENTIAL DISTRIBUTION

    A continuous random variable X is said to have a *negative exponential* or simply an *exponential distribution* with parameter λ, $\lambda > 0$, if its PDF is given by

$$\varphi_X(x) = \begin{cases} 0, & -\infty < x \le 0 \\ \lambda e^{-\lambda x}, & 0 < x < \infty \end{cases}$$

The Laplace transform of $\varphi_X(x)$ is

$$\begin{aligned} \bar{\varphi}_X(s) &= \int_{-\infty}^{\infty} e^{-sx} \varphi_X(x)\, dx \\ &= \int_0^{\infty} e^{-sx} \lambda e^{-\lambda x}\, dx \\ &= \int_0^{\infty} \lambda e^{-(\lambda+s)x}\, dx \\ &= \frac{\lambda}{\lambda + s}, \quad s > -\lambda \end{aligned}$$

From which we obtain

$$\bar{\varphi}'_X(s) = -\frac{\lambda}{(\lambda + s)^2}$$

$$\bar{\varphi}''_X(s) = \frac{2\lambda}{(\lambda + s)^3}$$

Hence

$$E[X] = -\bar{\varphi}'_X(0) = \frac{1}{\lambda}$$

and

$$\mathrm{Var}[X] = \bar{\varphi}''_X(0) - [\bar{\varphi}'_X(0)]^2$$

$$= \frac{2}{\lambda^2} - \frac{1}{\lambda^2} = \frac{1}{\lambda^2}$$

The negative exponential distribution plays an important role in elementary queuing theory and in reliability theory. It is the continuous counterpart of the geometric distribution defined for discrete variates.

c. Gamma Distribution

A continuous random variable X is said to have a *gamma distribution* of order n and with parameter λ if its PDF is given by

$$\varphi_X(x) = \begin{cases} 0, & -\infty < x \le 0 \\ \dfrac{(\lambda x)^{n-1}}{\Gamma(n)} \lambda e^{-\lambda x}, & 0 < x < \infty \end{cases}$$

where $\lambda > 0$ and $0 < n < \infty$. Note that when $n = 1$, the negative exponential distribution is recovered.

The Laplace transform of $\varphi_X(x)$ is given by

$$\bar{\varphi}_X(s) = \int_{-\infty}^{\infty} e^{-sx}\varphi_X(x)\,dx$$

$$= \int_0^{\infty} e^{-sx}\frac{(\lambda x)^{n-1}}{\Gamma(n)}\lambda e^{-\lambda x}\,dx$$

$$= \frac{\lambda^n}{\Gamma(n)}\int_0^{\infty} x^{n-1}e^{-x(s+\lambda)}\,dx$$

$$= \left(\frac{\lambda}{\lambda + s}\right)^n, \quad s > -\lambda$$

The first and second derivatives of $\bar{\varphi}_X(s)$ are given, respectively, by

$$\bar{\varphi}'_X(s) = -\frac{n\lambda^n}{(\lambda + s)^{n+1}}$$

and

$$\bar{\varphi}''_X(s) = \frac{n(n+1)\lambda^n}{(\lambda + s)^{n+2}}$$

Thus

$$E[X] = -\bar{\varphi}'_X(0) = \frac{n}{\lambda}$$

and

$$\operatorname{Var}[X] = \bar{\varphi}''_X(0) - [\bar{\varphi}'_X(0)]^2$$

$$= \frac{n(n+1)}{\lambda^2} - \frac{n^2}{\lambda^2} = \frac{n}{\lambda^2}$$

The gamma distribution is the continuous counterpart of the negative binomial distribution for discrete random variables. The χ^2 distribution encountered in statistics is a special form of the gamma distribution. We will in a later section show that if a sequence of k random variables $\{X_i\}$, $i = 1, 2, \ldots, k$, are independently distributed, each with a gamma distribution with parameter λ, their sum $\sum_{i=1}^{i=k} X_i$ has also a gamma distribution with parameter λ. Figure 1.11 illustrates possible shapes of the gamma density function for $n = \frac{1}{2}$, 1, and 10.

d. Normal or Gaussian Distribution

A continuous random variable X is said to have a *normal* or *Gaussian distribution* if its PDF is given by

$$\varphi_X(x) = \frac{1}{\sqrt{2\pi}\sigma} e^{-(x-\mu)^2/2\sigma^2}, \quad -\infty < x < \infty$$

where μ, $-\infty < \mu < \infty$, and σ, $\sigma > 0$, are the two parameters of the distribution. The Laplace transform of the PDF $\varphi_X(x)$ is

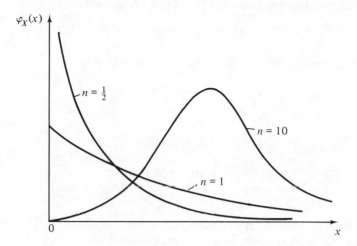

Figure 1.11. Gamma density functions of order $n = \frac{1}{2}$, 1 and 10.

$$\varphi_X(s) = \int_{-\infty}^{\infty} e^{-sx} \varphi_X(x)\, dx$$

$$= \int_{-\infty}^{\infty} e^{-sx} \frac{1}{\sqrt{2\pi}\sigma} e^{-(x-\mu)^2/2\sigma^2}\, dx$$

Making the change in variable $u = (x - \mu)/\sigma$, we obtain

$$\bar\varphi_X(s) = \int_{-\infty}^{\infty} e^{-s(\mu+\sigma u)} \cdot \frac{1}{\sqrt{2\pi}\sigma} e^{-u^2/2}\sigma\, du$$

$$= \frac{e^{-\mu s}}{\sqrt{2\pi}} \int_{-\infty}^{\infty} e^{-[(u^2/2) + \sigma s u]}\, du$$

$$= \frac{e^{-\mu s}}{\sqrt{2\pi}} \int_{-\infty}^{\infty} e^{-(1/2)(u^2 + 2\sigma s u + \sigma^2 s^2) + (1/2)\sigma^2 s^2}\, du$$

$$= \frac{e^{-\mu s + (1/2)\sigma^2 s^2}}{\sqrt{2\pi}} \int_{-\infty}^{\infty} e^{-(1/2)(u + \sigma s)^2}\, du$$

Let now

$$v = \frac{u + \sigma s}{\sqrt{2}}$$

Then

$$\bar\varphi_X(s) = \frac{e^{-\mu s + (1/2)\sigma^2 s^2}}{\sqrt{2\pi}} \int_{-\infty}^{\infty} \sqrt{2}\, e^{-v^2}\, dv$$

$$= \frac{e^{-\mu s + (1/2)\sigma^2 s^2}}{\sqrt{\pi}} \left(2 \int_{0}^{\infty} e^{-v^2}\, dv \right)$$

The last integral was encountered in Example 1.6, and we obtain immediately

$$\bar{\varphi}_X(s) = e^{-\mu s + (1/2)\sigma^2 s^2}$$

The first and second derivatives of $\bar{\varphi}_X(s)$ are, respectively,

$$\bar{\varphi}'_X(s) = (-\mu + \sigma^2 s)e^{-\mu s + (1/2)\sigma^2 s^2}$$
$$\bar{\varphi}''_X(s) = \sigma^2 e^{-\mu s + (1/2)\sigma^2 s^2} + (-\mu + \sigma^2 s)^2 e^{-\mu s + (1/2)\sigma^2 s^2}$$

Thus
$$E[X] = -\bar{\varphi}'_X(0) = \mu$$

and
$$\text{Var}[X] = \bar{\varphi}''_X(0) - [\bar{\varphi}'_X(0)]^2$$
$$= \sigma^2 + \mu^2 - \mu^2 = \sigma^2$$

The PDF of a normal distribution is symmetrical about $E[X]$ (see Figure 1.12). The importance of the normal distribution in the theory of statistics stems from results related to the theory of large numbers and the central limit theorem.

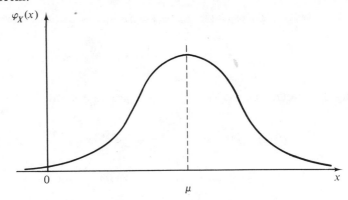

Figure 1.12. A normal density function.

v. *Jointly Distributed Continuous Random Variables*

a. Definition and Expectation

Let X_1 and X_2 be two continuous random variables defined as a pair over a given sample space Ω. By similarity to the theory of discrete variates, we define the *joint distribution function* of X_1 and X_2, $\Phi_{X_1, X_2}(x_1, x_2)$, to be

$$\Phi_{X_1, X_2}(x_1, x_2) = P\{X_1 \leq x_1, X_2 \leq x_2\}$$

The *joint probability density function*, joint PDF, of X_1 and X_2,

$\varphi_{X_1, X_2}(x_1, x_2)$, is defined by the relation

$$\Phi_{X_1, X_2}(x_1, x_2) = \int_{-\infty}^{x_2} \int_{-\infty}^{x_1} \varphi_{X_1, X_2}(u_1, u_2) \, du_1 \, du_2$$

from which we deduce that

$$\varphi_{X_1, X_2}(x_1, x_2) = \frac{\partial^2}{\partial x_1 \, \partial x_2} \Phi_{X_1, X_2}(x_1, x_2)$$

Also,

$$\int_{-\infty}^{\infty} \int_{-\infty}^{\infty} \varphi_{X_1, X_2}(x_1, x_2) \, dx_1 \, dx_2 = 1$$

For convenience, the subscripts X_1 and X_2 are sometimes omitted.

Let $g(X_1, X_2)$ be a function of the random variables X_1 and X_2. The expectation of $g(X_1, X_2)$, if it exists, is defined as

$$E[g(X_1, X_2)] = \int_{-\infty}^{\infty} \int_{-\infty}^{\infty} g(x_1, x_2) \varphi_{X_1, X_2}(x_1, x_2) \, dx_1 \, dx_2$$

b. Marginal and Conditional Probability Density Functions

Let $\varphi_{X_1, X_2}(x_1, x_2)$ be the joint PDF of the two random variables X_1 and X_2; the *marginal PDF* of the random variable X_1 is defined as

$$\varphi_{X_1}(x_1) = \int_{-\infty}^{\infty} \varphi_{X_1, X_2}(x_1, x_2) \, dx_2$$

One may similarly define the marginal PDF of the random variable X_2, $\varphi_{X_2}(x_2)$.

At any value of $X_1 = y_1$ for which $\varphi_{X_1}(y_1) > 0$, we define *the conditional PDF* of X_2, given that $X_1 = y_1$, as

$$\varphi_{X_2 | X_1}(x_2 | y_1) = \frac{\varphi_{X_1, X_2}(y_1, x_2)}{\varphi_{X_1}(y_1)}$$

Similarly, we can define for $\varphi_{X_2}(y_2) > 0$, the conditional PDF of X_1, given that $X_2 = y_2$, to be

$$\varphi_{X_1 | X_2}(x_1 | y_2) = \frac{\varphi_{X_1, X_2}(x_1, y_2)}{\varphi_{X_2}(y_2)}$$

c. Independent Random Variables

Two continuous random variables X_1 and X_2 are said to be *independent* if

$$\varphi_{X_1, X_2}(x_1, x_2) = \varphi_{X_1}(x_1) \varphi_{X_2}(x_2)$$

This definition implies that

$$\Phi_{X_1, X_2}(x_1, x_2) = \Phi_{X_1}(x_1)\Phi_{X_2}(x_2)$$
$$\varphi_{X_2|X_1}(x_2|x_1) = \varphi_{X_2}(x_2)$$

and

$$\varphi_{X_1|X_2}(x_1|x_2) = \varphi_{X_1}(x_1)$$

d. FUNCTIONS OF RANDOM VARIABLES AND SUMS OF RANDOM VARIABLES; CONVOLUTIONS; EXAMPLES

If $g(X_1, X_2)$ is a *function* of the two continuous random variables X_1 and X_2, then in general $g(X_1, X_2)$ will itself be a continuous random variable, say $X = g(X_1, X_2)$. The distribution function of X, $\Phi_X(x)$, may be found from the relation

$$\Phi_X(x) = P\{X \leq x\}$$
$$= P\{g(X_1, X_2) \leq x\}$$
$$= \iint\limits_{g(x_1, x_2) \leq x} \varphi_{X_1, X_2}(x_1, x_2) \, dx_1 \, dx_2$$

where the double integral is evaluated over all values of x_1 and x_2 for which $g(x_1, x_2) \leq x$.

A particular case of interest is when $X = X_1 + X_2$ and the random variables X_1 and X_2 are independently distributed; X then is a continuous random variable, and

$$\Phi_X(x) = \iint\limits_{x_1 + x_2 \leq x} \varphi_{X_1, X_2}(x_1, x_2) \, dx_1 \, dx_2$$
$$= \iint\limits_{x_1 + x_2 \leq x} \varphi_{X_1}(x_1)\varphi_{X_2}(x_2) \, dx_1 \, dx_2$$
$$= \int_{-\infty}^{\infty} \int_{-\infty}^{x - x_2} \varphi_{X_1}(x_1)\varphi_{X_2}(x_2) \, dx_1 \, dx_2$$
$$= \int_{-\infty}^{\infty} \varphi_{X_2}(x_2) \int_{-\infty}^{x - x_2} \varphi_{X_1}(x_1) \, dx_1 \, dx_2$$
$$= \int_{-\infty}^{\infty} \varphi_{X_2}(x_2)\Phi_{X_1}(x - x_2) \, dx_2$$

Differentiating this last expression with respect to x yields

$$\varphi_X(x) = \int_{-\infty}^{\infty} \varphi_{X_2}(x_2)\varphi_{X_1}(x - x_2) \, dx_2$$

Similarly, one can show that

$$\varphi_X(x) = \int_{-\infty}^{\infty} \varphi_{X_1}(x_1)\varphi_{X_2}(x - x_1) \, dx_1$$

These last two expressions are defined as the *convolution* of the PDF's $\varphi_{x_1}(x)$ and $\varphi_{x_2}(x)$ and are written as $\varphi_{x_1}(x) * \varphi_{x_2}(x)$ or $\varphi_{x_2}(x) * \varphi_{x_1}(x)$.

If the two random variables X_1 and X_2 are defined only over non-negative values, then the PDF of the random variable X is written more explicitly as

$$\varphi_X(x) = \int_0^x \varphi_{x_2}(x_2)\varphi_{x_1}(x - x_2)\, dx_2$$

$$= \int_0^x \varphi_{x_1}(x_1)\varphi_{x_2}(x - x_1)\, dx_1$$

In general, *if X_1 and X_2 are two continuous random variables independently distributed with respective PDF's $\varphi_{x_1}(x_1)$ and $\varphi_{x_2}(x_2)$, then the PDF $\varphi_X(x)$ of the continuous random variable $X = X_1 + X_2$ is obtained by forming the convolution of the two functions $\varphi_{x_1}(x)$ and $\varphi_{x_2}(x)$.*

Example 1.7

Let X_1 and X_2 be two continuous independent random variables, each uniformly distributed over the interval $(0, a]$. Find the PDF of the random variable X where

(a) $X = X_1 + X_2$.
(b) $X = X_1 X_2$.

Let $\varphi_{x_1}(x_1)$ and $\varphi_{x_2}(x_2)$ be the respective probability density functions of the random variables X_1 and X_2; then

$$\varphi_{x_1}(x_1) = \begin{cases} \dfrac{1}{a}, & 0 < x_1 \leq a \\ 0, & \text{otherwise} \end{cases}$$

$$\varphi_{x_2}(x_2) = \begin{cases} \dfrac{1}{a}, & 0 < x_2 \leq a \\ 0, & \text{otherwise} \end{cases}$$

Since X_1 and X_2 are independently distributed,

$$\varphi_{x_1, x_2}(x_1, x_2) = \varphi_{x_1}(x_1)\varphi_{x_2}(x_2)$$

(a) $X = X_1 + X_2$.

Because $\varphi_{x_1}(x_1) = 0$ for $-\infty < x_1 < 0$ and $\varphi_{x_2}(x_2) = 0$ for $-\infty < x_2 < 0$, it follows that $\varphi_X(x) = 0$ for $-\infty < x < 0$. Thus we only need to be concerned with the set of points $\{x_1, x_2 : x_1 > 0, x_2 > 0\}$ in Ω. Then

$$\Phi_X(x) = \iint\limits_{0 < x_1 + x_2 \leq x} \varphi_{x_1}(x_1)\varphi_{x_2}(x_2)\, dx_1\, dx_2$$

Referring to Figure 1.13, we consider the two cases when $0 < x \le a$ and when $a < x \le 2a$.

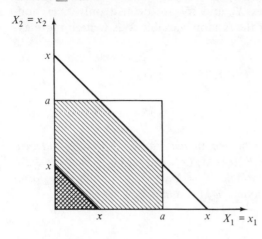

Figure 1.13. Sample space Ω when X_1 and X_2 are uniformly distributed and sets of points defining $P\{X_1 + X_2 \le x\}$.

Case 1. $0 < x \le a$

$$\Phi_X(x) = \int_0^x \int_0^{x-x_2} \frac{1}{a^2} \, dx_1 \, dx_2$$

$$= \frac{1}{a^2} \int_0^x (x - x_2) \, dx_2$$

$$= \frac{1}{a^2} \left| xx_2 - \frac{x_2^2}{2} \right|_0^x$$

$$= \frac{1}{a^2} \frac{x^2}{2}$$

Case 2. $a < x \le 2a$

$$\Phi_X(x) = \int_0^x \int_0^{x-x_2} \frac{1}{a^2} \, dx_1 \, dx_2 - 2 \int_a^x \int_0^{x-x_2} \frac{1}{a^2} \, dx_1 \, dx_2$$

$$= \frac{1}{a^2} \int_0^x (x - x_2) \, dx_2 - \frac{2}{a^2} \int_a^x (x - x_2) \, dx_2$$

$$= \frac{1}{a^2} \frac{x^2}{2} - \frac{2}{a^2} \left| xx_2 - \frac{x_2^2}{2} \right|_a^x$$

$$= \frac{1}{a^2} \frac{x^2}{2} - \frac{2}{a^2} \left[\frac{x^2}{2} - ax + \frac{a^2}{2} \right]$$

$$= \frac{1}{a^2} \left(2ax - \frac{x^2}{2} - a^2 \right)$$

We also note that $\Phi_X(x) = 1$ in the interval $2a < x < \infty$. By direct differentiation of $\Phi_X(x)$ in each of the intervals $0 < x \le a$, $a < x \le 2a$, and $2a <$

$x < \infty$, we obtain

$$\varphi_X(x) = \begin{cases} \dfrac{1}{a^2}x, & 0 < x \le a \\[2mm] \dfrac{1}{a^2}(2a - x), & a < x \le 2a \\[2mm] 0, & \text{otherwise} \end{cases}$$

The plot of the function $\varphi_X(x)$ is shown in Figure 1.14.

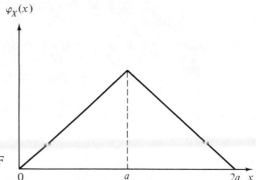

Figure 1.14. Plot of the PDF $\varphi_X(x)$ when $X = X_1 + X_2$.

(b) $X = X_1 X_2$

Here also we only need to be concerned with the set of points $\{x_1, x_2 : x_1 > 0,\ x_2 > 0\}$ in Ω, since $\varphi_X(x) = 0$ for $-\infty < x < 0$. We then have, for $0 < x \le a^2$,

$$\Phi_X(x) = \iint\limits_{0 < x_1 x_2 \le x} \varphi_{X_1}(x_1)\varphi_{X_2}(x_2)\, dx_1\, dx_2$$

From Figure 1.15,

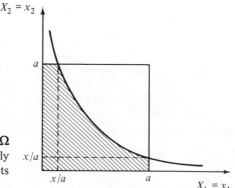

Figure 1.15. Sample space Ω when X_1 and X_2 are uniformly distributed and sets of points defining $P\{X_1 X_2 \le x\}$.

$$\Phi_X(x) = 1 - \int_{x/a}^{a} \int_{x/x_2}^{a} \frac{1}{a^2} \, dx_1 \, dx_2, \quad 0 < x \le a^2$$

$$= 1 - \frac{1}{a^2} \int_{x/a}^{a} \left(a - \frac{x}{x_2} \right) dx_2$$

$$= 1 - \frac{1}{a^2} \left| ax_2 - x \ln x_2 \right|_{x/a}^{a}$$

$$= 1 - \frac{1}{a^2} \left[a^2 - x \ln a - x + x \ln \frac{x}{a} \right]$$

$$= 1 - \frac{1}{a^2} [a^2 - 2x \ln a - x + x \ln x]$$

Differentiating this last expression, we obtain, for $0 < x \le a^2$,

$$\varphi_X(x) = -\frac{1}{a^2}[-2 \ln a + \ln x] = \frac{1}{a^2} \ln \frac{a^2}{x}$$

Since $\Phi_X(x) = 1$ for $a^2 < x < \infty$, therefore

$$\varphi_X(x) = \begin{cases} \frac{1}{a^2} \ln \frac{a^2}{x}, & 0 < x \le a^2 \\ 0, & \text{otherwise} \end{cases}$$

The plot of $\varphi_X(x)$ is shown in Figure 1.16.

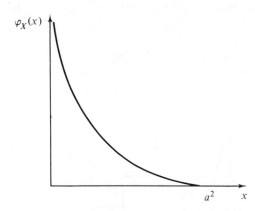

Figure 1.16. Plot of the PDF $\varphi_X(x)$ when $X = X_1 X_2$.

e. GENERALIZATION

The concept of joint probability distribution, joint probability density function, marginal density function, conditional density function, function of random variables, and independence can be extended from two continuous random variables to n continuous random variables, $n \geq 1$, in a way similar to the theory of discrete random variables. We shall not attempt to write the formal expressions; however, the reader is encouraged to introduce symbolically some of these concepts for his own benefit.

Suppose that we have a sequence $\{X_i\}$, $i = 1, 2, \ldots, n$, of n continuous random variables which are mutually independent, each having a PDF defined by $\varphi_{Xi}(x_i)$. Define now the sum

$$X = X_1 + X_2 + \cdots + X_n$$

of these random variables, and let $\varphi_X(x)$ be its PDF. Then it is easy to show that

$$\varphi_X(x) = \varphi_{X_1}(x) * \varphi_{X_2}(x) * \cdots * \varphi_{X_n}(x)$$

In particular, if each X_i has the same PDF $\varphi(x)$, then we simply write

$$\varphi_X(x) = \varphi*^{(n)}(x)$$

and call it the nth-fold convolution of $\varphi(x)$ with itself.

The reader should satisfy himself by showing that the convolution operation on density functions is commutative and associative.

vi. *Application to the Use of Laplace Transforms of Probability Density Functions*

For practical reasons, mostly based on the content of subsequent chapters, we shall in what follows restrict ourselves to nonnegative random variables, so that if X is a continuous random variable with CDF $\Phi_x(x)$ and PDF $\varphi_X(x)$, $0 \leq x < \infty$, its Laplace transform is

$$\varphi_X(s) = \int_0^\infty e^{-sx}\, \varphi_X(x)\, dx = \int_0^\infty e^{-sx}\, d\Phi_x(x)$$

However, the properties derived can easily be established for random variables defined over the entire real line. The reader will note that the methodology and some of the formal proofs are very similar to the ones used in the theory of discrete random variables, the concept of probability generating functions being replaced by that of Laplace transforms.

a. Sums of Independent Random Variables, Examples

If X_1 and X_2 are two continuous nonnegative random variables independently distributed with respective PDF's $\varphi_{X_1}(x_1)$ and $\varphi_{X_2}(x_2)$, then the sum $X = X_1 + X_2$ is a continuous nonnegative random variable whose PDF

$\varphi_X(x)$ is the convolution of $\varphi_{X_1}(x)$ and $\varphi_{X_2}(x)$; that is,

$$\varphi_X(x) = \varphi_{X_1}(x) * \varphi_{X_2}(x)$$

$$= \int_0^x \varphi_{X_1}(u)\varphi_{X_2}(x - u)\, du, \quad 0 < x < \infty$$

Let $\bar{\varphi}_X(s)$ be the Laplace transform of $\varphi_X(x)$, so that

$$\bar{\varphi}_X(s) = \int_0^\infty e^{-sx}\varphi_X(x)\, dx$$

Using the convolution form of $\varphi_X(x)$, we can write

$$\bar{\varphi}_X(s) = \int_0^\infty e^{-sx} \int_0^x \varphi_{X_1}(u)\varphi_{X_2}(x - u)\, du\, dx$$

Interchanging the order of integration, we obtain

$$\bar{\varphi}_X(s) = \int_0^\infty \int_u^\infty e^{-sx}\varphi_{X_1}(u)\varphi_{X_2}(x - u)\, dx\, du$$

Making the change in variable $x - u = v$, we obtain

$$\bar{\varphi}_X(s) = \int_0^\infty \int_0^\infty e^{-s(u+v)}\varphi_{X_1}(u)\varphi_{X_2}(v)\, dv\, du$$

$$= \left(\int_0^\infty e^{-su}\varphi_{X_1}(u)\, du \right)\left(\int_0^\infty e^{-sv}\varphi_{X_2}(v)\, dv \right)$$

$$= \bar{\varphi}_{X_1}(s) \cdot \bar{\varphi}_{X_2}(s)$$

Hence, if X_1 and X_2 are independent continuous random variables, then the Laplace transform of the PDF of $X = X_1 + X_2$ is the product of the Laplace transforms of the PDF's of X_1 and X_2.

Extending the above result to the sum of k continuous random variables, X_k, $k = 1, 2, \ldots$, independently distributed, we can say that if

$$X = X_1 + X_2 + \cdots + X_k$$

then

$$\bar{\varphi}_X(s) = \prod_{i=1}^{i=k} \bar{\varphi}_{X_i}(s)$$

In particular, if all the X_k, $k = 1, 2, \ldots$, are identically and independently distributed, whose common PDF $\varphi(x)$ has Laplace transform $\bar{\varphi}(s)$, then

$$\bar{\varphi}_X(s) = [\bar{\varphi}(s)]^k, \quad k = 1, 2, \ldots$$

Exercise 1.6

Using Laplace transforms, demonstrate the continuous counterpart of Example 1.2.

Example 1.8

Let $\{X_i\}$, $i = 1, 2, \ldots, n$, be a sequence of n independently distributed continuous random variables. Find the PDF of

$$X = \sum_{i=1}^{i=n} X_i$$

when each of the X_i has the following distribution:

(a) Exponential with parameter λ.
(b) Gamma with parameter λ and of order v_i.

    (a) If each X_i has a negative exponential distribution with PDF

$$\varphi(x) = \begin{cases} 0, & -\infty < x \leq 0 \\ \lambda e^{-\lambda x}, & 0 < x < \infty \end{cases}$$

then the Laplace transform of $\varphi(x)$ is $\bar{\varphi}(s) = \lambda/(\lambda + s)$; hence the Laplace transform of $\varphi_X(x)$ is

$$\bar{\varphi}_X(s) = \left(\frac{\lambda}{\lambda + s}\right)^n$$

which can be identified as the Laplace transform of a gamma density function; therefore,

$$\varphi_X(x) = \begin{cases} 0, & -\infty < x \leq 0 \\ \dfrac{(\lambda x)^{n-1}}{\Gamma(n)} \lambda e^{-\lambda x}, & 0 < x < \infty \end{cases}$$

    (b) If each X_i has the gamma distribution with PDF

$$\varphi_{X_i}(x) = \begin{cases} 0, & -\infty < x \leq 0 \\ \dfrac{(\lambda x)^{v_i-1}}{\Gamma(v_i)} \lambda e^{-\lambda x}, & 0 < x < \infty \end{cases}$$

then the Laplace transform of $\varphi_{X_i}(x)$, $i = 1, 2, \ldots, n$, is

$$\bar{\varphi}_{X_i}(s) = \left(\frac{\lambda}{\lambda + s}\right)^{v_i}$$

The Laplace transform of $\varphi_X(x)$ is

$$\bar{\varphi}_X(s) = \left(\frac{\lambda}{\lambda + s}\right)^{\sum_{i=1}^{i=n} v_i}$$

Thus X has a gamma density function with parameter λ and of order $\sum_{i=1}^{i=n} \nu_i$.

b. Mixed Distributions

Let $\{X_i\}$, $i = 1, 2, \ldots, N$, be a sequence of independent identically distributed continuous random variables with common PDF's $\varphi(x)$. Let $\bar{\varphi}(s)$ be the Laplace transform of $\varphi(x)$. Suppose now that N is itself a discrete random variable independently distributed from all X_i with PMF $\varphi_N(n)$, $n = 1, 2, \ldots$. The random variable

$$Y = X_1 + X_2 + \cdots + X_N$$

is said to be a *mixture* with a *mixed distribution*.

Define by $\varphi_Y(y)$ the PDF of the random variable Y, and let $\bar{\varphi}_Y(s)$ be the Laplace transform of $\varphi_Y(y)$. Following a reasoning similar to the case of discrete distribution, the reader can show that

$$\bar{\varphi}_Y(s) = \sum_{n=1}^{\infty} \varphi_N(n)[\bar{\varphi}(s)]^n$$

By introducing the PGF of the discrete random variable N to be

$$G_N(s) = \sum_{n=1}^{\infty} s^n \varphi_N(n)$$

it follows that

$$\bar{\varphi}_Y(s) = G_N[\bar{\varphi}(s)] \tag{1.17}$$

Exercise 1.7

Show that

$$E[Y] = E[N] \cdot E[X]$$
$$\text{Var}[Y] = \text{Var}[N] \cdot \{E[X]\}^2 + E[N] \cdot \text{Var}[X]$$

Example 1.9

In a given day, the number of orders for a product is a random variable N having a Poisson distribution with parameter λ. The size of each order, given that an order is placed, is a random variable having a negative exponential distribution with parameter μ. Assume that the size of each order is independent of the number of orders; find the distribution of the total quantity ordered in a given day.

Suppose that in the day at least one order is placed ($N \geq 1$), and let X_i be the size of the ith order. Then, the total quantity ordered is

$$X_1 + X_2 + \cdots + X_N$$

where each X_i has PDF $\varphi_X(x) = \mu e^{-\mu x}$, $x > 0$, whose Laplace transform is

$$\bar{\varphi}(s) = \frac{\mu}{\mu + s}$$

Also,

$$\varphi_N(n) = e^{-\lambda}\frac{\lambda^n}{n!}, \quad n = 1, 2, \ldots$$

When no order is placed $N = 0$; hence, the total quantity ordered in a given day is

$$Y = \begin{cases} 0, & N = 0 \\ X_1 + X_2 + \cdots + X_N, & N \geq 1 \end{cases}$$

Now, when $Y = 0$, the probability that no order is placed is

$$\varphi_N(0) = P\{Y = 0\} = e^{-\lambda}$$

Using expression (1.17) we have

$$\bar{\varphi}_Y(s) = \int_0^\infty e^{-sy}\, d\Phi(y) = e^{-\lambda} + \sum_{n=1}^\infty G_N\left(\frac{\mu}{\mu + s}\right)$$

$$= \sum_{n=0}^\infty e^{-\lambda}\frac{\lambda^n}{n!}\left(\frac{\mu}{\mu + s}\right)^n$$

$$= e^{-\lambda}\sum_{n=0}^\infty \frac{1}{n!}\left(\frac{\lambda\mu}{\mu + s}\right)^n$$

$$= e^{-\lambda}e^{\lambda\mu/(\mu+s)}$$

If we denote by $\Phi_Y(y)$ the distribution function of Y, then after inverting formally $\bar{\varphi}_Y(s)$ we have

$$d\Phi_Y(y) = \begin{cases} e^{-\lambda}, & y = 0 \\ e^{-(\lambda+\mu y)}\sqrt{\dfrac{\lambda\mu}{y}}\, I_1(2\sqrt{\lambda\mu y})\, dy, & 0 < y < \infty \end{cases}$$

where $I_1(\cdot)$ is the modified Bessel function of the first kind of order unity.

1.5. The Central Limit Theorem

The theory of probability studies to a considerable extent limit theorems related to sums of random variables, as the number of terms in the sum expression increases. One of the most important of these theorems is the *central limit theorem*, which we state without proof.

Let $X_1, X_2, \ldots, X_n$ be a sequence of independently distributed random variables (not necessarily with identical distributions). Suppose that each random variable X_i, $i = 1, 2, \ldots, n$, has a finite mean and a finite variance given respectively by

$$E[X_i] = \mu_i \quad \text{and} \quad \text{Var}[X_i] = \sigma_i^2$$

We know that the random variable

$$Y_n = X_1 + X_2 + \cdots + X_n$$

has mean

$$E[Y_n] = \sum_{i=1}^{i=n} \mu_i$$

and variance

$$\text{Var}[Y_n] = \sum_{i=1}^{i=n} \sigma_i^2$$

The central limit theorem states that under certain conditions, for large n, the random variable Y_n has approximately a normal distribution. More precisely,

$$\lim_{n \to \infty} P\left\{ \frac{Y_n - E[Y_n]}{\sqrt{\text{Var}[Y_n]}} \leq y \right\} = \int_{-\infty}^{y} \frac{1}{\sqrt{2\pi}} e^{-u^2/2} \, du$$

The use of the theorem necessitates only a knowledge of the mean and variance of each of the random variables X_i. Approximate expressions for calculating probabilities may then be obtained.

SELECTED REFERENCES

[1] CLARKE, A. B. and R. L. DISNEY, *Probability and Random Processes for Engineers and Scientists*, J. Wiley & Sons, Inc., New York, 1970.

[2] CRAMER, H., *Mathematical Methods of Statistics*, Princeton University Press, Princeton, New Jersey, 1946.

[3] CRAMER, H., *The Elements of Probability Theory*, John Wiley & Sons, Inc., New York, 1955.

[4] DUBES, R. C., *The Theory of Applied Probability*, Prentice-Hall, Inc., Englewood Cliffs, New Jersey, 1968.

[5] FELLER, W., *An Introduction to Probability Theory and its Applications*, Vol. I, J. Wiley & Sons, Inc., New York, 1957.

[6] FELLER, W., *An Introduction to Probability Theory and its Applications*, Vol. II, J. Wiley & Sons, Inc., New York, 1966.

[7] GNEDENKO, B. V., *The Theory of Probability*, Chelsea Publishing Company, New York, 1963.

[8] GRAY, J. R., *Probability*, Oliver & Boyd, London, 1967.

[9] PAPOULIS, A., *Probability, Random Variables and Stochastic Processes*, McGraw-Hill, New York, 1965.

[10] PARZEN, E., *Modern Probability and its Applications*, J. Wiley & Sons, Inc., New York, 1960.

[11] RÉNYI, A., *Probability Theory*, North Holland Publishing Company, Amsterdam, 1970.

[12] THOMASIAN, A. J., *The Structure of Probability Theory with Applications*, McGraw-Hill, New York, 1969.

[13] USPENSKY, J. V., *Introduction to Mathematical Probability*, McGraw-Hill, New York, 1937.

PROBLEMS

1. Let X_1 and X_2 be two independently and identically distributed random variables each with a geometric distribution. Find the PMF of the following random variables

   (a) $X = X_1 + X_2$ (c) max $\{X_1, X_2\}$

   (b) $X = |X_1 - X_2|$ (d) min $\{X_1, X_2\}$

2. In the experiment involving the throw of two symmetrical six-sided dice each numbered 1, 2, 3, 4, 5, 6, let X_1 and X_2 be random variables denoting, respectively, the outcome of the first and second die. Assume X_1 and X_2 to be independently distributed; find the PMF of the random variable X, where X is defined as

   (a) $X = X_1 X_2$ (c) $X = |X_1 - X_2|$

   (b) $X = \dfrac{X_1}{X_2}$ (d) $X = \min\{X_1, X_2\}$

3. Let X be a discrete random variable with PMF $\varphi_X(r)$. Let $\mu_X = E[X]$, and define a *median* of X, m_X, to be that value of r, $r = 0, 1, 2, \ldots$, such that

$$\sum_{r=0}^{r=m_X-1} \varphi_X(r) \leq \frac{1}{2}$$

and

$$\sum_{r=0}^{r=m_X} \varphi_X(r) \geq \frac{1}{2}$$

Show that

   (a) The mean μ_X is the value of τ, $0 \leq \tau < \infty$, which minimizes the expectation of the squared deviation of X from τ given by

$$E[(x - \tau)^2] = \sum_{r=0}^{\infty} (r - \tau)^2 \varphi_X(r)$$

   (b) The median m_X is the value of τ, $\tau = 0, 1, 2, \ldots$, which minimizes the expectation of the absolute deviation of X from τ given by

$$E[|X - \tau|] = \sum_{r=0}^{\infty} |r - \tau| \varphi_X(r)$$

4. In a particular day the number of customers arriving at a store is a random variable N having a Poisson distribution with parameter λ; that is,

$$\varphi_N(n) = e^{-\lambda} \frac{\lambda^n}{n!}, \quad n = 0, 1, 2, \ldots$$

Given that a customer is in the store, the probability that he will purchase a number of units X of a particular product is a random variable having the geometric distribution

$$\varphi_X(r) = qp^r, \quad r = 0, 1, 2, \ldots$$

This implies that a customer may be in the store but may not purchase the product.

Let Y be the total number of units of the product sold per day. Find the following:

(a) The probability that the product will not be sold on this particular day.

(b) The PGF of the random variable Y.

(c) $E[Y]$ and $\text{Var}[Y]$.

(*Hint:* The product may not be sold on this particular day if no customers arrive at the store or if customers in the store do not purchase it.)

5. Let $\{X_i\}$, $i = 1, 2, \ldots, k$, denote a sequence of k random variables independently distributed. Define the sequence of random variables $\{Y_i\}$, $i = 1, 2, \ldots, k$, where $Y_i = a_i X_i + b_i$, a_i and b_i being constants. Define the random variable $Z = \sum_{i=1}^{i=k} Y_i$. Show that

$$E[Z] = \sum_{i=1}^{i=k} [a_i E[X_i] + b_i]$$

and

$$\text{Var}[Z] = \sum_{i=1}^{i=k} a_i^2 \, \text{Var}[X_i]$$

(Consider the two cases when the Y_i is discrete and continuous.)

6. Show that if X is a random variable having a gamma distribution with parameter $\lambda = 1$ and of order $n = \nu/2$, $\nu = 1, 2, 3, \ldots$, then the random variable $Y = 2X$ has the χ^2 distribution with PDF

$$\varphi_Y(y) = \begin{cases} 0, & -\infty < y \le 0 \\ \dfrac{1}{2^{\nu/2} \Gamma\left(\dfrac{\nu}{2}\right)} y^{(\nu/2)-1} e^{-y/2}, & 0 < y < \infty \end{cases}$$

(In statistics, ν defines the *number of degrees of freedom*.)

7. Let X be a random variable having a gamma distribution of integer order r, $r = 1, 2, \ldots$, and with parameter λ. Show that

$$\Phi_X(x) = \int_0^x \frac{(\lambda u)^{r-1}}{\Gamma(r)} \lambda e^{-\lambda u}\, du = \sum_{i=r}^{\infty} e^{-\lambda x} \frac{(\lambda x)^i}{i!}$$

(*Hint:* Use integration by parts, and note the relationship between the gamma distribution and the Poisson distribution.)

8. Let X be a random variable having a continuous PDF $\varphi_X(x)$. We define the *median* of X, m_X, as

$$\int_{-\infty}^{m_X} \varphi_X(x)\, dx = \int_{m_X}^{\infty} \varphi_X(x)\, dx = \tfrac{1}{2}$$

where $-\infty < m_X < \infty$. Show that if X has the exponential distribution then the median is $m_X = (1/\lambda) \ln 2$. Show also that the mean value of X in the exponential distribution divides the sum of the probabilities of X approximately in the ratio $2:1$.

9. Let $\{X_i\}$, $i = 1, 2, \ldots, n$, be a sequence of independently distributed discrete random variables, each having a Poisson distribution with respective parameter λ_i. Find the PMF of $X = \sum_{i=1}^{i=n} X_i$.

10. Let $\{X_i\}$, $i = 1, 2, \ldots, n$, be a sequence of independently distributed continuous random variables, each having a normal distribution with respective mean μ_i and respective variance σ_i^2. Find the PDF of $X = \sum_{i=1}^{i=n} X_i$.

11. Let $\{X_i\}$, $i = 1, 2, \ldots, n$, be a sequence of real quantities. Consider the quantity

$$Y_n = \min\{X_1, X_2, \ldots, X_n\}$$

(a) Show that

$$Y_n = \min\{Y_{n-1}, X_n\}$$

(b) Suppose that the $\{X_i\}$, $i = 1, 2, \ldots, n$, form a sequence of random variables with joint probability density function $\varphi_{X_1, X_2, \ldots, X_n}(x_1, x_2, \ldots, x_n)$, $0 < x_i < \infty$. Prove that the distribution function of the random variable Y_n is

$$\Phi_{Y_n}(y) = 1 - \int_y^{\infty} \cdots \int_y^{\infty} \int_y^{\infty} \varphi_{X_1, X_2, \ldots, X_n}(x_1, x_2, \ldots, x_n)\, dx_1\, dx_2 \ldots dx_n$$

12. Let $\{X_i\}$, $i = 1, 2, \ldots, n$, be a sequence of real quantities. Consider the quantity

$$Y_n = \max\{X_1, X_2, \ldots, X_n\}$$

(a) Show that

$$Y_n = \max\{Y_{n-1}, X_n\}$$

(b) Suppose that the $\{X_i\}$, $i = 1, 2, \ldots, n$, form a sequence of random variables with joint probability density function $\varphi_{X_1, X_2, \ldots, X_n}(x_1, x_2, \ldots, x_n)$, $0 < x_i < \infty$. Prove that the distribution function of the random variable

Y_n is

$$\Phi_{Y_n}(y) = \int_0^y \cdots \int_0^y \int_0^y \varphi_{X_1, X_2, \ldots, X_n}(x_1, x_2, \ldots, x_n) \, dx_1 \, dx_2 \ldots dx_n$$

13. A continuous random variable X whose density function is ($\alpha > 0$, $\beta > 0$)

$$\varphi_X(x) = \begin{cases} \dfrac{\Gamma(\alpha + \beta)}{\Gamma(\alpha)\Gamma(\beta)} x^{\alpha-1}(1 - x)^{\beta-1}, & 0 < x \leq 1 \\ 0, & \text{otherwise} \end{cases}$$

is said to have a *beta distribution*. Find the mean and variance of X.

14. A continuous random variable X whose density function is

$$\varphi_X(x) = \begin{cases} 0, & -\infty < x \leq 0 \\ \dfrac{x}{\alpha^2} e^{-x^2/2\alpha^2}, & 0 < x < \infty \end{cases}$$

is said to have a *Rayleigh distribution*. Show that

$$E[X] = \sqrt{\frac{\pi}{2}} \alpha$$

15. Let $\varphi_X(x)$ and $H_X(x)$ be, respectively, the PDF and complementary cumulative function of a continuous random variable X, $0 < x < \infty$. Show that, for $n = 1, 2, \ldots$,

$$E[X^n] = \int_0^\infty x^n \varphi_X(x) \, dx = \int_0^\infty n x^{n-1} H_X(x) \, dx$$

16. Let X_1 and X_2 be two random variables with joint PDF $\varphi_{X_1, X_2}(x_1, x_2)$. The *covariance* of (X_1, X_2), written as $\text{Cov}(X_1, X_2)$, is defined to be

$$\text{Cov}(X_1, X_2) = E[(X_1 - E[X_1])(X_2 - E[X_2])]$$

The *correlation coefficient* between X_1 and X_2 is defined to be

$$\rho = \frac{\text{Cov}(X_1, X_2)}{\sqrt{\text{Var } X_1} \sqrt{\text{Var } X_2}}$$

Show that $-1 \leq \rho \leq 1$.

17. Let X be a continuous random variable with PDF $\varphi_X(x)$. The *mode* of the random variable X is defined as that value of x which maximizes the function $\varphi_X(x)$. Determine the mode when X has the following PDF:

(a) $\varphi_X(x) = \dfrac{(\lambda x)^{n-1}}{\Gamma(n)} \lambda e^{-\lambda x}$, $0 < x < \infty$, $\lambda > 0$, $n > 1$

(b) $\varphi_X(x) = \dfrac{\Gamma(\alpha + \beta)}{\Gamma(\alpha)\Gamma(\beta)} x^{\alpha-1}(1 - x)^{\beta-1}$, $0 < x \leq 1$, $\alpha > 0$, $\beta > 0$

18. Let X_1, X_2, and X_3 be three identically and independently distributed random variables each having the PDF

$$\varphi(x) = \begin{cases} 1, & 0 < x \le 1 \\ 0, & \text{otherwise} \end{cases}$$

Consider the random variable

$$Y = X_1 + X_2 + X_3$$

(a) Determine the PDF of Y and plot the resultant function on graph paper. Compute $P\{Y \le 1.75\}$.

(b) Consider now the random variable Z having a normal distribution with mean $E[Z] = E[Y]$ and with variance $\text{Var}[Z] = \text{Var}[Y]$. Plot the PDF of Z on the same graph paper as the PDF of Y, and compute $P\{Z \le 1.75\}$.

(c) Compare the results obtained from (a) and (b). What remarks can you make regarding the applicability of the central limit theorem to this particular example?

STOCHASTIC PROCESSES

2.1. Introduction

In real-life situations a large number of phenomena can be thought of as experiments involving a series of successive trials, rather than one single trial. In general, each trial in the experiment may depend on some or all past trials; furthermore, the outcome of each trial will in general be described according to some probability law. The study of this type of experiment constitutes the theory of *stochastic processes*.

It is possible that the succession of trials in an experiment be independent trials; that is, the outcome of a particular trial does not depend on any of the outcomes of past trials, and, in turn, does not affect the outcomes of future trials. This of course constitutes a special class of stochastic processes known as *independent processes*. In principle, it is possible to study the statistical characteristics of independent processes by making use of the theory underlying the probability laws of independent events without recourse to any special theory. In practice, however, this approach often leads to analytic difficulties. More elegant and easier techniques that rely on the theory of recurrence relations are available and will be discussed.

In a more complex experiment the outcome of a particular trial will

2

depend on the outcomes of previous trials and will, in turn, affect the outcomes of future trials. In the simplest form of dependency, the outcome of a particular trial depends *only* on the outcomes of the immediate previous trial, but not on the outcomes of any other past trials; in turn, the outcomes of a particular trial will affect only the outcomes of the next trial, but not those of any subsequent trials. A stochastic process possessing this type of dependency feature is known as a *Markov process;* this constitutes the next topic of discussion.

i. *Definition*

A stochastic process is a collection of random variables $\{X(t), t \in \mathbf{T}\}$ *defined on a given sample space, where each random variable is indexed by a parameter t that varies in an index set* $\mathbf{T}$. In general, t represents the time parameter.

We define the *state space* $\mathbf{I}$ of the process as the set of possible values that the random variables $X(t)$ can take. For example, if $X(t)$ is a discrete random variable taking on a denumerably infinite number of values over the set of nonnegative integers, then $\mathbf{I} = \{0, 1, 2, \ldots\}$. If $X(t)$ is a nonnegative continuous random variable, then $\mathbf{I} = \{x : 0 \leq x < \infty\}$.

ii. Classification

A stochastic process is said to be a *discrete-parameter process* if the index set $\mathbf{T}$ is defined as $\mathbf{T} = \{0, \pm 1, \pm 2, \ldots\}$, or as $\mathbf{T} = \{0, 1, 2, \ldots\}$.

A stochastic process is said to be a *continuous-parameter process* if the index set $\mathbf{T}$ is defined as $\mathbf{T} = \{t : -\infty < t < \infty\}$, or as $\mathbf{T} = \{t : 0 \leq t < \infty\}$.

A stochastic process is said to be a *discrete-valued process* if the state space $\mathbf{I}$ is discrete.

A stochastic process is said to be a *continuous-valued process* if the state space $\mathbf{I}$ is continuous.

From these definitions a stochastic process may fall into one of the four possible categories, 1, 2, 3, and 4, as represented in Table 2.1. When dealing with a discrete-parameter stochastic process, it is often convenient to denote the process by the collection of random variables $\{X_n, n \in \mathbf{T}\}$. It may be possible to illustrate a realization of the process by plotting a *sample function* on a graph with the index set represented on the horizontal axis and the

Table 2.1. CLASSIFICATION OF STOCHASTIC PROCESSES

| | | Index Set $\mathbf{T}$ | |
		Discrete	*Continuous*
	Discrete	1. Discrete-parameter discrete-valued process	2. Continuous-parameter discrete-valued process
State Space $\mathbf{I}$			
	Continuous	3. Discrete-parameter continuous-valued process	4. Continuous-parameter continuous-valued process

state space represented on the vertical axis. This is illustrated in the following example.

iii. An Example from Queuing Theory

Consider an example drawn from the theory of queues (see Chapter 4). Customers numbered $1, 2, \ldots, n, \ldots$ arrive one at a time in a random fashion and join a queue at a single service center. The amount of service provided to each customer when they start to be served is itself a random variable. The following are examples of stochastic processes with their sample functions.

1. *Discrete-parameter, discrete-valued process.* Let the index set $\mathbf{T}$ refer to customer numbers, $\mathbf{T} = \{1, 2, \ldots, n, \ldots\}$, and let the random variable X_n define the number of customers in the system at points of departure of the nth customer following service, $\mathbf{I} = \{0, 1, \ldots\}$ (Figure 2.1).

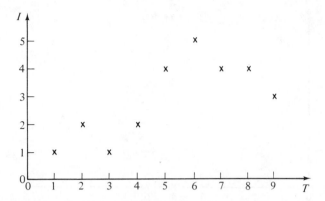

Figure 2.1. Sample function for a discrete-parameter, discrete-valued process.

2. *Continuous-parameter, discrete-valued process.* Let the index set **T** refer to continuous time t, $\mathbf{T} = \{t : 0 \leq t < \infty\}$, and let the random variable $X(t)$ define the number of customers in the system at time t, $\mathbf{I} = \{0, 1, 2, \ldots\}$ (Figure 2.2).

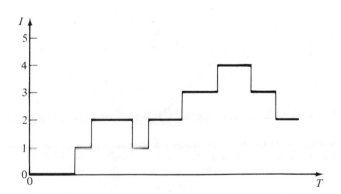

Figure 2.2. Sample function for a continuous-parameter, discrete-valued process.

3. *Discrete-parameter, continuous-valued process.* Let the index set **T** refer to customer numbers, $\mathbf{T} = \{1, 2, \ldots, n, \ldots\}$, and let the random variable W_n define the waiting time of the nth customer prior to service, $\mathbf{I} = \{x : 0 \leq x < \infty\}$ (Figure 2.3).

4. *Continuous-parameter, continuous-valued process.* Let the index set **T** refer to continuous time t, $\mathbf{T} = \{t : 0 \leq t < \infty\}$, and let the random variable $X(t)$ define the amount of service to be provided at time t for all the customers in the system, $\mathbf{I} = \{x : 0 < x < \infty\}$ (Figure 2.4).

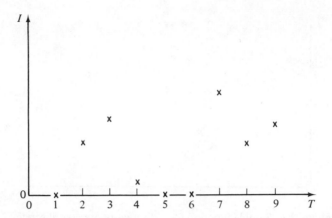

Figure 2.3. Sample function for a discrete-parameter, continuous-valued process.

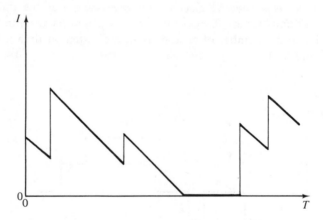

Figure 2.4. Sample function for a continuous-parameter, continuous-valued process.

iv. *Probability Law of a Stochastic Process*

Let $\{X(t), t \in \mathbf{T}\}$ define a stochastic process with index set $\mathbf{T}$. The stochastic process is completely described if for all n, $n = 0, 1, 2, \ldots$, and for any $t_0 < t_1 < t_2 < \cdots < t_n$, $t_0 = 0$, the joint distribution function of the $n + 1$ random variables $X(t_0), X(t_1), \ldots, X(t_n)$ defined by

$$\Phi_{X(t_0), X(t_1), \ldots, X(t_n)}(x_0, x_1, \ldots, x_n) = P\{X(t_0) \leq x_0, X(t_1) \leq x_1, \ldots, X(t_n) \leq x_n\}$$

can be determined. When dealing with a discrete-parameter, discrete-valued process $\{X_n, n \in \mathbf{T}\}$, $\mathbf{T} = \{0, 1, \ldots, n, \ldots\}$, $\mathbf{I} = \{0, 1, \ldots\}$, it is sufficient to determine for all n, $n = 0, 1, 2, \ldots$, the joint probability mass function of $X_0, X_1, \ldots, X_n$ given by

$$\varphi_{X_0, X_1, \ldots, X_n}(r_0, r_1, \ldots, r_n) = P\{X_0 = r_0, X_1 = r_1, \ldots, X_n = r_n\}$$

The problem of determining the probability law of a general stochastic process is an extremely difficult task. Fortunately, quite often, the class of stochastic processes encountered in practice possesses certain simple intrinsic features that make its study more amenable to analysis. Two important such processes we shall study in the next sections are the *independent processes* and the *Markov processes*.

In many practical situations the state of the system is completely specified initially; that is, $P\{X(t_0) \leq x_0\}$ or $\Phi_{X_0}(r_0)$ is known. These conditions, known as *initial conditions*, will be assumed to be given, if necessary. In general, the initial state of a process will affect and characterize its future behavior, thus necessitating its complete specification.

A final remark should be made. Although we have used the concept of random variables defined on a given sample space Ω to introduce the notion of stochastic process, certain types of discrete-valued processes have no particularly meaningful random variables attached to their sample or state space. The sample points in the state space are then identified by descriptive labels such as the letters A, B, C, . . . or the cardinal numbers 0, 1, 2, ...

2.2. Independent Processes

i. *Definition*

A stochastic process $\{X(t), t \subset \mathbf{T}\}$ is called an *independent process* if for $0 < t_1 < t_2 < \cdots < t_n$ and all $n = 0, 1, 2, \ldots$, and $t_0 = 0$, the probability law of the process is given by

$$P\{X(t_0) \leq x_0, X(t_1) \leq x_1, \ldots, X(t_n) \leq x_n\}$$
$$= P\{X(t_0) \leq x_0\}P\{X(t_1) \leq x_1\} \ldots P\{X(t_n) \leq x_n\}$$

In what follows we shall restrict our attention first to discrete-parameter discrete-valued independent processes. The process $\{X_n, n \in \mathbf{T}\}$, where $\mathbf{I} = \{0, 1, 2, \ldots\}$ and $\mathbf{T} = \{0, 1, \ldots, n, \ldots\}$, is then completely specified by the joint probability mass function of the sequence of random variables $\{X_n\}, n = 0, 1, \ldots,$

$$\varphi_{X_0, X_1, \ldots, X_n}(r_0, r_1, \ldots, r_n) = \varphi_{X_0}(r_0)\varphi_{X_1}(r_1) \ldots \varphi_{X_n}(r_n)$$

The elements $1, 2, \ldots$ of the index set $\mathbf{T}$ refer to the sequence of trials in the experiment. Since the trials are mutually independent, the initial condition factor is not relevant; that is, although the initial state of the system may be specified, it is not going to affect in any way the outcomes of the first trial, the second trial, or any subsequent trials. It is therefore natural to drop the element 0 in the set $\mathbf{T}$ and redefine it as $\mathbf{T} = \{1, 2, \ldots, n, \ldots\}$, thus establishing a one-to-one correspondence with the natural order of trials.

2.3. Independent Processes: Discrete Parameter

i. *Independent Bernoulli Trials: The Use of Recurrence Relations*

One of the simplest type of independent processes is the *independent Bernoulli trials* or *independent Bernoulli process* in which at any particular trial either of two mutually exclusive and exhaustive outcomes E_1 or E_2 can occur. It is assumed that the probability of occurrence of a particular outcome at a particular trial does not depend upon the order of the trial. For a given trial let $q, 0 < q < 1$, be the probability of occurrence of event E_1 and let $p, p + q = 1$, be the probability of occurrence of event E_2. For the nth trial, $n = 1, 2, \ldots$, define the random variable X_n such that

$$X_n = \begin{cases} 0, & \text{if event } E_1 \text{ occurs} \\ 1, & \text{if event } E_2 \text{ occurs} \end{cases}$$

Then $\{X_n\}$ is a sequence of identically and independently distributed random variables each having a Bernoulli distribution defined by the probability mass function

$$\varphi(r) = \begin{cases} p^r q^{1-r}, & r = 0, 1 \\ 0, & \text{otherwise}, \end{cases} \qquad (p + q = 1) \qquad (2.1)$$

A particular realization of the stochastic process $\{X_n, n \in \mathbf{T}\}$ is illustrated in Figure 2.5. The probability law of an independent Bernoulli process is completely specified by the following joint probability mass function:

$$\begin{aligned}
\varphi_{X_1, X_2, \ldots, X_n}(r_1, r_2, \ldots, r_n) &= \varphi_{X_1}(r_1)\varphi_{X_2}(r_2) \ldots \varphi_{X_n}(r_n) \\
&= \begin{cases} (p^{r_1}q^{1-r_1})(p^{r_2}q^{1-r_2}) \ldots (p^{r_n}q^{1-r_n}), & r_i = 0, 1; i = 1, 2, \ldots, n \\ 0, & \text{otherwise} \end{cases} \\
&= \begin{cases} \left(\dfrac{p}{q}\right)^{r_1 + r_2 + \cdots + r_n} q^n, & r_i = 0, 1; i = 1, 2, \ldots, n \\ 0, & \text{otherwise} \end{cases}
\end{aligned} \qquad (2.2)$$

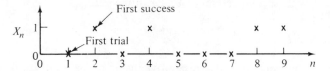

Figure 2.5. Sample function for a Bernoulli process.

An example of an experiment described by an independent Bernoulli process is the succession of trials involving the tossing of a single coin resulting in the appearance of either a head H or a tail T. The elements of the index set $\mathbf{T} = \{1, 2, \ldots\}$ correspond to the successive tosses of the coin. We may define for the nth toss the random variable X_n as

$$X_n = \begin{cases} 0, & \text{if the outcome is T} \\ 1, & \text{if the outcome is H} \end{cases}$$

Another example refers to the performance of independent successive tests on an object, each test resulting in either a success S or a failure F. We may then define for the nth test the random variable X_n as

$$X_n = \begin{cases} 0, & \text{if the outcome is F} \\ 1, & \text{if the outcome is S} \end{cases}$$

With a knowledge of the probability law (2.2), it is theoretically possible to obtain the probability of occurrence of a particular event associated with the process or the distribution function of any function of the random variables $X_1, X_2, \ldots, X_n$. The evaluation of these probabilities involves, in general, the computation of the multiple sum expression

$$\sum_R \cdots \sum \sum \left(\frac{p}{q}\right)^{r_1 + r_2 + \cdots + r_n} q^n$$

where the region R is a subset of the n-dimensional sample space on which the sequence of random variables $X_1, X_2, \ldots, X_n$ is defined. Despite the apparent elementary form of the function under the summation signs, considerable difficulty is often encountered in the analytic manipulation of this expression. However, it is often possible to derive the probability law of occurrences of particular events associated with the independent Bernoulli process by using the technique of recurrence relations We shall illustrate this method with special reference to Example 2.1.

Example 2.1

Suppose that each month a salesman calls on a customer to solicit orders. The probability that the customer will place an order in any particular month is p, $0 < p < 1$, and is independent of his actions in previous months. Thus the probability that the customer will not place an order in a particular month is $q = 1 - p$. We assume that the salesman is presently in month $n = 0$, and that he is interested in the customer's behavior in the month's ahead. For example, among other things the salesman may like to know the number of orders the customer will place over a given number of months; this interest may stem from the fact that the salesman is perhaps attempting to find out the commissions he expects to receive from sales to this particular customer. Because of the future uncertainty in the customer's behavior, the salesman can only provide a probabilistic assessment of the quantities he is evaluating.

We now select as the elements of the index set $\mathbf{T}$ the successive months period labeled $n = 1, 2, \ldots$. Next we construct a state or sample space consisting of two sample points or states: a state labeled 0 corresponding to the event "no order" placed, and a state labeled 1 corresponding to the event "an order" placed. If we now define the random variable X_n, $n = 1, 2, \ldots$, such that $X_n = 0$ if the process in month n is in state 0 and $X_n = 1$ if the process in month n is in state 1, then the process $\{X_n, n \in \mathbf{T}\}$ is an independent Bernoulli process (see Figure 2.5). One can imagine for example that the salesman is performing an experiment consisting of a single trial every month such that the outcome of the trial is a success if the customer places an order and a failure if the customer does not place an order.

In the next sections we provide examples illustrating the use of recurrence relations for determining the probability law of particular events associated with this process.

a. NUMBER OF SUCCESSES IN n TRIALS, $n \geq 1$. THE BINOMIAL PROCESS

Suppose that the salesman is interested in finding out the total number of orders Y_n that the customer will place during the first n months, $n \geq 1$ (see Figure 2.6). It is evident that Y_n is a random variable defined as the sum of the n independent random variables $X_1, X_2, \ldots, X_n$; that is,

$$Y_n = \sum_{i=1}^{i=n} X_i$$

It is known (see Example 1.4) that since each X_i, $i = 1, 2, \ldots, n$, has the same Bernoulli distribution (2.1), their sum Y_n will have a binomial distribution with PMF

$$\varphi_{Y_n}(r) = \begin{cases} \binom{n}{r} p^r q^{n-r}, & r = 0, 1, \ldots, n \\ 0, & \text{otherwise} \end{cases} \tag{2.3}$$

It is our purpose to illustrate the use of recurrence relations to arrive at this well-known solution. Let

$$P\{Y_n = r\} = P(r, n), \quad r = 0, 1, \ldots, n; n \geq 1$$

The notations $P_{r,n}$ or $P_r(n)$ or $P_n(r)$ are sometimes used for $P(r, n)$. If we relate the orders placed during the first n months to the orders placed during the first $n - 1$ months, $n \geq 2$, we note that

(1) r orders, $r = 1, 2, \ldots, n - 1$, in n months, $n = 2, 3, \ldots$, can be obtained in the following possible mutually exclusive and exhaustive ways: (a) if exactly r orders are placed in the first $n - 1$ months and no order is placed in the last month n; (b) if exactly $r - 1$ orders are placed in the first $n - 1$ months and one order is placed in the last month n. We can then write, for $r = 1, 2, \ldots, n - 1$ and $n = 2, 3, \ldots$,

$$P(r, n) = qP(r, n - 1) + pP(r - 1, n - 1) \tag{2.4}$$

(2) No orders, $r = 0$, in n months, $n = 2, 3, \ldots$, can be obtained only if no orders are placed during the first $n - 1$ months and no order is placed in the last month. Then for $r = 0$ and $n = 2, 3, \ldots$ we have

$$P(0, n) = qP(0, n - 1) \tag{2.5}$$

(3) Exactly $r = n$ orders in n months, $n = 2, 3, \ldots$, can be obtained only if exactly $r = n - 1$ orders are placed during the first $n - 1$ months and an order is placed in the last month. Thus, for $r = n$ and $n = 2, 3, \ldots$, we obtain

$$P(n, n) = pP(n - 1, n - 1) \tag{2.6}$$

(4) We know that the probability of placing exactly one order in the first month is p and the probability of placing no order is $q = 1 - p$. Hence, for $r = 0, 1$ and $n = 1$, we have

$$P(0, 1) = q$$
$$P(1, 1) = p \tag{2.7}$$

Relations (2.4), (2.5), (2.6), and (2.7) account for all the possible outcomes defined over the entire range of values of $r, r = 0, 1, \ldots, n$, and $n, n = 1, 2, \ldots$. These relations form a set of difference equations in the two variables r and n whose solution will yield the function $P(r, n)$ for the admissible values of r and n. This system of equations may be reduced to a difference equation in a single variable if we introduce the probability generating function

$$G(s, n) = \sum_{r=0}^{r=n} s^r P(r, n), \quad |s| \leq 1; n = 1, 2, \ldots \tag{2.8}$$

From relation (2.7), we obtain for $G(s, 1)$

$$G(s, 1) = \sum_{r=0}^{r=1} s^r P(r, 1)$$
$$= q + ps \tag{2.9}$$

For $n = 2, 3, \ldots$, multiply equation (2.4) by s^r, $r = 1, 2, \ldots, n - 1$, equation (2.5) by s^0, and equation (2.6) by s^n, and sum over all possible values of r to obtain

$$\sum_{r=0}^{r=n} s^r P(r, n) = q \sum_{r=0}^{r=n-1} s^r P(r, n - 1) + p \sum_{r=1}^{r=n} s^r P(r - 1, n - 1)$$

In the last sum expression on the right-hand side we perform the change in variable $i = r - 1$; thus

$$\sum_{r=0}^{r=n} s^r P(r, n) = q \sum_{r=0}^{r=n-1} s^r P(r, n - 1) + ps \sum_{i=0}^{i=n-1} s^i P(i, n - 1) \tag{2.10}$$

Using (2.8) in equation (2.10), the following difference equation in $G(s, n)$, $|s| \leq 1$, is obtained:

$$G(s, n) = qG(s, n - 1) + psG(s, n - 1)$$

or

$$= (q + ps)G(s, n - 1), \ n \geq 2$$

The expression for $G(s, n)$ can be obtained recursively starting from the expression for $G(s, 1)$ given by (2.9); thus

$$G(s, 2) = (q + ps)G(s, 1)$$
$$= (q + ps)^2$$

and so on; it is easy then to see that

$$G(s, n) = (q + ps)^n, \quad n \geq 1 \tag{2.11}$$

This probability generating function can be identified (Section 1.3-v-c) as that of a random variable possessing the binomial distribution; hence the PMF of Y_n, $\varphi_{Y_n}(r)$, is given by (2.3). The discrete-parameter discrete-valued process $\{Y_n, n \in \mathbf{T}\}$ defines the *binomial process*.

The statements that have been made so far about Y_n assumed an ordered sequence of n months or trials. This assumption is not necessary because of the independency of the occurrence of the events between any two trials. Hence the distribution of Y_n would still be binomial irrespective of the ordering or selection of the n months considered.

b. Number of Trials T_1 Until the First Success

We now consider another problem the salesman may be facing. Suppose that he attempts to find out the probability that the customer will place his first order in a given month n, $n \geq 1$; in other words, the salesman is trying to determine the probability that he will have to wait n months before the customer finally decides to place his order.

Let the random variable T_1 define the number of months elapsed until the occurrence of the first order (see Figures 2.5 and 2.6). For convenience, set $P\{T_1 = 0\} = 0$, and let

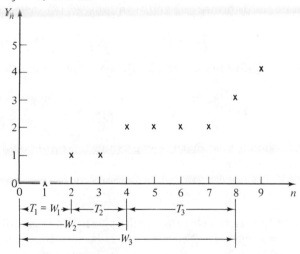

Figure 2.6. Sample function for a binomial process generated by the Bernoulli process of Figure 2.5.

$$\varphi_{T_1}(n) = P\{T_1 = n\}, \quad n = 0, 1, 2, \ldots$$

be the PMF of T_1. For $n = 2, 3, \ldots$, define the following:

$P_1(n)$ = probability that in n months the first $n - 1$ months result in no orders placed and the last month n results in an order placed

$P_0(n)$ = probability that in n months the first $n - 1$ months result in no orders placed and the last month n results also in no orders placed

Finally, for $n = 1$, we define

$P_1(1) =$ probability that an order is placed in the first month

$P_0(1) =$ probability that no order is placed in the first month

Then

$$P_1(1) = p \quad \text{and} \quad P_0(1) = q \tag{2.12}$$

Also, it is evident that

$$\varphi_{T_1}(n) = \begin{cases} 0, & n = 0 \\ P_1(n), & n = 1, 2, \ldots \end{cases}$$

Relating the events occurring during the first n months to the events occurring during the first $n - 1$ months, we obtain for $n = 2, 3, \ldots$ the following system of recurrence relations:

$$P_0(n) = qP_0(n - 1) \tag{2.13}$$

$$P_1(n) = pP_0(n - 1) \tag{2.14}$$

Solving equation (2.13) recursively and then using (2.12) yields

$$P_0(n) = q^{n-1}P_0(1) = q^n \tag{2.15}$$

Using expression (2.15) in equation (2.14), we finally obtain for $\varphi_{T_1}(n)$

$$\varphi_{T_1}(n) = \begin{cases} 0, & n = 0 \\ pq^{n-1}, & n = 1, 2, \ldots \end{cases} \tag{2.16}$$

The reader should note that the expression for $P_1(n)$ could have been obtained from (2.3) since

$$P_0(n - 1) = \varphi_{Y_{n-1}}(0) = q^{n-1}$$

Expression (2.16) can be identified as the alternative form of the geometric distribution encountered in Section 1.3-v-e. Because of the independence between trials, the distribution of the random variable T_1 will not depend on the selection of the origin month from which T_1 is counted. Thus, in our example, given that no orders were placed during the first 3 months, the probability that exactly $U = n$ months will elapse until the occurrence of the first customer order is

$$\varphi_U(n) = \begin{cases} 0, & n = 0 \\ pq^{n-1}, & n = 1, 2, \ldots \end{cases} \tag{2.17}$$

We shall express this property by saying that the geometric distribution characterizing the time T_1 until the occurrence of the first success has *no memory* or is *memoryless*.

c. NUMBER OF TRIALS T_i, $i = 2, 3, \ldots$, BETWEEN THE $(i - 1)$ ST AND iTH SUCCESS

The total number of months T_1 elapsed from the start of the process until the appearance of the first order was seen to have the geometric distribution defined by (2.16). Because the successive trials are mutually independent, the total number of months T_i, $i = 2, 3, \ldots$, elapsed between the appearance of the $(i - 1)$st order until the appearance of the ith order (see Figures 2.5 and 2.6) is such that $\{T_i\}$ forms a sequence of independently and identically distributed random variables each with PMF

$$\varphi_{T_i}(n) = \varphi(n) = \begin{cases} 0, & n = 0 \\ pq^{n-1}, & n = 1, 2, \ldots \end{cases} \tag{2.18}$$

Now, given that the total number of orders placed during the first x months, $x = 1, 2, \ldots$, is $i - 1$, $i = 1, 2, \ldots, x + 1$, the probability that exactly $U = n$ additional months will elapse until the occurrence of the ith order is still given by (2.17), since the distribution of T_i is geometric.

d. NUMBER OF TRIALS W_r UNTIL THE rTH SUCCESS, $r = 1, 2, \ldots$

As a further generalization, suppose that the salesman is interested in determining the probability distribution of the total number of months W_r to elapse until the customer places his rth order, $r = 1, 2, \ldots$ (see Figure 2.6). The special case of $r = 1$ was just analyzed. We shall not attempt, however, to formulate the general problem in terms of recurrence relations as the analysis becomes quite involved. Rather, we shall derive the distribution of W_r by making direct use of the results obtained in the previous section. Let the PMF of W_r be

$$\varphi_{W_r}(n) = P\{W_r = n\}, \quad n = 0, 1, 2, \ldots$$

The total number of months W_r elapsed from the start of the process until the appearance of the rth customer order is given by

$$W_r = T_1 + T_2 + \cdots + T_r$$

Since $\{T_i\}$, $i = 1, 2, \ldots, r$, form a sequence of r independently and identically distributed random variables, each with PMF given by (2.18), it follows that the PMF of W_r, $\varphi_{W_r}(n)$, is the rth-fold convolution of $\varphi(n)$; that is,

$$\varphi_{W_r}(n) = \varphi *^{(r)} (n)$$

From (2.18), the probability generating function of each T_i, $i = 1, 2, \ldots, r$, is

$$G(s) = \sum_{n=1}^{\infty} s^n pq^{n-1}$$

$$= \frac{ps}{1 - qs}$$

The probability generating function of W_r is thus

$$G_{W_r}(s) = \sum_{n=0}^{\infty} s^n \varphi_{W_r}(n) = \left(\frac{ps}{1 - qs}\right)^r, \quad |s| \leq 1$$

$$= p^r s^r (1 - qs)^{-r}$$

Using a binomial expansion, we obtain

$$G_{W_r}(s) = p^r s^r \left[1 + rqs + \frac{r(r + 1)}{2!} q^2 s^2 + \cdots \right.$$

$$+ \frac{r(r + 1) \ldots (n - 1)}{(n - r)!} q^{n-r} s^{n-r} + \cdots \right]$$

$$= p^r s^r \left[1 + rqs + \frac{r(r + 1)}{2!} q^2 s^2 + \cdots \right.$$

$$+ \binom{n - 1}{n - r} q^{n-r} s^{n-r} + \cdots \right]$$

The coefficient of s^n, $n = 0, 1, \ldots$, in this power series expansion of $G_{W_r}(s)$ identifies $\varphi_{W_r}(n)$; it can be seen that

$$\varphi_{W_r} = \begin{cases} 0, & n = 0, 1, \ldots, r - 1 \\ \binom{n - 1}{n - r} p^r q^{n-r}, & n = r, r + 1, \ldots \end{cases} \tag{2.19}$$

The reader can verify that the random variable $W_r - r$ has the negative binomial distribution (Section 1.3-v-f).

e. No Occurrence of a Run of Two Failures in n Trials, $n \geq 2$

Given any ordered sequence of elements of two types, each maximal subsequence of identical elements is called a *run*. For example, in an independent Bernoulli process in which the outcomes at each trial result in either a success S or a failure F, one may obtain the following sequence in six trials: SFFSSS. It is seen that this sequence has three subsequences, S, FF, and SSS, constituting, respectively, an S run of 1 followed by an F run of 2 followed by an S run of 3.

In our particular situation the sequence of trials performed every month will result in runs of "no orders" alternating with runs of "orders." The problem discussed in Section 2.3-i-b could have been formulated as follows: determine the probability that in n months an "order" run of 1 appears for the first time. The general theory of runs is beyond the scope of this book; in what follows we illustrate the use of recurrence relations in solving a special problem related to the theory of runs.

Suppose that the salesman would like to determine the probability that over an n-month period, $n \geq 2$, the customer will not place orders in any two successive months. If for any particular month we label 0 the event "no order placed" and we label 1 the event "order placed," then, for $n = 3$, out of the 2^3 mutually exclusive and exhaustive outcomes, the five outcomes

$$000, \quad 001, \quad 010, \quad 100, \quad 101$$

result in no runs in excess of two orders, while in each of the remaining three outcomes

$$111, \quad 110, \quad 011$$

a run equal to or exceeding two orders occurs.

For $n \geq 2$, let

$P_0(n)$ = probability that over an n-month period two orders in succession do not occur and that the last month n results in no order placed

$P_1(n)$ = probability that over an n-month period two orders in succession do not occur and that the last month n results in an order placed

Relating the events occurring during the first n-month period to the events occurring during the first $n - 1$ months, we see that, for $n \geq 3$,

• 1. The event "two orders in succession do not occur over n months and the last month n results in no order placed" can occur in the following two possible mutually exclusive and exhaustive ways: (a) "two orders in succession do not occur over $n - 1$ months, and the month $n - 1$ results in no order placed," and "no order is placed in the nth month"; (b) "two orders in succession do not occur over $n - 1$ months, and the last month $n - 1$ results in an order placed," and "no order is placed in the nth month." We thus obtain

$$P_0(n) = qP_0(n - 1) + qP_1(n - 1), \quad n = 3, 4, \ldots \tag{2.20}$$

2. The event "two orders in succession do not occur over n months, and the last month n results in an order placed" can occur only in the follow-

ing way: "two orders in succession do not occur over $n - 1$ months, and the month $n - 1$ results in no order placed," and "an order is placed in the last month n." Hence

$$P_1(n) = pP_0(n - 1), \quad n = 3, 4, \ldots \tag{2.21}$$

For $n = 2$, $P_1(2)$ is the probability of occurrence of the outcome 01, and $P_0(2)$ is the probability of occurrence of the outcomes 10 or 00. Therefore, for $n = 2$

$$P_0(2) = pq + q^2 \tag{2.22}$$
$$= q(p + q) = q$$

and

$$P_1(2) = qp \tag{2.23}$$

Let now

$P(n) =$ probability that over an n-month period two orders in succession do not occur

Since the outcome during the last month n is either an order placed or no order placed, it follows that

$$P(n) = P_0(n) + P_1(n), \quad n \geq 2 \tag{2.24}$$

The solution of the system of equations (2.20) to (2.24) will yield the function $P(n)$, $n \geq 2$.

Define by $G_0(s)$, $G_1(s)$, and $G(s)$ the generating functions of $P_0(n)$, $P_1(n)$, and $P(n)$, respectively; thus for $|s| \leq 1$,

$$G_0(s) = \sum_{n=2}^{\infty} s^n P_0(n) \tag{2.25}$$

$$G_1(s) = \sum_{n=2}^{\infty} s^n P_1(n) \tag{2.26}$$

$$G(s) = \sum_{n=2}^{\infty} s^n P(n)$$

$$= \sum_{n=2}^{\infty} s^n [P_0(n) + P_1(n)] \tag{2.27}$$

$$= G_0(s) + G_1(s)$$

Multiply equation (2.20) by s^n and sum over all $n = 3, 4, \ldots$; we obtain

$$\sum_{n=3}^{\infty} s^n P_0(n) = q \sum_{n=3}^{\infty} s^n P_0(n - 1) + q \sum_{n=3}^{\infty} s^n P_1(n - 1)$$

or

$$\sum_{n=2}^{\infty} s^n P_0(n) - s^2 P_0(2) = qs \sum_{m=2}^{\infty} s^m P_0(m) + qs \sum_{m=2}^{\infty} s^m P_1(m)$$

Expressed in terms of the generating functions (2.25) and (2.26) and using (2.22), this last relation can be written as

$$G_0(s) - qs^2 = qsG_0(s) + qsG_1(s) \tag{2.28}$$

Now multiply equation (2.21) by s^n and sum over all $n = 3, 4, \ldots$ to obtain

$$\sum_{n=3}^{\infty} s^n P_1(n) = p \sum_{n=3}^{\infty} s^n P_0(n-1)$$

or

$$\sum_{n=2}^{\infty} s^2 P_1(n) - s^2 P_1(2) = ps \sum_{m=2}^{\infty} s^m P_0(m)$$

Using (2.23) and the definition of $G_0(s)$ and $G_1(s)$, we can write for this last expression

$$G_1(s) - pqs^2 = psG_0(s) \tag{2.29}$$

Equations (2.28) and (2.29) form a system of simultaneous linear algebraic equations in $G_0(s)$ and $G_1(s)$. Substituting the expression for $G_1(s)$ in (2.29) into equation (2.28), we obtain

$$G_0(s) - qs^2 = qsG_0(s) + qs[pqs^2 + psG_0(s)]$$

Solving for $G_0(s)$, we get

$$G_0(s) = \frac{qs^2 + pq^2s^3}{1 - qs - pqs^2} \tag{2.30}$$

To derive an expression for $G(s)$, we note from equation (2.28) that

$$G_0(s) - qs^2 = qs[G_0(s) + G_1(s)]$$
$$= qsG(s)$$

Hence

$$G(s) = \frac{G_0(s)}{qs} - s \tag{2.31}$$

Using expression (2.30) in (2.31), we finally obtain

$$G(s) = \frac{s + pqs^2}{1 - qs - pqs^2} - s$$
$$= -1 - s + \frac{1 + ps}{1 - qs - pqs^2}, \quad |s| \leq 1 \tag{2.32}$$

This then is the expression for the generating function of $P(n)$, $n \geq 2$. The specific functional form of $P(n)$ may be obtained by writing relation (2.32) as a power series function of s and identifying the coefficient of s^n (see Problem 10). We shall show how to obtain $P(n)$ when p takes on a specific numerical value. Suppose that the probability that the customer will place an order in any given month is $p = \frac{2}{3}$; thus $q = \frac{1}{3}$. We then have

$$G(s) = -1 - s + \frac{1 + \frac{2}{3}s}{1 - \frac{1}{3}s - \frac{2}{9}s^2}$$

$$= -1 - s + \frac{1 + \frac{2}{3}s}{(1 - \frac{2}{3}s)(1 + \frac{1}{3}s)}$$

If we reduce into partial fractions the rational fraction on the right-hand side, we obtain

$$G(s) = -1 - s + \frac{4}{3}\frac{1}{1 - \frac{2}{3}s} - \frac{1}{3}\frac{1}{1 + \frac{1}{3}s}, \quad |s| \leq 1$$

Using a binomial expansion on the two fractions, we obtain

$$G(s) = -1 - s + \frac{4}{3}[1 + \frac{2}{3}s + (\frac{2}{3}s)^2 + \cdots + (\frac{2}{3}s)^n + \cdots]$$
$$- \frac{1}{3}[1 + (-\frac{1}{3}s) + (-\frac{1}{3}s)^2 + \cdots + (-\frac{1}{3}s)^n + \cdots]$$

We note that the coefficients of s^0 and s^1 are zero, as might be expected. The coefficient of s^n, $n \geq 2$, is thus

$$P(n) = \frac{4}{3}(\frac{2}{3})^n - \frac{1}{3}(-\frac{1}{3})^n$$

$$= \frac{4 \cdot 2^n}{3^{n+1}} + \frac{(-1)^{n+1}}{3^{n+1}}, \quad n \geq 2$$

Even for moderate values of n, the contribution of the second term on the right-hand side, $(-1)^{n+1}/3^{n+1}$, is negligible; thus, for large n,

$$P(n) \approx 2(\frac{2}{3})^{n+1}$$

ii. *Some Generalization of the Independent Bernoulli Process*

So far we have given special, but not detailed, attention to a discrete-parameter independent process consisting of a succession of trials, each resulting in the two possible outcomes, 0 or 1. The more elaborate theory underlying this independent Bernoulli process is particularly valuable in applied probability, because it is often possible to derive from this simple process rather interesting clues to formulate more general independent processes.

An immediate generalization is to assume that each trial may result in more than two possible outcomes; that is, the state space of the process contains more than two points with an associated positive probability. A particular realization of this more general process is shown in Figure 2.1. Another possible generalization would be to study the limiting behavior of the discrete process into a continuous process, and hence derive results for the continuous parameter and/or continuous state space independent process. In what follows we discuss some aspects of the first form of generalization. Section 2.4 will make use of the second form of generalization.

a. DISCRETE-PARAMETER INDEPENDENT PROCESSES WITH MORE THAN TWO STATES

In the example of the salesman soliciting orders every month from a customer, suppose that the orders refer to a particular product and that the size of each order is a discrete random variable. Let X_n denote the quantity ordered by the customer during month n, $n = 1, 2, \ldots$, and let the PMF of X_n be

$$P\{X_n = r\} = \varphi_{X_n}(r) = \varphi(r), \quad r = 0, 1, 2, \ldots$$

Thus $\varphi_{X_n}(0)$ is the probability that the customer will not place any orders during month n.

The probability law of this more general independent process is completely specified by the joint probability mass function of the sequence of independent random variables $\{X_n\}$ given by $\prod_{i=1}^{i=n} \varphi_{X_i}(r_i)$. Theoretically at least, one should be able to make any type of inferences about the probability structure of events associated with this process. Needless to say, at least the same type of difficulties are encountered as in the simpler independent Bernoulli process. Sometimes the use of recurrence relations presents a convenient avenue to overcome these difficulties. Other times it is possible to take direct advantage of the independence between trials to compute the probability of occurrences of particular events. For example, if the salesman is interested in determining the distribution of the total number of units Y_n to be ordered by the customer over an n-month period, $n = 1, 2, \ldots$, then

$$Y_n = X_1 + X_2 + \cdots + X_n$$

Thus the PMF of Y_n is the nth-fold convolution of $\varphi(r)$. Other characteristics of this process may be established in this direct fashion, as, for example, the distribution of the minimum or maximum number of units ordered over a given number of months. However, beyond these simple characterizations the formulation of the probability laws of more complex events, such as the occurrence of runs of particular length for a given order size, can be quite difficult and will not be treated here (see also Problem 11).

2.4. Independent Processes: Continuous Parameter

i. *Limit Form of the Independent Bernoulli Process*

For a continuous-parameter stochastic process where the index set is $\mathbf{T} = \{\theta : 0 \leq \theta < \infty\}$, we consider a given finite duration of the process of length t, $0 \leq \theta < t$. Suppose that the index set over this particular duration t is divided into small equal intervals of length $d\theta$ so that a total number of $n = t/d\theta$ intervals is generated.

Each of these intervals of length $d\theta$ may be assumed to correspond to a trial such that, at any particular trial either of two mutually exclusive and exhaustive outcomes F and S are possible. For a particular trial let

$\lambda\, d\theta$ = probability of occurrence of event S
$1 - \lambda\, d\theta$ = probability of occurrence of event F

It is assumed that the probability of occurrence of a particular outcome at a particular trial does not depend on the outcomes of past trials. For the ith trial, define the random variable X_i such that $(i = 1, 2, \ldots, n)$

$$X_i = \begin{cases} 0, & \text{if F occurs} \\ 1, & \text{if S occurs} \end{cases}$$

The discretized process is then an independent Bernoulli process consisting of n trials in which the probability of a success at a trial is $p = \lambda\, d\theta$. Within the context of this formulation it becomes possible to utilize the results of the analysis of the independent Bernoulli process and characterize the continuous-parameter process by letting $d\theta \longrightarrow 0$ and $n \longrightarrow \infty$, while the duration t of the process is kept constant (see Figure 2.7).

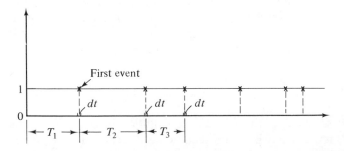

Figure 2.7. Sample function for the limit form of the Bernoulli process.

Referring in particular to the salesman problem, let time be measured over a continuous scale; then $\lambda\, d\theta$ is the probability that the customer will place an order in the time interval $(\theta, \theta + d\theta]^*$, while $(1 - \lambda\, d\theta)$ is the probability that he will not order in that time interval.

a. Number of Events over a Given Time t, $t \geq 0$: The Poisson Process

Suppose that we require to determine the total number of orders $X(t)$ that the customer will place over the entire time interval t (see Figure 2.8).

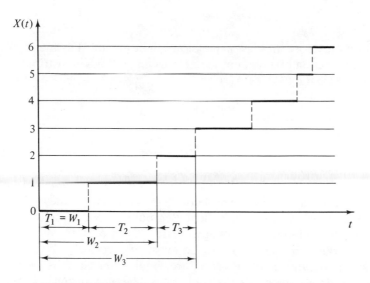

Figure 2.8. Sample function for the Poisson process generated by the process of Figure 2.7.

Let $\varphi_{X(t)}(r)$, $r = 0, 1, \ldots$, be the probability mass function of $X(t)$. From (2.11), the probability generating function of $X(t)$ is

$$G_{X(t)}(s, n) = [(1 - \lambda\, d\theta) + \lambda\, d\theta \cdot s]^n$$

Since $t = n\, d\theta$, we may write $G_{X(t)}(s, n)$ as

$$G_{X(t)}(s, n) = \left[1 - \frac{(1 - s)\lambda t}{n} \right]^n \tag{2.33}$$

To study the behavior of $G_{X(t)}(s, n)$ as $d\theta \to 0$, $n \to \infty$ while t is constant, we simply refer to a well-known transcendental function used in mathe-

---

*The notation ( ] refers to a half open interval; thus, the time interval $(\theta, \theta + d\theta]$ is equivalent to $\theta < \tau \leq \theta + d\theta$.

matics; this is the exponential function e^y defined by the relation

$$e^y = \lim_{n \to \infty} \left(1 + \frac{y}{n} \right)^n$$

Taking limits as $n \longrightarrow \infty$ on both sides of expression (2.33), we readily see that

$$G_{X(t)}(s) = \lim_{n \to \infty} G_{X(t)}(s, n) = \lim_{n \to \infty} \left[1 - \frac{(1 - s)\lambda t}{n} \right]^n$$

$$= e^{-(1-s)\lambda t}$$

From Section 1.3-v-d this last expression can be identified as the probability generating function of a random variable having a Poisson distribution with mean λt; hence the PMF of $X(t)$ is

$$\varphi_{X(t)}(r) = \frac{e^{-\lambda t}(\lambda t)^r}{r!}, \quad r = 0, 1, 2, \ldots \tag{2.34}$$

The continuous-parameter discrete-valued stochastic process $\{X(t), 0 \leq t < \infty\}$ defines the *homogeneous Poisson process* with *mean intensity* λ.

The Poisson process is the continuous-parameter counterpart of the binomial process describing the probability of occurrence of the total number of successes in successive independent Bernoulli trials discussed in Section 2.3-i-a. In the literature of continuous-parameter stochastic processes, the instants marking the occurrence of an event of specified character (in our example, the occurrence of an order) are simply referred to as *events*. Thus the Poisson process describes the probability law of occurrence of the total number of events happening according to some predefined conditions.

The Poisson process $\{X(t), 0 \leq t < \infty\}$ has the following important properties:

(1) It has *independent increments*; that is, for all $0 < t_1 < \cdots < t_n$, the n random variables $[X(t_1) - X(0)], \ldots, [X(t_n) - X(t_{n-1})]$ are independent. In words, the events occurring respectively during nonoverlapping intervals are mutually independent.

(2) It has *stationary increments*; that is, for all $h > 0$ and $0 \leq t_1 < t_2$, the random variable $[X(t_2 + h) - X(t_1 + h)]$ has the same distribution as the random variable $[X(t_2) - X(t_1)]$. In words, a translation in the origin of the index set does not affect the probability law of occurrence of events over a particular interval; the process is then said to be *homogeneous*.

We do not provide a formal proof to demonstrate these two properties; the reader should note, however, that underlying the generation of the Poisson process is the binomial process, which possesses similar type properties expressed over a discrete index set.

Referring to our particular example, it follows that the probability that the customer will place a given number of orders in the time interval $(t, t + h]$, $h > 0$, is independent of the starting time t from which the occurrence of orders is counted, and that it is a function of the elapsed length h only. Thus the total number of orders placed over $(t, t + h]$ can be expressed by the random variable $X(h)$. It also follows that $X(h)$ has a Poisson distribution with PMF

$$P\{X(h) = r\} = \varphi_{X(h)}(r) = e^{-\lambda h}\frac{(\lambda h)^r}{r!}, \quad r = 0, 1, 2, \ldots \qquad (2.35)$$

Finally, we note that given t, $t \geq 0$, the probability that no orders will be placed in the interval $(t, t + h]$ is

$$P\{X(h) = 0\} = \varphi_{X(h)}(0) = e^{-\lambda h} \qquad (2.36)$$

b. TIME ELAPSED T_1 UNTIL THE OCCURRENCE OF THE FIRST EVENT

As an extension of Section 2.3-i-b to continuous-parameter processes, let the random variable T_1 define the time elapsed from start of the process until the occurrence of the first order (see Figures 2.7 and 2.8). Let $\varphi_{T_1}(t)$, $t > 0$, be the PDF of T_1. Thus the probability that the first order occurs between t and $t + dt$ is

$$P\{t < T_1 \leq t + dt\} = \varphi_{T_1}(t)\, dt \qquad (2.37)$$

We note that the event "the first order occurs between t and $t + dt$" can only occur in the following way: "no orders occur between 0 and t" followed by "an order placed between t and $t + dt$." Since the occurrence of an order in the interval $(t, t + dt]$ is independent of the outcomes preceding time t, it follows that

$$P\{t < T_1 \leq t + dt\} = \text{probability that no orders occur before } t$$
$$\times \text{ probability that an order occurs in } (t, t + dt]$$

Using relation (2.36) and remembering that $\lambda\, dt$ is the probability that the customer will place an order in $(t, t + dt]$, it follows that

$$\varphi_{T_1}(t)\, dt = e^{-\lambda t} \cdot \lambda\, dt$$

Thus the PDF of T_1 is given by the negative exponential distribution

$$\varphi_{T_1}(t) = \lambda e^{-\lambda t}, \quad t > 0 \qquad (2.38)$$

The negative exponential distribution for T_1 in the continuous-parameter case plays the same role as the geometric distribution for T_1 in the discrete-parameter case. As such, it possesses the *memoryless* property.

Thus, referring to our example, given that the salesman is still waiting at time x, $x > 0$, for the occurrence of the first order, the distribution function of the additional time U he will have to wait until the first order is placed by the customer is negative exponential and is independent of the past elapsed time x. To show this property, we note that the probability that no orders are placed by the customer in the time interval $(x, x + h]$, $h > 0$, is the same as the probability that the salesman will have to wait an additional time U greater than h prior to the appearance of this first order. Thus

$$P\{X(h) = 0\} = P\{U > h\}$$

From this relation, we deduce that the distribution function of U is

$$\Phi_U(h) = P\{U \le h\} = 1 - P\{X(h) = 0\}$$

From relation (2.36) it follows that

$$\Phi_U(h) = 1 - e^{-\lambda h}, \quad h > 0$$

which is independent of x and can be recognized as the negative exponential distribution.

c. Interarrival Times $\{T_i\}$, $i = 1, 2, \ldots$, Between
   Successive Events

We now consider the more general problem of determining the distribution of the time T_i, $i = 1, 2, \ldots$, elapsed between the occurrence of the $(i - 1)$st order and the ith order, with T_1 defined as the time of occurrence of the first order (see Figures 2.7 and 2.8).

Let $\varphi_{T_i}(t)$ be the PDF of T_i. Using (2.36) and reasoning in a way similar to Section 2.4-i-b, it follows that

$$\varphi_{T_i}(t)dt = P\{t < T_i \le t + dt\} = e^{-\lambda t} \cdot \lambda \, dt$$

Thus, for all $i = 1, 2, \ldots$,

$$\varphi_{T_i}(t) = \varphi(t) = \lambda e^{-\lambda t}, \quad t > 0 \tag{2.39}$$

Therefore, $\{T_i\}$ form a sequence of identically distributed random variables each having the negative exponential distribution defined by expression (2.39).

We next show that $\{T_i\}$ form a sequence of independently distributed random variables. For $r = 1, 2, \ldots$, let $\varphi_{T_1, T_2, \ldots, T_r}(t_1, t_2, \ldots, t_r)$ define the joint probability density function of $T_1, T_2, \ldots, T_r$. Then, since the occurrence of an event over $(t, t + dt]$ does not depend on the occurrence of events prior to t, it follows that

$$\varphi_{T_1, T_2, \ldots, T_r}(t_1, t_2, \ldots, t_r) \, dt_1 \, dt_2 \, \cdots \, dt_r$$
$$= P\{t_1 < T_1 \leq t_1 + dt_1, t_2 < T_2 \leq t_2 + dt_2, \ldots, t_r < T_r \leq t_r + dt_r\}$$
$$= e^{-\lambda t_1} \lambda \, dt_1 \cdot e^{-\lambda t_2} \lambda \, dt_2 \, \cdots e^{-\lambda t_r} \lambda \, dt_r$$
$$= (\lambda e^{-\lambda t_1})(\lambda e^{-\lambda t_2}) \cdots (\lambda e^{-\lambda t_r}) \, dt_1 \, dt_2 \, \cdots dt_r$$
$$= \varphi_{T_1}(t_1) \cdot \varphi_{T_2}(t_2) \cdots \varphi_{T_r}(t_r) \, dt_1 \, dt_2 \, \cdots dt_r$$

and this last relation establishes the independence of the T_i.

The sequence of random variables $\{T_i\}$ defines the successive *interarrival times* $T_1, T_2, \ldots$ of events occurring in the interval 0 to ∞. We can then say that the homogeneous Poisson process with mean intensity λ is characterized by the fact that the interarrival times $\{T_i\}$ of events form a sequence of independent identically distributed random variables having a negative exponential distribution with mean $1/\lambda$.

It is also evident that given $i - 1$ events, $i = 1, 2, \ldots$, having occurred over the interval 0 to x, the additional time U elapsed till the occurrence of the ith event is a random variable having the negative exponential PDF

$$\varphi_U(t) = \lambda e^{-\lambda t}, \quad t > 0$$

d. Waiting Time W_r Until the Occurrence of the rth Event, $r = 1, 2, \ldots$

Finally, it is possible to obtain an expression for the total time W_r the salesman will have to wait before the customer places his rth order, $r = 1$, $2, \ldots$ (see Figure 2.8). The analysis proceeds in a way similar to the discrete case discussed in Section 2.3-i-d. Let $\varphi_{W_r}(t)$, $t > 0$, define the PDF of W_r; then $W_1 = T_1$, and

$$W_r = T_1 + T_2 + \cdots + T_r$$

Since $\{T_i\}$, $i = 1, 2, \ldots, r$, form a sequence of r independently and identically distributed random variables each with PDF given by expression (2.39), it follows that the PDF of W_r is the rth-fold convolution of $\varphi(t) = \lambda e^{-\lambda t}$. From Example 1.8, the rth-fold convolution of a negative exponential density function with parameter λ is a gamma density function of order r and with parameter λ. Hence

$$\varphi_{W_r}(t) = \frac{(\lambda t)^{r-1}}{\Gamma(r)} \lambda e^{-\lambda t}, \quad \lambda > 0, t > 0 \tag{2.40}$$

It is of interest to derive the distribution of W_r in yet another way. Referring to our example, we note that the probability that the salesman will have to wait a time W_r in excess of t for the occurrence of the rth order, $r = 1, 2, \ldots$, is the same as the probability that the customer will place a total number of orders $X(t)$ over $(0, t]$ less than or equal to $r - 1$. Hence

$$P\{W_r > t\} = P\{X(t) \le r - 1\} \tag{2.41}$$

The distribution function of W_r is then

$$\Phi_{W_r}(t) = P\{W_r \le t\} = P\{X(t) \ge r\}$$
$$= \sum_{i=r}^{\infty} e^{-\lambda t} \frac{(\lambda t)^i}{i!}$$

From Chapter 1, Problem 7, it follows that W_r has the gamma distribution function

$$\Phi_{W_r}(t) = \int_0^t \frac{(\lambda u)^{r-1}}{\Gamma(r)} \lambda e^{-\lambda u} \, du$$

The gamma distribution in the continuous-parameter process plays the same role as the negative binomial distribution in the discrete-parameter process encountered in Section 2.3-i-d.

Another important relationship between $X(t)$ and W_r is

$$P\{X(t) = 0\} = 1 - P\{w_1 \le t\}, \quad r = 0$$
$$P\{X(t) = r\} = P\{W_r \le t\} - P\{W_{r+1} \le t\}, \quad r = 1, 2, \ldots \tag{2.42}$$

Relations (2.41) and (2.42) are very general and are not restricted to the Poisson process or to continuous-parameter processes. Using formula (2.41), the reader should attempt to provide an alternative derivation for expression (2.19).

2.5. Markov Processes

i. *Definition*

A stochastic process $\{X(t), t \in \mathbf{T}\}$ with index set $\mathbf{T}$ and state space $\mathbf{I}$ is called a *Markov process* if for all $n, n = 0, 1, 2, \ldots$, and for any $t_0 < t_1 < t_2 < \cdots < t_n, t_0 = 0$,

$$P\{X(t_n) \le x_n \mid X(t_0) = x_0, X(t_1) = x_1, \ldots, X(t_{n-1}) = x_{n-1}\}$$
$$= P\{X(t_n) \le x_n \mid X(t_{n-1}) = x_{n-1}\}$$

In words, the previous relation asserts that

(1) The conditional probability distribution of $X(t_n)$ for given values of $X(t_0), X(t_1), \ldots, X(t_{n-1})$ depends only on the value taken by $X(t_{n-1})$,

the most recent value, and does not depend on the values taken by $X(t_0)$, $X(t_1)$, . . . , $X(t_{n-2})$.

(2) Or, given the present state of the system, the future state is independent of the past.

(3) Or, it is sufficient to know about the history of the stochastic process at time t_{n-1} to be able to deduce the properties of the process at time t_n.

In contradistinction to an independent process, which was previously discussed, a Markov process is an example of a stochastic process in which the most elementary form of dependency exists between successive trials. The probability law of a Markov process is then completely specified with a knowledge of (1) the *initial conditions* given by $P\{X(t_0) \leq x_0\}$, and (2) the set of conditional probability distributions given for all $0 \leq t_m < t_n, m, n = 0, 1, 2, \ldots$, by $P\{X(t_n) \leq x_n \mid X(t_m) = x_m\}$ and which defines the *transition probability distributions* of the Markov process.

The identification of the set of the transition probability distributions for all $0 \leq t_m < t_n$ for an arbitrary Markov process is a formidable task. Fortunately, many practical problems can be formulated as Markov processes in which the transition probability distribution is a function of the difference $t_n - t_m$ and is independent of t_n. We then say that the Markov process has *homogeneous* or *stationary* transition probability distributions. This homogeneity property introduces drastic simplifications when developing the probability law of the process. Even with this added advantage, the derivation of arbitrary statistical characteristics of the process from its probability law is, in general, a very difficult task. In practice, however, one is often interested in studying some special features of the process. Two particular problems of importance are (1) the determination of $P\{X(t_n) \leq x_n\}$, the (unconditional) probability distribution function of $X(t_n)$ for a given $t_n \in \mathbf{T}$, and (2) the characterization of the process for large t_n. By making use of the homogeneity property of the process, such problems can be solved without special reference to the probability law of the process.

ii. *Classification*

Markov processes are classified according to

(1) The nature of the index set $\mathbf{T}$ (discrete-parameter or continuous-parameter processes).

(2) The nature of the state space $\mathbf{I}$ (discrete-valued or continuous-valued processes).

The classification and corresponding terminologies are exhibited in Table 2.2. Whenever the state space $\mathbf{I}$ is discrete, a Markov process is known as a *Markov chain*. Markov chains constitute the subject we shall discuss in what follows.

Table 2.2 CLASSIFICATION OF MARKOV PROCESSES

		Index Set **T**	
		Discrete	*Continuous*
State Space **I**	*Discrete*	Discrete-parameter Markov chain	Continuous-parameter Markov chain
	Continuous	Discrete-parameter Markov process	Continuous-parameter Markov process

2.6. Markov Chains: Discrete Parameter

When dealing with discrete-parameter Markov chains, the probability law of the process $\{X_n, n \in \mathbf{T}\}$, where $\mathbf{I} = \{0, 1, 2, \ldots\}$ and $\mathbf{T} = \{0, 1, \ldots, n, \ldots\}$, is completely specified by the joint probability mass function of the sequence of random variables $\{X_n\}$, $n = 0, 1, 2, \ldots$, given by

$$\begin{aligned}
\varphi_{X_0, X_1, \ldots, X_n}&(r_0, r_1, \ldots, r_n) \\
&= P\{X_0 = r_0, X_1 = r_1, \ldots, X_n = r_n\} \\
&= P\{X_n = r_n \,|\, X_{n-1} = r_{n-1}\} \cdot P\{X_{n-1} = r_{n-1} \,|\, X_{n-2} = r_{n-2}\} \\
&\quad \cdots P\{X_2 = r_2 \,|\, X_1 = r_1\} \cdot P\{X_1 = r_1 \,|\, X_0 = r_0\} \cdot P\{X_0 = r_0\}
\end{aligned}$$

The elements of the index set $\mathbf{T} = \{0, 1, 2, \ldots, n, \ldots\}$ corresponding to the succession of trials are sometimes referred to as the *steps* of the process, while the elements of the state space $\mathbf{I} = \{0, 1, 2, \ldots\}$ are simply referred to as the *states* of the process.

It is then evident that the probability law of the process is completely specified knowing (1) The *initial state* of the system given by the PMF of X_0, $P\{X_0 = r\} = \varphi_{X_0}(r)$, $r = 0, 1, 2, \ldots$, and (2) the conditional probabilities or *transition probabilities* defined for all $0 \leq m < n$, $m, n = 0, 1, 2, \ldots$, by $P\{X_n = r_n \,|\, X_m = r_m\}$. For notational convenience we shall refer to the elements of the state space $\mathbf{I} = \{0, 1, 2, \ldots\}$ by the letters i, j, k. The transition probabilities are then written, for example, as

$$P\{X_n = j \,|\, X_m = i\}, \quad 0 \leq m < n$$

and this defines *the probability that the process is making a transition from state i in step m to state j in step n.*

Prior to delving into the general analysis of the discrete-parameter Markov chain, we shall study the simplest type of chains, known as the

Markov-dependent Bernoulli process, whose state space consists of the two points 0 and 1. We shall consider this process within the framework of an example already familiar to the reader.

i. *Markov-Dependent Bernoulli Process*

Example 2.2

Consider the following extension of Example 2.1: suppose that the salesman calls each month on another customer for order solicitation. Past experience shows that if the customer placed an order last month, the probability that he will place an order this month is $p_1, 0 < p_1 < 1$, and that if the customer did not place an order last month, the probability that he will place an order this month is $p_0, 0 < p_0 < 1$. We shall assume that the salesman is presently in month $n = 0$ with a knowledge of the customer's answer following his solicitation. Contrary to Example 2.1 in which the customer's behavior at a particular month in the future could be assessed independently of his past behavior, in the present example the customer's behavior at a particular month will depend, in general, on his past behavior. Thus, whereas in Example 2.1 the probability p that the customer will place an order at some future month $n, n \geq 1$, is a *given* known quantity, in Example 2.2 such probability will have *to be determined* and will, in general, depend on the particular value n assumes.

Suppose now that we select as the elements of the index set **T** the successive monthly periods labeled $n = 0, 1, 2, \ldots$. Next, as in Example 2.1, we construct a state space **I** consisting of the two points 0 and 1 such that the point 0 corresponds to the event "no order is placed" and the point 1 corresponds to the event "an order is placed." Define the sequence of random variables $\{X_n\}, n = 0, 1, 2, \ldots$, such that $X_n = 0$ if the process in month n is in state 0, and $X_n = 1$ if the process in month n is in state 1. The stochastic process $\{X_n, n \in \mathbf{T}\}$ is a discrete-parameter Markov chain characterized by the following:

1. *The initial conditions:* We have assumed that initially, at $n = 0$, the salesman knows what the customer has ordered. Of course, we can presume here one of the two possible outcomes, 0 or 1. However, to maintain some generalities in our level of discussions, we shall define the initial states as

$$P\{X_0 = 0\} = 1 - \alpha \tag{2.43}$$

$$P\{X_0 = 1\} = \alpha \tag{2.44}$$

where α is a specified quantity taking on the values of 0 or 1. Thus, if at $n = 0$ the customer has not placed an order, then $\alpha = 0$; while if he has placed an order, then $\alpha = 1$.

2. *The transition probabilities:* These may be written as

$$P\{X_n = 1 \mid X_{n-1} = 1\} = p_1 \tag{2.45}$$

$$P\{X_n = 1 \mid X_{n-1} = 0\} = p_0 \tag{2.46}$$

where it is implicitly assumed that the values of p_1 and p_0 are independent of n; hence expressions (2.45) and (2.46) are valid for $n = 1, 2, \ldots$. From the properties of conditional probabilities, it immediately follows that

$$P\{X_n = 0 \mid X_{n-1} = 1\} = q_1 = 1 - p_1 \tag{2.47}$$

$$P\{X_n = 0 \mid X_{n-1} = 0\} = q_0 = 1 - p_0 \tag{2.48}$$

This set of four conditional probabilities specify completely the transition from a particular month $n - 1$ to the immediately succeeding month n for any $n = 1, 2, \ldots$. At first, this information may seem to be insufficient to characterize the transition probabilities of the Markov chain, since, following our previous discussion, a knowledge of the conditional probabilities needs to be specified between any two steps m and n such that $0 \le m < n$. However, as we shall see later (and this may seem intuitive to some of the readers), it suffices to know for this situation the four relations (2.45) to (2.48) to be able to determine the transition probabilities between any two arbitrary steps.

Our objective is to determine for any particular value of $n, n = 0, 1, 2, \ldots$, the probability mass function $\varphi_{X_n}(r)$ of X_n, $r = 0, 1, 2, \ldots$. Since the customer either will order or will not order, it follows that $\varphi_{X_n}(r) = 0$ for $r = 2, 3, \ldots$. Hence we can write

$$\varphi_{X_n}(r) = P\{X_n = r\} = \begin{cases} P_r(n), & r = 0, 1 \\ 0, & r = 2, 3, \ldots \end{cases}$$

Therefore, the PMF of X_n will be completely specified if one can determine $P_0(n)$ and $P_1(n)$. It is evident that, for all $n = 0, 1, 2, \ldots$,

$$P_0(n) + P_1(n) = 1 \tag{2.49}$$

and
$$0 \le P_r(n) \le 1, \quad r = 0, 1 \tag{2.50}$$

For any particular month $n, n = 1, 2, \ldots$, we can write

$$P\{X_n = 0\} = P\{X_n = 0 \mid X_{n-1} = 0\} \cdot P\{X_{n-1} = 0\}$$
$$+ P\{X_n = 0 \mid X_{n-1} = 1\} \cdot P\{X_{n-1} = 1\} \tag{2.51}$$

$$P\{X_n = 1\} = P\{X_n = 1 \mid X_{n-1} = 0\} \cdot P\{X_{n-1} = 0\}$$
$$+ P\{X_n = 1 \mid X_{n-1} = 1\} \cdot P\{X_{n-1} = 1\} \tag{2.52}$$

We elaborate on the first of these relations: the probability that the customer will place no order in month n is the sum of the probabilities associated with two mutually exclusive and exhaustive events:

Event 1. No order placed in month $n - 1$ followed by no order placed in month n, and the probability of this event is

$$P\{X_{n-1} = 0, X_n = 0\} = P\{X_n = 0 \,|\, X_{n-1} = 0\} \cdot P\{X_{n-1} = 0\}$$

Event 2. An order placed in month $n - 1$ followed by no order placed in month n, and the probability of this event is

$$P\{X_{n-1} = 1, X_n = 0\} = P\{X_n = 0 \,|\, X_{n-1} = 1\} \cdot P\{X_{n-1} = 1\}$$

By a similar argument, relation (2.52) can be established. Using (2.47) and (2.48) in equation (2.51), and (2.45) and (2.46) in equation (2.52), we obtain, for $n = 1, 2, \ldots$,

$$P_0(n) = q_0 P_0(n - 1) + q_1 P_1(n - 1) \tag{2.53}$$

$$P_1(n) = p_0 P_0(n - 1) + p_1 P_1(n - 1) \tag{2.54}$$

From the initial conditions (2.43) and (2.44), we obtain, for $n = 0$, $\alpha = 0, 1$,

$$P_0(0) = 1 - \alpha \tag{2.55}$$

$$P_1(0) = \alpha \tag{2.56}$$

The functions $P_0(n)$ and $P_1(n)$, $n = 0, 1, 2, \ldots$, are therefore solutions to the system of equations (2.53) to (2.56). We shall solve this problem using two apparently different methods. We shall attempt to relate these two methods and simultaneously derive, sometimes intuitively, certain basic properties of the process.

a. Use of Difference Equations

Relations (2.53) and (2.54) form a system of simultaneous linear first-order difference equations in $P_0(n)$ and $P_1(n)$ subject to the initial conditions (2.55) and (2.56). We note that relations (2.49), (2.53), and (2.54) are not independent; for example, adding equation (2.53) to equation (2.54), relation (2.49) is obtained. It thus becomes possible to eliminate in a simple way either $P_0(n)$ or $P_1(n)$ from either of relations (2.53) or (2.54), and obtain a single difference equation. Solving for $P_0(n)$ in equation (2.49) and substituting for $P_0(n - 1)$ in equation (2.54), the following first-order linear difference equation in $P_1(n)$ is obtained

$$P_1(n) = p_0[1 - P_1(n - 1)] + p_1 P_1(n - 1), \quad n = 1, 2, \ldots \tag{2.57}$$

with $\qquad P_1(0) = \alpha, \qquad\qquad\qquad\qquad\qquad n = 0 \tag{2.58}$

Let $a = p_1 - p_0$; then equation (2.57) may be written as

$$P_1(n) = aP_1(n-1) + p_0, \quad n = 1, 2, \ldots \tag{2.59}$$

$$P_1(0) = \alpha \tag{2.60}$$

The solution of equation (2.59) with (2.60) as initial conditions will yield for any $n = 1, 2, \ldots$ the probability that the customer will place an order at some month n, knowing his most recent behavior, at $n = 0$. We consider the two cases depending on whether $a \neq 0$ or $a = 0$.

Case 1. $a = p_1 - p_0 \neq 0$

Since $0 < p_0 < 1$ and $0 < p_1 < 1$, then, clearly, $|a| < 1$. Define the generating function of $P_1(n)$ as

$$G_1(s) = \sum_{n=0}^{\infty} s^n P_1(n), \quad |s| < 1 \tag{2.61}$$

Multiply both sides of equation (2.59) by s^n and sum over all admissible values of n, $n = 1, 2, \ldots$; we obtain

$$\sum_{n=1}^{\infty} s^n P_1(n) = a \sum_{n=1}^{\infty} s^n P_1(n-1) + p_0 \sum_{n=1}^{\infty} s^n$$

or $\qquad \displaystyle\sum_{n=0}^{\infty} s^n P_1(n) - P_1(0) = as \sum_{i=0}^{\infty} s^i P_1(i) + p_0 s \sum_{i=0}^{\infty} s^i$

Using (2.60) and (2.61), this last relation may be written as

$$G_1(s) - \alpha = asG_1(s) + \frac{p_0 s}{1 - s}$$

Solving for $G_1(s)$, we obtain

$$G_1(s) = \frac{\alpha}{1 - as} + \frac{p_0 s}{(1 - as)(1 - s)}, \quad |a| < 1, |s| < 1 \tag{2.62}$$

$P_1(n)$ can then be identified as the coefficient of s^n in the power series expansion of $G_1(s)$. Expressing the right-hand side of relation (2.62) into partial fractions, we obtain

$$G_1(s) = \frac{\alpha}{1 - as} + \frac{p_0}{1 - a} \cdot \frac{1}{1 - s} - \frac{p_0}{1 - a} \cdot \frac{1}{1 - as}, \quad |a| < 1, |s| < 1$$

Since $|s| < 1$ and $|as| < 1$, expanding each fraction in s by the binomial theorem will yield

$$G_1(s) = \alpha(1 + as + a^2s^2 + \cdots + a^n s^n + \cdots)$$

$$+ \frac{p_0}{1-a}(1 + s + s^2 + \cdots + s^n + \cdots)$$

$$- \frac{p_0}{1-a}(1 + as + a^2s^2 + \cdots + a^n s^n + \cdots)$$

The expression for $P_1(n)$, $n = 0, 1, 2, \ldots$, is thus

$$P_1(n) = \alpha a^n + \frac{p_0}{1-a} - \frac{p_0}{1-a} a^n$$

Substituting for $a = p_1 - p_0$, we may write, for $n = 0, 1, 2, \ldots$,

$$P_1(n) = \left[\alpha - \frac{p_0}{1-(p_1-p_0)}\right](p_1 - p_0)^n + \frac{p_0}{1-(p_1-p_0)} \qquad (2.63)$$

The expression for $P_0(n)$ can then be obtained by substituting (2.63) in relation (2.49) to yield

$$1 - P_1(n) = P_0(n) = -\left[\alpha - \frac{p_0}{1-(p_1-p_0)}\right](p_1 - p_0)^n + \frac{1-p_1}{1-(p_1-p_0)}$$
$$(2.64)$$

The solution for $P_1(n)$ as given in (2.63) is an explicit expression for calculating the probability that the customer will place an order at a particular month n, in terms of the known parameters α, p_0, and p_1. We note that this solution is composed of two terms: the first term, representing the *transient component* of the solution,

$$\left[\alpha - \frac{p_0}{1-(p_1-p_0)}\right](p_1 - p_0)^n \qquad (2.65)$$

is a function of the initial condition α, the time parameter n, and p_0 and p_1; the second term, representing the *steady-state component* of the solution,

$$\frac{p_0}{1-(p_1-p_0)} \qquad (2.66)$$

is independent of the parameters α and n and is only a function of the transition probabilities p_0 and p_1. Since $|p_1 - p_0| < 1$, it follows that in absolute value the transient component takes on smaller and smaller values as n gets larger and larger. Ultimately, as $n \rightarrow \infty$,

$$\lim_{n\to\infty} \left[\alpha - \frac{p_0}{1-(p_1-p_0)}\right](p_1 - p_0)^n = 0$$

and
$$\lim_{n \to \infty} P_1(n) = P_1 = \frac{p_0}{1 - (p_1 - p_0)} \tag{2.67}$$

Thus the probability that the customer will place an order at some month in the far future is a constant independent of the particular month selected and whether or not the customer has presently placed an order. This probability, which depends only on the values of the transition probabilities p_0 and p_1, is equal to the steady-state component of the solution (2.63); it is denoted by the symbol P_1, which is not a function of n.

It is also easy to verify that

$$\lim_{n \to \infty} P_0(n) = P_0 = \frac{q_1}{1 - (p_1 - p_0)} \tag{2.68}$$

Case 2. $a = p_1 - p_0 = 0$

The system of equations (2.59) and (2.60) may be written as

$$P_1(n) = p_0, \quad n = 1, 2, \ldots \tag{2.69}$$
$$P_1(0) = \alpha, \quad n = 0 \tag{2.70}$$

It therefore follows that at all future months $n = 1, 2, \ldots$ the probability that the customer will place an order is a constant equal to p_0 and is independent of his present behavior or the particular month selected. This then is equivalent to an independent Bernoulli process. We also find that

$$P_0(n) = q_1 = 1 - p_1, \quad n = 1, 2, \ldots \tag{2.71}$$
$$P_0(0) = 1 - \alpha, \quad n = 0 \tag{2.72}$$

It should be noted that if in the system of equations (2.53) and (2.54) the parameter n is dropped the equations for the steady-state values of P_0 and P_1 are obtained:

$$P_0 = q_0 P_0 + q_1 P_1 \tag{2.73}$$
$$P_1 = p_0 P_0 + p_1 P_1 \tag{2.74}$$

Example 2.3

In Example 2.2, suppose that the customer has just placed an order; thus $P_1(0) = \alpha = 1$. Compute and plot $P_1(n)$, $n = 0, 1, 2, \ldots, 10$, for the following values of p_0 and p_1

(a) $p_0 = .30, p_1 = .70$.
(b) $p_0 = .70, p_1 = .30$.
(c) $p_0 = .40, p_1 = .40$.

Also find for each case $\lim_{n \to \infty} P_1(n)$.

For cases (a) and (b) we use relation (2.63); for case (c) we use relation (2.69). Table 2.3 provides the required numerical values for $P_1(n)$ for each of the three cases. The plot of $P_1(n)$ as a function of n is represented in Figures 2.9, 2.10, and 2.11 for cases (a), (b), and (c), respectively.

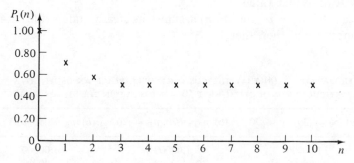

Figure 2.9. Plot of $P_1(n)$ as a function of $n(p_0 = .30,$ $p_1 = .70)$.

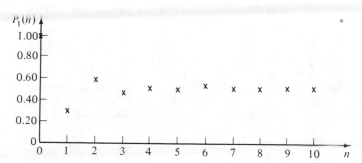

Figure 2.10. Plot of $P_1(n)$ as a function of $n(p_0 = .70,$ $p_1 = .30)$.

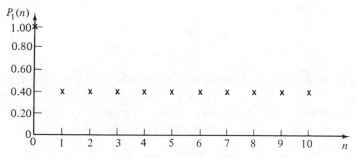

Figure 2.11. Plot of $P_1(n)$ as a function of $n(p_0 = p_1 = .40)$.

For case (a), $p_0 < p_1$; as n increases, $P_1(n)$ monotonically decreases and tends fairly rapidly (after 5 months for all practical purposes) toward its steady-state value of .50000.

For case (b), $p_0 > p_1$; as n increases, the value of $P_1(n)$ oscillates in a damped fashion about its steady-state value of .50000, and tends again fairly rapidly toward that value.

For case (c), $p_0 = p_1$, and $P_1(n)$ assumes its steady-state value of .40000 the first month and thereafter.

Table 2.3. PROBABILITY $P_1(n)$ THAT THE CUSTOMER WILL PLACE AN ORDER AT A PARTICULAR MONTH IN THE NEXT 10 MONTHS AND IN THE STEADY STATE

n	(a) $p_0 = .30, p_1 = .70$	(b) $p_0 = .70, p_1 = .30$	(c) $p_0 = .40, p_1 = .40$
0	1.00000	1.00000	1.00000
1	.70000	.30000	.40000
2	.58000	.58000	.40000
3	.53200	.46800	.40000
4	.51280	.51280	.40000
5	.50512	.49488	.40000
6	.50205	.50205	.40000
7	.50082	.49918	.40000
8	.50033	.50033	.40000
9	.50013	.49987	.40000
10	.50005	.50005	.40000
∞	.50000	.50000	.40000

b. USE OF MATRICES

For $n = 0, 1, 2, \ldots$, define the vector of probabilities

$$\mathbf{P}(n) = (P_0(n), P_1(n)) \tag{2.75}$$

and the square matrix

$$\mathbf{P} = \begin{pmatrix} q_0 & p_0 \\ q_1 & p_1 \end{pmatrix} \tag{2.76}$$

The matrix $\mathbf{P}$ identifies completely the probability of making a transition from a particular state i, $i = 0, 1$, to a particular state j, $j = 0, 1$, between two successive steps. It is known as the *one-step transition probability matrix*. Equations (2.53) and (2.54) may then be written as

$$\mathbf{P}(n) = \mathbf{P}(n-1) \cdot \mathbf{P}, \quad n = 1, 2, \ldots \tag{2.77}$$

with $\mathbf{P}(0)$ a given vector dictated by equations (2.55) and (2.56) to be

$$\mathbf{P}(0) = (1 - \alpha, \alpha) \qquad (2.78)$$

From equation (2.77) it is evident that, for any $0 \leq m < n$,

$$
\begin{aligned}
\mathbf{P}(n) &= [\mathbf{P}(n - 2)\cdot\mathbf{P}]\cdot\mathbf{P} = \mathbf{P}(n - 2)\cdot\mathbf{P}^2 \\
&= [\mathbf{P}(n - 3)\cdot\mathbf{P}]\cdot\mathbf{P}^2 = \mathbf{P}(n - 3)\cdot\mathbf{P}^3 \\
&\quad\cdots \\
&= \mathbf{P}(m)\cdot\mathbf{P}^{n-m}
\end{aligned}
$$

$\mathbf{P}^{n-m}$ is a 2×2 matrix that specifies the probability of making a transition from a particular state $i, i = 0, 1$, in step $m, m = 0, 1, \ldots, n - 1$, to a particular state $j, j = 0, 1$, in step n. As may be noted, $\mathbf{P}^{n-m}$ is independent of the step m and is a function only of the difference $n - m$; the transition probabilities are therefore *homogeneous*, and thus the Markov chain is completely characterized. $\mathbf{P}^{n-m}$ defines the $n - m$-*step transition probability matrix*.

From equation (2.77), we may compute $\mathbf{P}(n)$ recursively as follows:

$$
\begin{aligned}
\mathbf{P}(1) &= \mathbf{P}(0)\cdot\mathbf{P} \\
\mathbf{P}(2) &= \mathbf{P}(1)\cdot\mathbf{P} = \mathbf{P}(0)\cdot\mathbf{P}^2 \\
&\quad\cdots \\
\mathbf{P}(n) &= \mathbf{P}(n - 1)\cdot\mathbf{P} = \mathbf{P}(0)\cdot\mathbf{P}^n
\end{aligned}
$$

Thus, if the n-step transition probability matrix $\mathbf{P}^n$ can be determined, $\mathbf{P}(n)$ will be known. Let $p_{ij}^{(n)}$ denote the elements of $\mathbf{P}^n$. One may use mathematical induction to show that

$$
\begin{aligned}
\mathbf{P}^n &= \begin{pmatrix} p_{00}^{(n)} & p_{01}^{(n)} \\ p_{10}^{(n)} & p_{11}^{(n)} \end{pmatrix} \\
&= \frac{1}{1 - (p_1 - p_0)}\begin{pmatrix} q_1 + p_0(p_1 - p_0)^n & p_0 - p_0(p_1 - p_0)^n \\ q_1 - q_1(p_1 - p_0)^n & p_0 + q_1(p_1 - p_0)^n \end{pmatrix} \\
&= \frac{1}{1 - (p_1 - p_0)}\left[\begin{pmatrix} q_1 & p_0 \\ q_1 & p_0 \end{pmatrix} + (p_1 - p_0)^n\begin{pmatrix} p_0 & -p_0 \\ -q_1 & q_1 \end{pmatrix}\right]
\end{aligned} \qquad (2.79)
$$

Let now

$$\lim_{n\to\infty} \mathbf{P}(n) = (P_0, P_1) = \tilde{\mathbf{P}}$$

and

$$\lim_{n\to\infty} \mathbf{P}^n = \begin{pmatrix} \pi_{00} & \pi_{01} \\ \pi_{10} & \pi_{11} \end{pmatrix} = \boldsymbol{\pi}$$

From relation (2.79) we see that

$$\pi = \frac{1}{1 - (p_1 - p_0)}\begin{pmatrix} q_1 & p_0 \\ q_1 & p_0 \end{pmatrix}$$

The remarkable thing that we note if we compare π to relations (2.67) and (2.68) is that

$$\pi = \begin{pmatrix} P_0 & P_1 \\ P_0 & P_1 \end{pmatrix} = \begin{pmatrix} \tilde{\mathbf{P}} \\ \tilde{\mathbf{P}} \end{pmatrix}$$

Thus, in the steady state, for any $i, j = 0, 1$,

$$\pi_{ij} = \lim_{n \to \infty} p_{ij}^{(n)} = P_j$$

which is independent of i. Therefore, after a large number of transitions the probability of finding the system in a particular state is independent of (1) the number of transitions and (2) the state from which a transition has occurred. P_0 and P_1 are obtained from (2.73) and (2.74), which may be written as

$$(P_0, P_1) = (P_0, P_1)\begin{pmatrix} q_0 & p_0 \\ q_1 & p_1 \end{pmatrix}$$

or
$$\tilde{\mathbf{P}} = \tilde{\mathbf{P}} \cdot \mathbf{P}$$

ii. General Analysis of Discrete-Parameter Markov Chains

In Example 2.2, suppose that the orders refer to a particular product and that the size of each order is a discrete random variable (see Section 2.3-ii-a). Assume that the probability that a quantity $X_n = j, j = 0, 1, 2, \ldots$, ordered by the customer in month n, $n = 1, 2, \ldots$, depends on the amount $X_{n-1} = i$, $i = 0, 1, 2, \ldots$, he ordered in the previous month $n - 1$. Past experience has provided the salesman with a knowledge of the customer's behavior described by the conditional probabilities $P\{X_n = j \mid X_{n-1} = i\}$, $i, j = 0$, $1, 2, \ldots ; n = 1, 2, \ldots$. The salesman may then be interested in determining for all future months ahead the probability that the customer will place an order of a given size.

We thus come to extending the analysis of discrete-parameter Markov chains to cases involving an arbitrary number of points in their state space. We have seen that the probability law of the process is completely specified with a knowledge of the initial state of the system and the transition probabilities. In what follows we investigate the basic properties of the transition probabilities for homogeneous Markov chains. We show how such properties can be utilized to characterize (1) the (unconditional) probability mass function of X_n for a given $n, n = 1, 2, \ldots$, and (2) the long-run behavior of the Markov chain as $n \longrightarrow \infty$.

a. TRANSITION PROBABILITIES

(1) *Multiple-step Transition Probabilities*

Let

$$p_{jk}(m, n) = P\{X_n = k \,|\, X_m = j\}, \quad 0 \leq m < n$$

That is, $p_{jk}(m, n)$ is the probability that the process will go from state j to state k in $n - m$ steps, given that it was in state j at step m. Then $p_{jk}(m, n)$ is defined as the *transition probability* function of the Markov chain.

It is possible that the transition probability of the Markov chain depends only on the difference $n - m$; then we write

$$p_{jk}(m, n) = p_{jk}^{(n-m)}$$

and we say that the Markov chain possesses *stationary transition probabilities* or that it is time *homogeneous*. We shall consider only homogeneous Markov chains.

We define the *stationary transition probability function* as

$$p_{jk}^{(n)} = P\{X_{t+n} = k \,|\, X_t = j\}, \quad t, n \in \mathbf{T}$$

(2) *One-step Transition Probabilities*

Let

$$p_{jk} = P\{X_m = k \,|\, X_{m-1} = j\}$$

That is, $p_{jk} = p_{jk}^{(1)}$ is the probability that the process will go from j to k in exactly one step, given that it was in state j previously. Then p_{jk} is known as the *one-step transition probability function* of the Markov chain.

From the definition of $p_{jk}^{(n)}$ we note that

$$\sum_{k \in \mathbf{I}} p_{jk}^{(n)} = 1$$

(3) *Transition Probability Matrices*

It is convenient to express the transition probabilities of a Markov chain in terms of matrices. The one-step transition probability matrix is defined as

$$\mathbf{P} = \{p_{ij}\} = \begin{pmatrix} p_{00} & p_{01} & p_{02} & \cdots \\ p_{10} & p_{11} & p_{12} & \cdots \\ p_{20} & p_{21} & p_{22} & \cdots \\ \cdot & \cdot & \cdot & \cdots \end{pmatrix}$$

The n-step transition probability matrix of a homogeneous Markov chain is similarly expressed by

$$\mathbf{P}^{(n)} = \{p_{ij}^{(n)}\} = \begin{pmatrix} p_{00}^{(n)} & p_{01}^{(n)} & p_{02}^{(n)} & \cdots \\ p_{10}^{(n)} & p_{11}^{(n)} & p_{12}^{(n)} & \cdots \\ p_{20}^{(n)} & p_{21}^{(n)} & p_{22}^{(n)} & \cdots \\ \cdot & \cdot & \cdot & \cdots \end{pmatrix}$$

(4) *Chapman-Kolmogorov Equations*

The Chapman–Kolmogorov equations provide a method to relate the transition probabilities between successive steps. Let

$p_{ij}^{(m+n)} =$ probability that the Markov chain will go from state i to state j in exactly $m + n$ steps, given that it was in state i previously

$p_{ik}^{(m)} =$ probability that the Markov chain will go from state i to state k in exactly m steps, given that it was in state i previously

$p_{kj}^{(n)} =$ probability that the Markov chain will go from state k to state j in exactly n steps, given that it was in state k previously

The Chapman–Kolmogorov equations are then defined by Theorem 2.1.

Theorem 2.1

$$\boxed{p_{ij}^{(m+n)} = \sum_{k \in I} p_{ik}^{(m)} p_{kj}^{(n)}} \qquad \text{for all } i, j = 0, 1, \ldots$$

Proof: To establish these relations, suppose that it takes exactly u steps to get into state i. Then the total number of steps to reach state j is $u + m + n$; hence

$$P\{X_{u+m+n} = j \,|\, X_u = i\}$$
$$= \sum_{k \in I} P\{X_{u+m+n} = j \,|\, X_{u+m} = k, X_u = i\} \cdot P\{X_{u+m} = k \,|\, X_u = i\}$$

From the Markovian property,

$$P\{X_{u+m+n} = j \,|\, X_{u+m} = k, X_u = i\} = P\{X_{u+m+n} = j \,|\, X_{u+m} = k\}$$

Also, by definition,

$$P\{X_{u+m+n} = j \,|\, X_u = i\} = p_{ij}^{(m+n)}$$
$$P\{X_{u+m+n} = j \,|\, X_{u+m} = k\} = p_{kj}^{(n)}$$
$$P\{X_{u+m} = k \,|\, X_u = i\} = p_{ik}^{(m)}$$

Hence $p_{ij}^{(m+n)} = \sum_{k \in I} p_{kj}^{(n)} p_{ik}^{(m)}$

We next prove an important theorem, which uses the Chapman–Kolmogorov equations.

Theorem 2.2

The n-*step transition probability matrix* $\mathbf{P}^{(n)}$ *is equal to the* n*th power of the one-step transition probability matrix* $\mathbf{P}$.

$$\mathbf{P}^{(n)} = \mathbf{P}^n$$

Proof: The proof goes on by induction. For $n = 1$

$$\mathbf{P}^{(1)} = \mathbf{P}^1$$

by definition. Assume the theorem true for n; that is,

$$\mathbf{P}^{(n)} = \mathbf{P}^n$$

To prove that

$$\mathbf{P}^{(n+1)} = \mathbf{P}^{n+1}$$

we make use of the Chapman–Kolmogorov equation as well as the matricial notation for the transition probability

$$p_{ij}^{(n+1)} = \sum_{k \in \mathbf{I}} p_{ik}^{(n)} p_{kj}$$

$p_{ij}^{(n+1)}$ defines, of course, the (i, j) element of the matrix $\mathbf{P}^{(n+1)}$. Now, by definition of the multiplication of matrices, $p_{ij}^{(n+1)}$ is the i, j element of a matrix obtained by multiplying the matrix whose (i, k) element is $p_{ik}^{(n)}$ by a matrix whose (k, j) element is p_{kj}.

Since $p_{ik}^{(n)}$ is the (i, k) element of the n-step transition probability matrix $\mathbf{P}^{(n)}$, and p_{kj} is the (k, j) element of the one-step transition probability matrix $\mathbf{P}$, it follows that

$$\mathbf{P}^{(n+1)} = \mathbf{P}^{(n)} \cdot \mathbf{P} = \mathbf{P}^n \cdot \mathbf{P} = \mathbf{P}^{n+1}$$

which proves the theorem.

To illustrate the theorem, using matricial form, let

$$\mathbf{P} = \begin{pmatrix} p_{00} & p_{01} & p_{02} \\ p_{10} & p_{11} & p_{12} \\ p_{20} & p_{21} & p_{22} \end{pmatrix}$$

and

$$\mathbf{P}^{(n)} = \begin{pmatrix} p_{00}^{(n)} & p_{01}^{(n)} & p_{02}^{(n)} \\ p_{10}^{(n)} & p_{11}^{(n)} & p_{12}^{(n)} \\ p_{20}^{(n)} & p_{21}^{(n)} & p_{22}^{(n)} \end{pmatrix}$$

Then

$$\mathbf{P}^{(n)} \cdot \mathbf{P} = \begin{pmatrix} p_{00}^{(n)} & p_{01}^{(n)} & p_{02}^{(n)} \\ p_{10}^{(n)} & p_{11}^{(n)} & p_{12}^{(n)} \\ p_{20}^{(n)} & p_{21}^{(n)} & p_{22}^{(n)} \end{pmatrix} \cdot \begin{pmatrix} p_{00} & p_{01} & p_{02} \\ p_{10} & p_{11} & p_{12} \\ p_{20} & p_{21} & p_{22} \end{pmatrix}$$

$$= \begin{pmatrix} p_{00}^{(n)}p_{00} + p_{01}^{(n)}p_{10} + p_{02}^{(n)}p_{20} & p_{00}^{(n)}p_{01} + p_{01}^{(n)}p_{11} + p_{02}^{(n)}p_{21} & p_{00}^{(n)}p_{02} + p_{01}^{(n)}p_{12} + p_{02}^{(n)}p_{22} \\ p_{10}^{(n)}p_{00} + p_{11}^{(n)}p_{10} + p_{12}^{(n)}p_{20} & p_{10}^{(n)}p_{01} + p_{11}^{(n)}p_{11} + p_{12}^{(n)}p_{21} & p_{10}^{(n)}p_{02} + p_{11}^{(n)}p_{12} + p_{12}^{(n)}p_{22} \\ p_{20}^{(n)}p_{00} + p_{21}^{(n)}p_{10} + p_{22}^{(n)}p_{20} & p_{20}^{(n)}p_{01} + p_{21}^{(n)}p_{11} + p_{22}^{(n)}p_{21} & p_{20}^{(n)}p_{02} + p_{21}^{(n)}p_{12} + p_{22}^{(n)}p_{22} \end{pmatrix}$$

$$= \begin{pmatrix} \sum_{i=0}^{i=2} p_{0i}^{(n)} p_{i0} & \sum_{i=0}^{i=2} p_{0i}^{(n)} p_{i1} & \sum_{i=0}^{i=2} p_{0i}^{(n)} p_{i2} \\ \sum_{i=0}^{i=2} p_{1i}^{(n)} p_{i0} & \sum_{i=0}^{i=2} p_{1i}^{(n)} p_{i1} & \sum_{i=0}^{i=2} p_{1i}^{(n)} p_{i2} \\ \sum_{i=0}^{i=2} p_{2i}^{(n)} p_{i0} & \sum_{i=0}^{i=2} p_{2i}^{(n)} p_{i1} & \sum_{i=0}^{i=2} p_{2i}^{(n)} p_{i2} \end{pmatrix}$$

Using the Chapman–Kolmogorov equation, we obtain

$$\mathbf{P}^{(n)}\mathbf{P} = \begin{pmatrix} p_{00}^{(n+1)} & p_{01}^{(n+1)} & p_{02}^{(n+1)} \\ p_{10}^{(n+1)} & p_{11}^{(n+1)} & p_{12}^{(n+1)} \\ p_{20}^{(n+1)} & p_{21}^{(n+1)} & p_{22}^{(n+1)} \end{pmatrix}$$

$$= \mathbf{P}^{(n+1)}$$

Since by definition $\mathbf{P}^{(1)} = \mathbf{P}^1$, it follows that $\mathbf{P}^{(n+1)} = \mathbf{P}^{n+1}$.

The following are two important corollaries:

Corollary 1

In a homogeneous Markov chain it suffices to know the one-step transition probability matrix to be able to determine the n-step transition probability matrix, and thus completely specify the probability law of the process.

Corollary 2

For $j = 0, 1, 2, \ldots$ and $n = 0, 1, 2, \ldots$, define

$$P_j(n) = P\{X_n = j\}$$

and

$$\mathbf{P}(n) = (P_0(n), P_1(n), \ldots)$$

$\mathbf{P}(n)$ is known as the *unconditional probability vector* for determining the probability $P_j(n)$ that the system is in state j at step n. Clearly,

$$\mathbf{P}(n) = \mathbf{P}(n-1) \cdot \mathbf{P}$$

For a homogeneous Markov chain, it is easy to verify that, for any $0 \leq m < n$,

$$\mathbf{P}(n) = \mathbf{P}(m) \cdot \mathbf{P}^{n-m}$$

and
$$\mathbf{P}(n) = \mathbf{P}(0) \cdot \mathbf{P}^n$$

Consequently, *the unconditional probability vector of a homogeneous Markov chain is completely determined once one knows the one-step transition probability matrix and the unconditional probability vector at time 0.*

b. STEADY-STATE DISTRIBUTION

Perhaps one of the most important aspects in the study of Markov chains is the behavior of the n-step transition probabilities, $p_{ij}^{(n)}$ as $n \longrightarrow \infty$. The properties of this behavior have been rigorously established; the proofs, however, are beyond the scope of this book. It suffices to say that under certain conditions

$$\lim_{n \to \infty} p_{ij}^{(n)} = P_j = \lim_{n \to \infty} P_j(n) = \lim_{n \to \infty} P\{X_n = j\}$$

exists. That is, X_n has a limit distribution known as *steady-state* or *stationary distribution*. The *steady-state probabilities* P_j satisfy the following conditions:

$$P_j > 0$$
$$\sum_{j \in \mathbf{I}} P_j = 1$$

This means that after a large number of transitions the probability of finding the system in a particular state is independent of the number of transitions and the initial probability distributions. If such steady-state probabilities exist, the Chapman–Kolmogorov equations reduce to

$$P_j = \sum_{i \in \mathbf{I}} P_i p_{ij}, \quad \text{for all } j$$

Thus it becomes possible to determine the steady-state probabilities by solving a system of linear algebraic equations. Suppose that the Markov chain has exactly three states; then the $P_j, j = 0, 1, 2$, satisfy the following algebraic linear equations:

$$P_0 = P_0 p_{00} + P_1 p_{10} + P_2 p_{20}$$
$$P_1 = P_0 p_{01} + P_1 p_{11} + P_2 p_{21}$$
$$P_2 = P_0 p_{02} + P_1 p_{12} + P_2 p_{22}$$

Also
$$1 = P_0 + P_1 + P_2$$

Adding the first three equations yields the fourth one; hence at least one equation is redundant; since the solution $P_0 = 0 = P_1 = P_2$ satisfies the first three equations, but not the fourth one, it follows that the redundant equation cannot be the fourth one.

If steady state exists, the system of equations for the probabilities must always yield a unique solution. Although we have not discussed the conditions underlying the existence of a steady-state solution, nevertheless, often it is possible to find out about such existence by setting up the equations for the steady-state probabilities and analyzing the resultant system of simultaneous linear algebraic equations. If the equations have a unique solution, then steady state exists; if, on the other hand, the equations have no solution or a multiplicity of solutions, then steady state does not exist.

Example 2.4

Identify the binomial process discussed in Section 2.3-i-a as a discrete-parameter homogeneous Markov chain by defining an appropriate initial probability vector and the one-step transition probability matrix. Does a steady state exist?

Let Y_n denote the total number of successes in n trials, and, for $r = 0, 1, 2, \ldots$ and $n = 0, 1, 2, \ldots$, let

$$P\{Y_n = r\} = P_r(n)$$

and
$$\mathbf{P}(n) = (P_0(n), P_1(n), P_2(n), \ldots)$$

For $n = 0$, define

$$P_r(0) = \begin{cases} 1, & r = 0 \\ 0, & r = 1, 2, \ldots \end{cases}$$

For $n = 1, 2, \ldots$, we have, from Section 2.3-i-a,

$$P\{Y_n = r\} = P_r(n) = \begin{cases} P(r, n), & 0 \le r \le n \\ 0, & \text{otherwise} \end{cases}$$

where $P(r, n)$ is dictated by equations (2.4), (2.5), (2.6), and (2.7). The stochastic process $\{Y_n, n \in \mathbf{T}\}$ with index set $\mathbf{T} = \{0, 1, 2, \ldots\}$ and state space $\mathbf{I} = \{0, 1, 2, \ldots\}$ is a discrete-parameter homogeneous Markov chain with initial probability vector

$$\mathbf{P}(0) = (1, 0, 0, \ldots)$$

and one-step transition probability matrix

$$\mathbf{P} = \begin{pmatrix} q & p & 0 & 0 & 0 & \cdot \\ 0 & q & p & 0 & 0 & \cdot \\ 0 & 0 & q & p & 0 & \cdot \\ 0 & 0 & 0 & q & p & \cdot \\ \cdot & \cdot & \cdot & \cdot & \cdot & \cdot \end{pmatrix}$$

The steady-state equations are

$$P_0 = qP_0$$
$$P_1 = pP_0 + qP_1$$
$$P_2 = \quad\quad pP_1 + qP_2$$

$$\cdots$$

It is easy to verify that the only solution to this system of infinite equations has the form

$$P_0 = P_1 = P_2 = P_3 = \ldots = 0$$

However, since we must have

$$1 = P_0 + P_1 + P_2 + \cdots$$

it follows that a solution does not exist, and Y_n does not have a steady-state distribution.

2.7. Markov Chains: Continuous Parameter

The analysis of a continuous-parameter Markov chain is very similar to the discrete-parameter case, except that when dealing with transitions from a given state to another it is convenient to specify their characteristics over a small interval defined over the continuous-parameter set **T**.

i. *Definition*

The probability law of the continuous-parameter Markov chain $\{X(t),$ $t \in \mathbf{T}\}$, where $\mathbf{I} = \{0, 1, 2, \ldots\}$ and $\mathbf{T} = \{t : t \geq 0\}$, is completely specified by the joint probability mass function of the sequence of random variables $\{X(t_n)\}, n = 0, 1, 2, \ldots,$ and $0 = t_0 < t_1 < t_2 < t_n$, given by

$$P\{X(t_0) = r_0, X(t_1) = r_1, \ldots, X(t_n) = r_n\}$$
$$= P\{X(t_n) = r_n \mid X(t_{n-1}) = r_{n-1}\} \cdot P\{X(t_{n-1}) = r_{n-1} \mid X(t_{n-2}) = r_{n-2}\}$$
$$\ldots P\{X(t_2) = r_2 \mid X(t_1) = r_1\} \cdot P\{X(t_1) = r_1 \mid X(t_0) = r_0\} \cdot P\{X(t_0) = r_0\}$$

Thus the probability law of the process is characterized by (1) the initial state of the system given by the PMF of $X(t_o)$, $P\{X(t_0) = r\}, r = 0, 1, 2, \ldots$, and (2) the transition probabilities defined for all $0 \le m \le n, m, n = 0, 1, 2, \ldots$, by $P\{X(t_n) = r_n \mid X(t_m) = r_m\}$.

It is convenient to write the transition probabilities of the process as

$$p_{ij}(s, t) = P\{X(t) = j \mid X(s) = i\}$$

for $0 < s < t$ and $i, j = 0, 1, 2, \ldots$. The transition probabilities fulfill the following conditions:

(1) $p_{ij}(s, t) \ge 0$.
(2) $\sum_{j=0}^{j=1} p_{ij}(s, t) = 1$.
(3) The Chapman–Kolmogorov equations for continuous-parameter discrete state space Markov processes:

$$p_{ij}(s, u) = \sum_{k=0}^{\infty} p_{ik}(s, t) p_{kj}(t, u)$$

(4) $\lim_{t \to s} p_{ij}(s, t) = \begin{cases} 1, & \text{if } i = j \\ 0, & \text{if } i \ne j \end{cases}$

Quite often, the discrete-parameter Markov chain is specified by the transition probabilities over the interval $(t, t + dt]$; that is, the process is defined by

$$p_{ij}(t, t + dt) = P\{X(t + dt) = j \mid X(t) = i\}$$

The use of the Chapman–Kolmogorov equation is not altered, and all the previous properties still hold true.

As in the discrete-parameter case, we can define processes that are time homogeneous; thus the transition probabilities do not depend on the initial time but only on the elapsed time, and they may be written as

$$p_{ij}(\tau) = P\{X(s + \tau) = j \mid X(s) = i\}$$

for $i, j = 0, 1, 2, \ldots$ and $\tau > 0$. As an example, the homogeneous Poisson process discussed in Section 2.4-i-a may be defined as a continuous-parameter Markov chain with transition probability

$$p_{ij}(\tau) = \begin{cases} e^{-\lambda \tau} \dfrac{(\lambda \tau)^{j-i}}{(j-i)!}, & \text{if } j \geq i \\ 0, & \text{if } j < i \end{cases}$$

Equivalently, these transition probabilities may be expressed as

$$p_{ij}(dt) = \begin{cases} 1 - \lambda \, dt + o(dt), & \text{if } j = i \\ \lambda \, dt + o(dt), & \text{if } j = i + 1 \\ o(dt), & \text{otherwise} \end{cases}$$

where $o(dt)$ is a function of dt such that $\lim_{dt \to 0} [o(dt)/dt] = 0$

Let for $j = 0, 1, 2, \ldots$ and $t \geq 0$

$$P(j, t) = P\{X(t) = j\}$$

be the *unconditional probability* that the system is in state j at time t. Given the initial probabilities $\{P(j, 0)\}$, the $P(j, t)$'s are determined by the Chapman–Kolmogorov equations

$$P(j, t + dt) = \sum_{j=0}^{\infty} P(i, t) p_{ij}(t, t + dt)$$

For a time-homogeneous process, it can be shown that under certain conditions

$$\lim_{t \to \infty} P(j, t) = \lim_{t \to \infty} P\{X(t) = j\} = P_j$$

exists and is independent of t and the initial probability distribution. That is, $X(t)$ has a limit distribution known as *steady-state* or *stationary distribution*. The steady-state probabilities P_j satisfy the conditions $P_j > 0$ and $\sum_{j=0}^{\infty} P_j = 1$.

ii. *Transition Intensity*

We have seen that in a time-homogeneous discrete-parameter Markov chain, the transition probabilities were completely defined by the one-step transition probabilities. In a time-homogeneous continuous-parameter Markov chain, the role of the one-step transition probabilities is played by the *transition intensities* defined as

$$q_j = -\frac{d}{d\tau} p_{jj}(0) = \lim_{\tau \to 0} \frac{1 - p_{jj}(\tau)}{\tau}, \quad \text{for } j = 0, 1, 2, \ldots$$

and

$$q_{ij} = \frac{d}{d\tau} p_{ij}(0) = \lim_{\tau \to 0} \frac{p_{ij}(\tau)}{\tau}, \quad \text{for all } i \neq j$$

assuming that the limits exist. q_j is known as the *intensity of passage* for a process in a given state j; q_{ij} is known as the *intensity of transition* to state j from state i. The analysis of the transition intensities provides the foundation work for the study of discrete-parameter Markov chains. However, this analysis is beyond the scope of this book. We shall henceforth identify the processes by their transition probabilities and provide examples of such processes.

iii. *Birth and Death Processes*

In this section, we study several well-known and useful continuous-parameter Markov chains, which have found applications in such operations research areas as reliability theory, inventory theory, and queuing theory. Three processes that merit particular attention are

(a) The birth process.
(b) The death process.
(c) The birth and death process.

Originally, these processes were studied as stochastic models for population growth. Later, the models were found to be applicable to a wide variety of other natural processes.

Consider a population in which there is no interaction among individuals. New individuals may be added to the population through birth and individuals may be substracted from the population through death. Let $X(t)$ be the number of individuals in the population at time t, $t \geq 0$; $X(t)$ is sometimes referred to as the population size. The process $\{X(t), t \in \mathbf{T}\}$ is a continuous-parameter discrete state space stochastic process. We shall primarily be interested in finding the probability that at time t the population size is exactly n; that is,

$$P(n, t) = P\{X(t) = n\}$$

In a population model involving simultaneously birth and death, consider a time interval $(t, t + dt]$. Suppose that at time t, $t \geq 0$, the population size, that is, the state of the system, is i, $i = 0, 1, 2, \ldots$. Assume now that in the time interval dt either of the following takes place:

(1) A birth, which may occur with probability $\lambda_i(t) \, dt + o(dt)$, thus bringing the population size at time $t + dt$ to $i + 1$; $\lambda_i(t)$ is known as the *birth rate* at time t when population size is i.
(2) A death, if $i \neq 0$, which may occur with probability $\mu_i(t) \, dt + o(dt)$, thus bringing the population size at time $t + dt$ to $i - 1$; $\mu_i(t)$ is known as the *death rate* at time t when population size is i.

Any other events that may occur, resulting in a change in the population size at time $t + dt$, are assumed to have a probability of occurrence of $o(dt)$. For example, two births and a death may occur in dt with probability $o(dt)$. It is then evident that the probability that the population size will not be altered at time $t + dt$, that is, will be i, is

$$1 - [\lambda_i(t) \, dt + \mu_i(t) \, dt] + o(dt)$$

Under these assumptions, the process describing the population growth constitutes a continuous-parameter Markov chain with transition probabilities

$$p_{ii}(t, t + dt) = 1 - [\lambda_i(t) \, dt + \mu_i(t) \, dt] + o(dt)$$

$$p_{ij}(t, t + dt) = \begin{cases} \lambda_i(t) \, dt + o(dt), & \text{if } j = i + 1; i = 0, 1, \ldots \\ \mu_i(t) \, dt + o(dt), & \text{if } j = i - 1; i = 1, 2, \ldots \\ o(dt), & \text{otherwise} \end{cases}$$

The population size at time origin will be assumed to be given by

$$P(n, 0) = \begin{cases} 1, & n = N; N = 0, 1, 2, \ldots \\ 0, & n \neq N \end{cases}$$

Prior to attempting the formulation of stochastic models for this general population growth problem, we shall analyze two important special cases. In the first case death is absent; the population growth is solely affected by births, thus increasing in size as time elapses. In the second case birth is absent; the population is subject to attrition as a result of death from individuals; thus the size of the population is steadily reduced, ultimately reaching a zero level (population extinction).

a. BIRTH PROCESS

In the special case when $\mu_i(t) = 0$ for all $i = 1, 2, \ldots$ and all $t \geq 0$, that is, when no death occurs, the population growth is a birth process. We consider now three special forms of the birth rate $\lambda_i(t)$.

(1) *Homogeneous Poisson Process:* $\lambda_i(t) = \lambda = $ *Constant*

Suppose that the initial state of the process is such that $N = 0$, and assume that the birth rate is a constant λ, independent of the time parameter t or the population size. Then the process is time homogeneous and the transition probabilities take the form

$$p_{ij}(dt) = \begin{cases} \lambda \, dt + o(dt), & \text{if } j = i + 1 \\ 1 - \lambda \, dt + o(dt), & \text{if } j = i \\ o(dt), & \text{otherwise} \end{cases} \tag{2.80}$$

A little reflection will show that this particular process is the continuous-parameter counterpart of the binomial process previously discussed and that its axiomatic basis is the same as the continuous-parameter independent process. Clearly, we have the following initial conditions:

$$P(n, 0) = \begin{cases} 1, & n = 0; t = 0 \\ 0, & n \geq 1; t = 0 \end{cases} \tag{2.81}$$

The equations for $P(n, t)$, $t \geq 0$, are obtained as follows:
For $n = 0, t \geq 0$,

$$P(0, t + dt) = P(0, t) \cdot p_{0,0}(dt) \tag{2.82}$$

For $n \geq 1, t \geq 0$,

$$P(n, t + dt) = P(n - 1, t) \cdot p_{n-1,n}(dt) + P(n, t) \cdot p_{n,n}(dt) + o(dt) \tag{2.83}$$

We elaborate on the second of these equations: the probability that the population size at time $t + dt$ is n is the sum of the probabilities associated with the two following mutually exclusive and exhaustive events

Event 1. The population size at time t is $n - 1$ and a birth occurs during $(t, t + dt]$; the probability of this event is

$$P\{X(t) = n - 1; X(t + dt) = n\}$$
$$= P\{X(t + dt) = n \mid X(t) = n - 1\} \cdot P\{X(t) = n - 1\}$$
$$= p_{n-1,n}(dt) \cdot P(n - 1, t)$$

Event 2. The population size at time t is n and no birth occurs during $(t, t + dt]$; the probability of this event is

$$P\{X(t) = n; X(t + dt) = n\} = P\{X(t + dt) = n \mid X(t) = n\} \cdot P\{X(t) = n\}$$
$$= p_{n,n}(dt) \cdot P(n, t)$$

Any other event has a probability of occurrence of $o(dt)$. Using (2.80) in (2.82) and (2.83), we obtain, respectively, for $n = 0, t \geq 0$,

$$P(0, t + dt) = P(0, t)[1 - \lambda \, dt + o(dt)]$$

and for $n \geq 1, t \geq 0$,

$$P(n, t + dt) = P(n - 1, t)[\lambda \, dt + o(dt)]$$
$$+ P(n, t)[1 - \lambda \, dt + o(dt)] + o(dt)$$

These equations may be written as

$$\frac{P(0, t + dt) - P(0, t)}{dt} = -\lambda P(0, t) + \frac{o(dt)}{dt}, \quad n = 0; t \geq 0$$

$$\frac{P(n, t + dt) - P(n, t)}{dt} = \lambda P(n - 1, t) - \lambda P(n, t) + \frac{o(dt)}{dt}, \quad n \geq 1; t \geq 0$$

Let $dt \to 0$; then, since

$$\lim_{dt \to 0} \frac{o(dt)}{dt} = 0$$

we obtain the following system of differential difference equations for $t \geq 0$:

$$\frac{dP(0, t)}{dt} = -\lambda P(0, t), \quad n = 0 \tag{2.84}$$

$$\frac{dP(n, t)}{dt} = \lambda P(n - 1, t) - \lambda P(n, t), \quad n \geq 1 \tag{2.85}$$

subject to the initial conditions (2.81). These equations may be solved recursively as follows. From equation (2.84) we obtain

$$P(0, t) = A_0 e^{-\lambda t}, \quad t \geq 0 \tag{2.86}$$

where A_0 is an arbitrary constant whose value can be specified from the initial conditions (2.81). Setting $t = 0$, we immediately see that $A_0 = 1$; hence

$$P(0, t) = e^{-\lambda t}, \quad t \geq 0 \tag{2.87}$$

Having obtained the explicit expression for $P(0, t)$, we consider equation (2.85) for $n = 1$ and use expression (2.87) to obtain

$$\frac{dP(1, t)}{dt} = \lambda P(0, t) - \lambda P(1, t)$$

or

$$\frac{dP(1, t)}{dt} + \lambda P(1, t) = \lambda e^{-\lambda t}$$

Multiplying both sides by $e^{\lambda t}$, we obtain

$$\frac{d}{dt}[e^{\lambda t} P(1, t)] = \lambda$$

Integration yields

$$P(1, t) = \lambda t e^{-\lambda t} + A_1 e^{-\lambda t}, \quad t \geq 0 \tag{2.88}$$

where A_1 is an arbitrary constant whose value can again be obtained from the initial conditions (2.81). Set $t = 0$ in expression (2.88); we get

$$P(1, 0) = A_1 = 0$$

Hence we can finally write

$$P(1, t) = e^{-\lambda t}\frac{(\lambda t)}{1!}, \quad t \geq 0 \tag{2.89}$$

The reader may verify by induction or otherwise that, in general,

$$P(n, t) = e^{-\lambda t}\frac{(\lambda t)^n}{n!}, \quad n = 0, 1, 2, \ldots ; t \geq 0$$

This can be recognized as the Poisson distribution with mean λt.

Relating this analysis to the discussion of Section 2.4-i-a, we note that the population growth under no death rate and constant birth rate is a *homogeneous Poisson process* with mean intensity λ (see Figure 2.8).

(2) *Nonhomogeneous Poisson Process:* $\lambda_i(t) = \lambda(t)$

We assume again that at time origin the population size is zero; however, the birth rate is assumed to be an arbitrary function of the time parameter t, but independent of the population size. The process is not time homogeneous; the transition probabilities are

$$p_{ij}(t, t + dt) = \begin{cases} \lambda(t) \, dt + o(dt), & \text{if } j = i + 1 \\ 1 - \lambda(t) \, dt + o(dt), & \text{if } j = i \\ o(dt), & \text{otherwise} \end{cases}$$

The initial conditions are the same as (2.81). The equations for $P(n, t)$ are similar to equations (2.84) and (2.85) except for λ being replaced by $\lambda(t)$. We thus obtain for $t \geq 0$

$$\frac{dP(0, t)}{dt} = -\lambda(t)P(0, t), \quad n = 0 \tag{2.90}$$

$$\frac{dP(n, t)}{dt} = \lambda(t)P(n - 1, t) - \lambda(t)P(n, t), \quad n \geq 1 \tag{2.91}$$

These equations may again be solved recursively. From equation (2.90) we obtain

$$P(0, t) = A_0 e^{\int_0^t \lambda(\theta) \, d\theta}, \quad t \geq 0$$

Setting $t = 0$, we find $A_0 = P(0, 0) = 1$.

Let

$$\Lambda(t) = \int_0^t \lambda(\theta) \, d\theta$$

then
$$P(0, t) = e^{-A(t)}$$

In equation (2.91), set $n = 1$; we obtain

$$\frac{dP(1, t)}{dt} + \lambda(t)P(1, t) = \lambda(t)e^{-A(t)}$$

Multiplying both sides by $e^{A(t)}$ yields

$$\frac{d}{dt}[e^{A(t)}P(1, t)] = \lambda(t)$$

Integrating both sides with respect to t, we obtain

$$e^{A(t)}P(1, t) = \int_0^t \lambda(\theta)\, d\theta + A_1$$

where A_1 is an arbitrary constant; thus

$$P(1, t) = e^{-A(t)}[A(t)] + A_1 e^{-A(t)}$$

At $t = 0$, $P(1, 0) = 0$; therefore, $A_1 = 0$; hence

$$P(1, t) = e^{-A(t)}\frac{[A(t)]}{1!}$$

In general, it can be verified that

$$P(n, t) = e^{-A(t)}\frac{[A(t)]^n}{n!}, \quad n = 0, 1, 2, \ldots; t \geq 0 \qquad (2.92)$$

The process is known as the *nonhomogeneous Poisson process*. The form of $P(n, t)$ is similar to the homogeneous case that results from expression (2.92) if we set $\lambda(t) = \lambda$.

(3) *Yule–Furry Process*: $\lambda_i(t) = i\lambda$, $i \geq 1$

    In the *Yule–Furry birth process*, the population consists of exactly N, $N = 1, 2, \ldots$, individuals at time origin, and during a time interval of length dt, each individual has a probability $\lambda\, dt + o(dt)$ of creating a new one. Thus, if at time t the population consists of i individuals, the probability of a birth in the time interval $(t, t + dt)$ is $i\lambda\, dt + o(dt)$. The transition probabilities are time homogeneous and are given by $(i \geq N)$

$$p_{ij}(dt) = \begin{cases} i\lambda\, dt + o(dt), & \text{if } j = i + 1 \\ 1 - i\lambda\, dt + o(dt), & \text{if } j = i \\ o(dt), & \text{otherwise} \end{cases}$$

The initial conditions for this problem are

$$P(n, 0) = \begin{cases} 1, & \text{if } n = N \\ 0, & \text{otherwise} \end{cases}$$

The equations for $P(n, t), n = 1, 2, \ldots, t \geq 0$, are obtained as follows:
For $n = N$ and $t \geq 0$,

$$P(N, t + dt) = P(N, t)[1 - N\lambda \, dt + o(dt)]$$

For $n \geq N + 1$ and $t \geq 0$,

$$P(n, t + dt) = P(n - 1, t)[(n - 1)\lambda \, dt + o(dt)] \\ + P(n, t)[1 - n\lambda \, dt + o(dt)] + o(dt)$$

Transposing and setting $dt \to 0$, we obtain for $t \geq 0$

$$\frac{dP(N, t)}{dt} = -\lambda N P(N, t), \quad n = N$$

$$\frac{dP(n, t)}{dt} = (n - 1)\lambda P(n - 1, t) - n\lambda P(n, t), \quad n \geq N + 1$$

These equations, together with the initial conditions, may again be solved recursively, and it may be verified that

$$P(n, t) = \binom{n - 1}{n - N} e^{-N\lambda t}(1 - e^{-\lambda t})^{n-N}, \quad n \geq N, t \geq 0$$

This distribution is the negative binomial distribution which appears in (2.19). It follows that the population size at time t is the sum of N independent random variables, each having the geometric distribution $e^{-\lambda t}(1 - e^{-\lambda t})^{n-1}$, $n = 1, 2, \ldots$.

b. DEATH PROCESS

Another special case of interest arising in the process describing population growth stems by setting $\lambda_i(t) = 0$ for all $i = 0, 1, 2, \ldots$ and all $t \geq 0$. Thus there are no births and the population size decays with time through death of individuals (see Figure 2.12). We shall study three special forms of death processes, corresponding to death rates whose expressions are similar to the birth rates previously discussed.

(1) Homogeneous Death Process with Constant Rate: $\mu_i(t) = \mu = Constant$

We consider a population of initial size N whose individual members die at a constant rate μ, independent of the time parameter t and the population size. This time-homogeneous process has transition probabilities

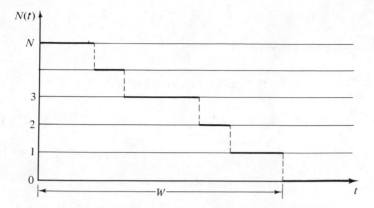

Figure 2.12. Sample function for a death process.

defined by

$$p_{ij}(dt) = \begin{cases} \mu\, dt + o(dt), & \text{if } j = i - 1;\ i = 1, 2, \ldots, N \\ 1 - \mu\, dt + o(dt), & \text{if } j = i;\ i = 1, 2, \ldots, N \\ o(dt), & \text{otherwise} \end{cases}$$

Once the population reaches the state $i = 0$ (population extinction), it remains in that state thereafter. The initial conditions to this problem are

$$P(n, 0) = \begin{cases} 1, & \text{if } n = N \neq 0 \\ 0, & \text{if } n \neq N \end{cases} \tag{2.93}$$

Expressions for $P(n, t)$, $n = 0, 1, 2, \ldots$ and $t \geq 0$, can be obtained as follows:
   For $n \geq N + 1$, $t \geq 0$,

$$P(n, t) = 0$$

For $n = N$, $t \geq 0$,

$$P(N, t + dt) = P(N, t)[1 - \mu\, dt + o(dt)] + o(dt)$$

For $1 \leq n \leq N - 1$, $t \geq 0$,

$$P(n, t + dt) = P(n, t)[1 - \mu\, dt + o(dt)]$$
$$+ P(n + 1, t)[\mu\, dt + o(dt)] + o(dt)$$

For $n = 0$, $t \geq 0$,

$$P(0, t + dt) = P(0, t) + P(1, t)[\mu\, dt + o(dt)] + o(dt)$$

Transposing, dividing by dt, and letting $dt \to 0$, we obtain for $t \geq 0$

$$P(n, t) = 0, \quad n \geq N + 1 \tag{2.94}$$

$$\frac{dP(N, t)}{dt} = -\mu P(N, t), \quad n = N \tag{2.95}$$

$$\frac{dP(n, t)}{dt} = -\mu P(n, t) + \mu P(n + 1, t), \quad 1 \leq n \leq N - 1 \tag{2.96}$$

$$\frac{dP(0, t)}{dt} = \mu P(1, t), \quad n = 0 \tag{2.97}$$

These equations may be solved recursively. From equation (2.95) we have

$$P(N, t) = A_N e^{-\mu t}, \quad t \geq 0$$

where A_N is an arbitrary constant; from (2.93) we have $P(N, 0) = 1$; hence $A_N = 1$ and

$$P(N, t) = e^{-\mu t} \tag{2.98}$$

Now, in equation (2.96) let $n = N - 1$; then

$$\frac{dP(N - 1, t)}{dt} = -\mu P(N - 1, t) + \mu P(N, t)$$

Transposing and using expression (2.98), we obtain

$$\frac{dP(N - 1, t)}{dt} + \mu P(N - 1, t) = \mu e^{-\mu t}$$

Multiplying both sides of this equation by $e^{\mu t}$ yields

$$\frac{d}{dt}[e^{\mu t} P(N - 1, t)] = \mu$$

Integrating yields

$$P(N - 1, t) = e^{-\mu t}(\mu t) + A_{N-1} e^{-\mu t}, \quad t \geq 0$$

where A_{N-1} is an arbitrary constant. From the initial conditions (2.93), $P(N - 1, 0) = 0$; hence $A_1 = 0$ and

$$P(N - 1, t) = e^{-\mu t} \frac{(\mu t)}{1!}$$

An inductive proof will show that, in general, we have

$$P(n, t) = e^{-\mu t} \frac{(\mu t)^{N-n}}{(N - n)!}, \quad n = 1, 2, \ldots, N - 1; t \geq 0 \tag{2.99}$$

Substituting the expression for $P(1, t)$ in equation (2.97) and integrating both sides yields

$$P(0, t) = \int_0^t \mu e^{-\mu\theta} \frac{(\mu\theta)^{N-1}}{(N-1)!} \, d\theta + A_0, \quad t \geq 0$$

From (2.93), $P(0, 0) = 0$; hence the value of the arbitrary constant $A_0 = 0$ and

$$P(0, t) = \int_0^t \mu e^{-\mu\theta} \frac{(\mu\theta)^{N-1}}{(N-1)!} \, d\theta \qquad (2.100)$$

Alternatively, the expression for $P(0, t)$ could have been obtained by noting that, for all $t \geq 0$,

$$\sum_{n=0}^{\infty} P(n, t) = 1$$

Thus
$$P(0, t) = 1 - \sum_{n=1}^{n=N} P(n, t)$$

$$= 1 - \sum_{n=1}^{n=N} e^{-\mu t} \frac{(\mu t)^{N-n}}{(N-n)!}$$

$$= 1 - \sum_{n=0}^{n=N-1} e^{-\mu t} \frac{(\mu t)^n}{n!}$$

Note that for $1 \leq n \leq N - 1$, $P(n, t)$ has the same structure as the Poisson distribution.

Expressions for the expectation and variance of the population size at any given time t cannot be obtained in closed form, but may be presented in terms of incomplete gamma functions.

An important question that arises in a death process relates to the time elapsed till population extinction. Let the random variable W define the time necessary for the population to reach a size zero for the first time (see Figure 2.12); then

$$P(0, t) = P\{W \leq t\}$$

The probability density function of W is then

$$\varphi_W(t) = \frac{dP(0, t)}{dt}$$

$$= \mu e^{-\mu t} \frac{(\mu t)^{N-1}}{(N-1)!}, \quad t > 0$$

which can be identified as a gamma distribution with parameter μ and of order N. W is thus the sum of N independent random variables each having the identical negative exponential PDF $\mu e^{-\mu t}$.

(2) *Nonhomogeneous Death Process:* $\mu_i(t) = \mu(t)$

This process is similar to the homogeneous death process except that the death rate at time t, $t \geq 0$, is $\mu(t)$. The transition probabilities are

$$p_{ij}(t, t + dt) = \begin{cases} \mu(t)\, dt + o(dt), & \text{if } j = i - 1; i = 1, 2, \ldots, N \\ 1 - \mu(t)\, dt + o(dt), & \text{if } j = i; i = 1, \ldots, N \\ o(dt), & \text{otherwise} \end{cases}$$

Let
$$\mathfrak{M}(t) = \int_0^t \mu(\theta)\, d\theta$$

Then, proceeding as before, it can be shown that

$$P(n, t) = \begin{cases} 0, & n > N; t \geq 0 \\ e^{-\mathfrak{M}(t)} \dfrac{[\mathfrak{M}(t)]^{N-n}}{(N - n)!}, & 1 \leq n \leq N; t \geq 0 \\ \displaystyle\int_0^t \mu(\theta) e^{-\mathfrak{M}(\theta)} \dfrac{[\mathfrak{M}(\theta)]^{N-1}}{(N - 1)!}\, d\theta, & n = 0; t \geq 0 \end{cases}$$

It is easy to verify that the time W till population extinction has the PDF

$$\varphi_W(t) = \mu(t) e^{-\mathfrak{M}(t)} \frac{[\mathfrak{M}(t)]^{N-1}}{(N - 1)!}, \quad t > 0$$

(3) Homogeneous Death Process with Linear Rate: $\mu_i(t) = i\mu$

As a final example of a death process, we consider a situation where the death rate at any particular time is proportional to the population size i at that particular time; that is, $\mu_i(t) = i\mu$. We assume again that at time origin the population size is $N > 0$. For this time-homogeneous process, the transition probabilities are

$$p_{ij}(dt) = \begin{cases} i\mu\, dt + o(dt), & \text{if } j = i - 1; i = 1, 2, \ldots, N \\ 1 - i\mu\, dt + o(dt), & \text{if } j = i; i = 1, 2, \ldots, N \\ o(dt), & \text{otherwise} \end{cases}$$

The state of the process does not change once it reaches state $i = 0$. The initial conditions are the same as (2.93), and the equations for $P(n, t)$ can be verified to be for $t \geq 0$

$$P(n, t) = 0, \quad n \geq N + 1$$

$$\frac{dP(N, t)}{dt} = -N\mu P(N, t), \quad n = N$$

$$\frac{dP(n, t)}{dt} = -n\mu P(n, t) + (n + 1)\mu P(n + 1, t), \quad 1 \leq n \leq N - 1$$

$$\frac{dP(0, t)}{dt} = \mu P(1, t), \quad n = 0,$$

Solving these equations recursively yields

$$P(n, t) = \begin{cases} 0, & n \geq N + 1; t \geq 0 \\ \binom{N}{n}(e^{-\mu t})^n(1 - e^{-\mu t})^{N-n}, & 0 \leq n \leq N; t \geq 0 \end{cases}$$

This distribution can be recognized as a binomial distribution. It follows that the population size at time t is the sum of N independent random variables each having the Bernoulli distribution $(e^{-\mu t})^r(1 - e^{-\mu t})^{1-r}, r = 0, 1$; $t \geq 0$.

The expected population size at time t is

$$E[X(t)] = Ne^{-\mu t}$$

and its variance is

$$\text{Var}[X(t)] = Ne^{-\mu t}(1 - e^{-\mu t})$$

The time elapsed W till population extinction has the PDF

$$\psi_W(t) \quad \mu N e^{-\mu t}(1 - e^{-\mu t})^{N-1}, \quad t \geq 0$$

Exercise

Find $E[W]$ and $\text{Var}[W]$.

c. Birth and Death Process

In a general stochastic model describing population growth, new individuals may be added to the population through birth, thus contributing to an increase in the population size, while simultaneously individuals may be removed through death, thus inducing a population attrition (see Figure 2.13).

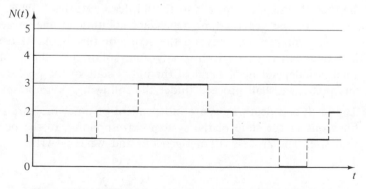

Figure 2.13. Sample function for a birth and death process.

When the birth and death rates are, respectively, $\lambda_i(t)$ and $\mu_i(t)$, the equations for $P(n, t)$ are

For $n = 0$, $t \geq 0$,

$$P(0, t + dt) = P(0, t)[1 - \lambda_0(t)\, dt + o(dt)]$$
$$+ P(1, t)[1 - \lambda_1(t)dt + o(dt)][\mu_1(t)\, dt + o(dt)]$$

For $n \geq 1$, $t \geq 0$,

$$P(n, t + dt) = P(n - 1, t)[\lambda_{n-1}(t)\, dt + o(dt)][1 - \mu_{n-1}(t)\, dt + o(dt)]$$
$$+ P(n, t)[1 - \lambda_n(t)\, dt + o(dt)][1 - \mu_n(t)\, dt + o(dt)]$$
$$+ P(n + 1, t)[1 - \lambda_{n+1}(t)\, dt + o(dt)][\mu_{n+1}(t)\, dt + o(dt)]$$
$$+ o(dt)$$

It is assumed that at time origin $t = 0$, the population consists of exactly N individuals, $N \geq 0$. Transposing terms in these equations, dividing both sides by dt, and passing to the limit as $dt \to 0$, we obtain, for $t \geq 0$,

$$\frac{dP(0, t)}{dt} = -\lambda_0(t)P(0, t) + \mu_1(t)P(1, t), \quad n = 0$$

$$\frac{dP(n, t)}{dt} = \lambda_{n-1}(t)P(n - 1, t) - [\lambda_n(t) + \mu_n(t)]P(n, t) \quad\quad (2.101)$$
$$+ \mu_{n+1}(t)P(n + 1, t), \quad n \geq 1$$

subject to the initial conditions

$$P(n, 0) = \begin{cases} 1, & n = N \\ 0, & \text{otherwise} \end{cases}$$

Even in the special case of a constant birth and death rate that is when $\lambda_i(t) = \lambda$ and $\mu_i(t) = \mu$, the general time-dependent solution of the preceding equations is very complex and beyond the scope of this book. It can be shown then that when $\lambda/\mu < 1$, a steady-state solution to the problem exists, which is independent of t and of the initial state of the process.

At this point, we shall proceed under the following assumptions:

(1) The birth and death rates are independent of time; they are solely functions of the state of the system corresponding to the population size. The process is homogeneous and we may write

$$\lambda_i(t) = \lambda_i, \quad i = 1, 2, \ldots$$
$$\mu_i(t) = \mu_i, \quad i = 0, 1, \ldots$$

(2) Steady state conditions are fulfilled and have been reached. Let then

$$P_n = \lim_{t \to \infty} P(n, t), \quad n = 0, 1, 2, \dots$$

denote the probability that the population size is n under steady state conditions.

The system of differential-difference equations (2.101) reduces then to

$$\begin{aligned} 0 &= -\lambda_0 P_0 + \mu_1 P_1, \quad n = 0 \\ 0 &= \lambda_{n-1} P_{n-1} - (\lambda_n + \mu_n) P_n + \mu_{n+1} P_{n+1}, \quad n \geq 1 \end{aligned} \tag{2.102}$$

with the normalizing condition

$$\sum_{n=0}^{\infty} P_n = 1 \tag{2.103}$$

The system of equations (2.102) may alternatively be written as

$$\begin{aligned} \lambda_0 P_0 - \mu_1 P_1 &= 0, \quad n = 0 \\ \lambda_n P_n - \mu_{n+1} P_{n+1} &= \lambda_{n-1} P_{n-1} - \mu_n P_n, \quad n \geq 1 \end{aligned} \tag{2.104}$$

From these, we simply deduce that

$$\lambda_n P_n - \mu_{n+1} P_{n+1} = 0, \quad n \geq 0 \tag{2.105}$$

or equivalently

$$\lambda_n P_n = \mu_{n+1} P_{n+1}, \quad n \geq 0 \tag{2.106}$$

Relation (2.106) may also be written

$$P_{n+1} = \frac{\lambda_n}{\mu_{n+1}} P_n, \quad n \geq 0$$

By substituting $n = 0, 1, 2, \dots$ we obtain

$$P_1 = \frac{\lambda_0}{\mu_1} P_0$$

$$P_2 = \frac{\lambda_1}{\mu_2} P_1$$

$$= \frac{\lambda_1}{\mu_2} \frac{\lambda_0}{\mu_1} P_0$$

and in general

$$P_n = \frac{\lambda_0 \lambda_1 \cdots \lambda_{n-1}}{\mu_1 \mu_2 \cdots \mu_n} P_0, \quad n \geq 1 \tag{2.107}$$

Finally, using relation (2.107) in the normalizing condition (2.103), the value of P_0 may be determined from the relation

$$P_0\left(1 + \frac{\lambda_0}{\mu_1} + \frac{\lambda_0\lambda_1}{\mu_1\mu_2} + \cdots + \frac{\lambda_0\lambda_1 \cdots \lambda_{n-1}}{\mu_1\mu_2 \cdots \mu_n} + \cdots\right) = 1$$

provided the series

$$S = 1 + \frac{\lambda_0}{\mu_1} + \frac{\lambda_0\lambda_1}{\mu_1\mu_2} + \cdots + \frac{\lambda_0\lambda_1 \cdots \lambda_{n-1}}{\mu_1\mu_2 \cdots \mu_n} + \cdots$$

converges. In fact, the conditions for the existence of a steady state solution are exactly the one for the convergence of the series S.

It is often convenient to use a flow diagram such as Figure 2.14 to set up the equations corresponding to steady state regimes in a birth and death

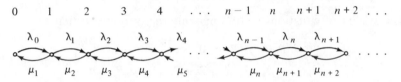

Figure 2.14. Flow diagram for the steady state birth and death process.

process. In this diagram, the node points refer to the state of the system, in this case the number of units in the system. The arrows indicate the possible transitions to other states and are labeled with the transition intensities. The quantities $\lambda_n P_n$ and $\mu_{n+1}P_{n+1}$, $n = 0, 1, 2, \ldots$, are interpreted as average rates. The following law relating to the conservation of flow is used at each node: *the average rate at which the system enters a state equals the average rate at which the system leaves that state.*

To see that the system of equations (2.102) implies this law, we simply write it as

$$\mu_1 P_1 = \lambda_0 P_0$$
$$\lambda_0 P_0 + \mu_2 P_2 = \lambda_1 P_1 + \mu_1 P_1$$
$$\lambda_1 P_1 + \mu_3 P_3 = \lambda_2 P_2 + \mu_2 P_2$$
$$\cdots$$

Thus, referring to node 1, the average rate at which the system enters state 1 is $(\lambda_0 P_0 + \mu_2 P_2)$, while the average rate at which the system leaves state 1 is $(\lambda_1 P_1 + \mu_1 P_1)$, and these two quantities are equal.

The law of conservation of flow may also be expressed equivalently by stating that *the average rate at which the system enters a higher state from a lower one equals the average rate at which it leaves the higher state to the lower one.*

If we make a cut across the two arcs between states 1 and 2 in Figure 2.14, then $\lambda_1 P_1$ is the rate at which the system moves up from state 1 to state 2, while $\mu_2 P_2$ is the rate at which the system moves down from state 2 to state 1. These two rates are equal as implied by equation (2.106). Application of the law to successive cuts will yield the entire system of equations (2.106).

2.8. Renewal Theory

i. *Counting Process*

In Section 2.4-i-c we found that one way of characterizing the Poisson process is to describe the time interval between successive events by a sequence of random variables $\{T_i\}$, $i = 1, 2, \ldots$, independently and identically distributed, each having a negative exponential distribution. A natural generalization is to consider the continuous-parameter discrete state space process $\{N(t), t \geq 0\}$, which counts the number of points occurring over the interval $(0, t]$ and such that these points are distributed over the parameter set in such a way that their successive interarrival times $\{T_i\}$, $i = 1, 2, \ldots$, are random variables. The process $\{N(t), t \geq 0\}$ is called the *counting process* of the series of points (see Figure 2.15).

Typically, referring to the salesman problem, the points in the counting process would correspond to the event of the customer placing an order. The occurrence of such orders over a continuous time scale would then be specified by the distribution of the time elapsed T_i between the $(i - 1)$th and ith order. Thus, alternatively, it is possible to characterize the customer's behavior by defining the sequence of random variables $\{T_i\}$. The process $\{N(t), t \geq 0\}$ would then specify the total number of orders placed by the customer in the time interval $(0, t]$.

The theory of counting processes, of which the Poisson process is a special case, plays an important role in stochastic processes. Specific applications arise in the recording of pulses by counters, in astrophysics, in actuarial work, and in operations research. In what follows, we study a special type of counting process in which the sequence of random variables $\{T_i\}$ is independently and identically distributed.

ii. *Definitions*

Consider a *counting process* $\{N(t), t \geq 0\}$, where $N(t)$ represents the number of events that have happened up to time t. Let $\{T_i\}$, $i = 1, 2, \ldots$,

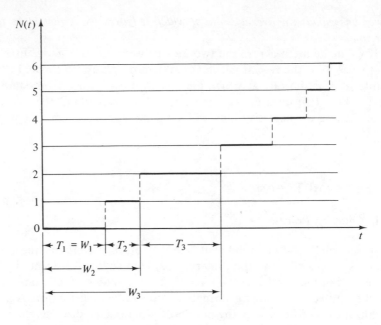

Figure 2.15. Sample function for a counting process or a renewal process.

denote the sequence of random variables for the *interarrival times*, that is, the time interval between two successive event occurrences. Let $\{W_i\}$, $i = 1, 2, \ldots$, denote the sequence of random variables for the *waiting times* until the occurrence of the ith event so that

$$W_i = \sum_{r=1}^{r=i} T_r$$

The three processes, counting, interarrival times, and waiting times, characterize a renewal process (see Figure 2.15). We introduce next a number of concepts useful in renewal theory.

(1) The sequence of nonnegative random variables $\{T_i\}$, $i = 1, 2, \ldots$, is called an *ordinary renewal process* if the random variables T_i are *independently* and *identically* distributed, possessing a common distribution function $\Phi(\cdot)$.

(2) The random variable $W_i = \sum_{j=1}^{j=i} T_j$ is called the *time up to the ith renewal*.

(3) The quantity $N(t)$ is called *the number of renewals up to time t*. By definition $N(0) = 0$.

(4) The expectation of $N(t)$ denoted by $M(t)$

$$M(t) = E[N(t)]$$

is the *renewal function*, sometimes known as the *renewal quantity*.
(5) The first derivative of $M(t)$

$$m(t) = \frac{dM(t)}{dt} = \lim_{dt \to 0} \frac{E[N(t + dt)] - E[N(t)]}{dt}$$

is called the *renewal density* and is the expected number of renewals per unit time.

(6) A renewal process where $T_1 \neq T_2, T_3, \ldots$, that is, where the distribution function of T_1, $\Phi_1(\cdot)$ is different from the distribution function $\Phi(\cdot)$ of $T_2, T_3, \ldots$, is called a *modified renewal process*.

In what follows we shall study the renewal function $M(t)$, the renewal density $m(t)$, and the number of renewals $N(t)$.

iii. *Expected Number of Renewals*

Theorem 2.3

The *renewal function* $M(t)$ in the time interval $(0, t]$ satisfies the integral equation

$$M(t) = \Phi(t) + \int_0^t M(t - x)\, d\Phi(x), \quad t > 0$$

where $\Phi(\cdot)$ is the distribution function of the time interval between renewals.

Proof: Using the concept of conditional expectation, we may write

$$M(t) = \int_0^\infty E[N(t)\,|\,T_1 = x]\, d\Phi(x)$$

where T_1 is the interarrival time to the first renewal. We can break this integral into two parts as follows (see Figure 2.16):

$$M(t) = \int_0^t E[N(t)\,|\,T_1 = x]\, d\Phi(x) + \int_t^\infty E[N(t)\,|\,T_1 = x]\, d\Phi(x)$$

Now $$E[N(t)\,|\,T_1 = x \leq t] = 1 + E[N(t - x)]$$

and $$E[N(t)\,|\,T_1 = x > t] = 0$$

Figure 2.16. $M(t) = \int_0^t E[N(t)\,|\,T_1 = x]\, d\Phi(x)$.

Hence
$$M(t) = \int_0^t [1 + E[N(t - x)]] \, d\Phi(x)$$
$$= \Phi(t) + \int_0^t M(t - x) \, d\Phi(x), \quad t > 0$$

This equation is known as the *renewal equation*. Henceforth, we shall assume that the interarrival times can be described by a probability density function $\varphi(\cdot)$, so that $d\Phi(t) = \varphi(t) \, dt$, and the integral equation becomes

$$M(t) = \Phi(t) + \int_0^t M(t - x)\varphi(x) \, dx$$

We may solve this equation using Laplace transforms (see Appendix). Let

$$\bar{M}(s) = \int_0^\infty e^{-st} M(t) \, dt$$

$$\bar{\Phi}(s) = \int_0^\infty e^{-st} \Phi(t) \, dt$$

$$\bar{\varphi}(s) = \int_0^\infty e^{-st} \varphi(t) \, dt$$

We assume that each of the quantities $\bar{M}(s)$, $\bar{\Phi}(s)$, and $\bar{\varphi}(s)$ exists for some $s \geq 0$. Clearly,

$$\bar{\Phi}(s) = \frac{\bar{\varphi}(s)}{s}$$

We shall make use of the convolution properties in Laplace transforms and obtain an explicit expression for $\bar{M}(s)$ in terms of $\bar{\varphi}(s)$. Inversion of $\bar{M}(s)$ will result in an expression for $M(t)$ involving convolutions of $\varphi(t)$. Remembering that the Laplace transform of the convolution of two functions is the product of the Laplace transforms of the two functions, by taking the Laplace transform on both sides of the renewal equation, we obtain

$$\bar{M}(s) = \bar{\Phi}(s) + \bar{M}(s) \cdot \bar{\varphi}(s)$$
$$= \bar{\Phi}(s) + \bar{M}(s) \cdot s\bar{\Phi}(s)$$

Hence, solving for $\bar{M}(s)$,

$$\bar{M}(s) = \frac{\bar{\Phi}(s)}{1 - s\bar{\Phi}(s)}$$

or
$$\bar{M}(s) = \frac{\bar{\varphi}(s)}{s[1 - \bar{\varphi}(s)]} \qquad (2.108)$$

Using the binomial theorem, we can write, for $s > 0$,

$$\frac{1}{1 - \bar{\varphi}(s)} = 1 + \bar{\varphi}(s) + [\bar{\varphi}(s)]^2 + [\bar{\varphi}(s)]^3 + \cdots$$

so that

$$\bar{M}(s) = \frac{\bar{\varphi}(s)}{s}\left[\sum_{r=0}^{\infty} [\bar{\varphi}(s)]^r\right]$$

$$= \sum_{r=1}^{\infty} \frac{[\bar{\varphi}(s)]^r}{s}$$

$$= \frac{\bar{\varphi}(s)}{s} + \frac{[\bar{\varphi}(s)]^2}{s} + \frac{[\bar{\varphi}(s)]^3}{s} + \cdots$$

Inverting term by term, we obtain

$$M(t) = \sum_{r=1}^{\infty} \int_0^t \varphi *^{(n)}(x)\, dx$$

where $\varphi *^{(n)}(x)$ is the nth-fold convolution of $\varphi(x)$ with itself.

To obtain an expression for $m(t)$, we recall that

$$m(t) = \frac{dM(t)}{dt}$$

Hence

$$\bar{m}(s) = s\bar{M}(s) - M(0) = \frac{\bar{\varphi}(s)}{1 - \bar{\varphi}(s)}$$

since $M(0) = 0$. Therefore

$$\bar{m}(s) = \sum_{r=1}^{\infty} [\bar{\varphi}(s)]^r$$

and by direct inversion or by term-by-term differentiation of $M(t)$, we obtain

$$m(t) = \sum_{r=1}^{\infty} \varphi *^{(n)}(t)$$

It is easily verified that $m(t)$ satisfies the equation

$$m(t) = \varphi(t) + \int_0^t m(t - x)\varphi(x)\, dx \qquad (2.109)$$

This is known as the *integral equation for the renewal density*.

Example 2.5

Assume that $\varphi(t) = \lambda e^{-\lambda t}$, $\lambda > 0$, $t > 0$.

$$\bar{\varphi}(s) = \int_0^{\infty} e^{-st}\lambda e^{-\lambda t}\, dt = \frac{\lambda}{\lambda + s}$$

Substituting in the expression for $\bar{m}(s)$, we obtain

$$\bar{m}(s) = \frac{\lambda/(\lambda + s)}{1 - [\lambda/(\lambda + s)]} = \frac{\lambda}{s}$$

Taking the inverse, we find for the renewal density function

$$m(t) = \lambda, \quad t > 0$$

The expected interarrival time is

$$\frac{1}{m(t)} = \frac{1}{\lambda}$$

iv. *An Asymptotic Result*

Often we are interested in the asymptotic behavior of $m(t)$ as t approaches infinity. From the final value theorem (see Appendix) we know that

$$\lim_{t \to \infty} m(t) = \lim_{s \to 0} s\bar{m}(s)$$

Now

$$\lim_{s \to 0} s\bar{m}(s) = \lim_{s \to 0} \frac{s\bar{\varphi}(s)}{1 - \bar{\varphi}(s)}$$

Using L'Hospital's rule, we obtain

$$\lim_{s \to 0} s\bar{m}(s) = \frac{\lim_{s \to 0}(s\bar{\varphi}'(s) + \bar{\varphi}(s))}{\lim_{s \to 0} -\bar{\varphi}'(s)}$$

Now

$$\lim_{s \to 0} \bar{\varphi}(s) = 1$$

and

$$\lim_{s \to 0} \bar{\varphi}'(s) = \int_0^\infty -t\varphi(t)\, dt = -\mu$$

where $\mu = E[T_i]$ is the expected interarrival time. Thus

$$\lim_{s \to 0} s\bar{m}(s) = \frac{1}{\mu}$$

Therefore,

$$\lim_{t \to \infty} m(t) = \frac{1}{\mu}$$

In a similar way, we could have shown that

$$\lim_{t \to \infty} M(t) \approx \frac{t}{\mu}$$

From this it follows that for large values of t the expected number of renewals per unit time depends only on the expected interarrival time equaling its inverse.

v. The Number of Renewals N(t)

The distribution of $N(t)$ in the stochastic process $\{N(t),\ t \geq 0\}$ is [see (2.42)]

$$P(0, t) = P\{N(t) = 0\} = 1 - P\{W_1 \leq t\} = 1 - \int_0^t \varphi(x)\, dx$$

$$P(n, t) = P\{N(t) = n\} = P\{W_n \leq t\} - P\{W_{n+1} \leq t\}$$

$$= P\{T_1 + \cdots + T_n \leq t\} - P\{T_1 + \cdots + T_{n+1} \leq t\}$$

$$= \int_0^t \varphi *^{(n)}(x)\, dx - \int_0^t \varphi *^{(n+1)}(x)\, dx, \quad n = 1, 2, \ldots$$

Let
$$\bar{P}(n, s) = \int_0^\infty e^{-st} P(n, t)\, dt$$

denote the Laplace transform of $P(n, t)$ with respect to the variable t, and let

$$\bar{P}(z, s) = \sum_{n=0}^\infty z^n \bar{P}(n, s)$$

denote the generating function of $\bar{P}(n, s)$ with respect to the variable n. Then

$$\bar{P}(n, s) = \frac{1}{s}[\bar{\varphi}(s)]^n - \frac{1}{s}[\bar{\varphi}(s)]^{n+1}$$

$$= \frac{[\bar{\varphi}(s)]^n[1 - \bar{\varphi}(s)]}{s}, \quad n = 1, 2, \ldots$$

and

$$\bar{P}(z, s) = \frac{1 - \bar{\varphi}(s)}{s} \cdot \frac{1}{1 - z\bar{\varphi}(s)}$$

The Laplace transform of the expected number of renewals is (see (2.108))

$$\bar{M}(s) = \lim_{z \to 1} \frac{\partial \bar{P}(z, s)}{\partial z} = \lim_{z \to 1} \frac{1 - \bar{\varphi}(s)}{s} \cdot \frac{z\bar{\varphi}(s)}{[1 - z\bar{\varphi}(s)]^2}$$

$$= \frac{\bar{\varphi}(s)}{s[1 - \bar{\varphi}(s)]}$$

Example 2.6

Assume that $\varphi(t) = \lambda e^{-\lambda t}$, $\lambda > 0$, $t > 0$. Then

$$P\{N(t) = 0\} = 1 - \int_0^t \lambda e^{-\lambda x}\, dx$$

$$P\{N(t) = n\} = \int_0^t \lambda e^{-\lambda x} \frac{(\lambda x)^{n-1}}{(n-1)!}\, dx - \int_0^t \lambda e^{-\lambda x} \frac{(\lambda x)^n}{n!}\, dx, \quad n = 1, 2, \ldots$$

Using the sum expression for the incomplete gamma functions we obtain

$$P\{N(t) = n\} = \sum_{i=n}^{\infty} e^{-\lambda t} \frac{(\lambda t)^i}{i!} - \sum_{i=n+1}^{\infty} e^{-\lambda t} \frac{(\lambda t)^i}{i!}$$

$$= e^{-\lambda t} \frac{(\lambda t)^n}{n!}, \quad n = 0, 1, 2, \ldots$$

SELECTED REFERENCES

[1] BAILEY, N. T. J., *The Elements of Stochastic Processes with Applications to the Natural Sciences*, J. Wiley & Sons, Inc., New York, 1964.

[2] BHARUCHA-REID, A. T., *Elements of the Theory of Markov Processes and their Applications*, McGraw-Hill, New York, 1960.

[3] BHAT, U.N., *Elements of Applied Stochastic Processes,* J. Wiley & Sons, Inc., New York, 1972.

[4] CHUNG, K. L., *Markov Chains*, Springer-Verlag, New York, 1967.

[5] CLARKE, A. B. and RALPH L. DISNEY, *Probability and Random Processes for Engineers and Scientists*, J. Wiley & Sons, Inc., New York, 1970.

[6] COX, D. R. and H. D. MILLER, *The Theory of Stochastic Processes*, J. Wiley & Sons, Inc., New York, 1965.

[7] FELLER, W., *An Introduction to Probability Theory and Its Applications*, Vol. I, J. Wiley & Sons, Inc., New York, 1957.

[8] FELLER, W., *An Introduction to Probability Theory and Its Applications*, Vol. II, J. Wiley & Sons, Inc., New York, 1966.

[9] GNEDENKO, B. V., *The Theory of Probability*, Chelsea Publishing Company, New York, 1963.

[10] GRAY, J. R., *Probability*, Oliver & Boyd, London, 1967.

[11] PARZEN, E., *Stochastic Processes*, Holden-Day, Inc., San Francisco, 1962.

[12] PRABHU, N. V., *Stochastic Processes*, The Macmillan Company, New York, 1965.

[13] USPENSKY, J. V., *Introduction to Mathematical Probability*, McGraw-Hill, New York, 1937.

PROBLEMS

1. In the generalized Bernoulli independent process, the probability of success p_n at the nth trial is a function of the order n of trials, $n = 1, 2, \ldots$. Formulate the system of difference equations for $P(r, n)$, the probability of r successes in *n* trials, and show that

$$G(s, n) = \sum_{r=0}^{r=n} s^r P(r, n) = \prod_{i=1}^{i=n} (q_i + p_i s)$$

2. In Example 2.1 suppose that the total number of months N the salesman will have the customer's account is a random variable with PMF

$$\varphi_N(n) = e^{-\alpha} \frac{\alpha^{n-1}}{(n-1)!}, \quad n = 1, 2, \ldots$$

Let X denote the total number of orders placed by the customer during his entire business association with the salesman. Find $G_X(s)$, $E[X]$, and $\mathrm{Var}[X]$.

3. A meteorological group is attempting to forecast the winter weather at a particular location. Past experience shows that they never have two nice days in a row. If they have a nice day, they are just as likely to have snow as rain the next day. If they have snow (or rain), they have an even chance of having the same the next day. If there is a change from snow or rain, only half the time is this change to a nice day. Define by 0, 1, and 2 the states corresponding to "nice day," "snowy day," and "rainy day," respectively.

   (a) Show that the weather phenomenon over successive days is a homogeneous Markov chain with one-step transition probability matrix

   $$\mathbf{P} = \begin{pmatrix} p_{00} & p_{01} & p_{02} \\ p_{10} & p_{11} & p_{12} \\ p_{20} & p_{21} & p_{22} \end{pmatrix} = \begin{pmatrix} 0 & \frac{1}{2} & \frac{1}{2} \\ \frac{1}{4} & \frac{1}{2} & \frac{1}{4} \\ \frac{1}{4} & \frac{1}{4} & \frac{1}{2} \end{pmatrix}$$

   (b) Find the steady-state probabilities.

4. Consider the problem of the wartime operation of a bomber squadron. Suppose that at the end of each day the squadron has 0, 1, 2, or 3 bombers. If it has less than 3 bombers, it requests replacement planes to be flown to bring the squadron up to full strength; planes are flown in overnight and are available for the morning. Unfortunately, the replacement bombers are flown over enemy lines, and there is a probability p_1 that a replacement plane will be shot down on the way to the squadron's base of operations. In the morning, if 1 or 2 or 3 planes are available, then they are sent on a bombing mission; otherwise, no mission is flown (since no planes are available at the base). If planes are sent out, each plane has a probability p_2 of being shot down. Any plane that returns to base is made serviceable by the next day. At the end of the day, if the squadron is below full strength, additional airplanes are requested to be flown up. Let $X(n)$ denote the number of bombers at the squadron base on the *n*th morning, $n = 0, 1, 2, \ldots$. Show that $\{X(n), n \in \mathbf{T}\}$ is a discrete-parameter homogeneous Markov chain whose one-step transition probability matrix can be expressed as $\mathbf{P} = \mathbf{A} \cdot \mathbf{B}$, where $\mathbf{A}$ and $\mathbf{B}$ are given by

$$\mathbf{A} = \begin{pmatrix} 1 & 0 & 0 & 0 \\ p_2 & q_2 & 0 & 0 \\ p_2^2 & 2p_2 q_2 & q_2^2 & 0 \\ p_2^3 & 3p_2^2 q_2 & 3p_2 q_2^2 & q_2^3 \end{pmatrix}$$

$$\mathbf{B} = \begin{pmatrix} p_1^3 & 3p_1^2 q_1 & 3p_1 q_1^2 & q_1^3 \\ 0 & p_1^2 & 2p_1 q_1 & q_1^2 \\ 0 & 0 & p_1 & q_1 \\ 0 & 0 & 0 & 1 \end{pmatrix}$$

Give an interpretation of $\mathbf{A}$ and $\mathbf{B}$.

5. The marketing division of a large corporation introduces on a pilot basis at a specific location a new product. Assume that the profit and loss in successive months are independent events such that the probability of a month turning out to be a profit is $p = \frac{2}{3}$. If over a period of a year the product does not show monthly losses over 3 consecutive months, the corporation will market the product on a nationwide basis. Determine the probability that the product will not appear on the national market.

6. For a nonhomogeneous death process with death rate $\mu_i(t) = i\mu(t)$, suppose that the entire population may be subject independently, at any time t to a sudden annihilation. Let $v(t)\,dt + o(dt)$ be the probability that a population of size i, $i = 1, 2, \ldots, N$, will be annihilated in the time interval $(t, t + dt]$. Obtain the differential difference equations for $P(n, t)$, the probability that at time t the population consists of exactly n individuals. Show that if

$$\mathfrak{M}(t) = \int_0^t \mu(\theta)\,d\theta$$

and
$$\mathfrak{N}(t) = \int_0^t v(\theta)\,d\theta$$

then for $1 \leq n \leq N$, $t \geq 0$,

$$P(n, t) = e^{-\mathfrak{N}(t)} \binom{N}{n} [e^{-\mathfrak{M}(t)}]^n [1 - e^{-\mathfrak{M}(t)}]^{N-n}$$

Find the PDF of the time elapsed W till population extinction. Determine $E[W]$ and $\text{Var}[W]$. [Note that $P(n, t)$ is the product of two quantities, one being $e^{-\mathfrak{N}(t)}$.]

An interesting military application to this problem would be the following:

A convoy of N cargo ships is carrying supplies to a given destination. If convoy protection is ineffective, enemy attacks on the convoy may cause attrition or total annihilation such that if there are n units in the convoy at time t the probability that a single unit will be lost in the time interval $(t, t + dt]$ is $n\mu(t)\,dt + o(dt)$. The probability of total annihilation in the time interval $(t, t + dt]$ is $v(t)\,dt + o(dt)$. Analysis of World War II data has shown that effective convoy protection would result in an attrition rate which is independent of the size of the convoy at a given time.

7. An equipment can be in either of two states: operating or failed. The time period X during which the equipment is in an operating state is a random variable having a negative exponential distribution with mean $1/\lambda$. As soon as the equipment fails, repair starts. Repair time Y is a random variable with a negative exponential distribution with mean $1/\mu$. Immediately upon termination of repair, operation of the equipment starts up again. The

sequence of random variables X_1, Y_1, X_2, Y_2, ... describing the succession of times spent in each of the two states is mutually independent. Initially, assume that the equipment is in an operating state.

(a) Find the probability that the equipment is in an operating state at time t, $t \geq 0$.

(b) Determine the PDF of the time elapsed between two successive start-ups.

(c) Derive the equation and obtain an expression for the expected number of start-ups in the time interval $(0, t]$.

[*Hint:* Define $P_0(t)$ as the probability that the equipment is in a state of repair at time t and $P_1(t)$ as the probability that the equipment is operating at time t. Also note that $[\lambda \, dt + o(dt)]$ is the probability that the operating equipment at time t will fail in the time interval $(t, t + dt]$, and that $[\mu \, dt + o(dt)]$ is the probability that the nonoperating equipment at time t will be repaired in the time interval $(t, t + dt]$.]

8. Consider N automatic machines which when subject to breakdown call for service to be repaired by a single repairman. The time required to service a machine is a random variable having a negative exponential distribution with parameter μ. A machine that breaks down is serviced immediately if the repairman is free; otherwise, machines in need of repair form a waiting line for service. If at time t the machine is in a working state, the probability that it will break down is $\lambda \, dt + o(dt)$. Let $N(t)$ denote the number of machines working at time t. For the continuous-parameter Markov chain $\{N(t), t \in \mathbf{T}\}$,

(a) Find the transition probabilities.

(b) Derive the steady-state equations for $N(t)$ using the analytic method and the flow-diagram method.

(c) Find the steady-state distribution of $N(t)$.

9. In an ordinary renewal process, the interarrival times T_i, $i = 1, 2, \ldots$, for the first, second, ... renewals form a sequence of independent identically distributed random variables with common PDF given by

$$\varphi(x) = (\lambda x)\lambda e^{-\lambda x}, \quad \lambda > 0, x > 0$$

Determine the renewal density function $m(t)$, $t > 0$. Using the explicit form of $m(t)$, show that

$$\lim_{t \to \infty} m(t) = \frac{2}{\lambda}$$

10. Starting from expression (2.32), show that in the general case,

$$P(n) = A_1 \alpha_1^n + B_1 \alpha_2^n, \quad n \geq 2$$

where

$$\alpha_1 = \frac{-2pq}{q + \sqrt{q^2 + 4pq}}$$

$$\alpha_2 = \frac{-2pq}{q - \sqrt{q^2 + 4pq}}$$

Determine the values of A_1 and B_1.

11. Characterize a discrete-parameter independent process as a Markov chain.

RELIABILITY THEORY

3.1. Introduction

The theory of reliability studies the failure laws of systems that are operating within predetermined limits under given operating conditions. Complex systems, even if properly designed, are prone to failure; it is therefore important to assure the proper functioning of systems not only from a strict technological standpoint, but simultaneously by considering the failure properties of its components. Reliability is an internal property of operating systems, and, as such, it has to be specified as part of the systems' characteristics. Thus it becomes imperative to develop a quantitative basis for describing the failure process of systems to measure unambiguously their reliability; this then enables one to compare the performance of similar systems on a reliability basis and to develop methods for improving the reliability of a particular system.

In this chapter we develop the mathematical basis of the theory of reliability, and assume that the reader is familiar with the content of Chapters 1 and 2. Following a discussion of the failure phenomena of systems and their statistical characteristics, a number of mathematical models are presented to describe the failure process. Next, a basis for assessing the failure of complex

3

systems, given the failure characteristics of its components, is presented. Finally, various methods for improving the reliability of systems are discussed; these include the introduction of parallel and standby redundancies, the use of repair and/or replacement, and specific design considerations.

3.2. Failure Phenomenon: Exogenous- and Endogenous-Type Failures

We define *failure* as the event associated with a shift in the operating characteristics of a system from its permissible limits. It is convenient to classify the causes of failure in a system as being either of an *exogenous* type or of an *endogenous* type.

Exogenous-type failures are extrinsic or extraneous in character; that is, they originate from causes external to the system. Exogenous-type failures are usually associated with the environmental conditions under which the equipment is operating. They are the results of severe, unpredictable stresses arising from such environmental factors as sudden shocks. The rate of incidence of this type of failure is determined by the severity of the environmental conditions: the more severe the environment, the more frequent the occurrence of failure. An exogenous-type failure is sometimes called a *chance failure* or *random failure*.

Endogenous-type failures originate from internal causes. They are usually associated with certain inherent characteristics of the system and are dominant in the initial and final phase of operation of a system. During the initial phase of operation of a system, an endogenous-type failure, known also as *initial failure*, results from inherent defects in the system attributed to faulty design, manufacturing, or assembly. As the system operates, it wears out; hence, during its final phase of operation a second type of endogenous failure, known as *wear-out failure*, may be experienced by the system. The wear-out failure is the outcome of accumulated wear and tear; a depletion process occurs through abrasion, fatigue, creep, and the like, which results in the loss of specified physical, mechanical, chemical, or other properties characterizing the initial design features of the system. Ultimately, an operating system will fail as a result of wear and tear.

At this stage, it is useful to provide a heuristic treatment of an important concept, which will be defined later in an exact quantitative way. This concept is that of *hazard rate* or *failure rate* or *force of mortality* or *failure intensity*. It is the chance of failure of a system at a particular time, given that the system has survived up to that time. The behavior of the hazard rate as a function of time is sometimes known as the *hazard function* or *life characteristic* or *lambda characteristic* of the system. For a typical system that may experience any of the three previously described types of failure, the life characteristic will appear as in Figure 3.1.

This graph is to be interpreted in the following way: at the very beginning, the possibility of defective design, manufacturing, or assembly produces a high hazard on the system and is the significant cause of failure; during the second period, failure due to chance is predominant; at the final stage, a

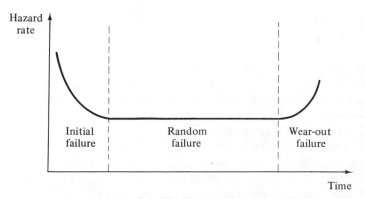

Figure 3.1. Typical life characteristics of a system.

system is more likely to fail through wear and tear. It must be realized that the three causes of failure are present at any particular time; nevertheless, their significance is dominant only at particular phases of the operation of the system.

From an actuarial point of view, the failure phenomenon in systems is very analogous to the mortality or death phenomenon in human beings. This is illustrated in Table 3.1.

Table 3.1. CAUSES OF MORTALITY FOR SYSTEMS
AND HUMAN BEINGS

Causes for Systems Failure	*Causes for Human Death*
Original defects	Birth defects
Chance	Accident
Wear out	Age

3.3. Statistical Characteristics of a System Subject to Failure

In what follows we shall provide a quantitative basis for developing the theory of reliability. Since, in general, the failure process is random, it follows that the characterization of an operating system subject to failure is probabilistic in nature. We shall mainly consider "one-shot-type" operations; that is, we shall assume that the system has to perform a specific function for a specific period starting from some time origin, and we shall be interested in determining the statistical features underlying the survival of the system. The basic quantity that will play an important role in this chapter is the *time to failure* of the system regarded as a random variable. It is defined as the time elapsed from the start of operation of the system until failure occurs for the first time.

We shall provide a quantitative definition for such important concepts as failure density function, life distribution, reliability function, mean time to failure, and hazard functions. Finally, the following well-known time to failure distributions will be discussed:

(1) Exponential distribution.
(2) Two-parameter Weibull distribution.
(3) Extreme value distribution.
(4) Gamma distribution.
(5) Lognormal distribution.

i. *Reliability*

The reliability of an operating system is defined as the probability that the system will perform satisfactorily within specified conditions over a given future time period when the system starts operating at some time origin.

ii. *Failure Density Function, Life Distribution, Reliability Function, and Mean Time to Failure*

Define by X the random variable for the time to failure of a system that has started operating at some time origin.

The PDF of the random variable X, $f_X(x)$, $0 < x < \infty$, is known as the *failure density function* of the system; thus $f_X(x)\,dx$ is the probability that the system will fail in the time interval $(x, x + dx]$.

The CDF of the random variable X, $F_X(x)$, $0 \le x < \infty$, is known as the *life distribution* of the system; that is, the probability that a system which has started operating at time origin $t = 0$ will fail on or before some time t, $0 \le t < \infty$, is

$$P\{X \le t\} = F_X(t) = \int_0^t f_X(x)\,dx$$

The complementary CDF of X, $R_X(x)$, $0 \le x < \infty$, is known as the *reliability function* or *survival function* of the system. The probability that a system that has started operating at time origin $t = 0$ will survive up to time t (i.e., will not fail on or before time t), $0 \le t < \infty$, is

$$P\{X > t\} = R_X(t) = 1 - F_X(t) = \int_t^\infty f_X(x)\,dx$$

The expectation of the random variable X, $E[X]$, defines the *mean time to failure* of a system that has started operating from some time origin. Clearly,

$$E[X] = \int_0^\infty x f_X(x)\,dx = \int_0^\infty R_X(x)\,dx$$

Example 3.1

A particular component in a complex mechanical system has a negative exponential time-to-failure distribution with mean time to failure of 1000 hours. Find

(a) The reliability function.

(b) The reliability of the component if it has to operate for 2000 hours.

(c) The maximum time of operation of the component if a reliability level of at least (1) .99 (2) .999 is to be maintained.

(a) Let X denote the time to failure of the component. Its PDF is given by

$$f(x) = \lambda e^{-\lambda x}, \quad \lambda > 0, x > 0$$

where $\lambda = 1/1000 = .001$. Thus

$$R_X(t) = e^{-\lambda t} = e^{-.001t}, \quad t \geq 0$$

(b) If the component has to operate for 2000 hours, its reliability, which is its probability of survival after 2000 hours of operation, is

$$R_X(2000) = e^{-.001(2000)} = e^{-2} = .1352$$

(c) Theoretically, one can achieve any reliability level by controlling the time of operation of the component; for this particular component, the reliability is a strictly decreasing function of the time of operation. Thus a high reliability may be achieved by reducing the time of operation of the component. Now, since

$$R_X(t) = e^{-.001t}$$

then

$$t = 1000 \ln \frac{1}{R_X(t)}$$

If $R_X(t) = .99$,

$$t = 1000 \ln \frac{1}{.99} \approx 10 \text{ hours}$$

If $R_X(t) = .999$,

$$t = 1000 \ln \frac{1}{.999} \approx 1 \text{ hour}$$

iii. *Hazard Function*

a. DEFINITION

The *hazard function* or *hazard rate* $h_X(t)$ of a system is so defined that $h_X(t) \, dt$ is the conditional probability of time to failure of the system in the time interval $(t, t + dt]$, given that the system has not failed from time origin up to time t, $t \geq 0$. Thus

$$h_X(t) \, dt = \frac{P\{t < X \leq t + dt, X > t\}}{P\{X > t\}}$$

The numerator in the right-hand expression is the joint probability that the system has survived to time t and will fail in the time interval $(t, t + dt]$; this is equivalent to the probability that the system will fail in the time interval $(t, t + dt]$, since this event presumes survival of the system to time t. Hence

$$h_X(t) \, dt = \frac{P\{t < X \le t + dt\}}{P\{X > t\}} = \frac{f_X(t) \, dt}{R_X(t)}$$

$$= -\frac{dR_X(t)}{R_X(t)}$$

$$= \frac{dF_X(t)}{1 - F_X(t)} \tag{3.1}$$

Suppose that we have a large number of identical components which start operating at exactly the same time under identical conditions; then $h_X(t) \, dt$ denotes the proportion of components that have survived up to time t, but that will fail in the time interval $(t, t + dt]$.

Given the expression for the hazard function, it is possible to obtain expressions for the life distribution, the reliability function, and the density function for the time to failure. We are interested in solving for $F_X(t)$ in expression (3.1), identified as a first-order differential equation. Integrating both sides of equation (3.1), we obtain

$$\int_0^t h_X(x) \, dx = \int_0^t \frac{dF_X(x)}{1 - F_X(x)}$$

$$= -\ln\left[1 - F_X(t)\right] + \ln\left[1 - F_X(0)\right]$$

or $$\frac{1 - F_X(t)}{1 - F_X(0)} = e^{-\int_0^t h_X(x) \, dx}$$

or $$1 - F_X(t) = [1 - F_X(0)]e^{-\int_0^t h_X(x) \, dx} \tag{3.2}$$

Since $F_X(t)$ is a distribution function with the property that $\lim_{t \to \infty} F_X(t) = 1$, it follows that we must have the condition $\lim_{t \to \infty} \int_0^t h_X(x) \, dx = \infty$ satisfied for all hazard functions.

Expression (3.2) is the most general relation between the life distribution function and the hazard function. This expression involves the quantity $F_X(0)$, which is the probability of failure of the system at $t = 0$; clearly, $0 \le F_X(0) \le 1$. In what follows we shall assume, unless specified otherwise, that $F_X(0) = 0$; then, from relation (3.2), we obtain the following expressions for $F_X(t)$, $R_X(t)$, and $f_X(t)$:

$$F_X(t) = 1 - e^{-\int_0^t h_X(x) \, dx}, \quad t \ge 0 \tag{3.3}$$

$$R_X(t) = e^{-\int_0^t h_X(x) \, dx}, \quad t \ge 0 \tag{3.4}$$

$$f_X(t) = h_X(t)e^{-\int_0^t h_X(x) \, dx}, \quad t > 0 \tag{3.5}$$

An expression for the mean time to failure in terms of the hazard function is

$$E[X] = \int_0^\infty e^{-\int_0^t h_X(x)\,dx}\,dt \tag{3.6}$$

b. EXAMPLES OF HAZARD FUNCTIONS AND
    RELATED FAILURE DENSITY FUNCTIONS

(1) Exponential Distribution

The simplest form for the hazard function is a constant function; that is,

$$h_X(t) = \lambda \text{ (a constant)}, \quad t \geq 0$$

Then, from expression (3.5),

$$f_X(t) = \lambda e^{-\lambda t}, \quad t > 0$$

which is the *exponential distribution*. The physical interpretation of a constant hazard function is that irrespective of the time t elapsed since the start of operation of a system the probability that the system will fail in the next time interval dt, given that it has survived to time t, is independent of the elapsed time t. A situation of this sort corresponds in practice to chance or random failure.

(2) Two-Parameter Weibull Distribution

In general, the hazard function depends on time. Suppose that

$$h_X(t) = \alpha \lambda t^{\alpha-1}, \quad \alpha > 0, \lambda > 0, t \geq 0$$

where α and λ are constants; from (3.4) and (3.5) we obtain, respectively, for the reliability function and the failure density function

$$R_X(t) = e^{-\lambda t^\alpha}, \quad t \geq 0$$
$$f_X(t) = \alpha \lambda t^{\alpha-1} e^{-\lambda t^\alpha}, \quad t > 0$$

This is the *two-parameter Weibull distribution*, which reduces to the exponential distribution when $\alpha = 1$. We note in particular that for the Weibull distribution the hazard function is

decreasing if $0 < \alpha < 1$

constant if $\alpha = 1$

increasing if $1 < \alpha < \infty$

This class of hazard function provides a practical means for fitting the various portions of the life characteristic curve of a system. The initial-failure portion would be represented by a hazard function with $\alpha < 1$, the random-failure portion would correspond to a hazard function with $\alpha = 1$, and the wear-out failure portion would be represented by a hazard function with $\alpha > 1$; the quantities α and λ would have to be empirically estimated.

The parameter λ is known as the scale parameter. Typical shapes for the Weibull distribution for a given λ and $\alpha = 1, 2$, and 3 are shown in Figure 3.2.

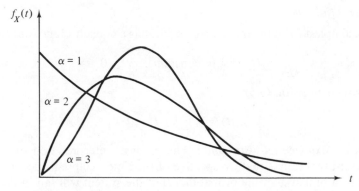

Figure 3.2. The Weibull distribution $f_X(t) = \alpha \lambda t^{\alpha-1} e^{-\lambda t^\alpha}$ $(\lambda = 1)$.

An expression for the kth moment about the origin for the two-parameter Weibull distribution can be obtained by evaluating the integral

$$E[X^k] = \int_0^\infty x^k \alpha \lambda x^{\alpha-1} e^{-\lambda x^\alpha}\, dx$$

Let $u = x^\alpha$; then

$$E[X^k] = \int_0^\infty \lambda u^{k/\alpha} e^{-\lambda u}\, du$$
$$= \lambda^{-k/\alpha} \Gamma\left(\frac{k}{\alpha} + 1\right)$$

Expressions for the mean and variance of X are

$$E[X] = \lambda^{-1/\alpha} \Gamma\left(\frac{1}{\alpha} + 1\right)$$
$$\text{Var}\,[X] = \lambda^{-2/\alpha}\left\{\Gamma\left(\frac{2}{\alpha} + 1\right) - \left[\Gamma\left(\frac{1}{\alpha} + 1\right)\right]^2\right\}$$

Referring to the expression for the reliability function, we obtain, by taking logarithms twice,

$$\ln\left[-\ln R_X(t)\right] = \alpha \ln t + \ln \lambda$$

Thus $\ln[-\ln R_X(t)]$ plotted against $\ln t$ is a straight line with slope α. Kao [6] has designed and produced a graph paper known as the Weibull probability paper based on this double logarithmic transformation. For a system having a Weibull time-to-failure distribution, a plot of the survival function versus time on this special graph paper will yield a straight line. The slope of the line is an estimate of the parameter α. The scale parameter λ may be estimated by noting that at $t = 1$

$$\ln[-\ln R_X(1)] = \ln \lambda$$

The statistical procedure for obtaining the survival function empirically is a subject in life testing statistics and will not be discussed here.

We have seen that when $\alpha = 1$, the hazard function is a constant, thus generating the *exponential distribution* as a special case of the Weibull distribution.

Another important special case is when $\alpha = 2$; then the expression for $h_X(t)$ is

$$h_X(t) = 2\lambda t$$

which is a linear function of t. The failure density function becomes

$$f_X(t) = 2\lambda t e^{-\lambda t^2}$$

which can be recognized as the *Rayleigh distribution* (see Chapter 1, Problem 14).

(3) *Extreme Value Distribution*

Another form of the hazard function is

$$h_X(t) = \alpha \gamma e^{\gamma t}, \quad \alpha > 0, \gamma > 0, t \geq 0$$

The corresponding failure density function is

$$f_X(t) = \alpha \gamma e^{\gamma t} e^{-\alpha(e^{\gamma t}-1)}, \quad t > 0$$

This is a two-parameter distribution (see Figure 3.3); it can be reduced to the standardized *extreme value distribution* encountered in the statistical theory of extreme values investigated by Gumbel [5]. From (3.6) we obtain for $E[X]$

$$E[X] = \int_0^\infty e^{-\int_0^t \alpha \gamma e^{\gamma x}\,dx}\,dt$$

$$= \int_0^\infty e^{-\alpha(e^{\gamma t}-1)}\,dt$$

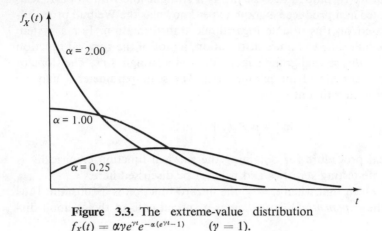

Figure 3.3. The extreme-value distribution
$f_X(t) = \alpha\gamma e^{\gamma t}e^{-\alpha(e^{\gamma t}-1)}$ $(\gamma = 1)$.

Let $e^{\gamma t} = u$; then $\gamma t = \ln u$ and $dt = du/\gamma u$. Hence

$$E[X] = \frac{e^\alpha}{\gamma} \int_1^\infty \frac{e^{-\alpha u}}{u}\, du$$

The right-hand side integral is the *exponential integral* function, which can be evaluated from tables.

(4) Gamma Distribution

This distribution was already investigated in Chapter 1. The form of the failure density function is

$$f_X(t) = \frac{(\lambda t)^{n-1}}{\Gamma(n)} \lambda e^{-\lambda t}, \quad \lambda > 0, n > 0, t > 0$$

The reliability function is

$$R_X(t) = \int_t^\infty \frac{(\lambda x)^{n-1}}{\Gamma(n)} \lambda e^{-\lambda x}\, dx$$

$$= \frac{1}{\Gamma(n)} \int_{\lambda t}^\infty u^{n-1}e^{-u}\, du, \quad \lambda > 0, n > 0, t \geq 0$$

which can be recognized as the incomplete gamma function. The hazard function is

$$h_X(t) = \frac{(\lambda t)^{n-1}\lambda e^{-\lambda t}}{\displaystyle\int_{\lambda t}^\infty u^{n-1}e^{-u}\, du}$$

The reader can verify that $h_X(t)$ is an increasing function of t when $n > 1$, and is a decreasing function of t when $n < 1$. When $n = 1$, $h_X(t) = \lambda$; furthermore, it may be shown that $\lim_{t \to \infty} h_X(t) = \lambda$.

Expressions for the mean and variance of the time to failure are

$$E[X] = \frac{n}{\lambda}$$

$$\text{Var}\,[X] = \frac{n}{\lambda^2}$$

(5) Lognormal Distribution

The time-to-failure X has the lognormal distribution if its PDF is

$$f_X(t) = \begin{cases} \dfrac{1}{t\sigma\sqrt{2\pi}}e^{-(\ln t - \mu)^2/2\sigma^2}, & t > 0 \\ 0, & \text{otherwise} \end{cases}$$

We note that if Z is a normally distributed random variable with density function

$$f_Z(t) = \frac{1}{\sigma\sqrt{2\pi}}e^{-(t-\mu)^2/2\sigma^2}, \quad -\infty < t < \infty$$

then the random variable $X = e^Z$ has the lognormal distribution. It is easy to show that the reliability function is

$$R_X(t) = \int_{\ln t}^{\infty} \frac{1}{\sigma\sqrt{2\pi}}e^{-(x-\mu)^2/2\sigma^2}\,dx, \quad t \geq 0$$

Expressions for the mean and variance of the time to failure are, respectively,

$$E[X] = e^{\mu + (\sigma^2/2)}$$
$$\text{Var}\,[X] = e^{2\mu}(e^{2\sigma^2} - e^{\sigma^2})$$

The hazard function for the lognormal distribution has the property that it increases for a while and eventually decreases.

The lognormal distribution possesses the same properties under multiplication of positive random variables as does the normal distribution under addition of random variables. The reader should verify, for example, that if X_1 and X_2 are two independently distributed random variables with a lognormal distribution, then the random variable $X = X_1 X_2$ has a lognormal distribution.

A plot of $f_X(t)$ is shown in Figure 3.4.

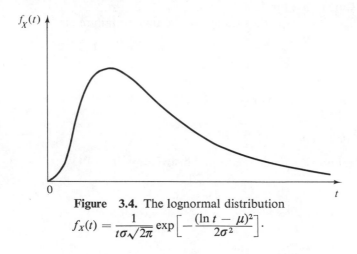

Figure 3.4. The lognormal distribution

$$f_X(t) = \frac{1}{t\sigma\sqrt{2\pi}} \exp\left[-\frac{(\ln t - \mu)^2}{2\sigma^2}\right].$$

3.4. Stochastic Processes Underlying the Failure Phenomenon

In general the causes underlying the failure of a system can be regarded as random phenomena, which can be formulated as a stochastic model relating the time-to-failure distribution of the system to the particular causal phenomenon. Consider a simplified situation in which a system is free from initial-type failure. We may assume that the system possesses an internal characteristic with a threshold magnitude known as *strength* such that the imposition or generation of *stresses* with magnitude greater than strength causes system failure.

When the system, whether operating or not, is placed in an adverse environment, it may be subject to exogenous stresses or shocks occurring in the form of pulses or peaks, whose occurrence and magnitude can be described as a random process. Failure of the system occurs whenever a peak stress exceeds the system's strength. A realization of this particular process is represented in Figure 3.5.

When the system is in an operating state, a process of deterioration or damage takes place that is cumulative in nature. Here again the amount of damage and the time interval between its occurrence are in general random variables. We may theorize that whenever the amount of damage grows, the *strength* of the system diminishes as a function of the time of operation of the system. Thus, even if the system is not subject to an adverse environment,

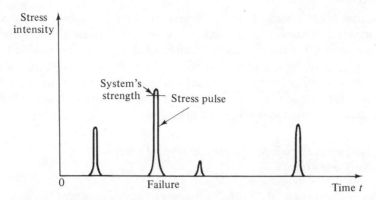

Figure 3.5. A sample function of a stress generating environment.

a physical condition is ultimately reached corresponding to the wear-out failure. On the other hand, for a system operating under adverse environmental conditions, failure may also be due to the event "stress greater than strength"; thus system failure will occur as a result of either chance or wear-out failure. Figure 3.6 is a realization of a wear-out process.

During the initial state of operation the amount of deterioration is negligible, so that as a first approximation one may assume that the system is subject only to an adverse environment but no wear out. This then forms the basis of the stochastic models for exogenous-type failures, where it is assumed that no wear out is present and hence that the system's strength remains constant. The next set of stochastic models assumes that the operating system is not subject to exogenous stresses but that it wears out with a consequent failure; the system's strength is a decreasing function of the time

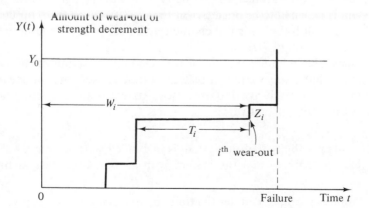

Figure 3.6. A sample function of a wear-out process in a system.

of operation of the system. Finally, the last set of models considers the combined presence of exogenous stresses and wear out. For each model appropriate expressions for the reliability function of the system are obtained. We have distinguished between the two states *operating* and *operative*: a system is in an operating state if it is functioning; a system is in an operative state if it is functioning or if it is idle but in a ready state of functioning. A failed system is neither operating nor operative.

i. *Stochastic Models for Exogenous-Type Failures— No Wear Outs*

The phenomenon of failure of systems operative under a particular environmental condition can often be formulated as a stochastic process relating the causes and frequencies of failure-inducing factors to the time to failure of the system. This approach, which provides a rational basis to characterize certain exogenous-type failures, was first undertaken by Epstein [2]. In what follows we shall consider three stochastic models, which can be considered as examples of systems subject to random extreme conditions of instantaneous duration.

a. SYSTEM SUBJECT TO A SINGLE ENVIRONMENT

Suppose that a system is operative in an environment ε that can be simulated as a stochastic process consisting of an alternation of normal and peak events. When a peak event occurs, the system may fail; otherwise, the system will not fail. Suppose that these peak events have an instantaneous duration, and let the probability of a peak occurring in the time interval $(t, t + dt]$ be $\lambda(t) \, dt + o(dt)$, $t \geq 0$. Suppose now that conditional on the occurrence of a peak at time t, the probability that the system will fail is $p(t)$. The system is assumed to be operative at time $t = 0$. Given these inputs, it is possible to establish the statistical characteristics of the time to failure of the system as a function of $\lambda(t)$ and $p(t)$.

We may obtain an expression for the reliability function of the system by imagining that at any particular instant of time the system is in one of the following mutually exclusive and exhaustive states: state 0 denoting failure and state 1 denoting operative.
Let

$P_0(t) =$ probability that the system is in a failed state at time t, $t \geq 0$

$P_1(t) =$ probability that the system is in an operative state at time t, $t \geq 0$

If the random variable X denotes the time to failure of the system, given that it is operative at time origin $t = 0$, then it is evident that $P_0(t)$ is the life distribution $F_X(t)$ of the system, while $P_1(t)$ is the reliability function $R_X(t)$ of the system; thus

$$P_0(t) = F_X(t) = P\{X \le t\}$$
$$P_1(t) = R_X(t) = P\{X > t\}$$

and, of course,

$$P_0(t) + P_1(t) = 1, \quad \text{for all } t \ge 0$$

Because of this last identity, it is only necessary to obtain an expression for either $P_0(t)$ or $P_1(t)$. We select $P_1(t)$ and write for $t \ge 0$

$$P_1(t + dt) = P_1(t)[\lambda(t)\,dt + o(dt)][1 - p(t)]$$
$$+ P_1(t)[1 - \lambda(t)\,dt + o(dt)]$$

which can be rewritten as

$$\frac{P_1(t + dt) - P_1(t)}{dt} = -p(t)\lambda(t)P_1(t) + \frac{o(dt)}{dt}$$

Taking limits on both sides as $dt \longrightarrow 0$, we obtain

$$\frac{dP_1(t)}{dt} = -p(t)\lambda(t)P_1(t), \quad t \ge 0$$

Since the system is operative at time $t = 0$, we can write for the initial condition

$$P_1(0) = 1, \quad t = 0$$

and obtain for $P_1(t)$ or $R_X(t)$

$$R_X(t) = P_1(t) = e^{-\int_0^t p(x)\lambda(x)\,dx}, \quad t \ge 0$$

The failure density function is

$$f_X(t) = p(t)\lambda(t)e^{-\int_0^t p(x)\lambda(x)\,dx}, \quad t > 0$$

and the hazard function is

$$h_X(t) = p(t)\lambda(t), \quad t \ge 0$$

It is easy to see that the exponential and the Weibull cases are special forms of $p(t)\lambda(t)$.

b. SYSTEM SUBJECT TO A CERTAIN PRESENCE MULTIPLE ENVIRONMENT

We can extend the previous model and assume that a system is exposed simultaneously to k environments $\mathcal{E}_1, \mathcal{E}_2, \ldots, \mathcal{E}_k$. Each environment is acting independently of any other environment and is characterized by the occurrence of peak events inducing shocks on the system. Assume that the system is operative at time origin, and for $i = 1, 2, \ldots, k$, let $\lambda_i(t) \, dt + o(dt)$ be the probability that a peak event in environment $\mathcal{E}_i$ occurs in the time interval $(t, t + dt]$, $t \geq 0$. Let also $p_i(t)$ be the conditional probability that the system fails at time t, given that a peak event in environment $\mathcal{E}_i$ occurs, $t \geq 0$. Then we may write for $P_1(t)$, the probability that the system is operative at time t, $t \geq 0$,

$$P_1(t + dt) = P_1(t) \sum_{i=1}^{i=k} [\lambda_i(t) \, dt + o(dt)][1 - p_i(t)]$$
$$+ P_1(t)\left\{1 - \sum_{i=1}^{i=k} [\lambda_i(t) \, dt + o(dt)]\right\}$$

or

$$\frac{P_1(t + dt) - P_1(t)}{dt} = -P_1(t) \sum_{i=1}^{i=k} p_i(t)\lambda_i(t) + \frac{o(dt)}{dt}$$

This reduces to the differential equation

$$\frac{dP_1(t)}{dt} = -P_1(t) \sum_{i=1}^{i=k} p_i(t)\lambda_i(t), \quad t \geq 0$$

subject to the initial condition

$$P_1(0) = 1, \quad t = 0$$

We then obtain, for $P_1(t)$ or $R_X(t)$,

$$R_X(t) = P_1(t) = \exp\left\{-\int_0^t \left[\sum_{i=1}^{i=k} p_i(x)\lambda_i(x)\right] dx\right\}, \quad t \geq 0$$

The corresponding hazard function is

$$h_X(t) = \sum_{i=1}^{i=k} p_i(t)\lambda_i(t), \quad t \geq 0$$

c. System Subject to an Uncertain Presence of Multiple Mutually Exclusive Environments

Suppose now that the system is exposed to k possible environments $\mathcal{E}_1, \mathcal{E}_2, \ldots, \mathcal{E}_k$, considered as mutually exclusive and exhaustive events occurring with respective probabilities $\pi_1, \pi_2, \ldots, \pi_k$; $\pi_i \geq 0$ and $\sum_{i=1}^{i=k} \pi_i = 1$. The ith environment is characterized by the occurrence of peaks such that $\lambda_i(t) \, dt + o(dt)$ is the probability that a peak event occurs in the time interval $(t, t + dt]$, given that the ith environment prevails. Let also $p_i(t)$ be the conditional probability that the system fails at time t, given that a peak event in environment i occurs. The expression for the reliability function of the system is

$$R_X(t) = P\{X > t\} = \sum_{i=1}^{i=k} \pi_i e^{-\int_0^t p_i(x)\lambda_i(x)\, dx}$$

The corresponding failure density function is

$$f_X(t) = \sum_{i=1}^{i=k} \pi_i p_i(t)\lambda_i(t) e^{-\int_0^t p_i(x)\lambda_i(x)\, dx}, \quad t > 0$$

an expression involving the sum of exponentials.

Epstein [2] suggests a continuous analogue expression for $f_X(t)$. Let $G(x)$ be a distribution function, $0 \le x < \infty$, and define

$$f_X(t) = \int_0^\infty p(t)\lambda(t) e^{-\int_0^t p(x)\lambda(x)\, dx}\, dG[p(t)\lambda(t)], \quad t > 0$$

Let $v(t) = p(t)\lambda(t)$; then

$$f_X(t) = \int_0^\infty v(t) e^{-\int_0^t v(x)\, dx}\, dG[v(t)] \quad t > 0$$

In the special case when $v(t) = v = $ constant, and

$$g(v) = \frac{dG(v)}{dv} = \frac{(Av)^r}{\Gamma(r+1)} A e^{-Av}, \quad A > 0$$

we have

$$f_X(t) = \int_0^\infty \frac{v^{r+1} A^{r+1} e^{-v(A+t)}}{\Gamma(r+1)}\, dv, \quad t > 0$$

The reader may verify that $f_X(t)$ is a proper density function.

ii. *Stochastic Models for Endogenous-Type Failures*

We now illustrate by two models a situation involving an operating system subject to no exogenous conditions whose strength deteriorates with time as a result of wear out. The first model assumes that the time intervals between strength decrements are random, but the magnitude of the decrements is a constant. The second model assumes that both the time interval between strength decrements and the magnitude of these decrements are random. In both models we postulate that initially the system's strength is a given known quantity $Y_0 > 0$ and that failure at time t occurs once the cumulative wear out $Y(t)$ exceeds the value of Y_0 (see Figure 3.6).

a. STOCHASTIC MODEL FOR WEAR OUT WITH CONSTANT DAMAGE RATE

For $t \geq 0$, let

$\mu(t) \, dt + o(dt) =$ probability that in the time interval $(t, t + dt]$ the system strength decreases by one unit

$P(y, t) =$ probability that the system strength is y at time t; that is, $P\{Y_0 - Y(t) = y\} = P(y, t)$

It is evident that $P(y, t) = 0$ for all $y > Y_0$ and all $t \geq 0$. The initial conditions for the problem are

$$P(Y_0, 0) = 1, \quad y = Y_0$$
$$P(y, 0) = 0, \quad \text{otherwise}$$

It is not then difficult to identify the wear-out process as a nonhomogeneous death process (Section 2.7-iii-b). The differential difference equations governing $P(y, t)$ for $y = 0, 1, \ldots, Y_0$ and $t \geq 0$ are

$$\frac{dP(Y_0, t)}{dt} = -\mu(t)P(Y_0, t), \quad y = Y_0$$

$$\frac{dP(y, t)}{dt} = -\mu(t)P(y, t) + \mu(t)P(y + 1, t), \quad y = 1, 2, \ldots, Y_0 - 1$$

$$\frac{dP(0, t)}{dt} = \mu(t)P(1, t), \quad y = 0$$

Let

$$\tilde{M}(t) = \int_0^t \mu(x) \, dx$$

Then we obtain for $t \geq 0$

$$P(y, t) = e^{-\tilde{M}(t)} \frac{[\tilde{M}(t)]^{Y_0-y}}{(Y_0 - y)!}, \quad y = 1, 2, \ldots, Y_0$$

$$P(0, t) = \int_0^t \mu(x)e^{-\tilde{M}(x)} \frac{[\tilde{M}(x)]^{Y_0-1}}{(Y_0 - 1)!} \, dx$$

The distribution function for the time to failure X of the system is the probability that the system is completely worn out at time t. Hence

$$F_X(t) = P\{X \leq t\}$$
$$= P\{Y(t) \geq Y_0\} = P(0, t)$$

The failure density function is then

$$f_X(t) = \frac{dP(0, t)}{dt} = \mu(t)e^{-\tilde{M}(t)}\frac{[\tilde{M}(t)]^{Y_0-1}}{(Y_0 - 1)!}, \quad Y_0 = 1, 2, \ldots$$

In the special case when $\mu(t) = \mu$ for all $t \geq 0$, the failure density function will have the gamma distribution

$$f_X(t) = \mu e^{-\mu t}\frac{(\mu t)^{Y_0-1}}{(Y_0 - 1)!}$$

and the reliability function of the system is the incomplete gamma function.

b. Stochastic Model for Wear Out with Random Damage Rate

Suppose now that both the time intervals between wear outs and the amount of wear out are random variables. For $i = 1, 2, \ldots$ and $t \geq 0$ (Figure 3.6), let

T_i = time interval between the two successive wear outs $i - 1$ and i

W_i = waiting time till the ith wear out; thus $W_i = T_1 + T_2 + \cdots + T_i$

Z_i = amount of the ith wear out

$N(t)$ = total number of wear outs over the interval $(0, t]$, $N(t) = 0, 1, 2, \ldots$

$Y(t)$ = cumulative wear out up to time t

Assume that $\{T_i\}$ and $\{Z_i\}$ form each a sequence of independently and identically distributed random variables with respective PDF's $f_T(t)$ and $\varphi_Z(z)$. $\{T_i\}$ and $\{Z_i\}$ are ordinary renewal processes. We further assume that $\{T_i\}$ and $\{Z_i\}$ are mutually independent. Our objective is to obtain an expression for the reliability function in terms of the two functions $f_T(t)$ and $\varphi_Z(z)$ assumed known, and the initial strength Y_0 of the system. It is evident that

$$Y(t) = 0, \quad \text{if } N(t) = 0$$

and

$$Y(t) = X_1 + X_2 + \cdots + X_{N(t)}, \quad \text{if } N(t) = 1, 2, \ldots$$

Denote by $F_{Y(t)}(y)$, $0 \leq y < \infty$, the distribution function of $Y(t)$; then $F_{Y(t)}(y)$ can be recognized as a mixed distribution (Section 1.4-vi-b). We can write

$$F_{Y(t)}(0) = P\{N(t) = 0\}, \quad y = 0$$

$$dF_{Y(t)}(y) = \sum_{n=1}^{\infty} \varphi_Z *^{(n)}(y) \cdot P\{N(t) = n\}\,dy, \quad y > 0$$

From equation (2.42), we have

$$P\{N(t) = 0\} = 1 - P\{W_1 \le t\} = 1 - \int_0^t f_{W_1}(\theta)\, d\theta$$

$$P\{N(t) = n\} = P\{W_n \le t\} - P\{W_{n+1} \le t\}$$

$$= \int_0^t f_{W_n}(\theta)\, d\theta - \int_0^t f_{W_{n+1}}(\theta)\, d\theta$$

$$= \int_0^t f_T *^{(n)}(\theta)\, d\theta - \int_0^t f_T *^{(n+1)}(\theta)\, d\theta, \quad n = 1, 2, \ldots$$

Substituting in the previous expression for $dF_{Y(t)}(y)$, we obtain, for $y > 0$,

$$dF_{Y(t)}(y) = \sum_{n=1}^{\infty} \varphi_Z *^{(n)}(y)\left[\int_0^t f_T *^{(n)}(\theta)\, d\theta - \int_0^t f_T *^{(n+1)}(\theta)\, d\theta\right] dy$$

Failure due to wear out will occur if and only if the cumulative wear out or strength decrement $Y(t)$ is greater than or equal to Y_0. If X denotes the time to failure of the system, then

$$P\{X \le t\} = P\{Y(t) \ge Y_0\}$$

The reliability function of the system is thus

$$R_X(t) = P\{X > t\} = P\{Y(t) < Y_0\}$$

$$= F_{Y(t)}(0) + \int_0^{Y_0} dF_{Y(t)}(y)$$

Now $\qquad F_{Y(t)}(0) = P\{N(t) = 0\} = 1 - P\{W_1 \le t\}$

$$= 1 - \int_0^t f_T(\theta)\, d\theta$$

Hence the explicit expression for the reliability function for $t \ge 0$ is

$$R_X(t) = 1 - \int_0^t f_T(\theta)\, d\theta$$

$$+ \int_0^{Y_0} \sum_{n=1}^{\infty} \varphi_Z *^{(n)}(y)\left[\int_0^t f_T *^{(n)}(\theta)\, d\theta - \int_0^t f_T *^{(n+1)}(\theta)\, d\theta\right] dy$$

An expression for the mean time to failure can be obtained by forming first the Laplace transform of $R_X(t)$ with respect to t. Using the conventional notation $\tilde{g}(s)$ to denote the Laplace transform of the function $g(t)$, $t \ge 0$ (see Appendix), we obtain

$$\tilde{R}_X(s) = \frac{1}{s} - \frac{1}{s}\tilde{f}_T(s) + \int_0^{Y_0} \sum_{n=1}^{\infty} \varphi_Z *^{(n)}(y)\left[\frac{1}{s}\tilde{f}_T *^{(n)}(s) - \frac{1}{s}\tilde{f}_T *^{(n+1)}(s)\right] dy$$

Remembering that the Laplace transform of the nth-fold convolution of a function is the nth power of the convolution of the function, it follows that

$$\bar{R}_X(s) = \frac{1}{s} - \frac{1}{s}\tilde{f}_T(s) + \frac{1}{s}\int_0^{Y_0}\sum_{n=1}^{\infty}\varphi_Z*^{(n)}(y)\{[\tilde{f}_T*(s)]^n - [\tilde{f}_T*(s)]^{n+1}\}\,dy$$

$$= \frac{1-\tilde{f}_T(s)}{s} + \frac{1-\tilde{f}_T(s)}{s}\int_0^{Y_0}\sum_{n=1}^{\infty}\varphi_Z*^{(n)}(y)[\tilde{f}_T*(s)]^n\,dy$$

$$= \frac{1-\tilde{f}_T(s)}{s}\left\{1 + \int_0^{Y_0}\sum_{n=1}^{\infty}\varphi_Z*^{(n)}(y)[\tilde{f}_T*(s)]^n\,dy\right\}$$

The mean time to failure of the system is

$$E[X] = \lim_{s\to 0}\bar{R}_X(s)$$

$$= \lim_{s\to 0}\left\{\frac{1-\tilde{f}_T(s)}{s}\right\}\cdot\lim_{s\to 0}\left\{1 + \int_0^{Y_0}\sum_{n=1}^{\infty}\varphi_Z*^{(n)}(y)[\tilde{f}_T*(s)]^n\,dy\right\}$$

Applying l'Hospital's rule once to the first limit on the right-hand side, we obtain

$$E[X] = \left[\lim_{s\to 0} - \tilde{f}_T'(s)\right]\left[1 + \int_0^{Y_0}\sum_{n=1}^{\infty}\varphi_Z*^{(n)}(y)\,dy\right]$$

$$= E[T]\left[1 + \int_0^{Y_0}\sum_{n=1}^{\infty}\varphi_Z*^{(n)}(y)\,dy\right]$$

In the ordinary renewal process $\{Z_i\}$, the renewal function $M(Y_0)$ satisfies the integral equation (see Section 2.8-iii)

$$M(Y_0) = \Phi_Z(Y_0) + \int_0^{Y_0}M(Y_0 - x)\,d\Phi_Z(x)$$

and has for solution the quantity

$$M(Y_0) = \sum_{n=1}^{\infty}\int_0^{Y_0}\varphi_Z*^{(n)}(y)\,dy$$

Therefore, the mean time to failure of the system is

$$E[X] = E[T][1 + M(Y_0)]$$

Also, for large values of Y_0, we have (Section 2.8-iv)

$$M(Y_0) \approx \frac{Y_0}{E[Z]}$$

Thus, for large Y_0, we have approximately

$$E[X] \approx E[T]\left(1 + \frac{Y_0}{E[Z]}\right)$$

In the special case when $\varphi_Z(y) = \mu e^{-\mu y}$, $0 < y < \infty$, $M(Y_0) = Y_0/\mu$ and

$$E[X] = E[T]\left(1 + \frac{Y_0}{\mu}\right)$$

Other models for wear out can be found in Patterson's work [10].

iii. *Stochastic Models for Mixed-Type Failures*

Mixed-type failures result from the combined presence of exogenous stresses and endogenous wear out. A realization of the phenomenon could simply be represented by superimposing Figures 3.5 and 3.6. We shall assume that the exogenous events occur independently from the endogenous events and that failure occurs at a time when stress is greater than or equal to strength.

The characterization of this mixed-type form of failure necessitates a knowledge of

(1) The type of environment the system is operating under.
(2) The frequency and magnitude of the stresses imposed upon the system.
(3) The wear out process.
(4) The initial strength Y_0 of the system.

As an illustration, consider the situation in Section 3.4-i-a where an operating system is subject to a single environment generating peak events. Let

$\lambda(t)\, dt + o(dt) =$ probability of a peak occurring in the time interval $(t,\, t + dt]$

$p(t) =$ probability that the system will fail at time t, given that a peak has occurred at time t

Assume now that the magnitude of a peak occurring at time t is a random variable $U(t)$ with PDF $g_{U(t)}(x)$, $x > 0$. In addition, the system is subject to wear out such that $Y(t)$ represents the cumulative strength decrement at time t. The reliability of the system as given in Section 3.4-i-a is

$$R_X(t) = e^{-\int_0^t p(x)\lambda(x)\, dx}, \quad t \geq 0$$

The expression for $p(t)$ is simply obtained by noting that

$$\begin{aligned} p(t) &= P\{\text{stress at } t \geq \text{strength at } t\} \\ &= P\{U(t) \geq Y_0 - Y(t)\} \\ &= 1 - P\{U(t) + Y(t) < Y_0\} \end{aligned}$$

Since the two random variables $U(t)$ and $Y(t)$ are independent, it follows that

$$p(t) = 1 - \int_0^{Y_0} g_{U(t)}(y) * dF_{Y(t)}(y)$$

The appropriate expression for $F_{Y(t)}(y)$ will be dictated by the analysis of the wear-out process as performed, for example, in Section 3.4-ii-a. A final expression for the system reliability function can thus be obtained.

The importance of a mathematical analysis of the failure phenomena of a system stems from the fact that it is possible to obtain theoretical explicit expressions for the reliability function of the system in terms of the various complex interacting factors that are conducive to failure. This, then, enables one to measure the significance of each factor upon the system reliability, and ultimately, at the design level, to develop procedures for improving such reliability.

3.5. Determination of the Failure Characteristics of a System, Given the Failure Characteristics of Its Components Serial Systems

In its simplest design form, a complex system can be considered as made up of a number of components whose simultaneous satisfactory performance determines the satisfactory performance of the entire system. Thus the survival of the system hinges upon the joint survival of all its components; in other words, the system fails if and only if at least one of its components fails. Such a system is said to have a *serial configuration*, or the system is said to be a *serial system* or *series system*, or to have n, $(n = 1, 2, \dots)$, *components in series*. A serial system consisting of n components is schematically represented as in Figure 3.7.

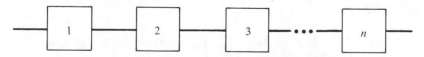

Figure 3.7. Serial system with n components.

Suppose that a serial system consisting of exactly n components starts operating at some time origin. Let $X_1, X_2, \dots, X_n$ be random variables denoting, respectively, the time to failure of components $1, 2, \dots, n$. Denote, respectively, by $f_{X_1, X_2, \dots, X_n}(x_1, x_2, \dots, x_n)$ and $F_{X_1, X_2, \dots, X_n}(x_1, x_2, \dots, x_n)$ the joint PDF and joint distribution function of $X_1, X_2, \dots, X_n$. It is required to obtain an expression for the reliability function of the system. Let

Y denote the time to failure of the system. Since, by definition, the system fails at the time when the first failure among the n components occurs, then

$$Y = \min \{X_1, X_2, \ldots, X_n\}$$

The distribution function of Y would then be given by (see Chapter 1, Problem 11)

$$F_Y(y) = \int \cdots \int\int_{0 \le \min\{X_1, X_2, \ldots, X_n\} \le y} f_{X_1, X_2, \ldots, X_n}(x_1, x_2, \ldots, x_n) \, dx_1 \, dx_2 \ldots dx_n$$

$$= 1 - \int_y^\infty \cdots \int_y^\infty \int_y^\infty f_{X_1, X_2, \ldots, X_n}(x_1, x_2, \ldots, x_n) \, dx_1 \, dx_2 \ldots dx_n$$

Hence the expression for the reliability function $R_Y(t)$ of the system is

$$R_Y(t) = \int_y^\infty \cdots \int_y^\infty \int_y^\infty f_{X_1, X_2, \ldots, X_n}(x_1, x_2, \ldots, x_n) \, dx_1 \, dx_2 \ldots dx_n$$

This constitutes the most general expression for the reliability function of a serial system with n components, given in terms of the joint density function of the time to failure of the components. Theoretically, then, expressions for the failure density function, the hazard function, and the mean time to failure of the system could be derived; unfortunately, closed-form expressions cannot be obtained unless particular cases are investigated.

An important special case arises if the survival and failure of a particular component in the system do not depend on the operation of any of the other components. The random variables $X_1, X_2, \ldots, X_n$ are then mutually independent, and the system is known as an *independent serial system*.

For $i = 1, 2, \ldots, n$, let $f_{X_i}(x_i)$ and $F_{X_i}(x_i)$ denote, respectively, the marginal PDF and CDF of the random variable X_i. Let $R_{X_i}(t)$ denote the reliability function of component i, $i = 1, 2, \ldots, n$. Then, for an independent serial system, the expressions for the life distribution and reliability function become

$$F_Y(y) = 1 - \left[\int_y^\infty f_{X_1}(x_1) \, dx_1 \right] \left[\int_y^\infty f_{X_2}(x_2) \, dx_2 \right] \cdots \left[\int_y^\infty f_{X_n}(x_n) \, dx_n \right]$$

$$= 1 - \prod_{i=1}^{i=n} [1 - F_{X_i}(y)], \quad y \ge 0$$

$$R_Y(t) = \prod_{i=1}^{i=n} [1 - F_{X_i}(t)]$$

$$= \prod_{i=1}^{i=n} R_{X_i}(t), \quad t \ge 0$$

Thus *the reliability function of an independent serial system is the product of the reliability functions of its components.*

To obtain an expression for the hazard function of an independent serial system, we first note that an expression for the failure density function is

$$f_Y(y) = \frac{dF_Y(y)}{dy} = \left\{ \prod_{i=1}^{i=n} [1 - F_{X_i}(y)] \right\} \sum_{i=1}^{i=n} \frac{f_{X_i}(y)}{1 - F_{X_i}(y)}$$

It immediately follows that the hazard function of the system is

$$h_Y(t) = \frac{f_Y(t)}{R_Y(t)} = \sum_{i=1}^{i=n} \frac{f_{X_i}(t)}{1 - F_{X_i}(t)}$$

Since the hazard function of the ith component is

$$h_{X_i}(t) = \frac{f_{X_i}(t)}{1 - F_{X_i}(t)}$$

it follows that

$$h_Y(t) = \sum_{i=1}^{i=n} h_{X_i}(t)$$

Hence *the hazard function of an independent serial system is the sum of the hazard functions of its components.*

Example 3.2

An independent serial system consists of n components each with an exponential failure density function. Let λ_i be the hazard function of component i, $i = 1, 2, \ldots, n$. Show that the system has an exponential failure density function with hazard function $\sum_{i=1}^{i=n} \lambda_i$.

The reliability function of component i being $R_{X_i}(t) = e^{-\lambda_i t}$, it follows that the system reliability is

$$R_Y(t) = \prod_{i=1}^{i=n} e^{-\lambda_i t} = \exp\left[-\sum_{i=1}^{i=n} \lambda_i \right] t, \quad t \geq 0$$

and the result follows immediately.

3.6. Methods for Improving the Reliability of a System

There are four basic methods for improving the reliability of a system:

(i) By decreasing the failure intensity of the system through improved technology at the component or system level.

(ii) By introducing redundancy at the component or system level.

(iii) By supplementing redundancy with repair or replacement.
(iv) By improving the overall design of the system to achieve maximum reliability within given limitations.

i. *Technological Considerations for Decreasing the Failure Intensity of a System*

The following methods are cited for achieving improved system reliability by decreasing the failure intensity of the system through technological consideration.

(1) Simplification of the system.
(2) Selection of the most reliable components.
(3) Lightening the operating conditions of the system.
(4) Rejection of components of low reliability.
(5) Standardization of components.
(6) Improvement of production technology.
(7) Statistical control of component quality.
(8) Elimination of initial-type failures using break-ins.
(9) Minimization of wear-out-type failures through an effective preventive-maintenance program.
(10) Reducing the hazard contribution of human errors by proper automation and by increasing the skill of the servicing personnel.

For additional discussion, the reader is referred to Polovko [11].

ii. *Introduction of Redundancy*

A failed component could be repaired or replaced by a new component. Repair, if possible, is assumed to bring back a failed component into a state of "as good as new." Conceptually, therefore, the process of repair and that of replacement are functionally equivalent.

It is evident that if in a system failed components are *immediately* replaced or repaired, their reliability and that of the system would be maintained at unity at all times. In practice, however, the time factor is an important element to consider: the physical process of removing and replacing a failed component, or repairing it, will invariably cause a disruption in the operation of the system.

One possible way for improving the reliability of a system is to introduce alternative components to help the system operate successfully in case of failure of one or more components. The term *redundancy* is used to describe the property of systems having this improved reliability capability. A *redundant system* is one that incorporates additional components, known as *redundancies*, whose purpose is nonfunctional from a conventional engineering design viewpoint, but whose presence increases the chances of successful operation of the system.

There exist two basic types of redundancies that have received considerable attention both in theory and in practice; these are the *parallel redundancy* and the *standby redundancy*. Roughly speaking, a parallel redundancy starts its operation at the time when the system begins to operate. A standby redundancy starts its operation in case of failure only, by taking over instantaneously the function performed by the failed component; a standby redundancy is thus equivalent to an instantaneous repair or replacement process. In a standby redundant system the backup components may deteriorate while in storage, in which case one has a *warm* standby system, or they may not, in which case one has a *cold* standby system. In practice, it is also possible that the backup elements may fail while idle. All these cases have been investigated in the literature. We shall only consider the case of a (cold) standby system in which the backup components do not deteriorate or fail while idle. The practical implementation of a system operating with standby redundancy requires that the operation of the components be continuously monitored, that failure be immediately detected, and that switching over into a backup component occurs instantaneously. These requirements are of necessity if disruption in the operation of the system is to be avoided.

We shall now develop theoretical models of reliability for parallel systems and (cold) standby systems involving no repair and/or replacement of failed components.

a. Parallel System—No Repair or Replacement
   of Failed Components

Consider a system consisting of exactly n components, $n = 1, 2, \ldots$. All n components start operating simultaneously. A component is not repaired or replaced when it fails and the system is operating when at least one component is operating; that is, the system fails if and only if all n components fail. Such a system is said to have a *parallel configuration* or to be a *parallel system*, or the components of the system are said to be in *parallel*, or the system is said to have $n - 1$ *parallel redundancies*. In general, the components need not possess the same time-to-failure characteristics, although they must be functionally equivalent. A parallel system with n components is schematically represented as in Figure 3.8.

Consider a parallel system consisting of exactly n components, which start operating simultaneously at some time origin. Let $X_1, X_2, \ldots, X_n$ be random variables denoting, respectively, the time to failure of components $1, 2, \ldots, n$. Denote, respectively, by $f_{X_1, X_2, \ldots, X_n}(x_1, x_2, \ldots, x_n)$ and $F_{X_1, X_2, \ldots, X_n}(x_1, x_2, \ldots, x_n)$ the joint PDF and joint distribution function of $X_1, X_2, \ldots, X_n$. Let Y denote the time to failure of the system. Since the system fails at the time when the last failure among the n components occurs,

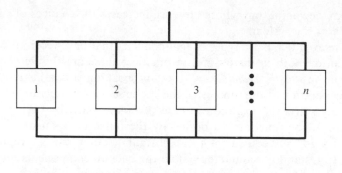

Figure 3.8. Parallel system with n components.

it follows that

$$Y = \max \{X_1, X_2, \ldots, X_n\}$$

From Chapter 1, Problem 12, the distribution function of Y is $(y \geq 0)$

$$F_Y(y) = \int \cdots \int\int_{0 \leq \max\{X_1, X_2, \ldots, X_n\} \leq y} f_{X_1, X_2, \ldots, X_n}(x_1, x_2, \ldots, x_n) \, dx_1 \, dx_2 \ldots dx_n$$

$$= \int_0^y \cdots \int_0^y \int_0^y f_{X_1, X_2, \ldots, X_n}(x_1, x_2, \ldots, x_n) \, dx_1 \, dx_2 \ldots dx_n$$

The reliability function of the system is then $(t \geq 0)$

$$R_Y(t) = 1 - \int_0^t \cdots \int_0^t \int_0^t f_{X_1, X_2, \ldots, X_n}(x_1, x_2, \ldots, x_n) \, dx_1 \, dx_2 \ldots dx_n$$

We consider now the special case of an *independent parallel system* in which the operation of a particular component does not depend on the operation of the other components. The random variables $X_1, X_2, \ldots, X_n$ are then mutually independent. Let $f_{X_i}(x_i)$ be the marginal PDF of X_i and $F_{X_i}(x_i)$ be its marginal CDF. Let again $R_{X_i}(t)$ be the reliability function of component i, $i = 1, 2, \ldots, n$. The expression for the distribution function of the system becomes

$$F_Y(y) = \left[\int_0^y f_{X_1}(x_1) \, dx_1 \right] \left[\int_0^y f_{X_2}(x_2) \, dx_2 \right] \cdots \left[\int_0^y f_{X_n}(x_n) \, dx_n \right]$$

$$= \prod_{i=1}^{i=n} F_{X_i}(y), \quad y \geq 0$$

Hence, *in an independent parallel system, the life distribution function of the system is the product of the life distribution functions of its components.*

The system reliability function is

$$R_Y(t) = 1 - \prod_{i=1}^{i=n} F_{X_i}(t) = 1 - \prod_{i=1}^{i=n} [1 - R_{X_i}(t)], \quad t \geq 0$$

By differentiating $F_Y(y)$ with respect to y, we obtain, for the failure density function,

$$f_Y(y) = \left[\prod_{i=1}^{i=n} F_{X_i}(y) \right] \sum_{i=1}^{i=n} \frac{f_{X_i}(y)}{F_{X_i}(y)}$$

The reader can verify that an expression relating the hazard function of the system to the hazard function of the components is

$$e^{-\int_0^t h_Y(x)\,dx} = 1 - \prod_{i=1}^{i=n} [1 - e^{-\int_0^t h_{X_i}(x)\,dx}]$$

When all n components are identical, then dropping the subscript X_i from $f_{X_i}(\cdot)$, $F_{X_i}(\cdot)$, and $R_{X_i}(\cdot)$, $i = 1, 2, \ldots, n$, we obtain

$$F_Y(y) = [F(y)]^n$$
$$R_Y(t) = 1 - [1 - R(t)]^n$$
$$f_Y(y) = nf(y)[F(y)]^{n-1}$$

Example 3.3

An independent parallel system consists of n identical components having an exponential failure density with rate λ. Find the reliability function of the system; hence show that its mean time to failure is

$$\frac{1}{\lambda} \sum_{i=1}^{i=n} \frac{1}{i}$$

The reliability function of the system is

$$R_Y(t) = 1 - (1 - e^{-\lambda t})^n, \quad t \geq 0$$

The plot of $R_Y(t)$ as a function of λt for $n = 0, 1, 2, 3$ is shown in Figure 3.9. The mean time to failure of the system is

$$E[Y] = \int_0^\infty R_Y(t)\,dt = \int_0^\infty [1 - (1 - e^{-\lambda t})^n]\,dt$$

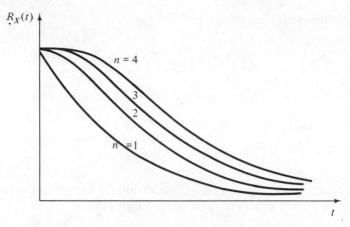

$R_X(t)$

$n = 4$

3

2

$n = 1$

t

Figure 3.9. The reliability function for independent parallel system with $n = 1, 2, 3, 4$ identical components.

Making the change in variable $Z = 1 - e^{-\lambda t}$ in the last integral, we obtain

$$E[Y] = \frac{1}{\lambda} \int_0^1 \frac{1 - Z^n}{1 - Z} \, dZ$$

$$= \frac{1}{\lambda} \int_0^1 (1 + Z + Z^2 + \cdots + Z^{n-1}) \, dZ$$

$$= \frac{1}{\lambda} \sum_{i=1}^{i=n} \frac{Z^i}{i} \Big|_0^1$$

$$= \frac{1}{\lambda} \sum_{i=1}^{i=n} \frac{1}{i} = \frac{1}{\lambda} \left(1 + \frac{1}{2} + \cdots + \frac{1}{n} \right)$$

For large n, an approximate expression for $E[Y]$ is

$$E[Y] \approx \frac{1}{\lambda} \ln n$$

The plot of $\lambda E[Y]$ as a function of n is shown in Figure 3.10.

Example 3.4

For a system with independent parallel configuration consisting of n components, $n = 1, 2, \ldots$, it may be of interest to determine at any particular time of operation the probability $P(r, t)$ that exactly r components, $r = 0, 1, 2, \ldots, n$, are operating ($t \geq 0$). Consider for simplicity the case when all components have identical hazard rate function $h(t)$ and assume that initially, at time $t = 0$, all components are operating. Find $P(r, t)$, and determine the reliability function of the system.

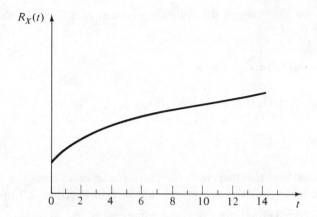

Figure 3.10. The dependence of mean time-to-failure of an independent parallel system as a function of the number of its identical components.

The initial conditions for this problem are

$$P(n, 0) = 1, \quad r = n, \, t = 0$$
$$P(r, 0) = 0, \quad 0 \leq r \leq n - 1, \, t = 0$$

For $r = n$ and $t \geq 0$,

$$P(n, t + dt) = P(n, t)[1 - nh(t) \, dt] + o(dt)$$

For $r = 1, 2, \ldots, n - 1$ and $t \geq 0$,

$$P(r, t + dt) = P(r, t)[1 - rh(t) \, dt] + P(r + 1, t)[(r + 1)h(t) \, dt] + o(dt)$$

and for $r = 0$ and $t \geq 0$,

$$P(0, t + dt) = P(0, t) + P(1, t)h(t) \, dt + o(dt)$$

These three equations may be written as, $t \geq 0$,

$$\frac{P(n, t + dt) - P(n, t)}{dt} = -nh(t)P(n, t) + \frac{o(dt)}{dt}, \quad r = n$$

$$\frac{P(r, t + dt) - P(r, t)}{dt} = -rh(t)P(r, t) + (r + 1)h(t)P(r + 1, t) + \frac{o(dt)}{dt},$$
$$1 \leq r < n - 1$$

$$\frac{P(0, t + dt) - P(0, t)}{dt} = h(t)P(1, t) + \frac{o(dt)}{dt}, \quad r = 0$$

Let $dt \rightarrow 0$; we then obtain the following system of differential difference equations for $t \geq 0$:

$$\frac{dP(n, t)}{dt} = -nh(t)P(n, t), \quad r = n$$

$$\frac{dP(r, t)}{dt} = -rh(t)P(r, t) + (r + 1)h(t)P(r + 1, t), \quad 1 \leq r \leq n - 1$$

$$\frac{dP(0, t)}{dt} = h(t)P(1, t), \quad r = 0$$

These equations together with the given initial conditions are those of a nonhomogeneous death process with linear rate (Section 2.7-iii-b). Let $\tilde{H}(t) = \int_0^t h(\theta) \, d\theta$. The solution of these equations is

$$P(r, t) = \binom{n}{r} [e^{-\tilde{H}(t)}]^r [1 - e^{-\tilde{H}(t)}]^{n-r}, \quad r = 0, 1, 2, \ldots, n$$

The reliability function of the system is $R_Y(t) = 1 - P(0, t) = 1 - [1 - e^{-\tilde{H}(t)}]^n$.

Example 3.5

Consider a system consisting of n components, $n = 1, 2, 3, \ldots$. Let X_i be the time to failure of the ith component, $i = 1, 2, \ldots, n$, and let $R_{X_i}(t)$ and $E[X_i]$ be, respectively, its reliability function and its mean time to failure. Let Y be the time to failure of the system and $R_Y(t)$ its reliability function.

(1) If the system has an independent serial configuration, show that

$$0 \leq E[Y] \leq \min_i \{E[X_i]\}$$

(2) If the system has an independent parallel configuration, show that

$$\max_i \{E[X_i]\} \leq E[Y] \leq \sum_{i=1}^{i=n} E[X_i]$$

1. For an independent serial system

$$E[Y] = \int_0^\infty R_Y(t) \, dt = \int_0^\infty \left\{ \prod_{i=1}^{i=n} R_{X_i}(t) \right\} dt$$

Since for all $i = 1, 2, \ldots, n$ and all $t \geq 0$,

$$0 \leq R_{X_i}(t) \leq 1$$

Then
$$\prod_{i=1}^{i=n} R_{X_i}(t) \leq R_{X_i}(t)$$

It follows that, for all $i = 1, 2, \ldots, n$,

$$\int_0^\infty \left\{ \prod_{i=1}^{i=n} R_{X_i}(t) \right\} dt \leq \int_0^\infty R_{X_i}(t) \, dt$$

and in particular

$$\int_0^\infty \left\{ \prod_{i=1}^{i=n} R_{X_i}(t) \right\} dt \leq \min_i \left\{ \int_0^\infty R_{X_i}(t) \, dt \right\}$$

Therefore, for an independent serial system

$$0 \leq E[Y] \leq \min_i \{E[X_i]\}$$

In the special case of identical components, where $E[X]$ is the mean time to failure of all components, we have

$$0 \leq E[Y] \leq E[X]$$

2. For an independent parallel system

$$E[Y] = \int_0^\infty R_Y(t) \, dt = \int_0^\infty \left\{ 1 - \prod_{i=1}^{i=n} [1 - R_{X_i}(t)] \right\} dt$$

Since for all $i = 1, 2, \ldots, n$ and all $t \geq 0$,

$$0 \leq 1 - R_{X_i}(t) \leq 1$$

then

$$1 - \prod_{i=1}^{i=n} [1 - R_{X_i}(t)] \geq 1 - [1 - R_{X_i}(t)]$$

It follows that, for all $i = 1, 2, \ldots, n$,

$$\int_0^\infty \left\{ 1 - \prod_{i=1}^{i=n} [1 - R_{X_i}(t)] \right\} dt \geq \int_0^\infty R_{X_i}(t) \, dt$$

and in particular

$$\int_0^\infty \left\{ 1 - \prod_{i=1}^{i=n} [1 - R_{X_i}(t)] \right\} dt \geq \max_i \left\{ \int_0^\infty R_{X_i}(t) \, dt \right\}$$

Hence $$E[Y] \geq \max_i E[X_i]$$

To demonstrate that $E[Y] \leq \sum_{i=1}^{i=n} E[X_i]$, we first show that, for all $t \geq 0$,

$$1 - \prod_{i=1}^{i=n} [1 - R_{X_i}(t)] \leq \sum_{i=1}^{i=n} R_X(t)$$

The proof is by induction. For all $t \geq 0$,

$$1 - [1 - R_{X_1}(t)][1 - R_{X_2}(t)] = 1 - [1 - R_{X_1}(t) - R_{X_2}(t) + R_{X_1}(t)R_{X_2}(t)]$$
$$= R_{X_1}(t) + R_{X_2}(t) - R_{X_1}(t)R_{X_2}(t)$$
$$\leq R_{X_1}(t) + R_{X_2}(t)$$

Suppose now that, for $t \geq 0$,

$$1 - \prod_{i=1}^{i=n-1} [1 - R_{X_i}(t)] \leq \sum_{i=1}^{i=n-1} R_{X_i}(t)$$

Then
$$1 - \sum_{i=1}^{i=n-1} R_{X_i}(t) \leq \prod_{i=1}^{i=n-1} [1 - R_{X_i}(t)]$$

Multiplying both sides by $[1 - R_{X_n}(t)]$, we obtain

$$\left[1 - \sum_{i=1}^{i=n-1} R_{X_i}(t) \right][1 - R_{X_n}(t)] \leq \prod_{i=1}^{i=n} [1 - R_{X_i}(t)]$$

or
$$1 - \sum_{i=1}^{i=n} R_{X_i}(t) + R_{X_n}(t) \sum_{i=1}^{i=n-1} R_{X_i}(t) \geq \prod_{i=1}^{i=n} [1 - R_{X_i}(t)]$$

Transposing, we obtain

$$1 - \prod_{i=1}^{i=n} [1 - R_{X_i}(t)] \leq \sum_{i=1}^{i=n} R_{X_i}(t) - R_{X_n}(t) \sum_{i=1}^{i=n-1} R_{X_i}(t)$$

$$\leq \sum_{i=1}^{i=n} R_{X_i}(t), \quad \text{for all } t \geq 0$$

Hence
$$\int_0^\infty \left\{ 1 - \prod_{i=1}^{i=n} [1 - R_{X_i}(t)] \right\} dt \leq \int_0^\infty \left\{ \sum_{i=1}^{i=n} R_{X_i}(t) \right\} dt$$

or
$$\int_0^\infty R_Y(t)\, dt \leq \sum_{i=1}^{i=n} \left\{ \int_0^\infty R_{X_i}(t)\, dt \right\}$$

Hence
$$E[Y] \leq \sum_{i=1}^{i=n} E[X_i]$$

Combining the two results, we obtain, for a system having an independent parallel configuration with n components,

$$\max_i \{E[X_i]\} \leq E[Y] \leq \sum_{i=1}^{i=n} E[X_i]$$

When all components are identical with mean time to failure equal to $E[X]$, we have

$$E[X] \leq E[Y] \leq nE[X]$$

The inequalities obtained for independent serial and parallel system provide an upper and lower bound for the mean time to failure of the system in terms of the mean time to failure of its individual components. These inequalities are particularly useful since explicit expressions relating $E[Y]$ to $E[X_i]$, $i = 1, 2, \ldots, n$ cannot be obtained unless the functional form of the failure density function of each component is known.

b. STANDBY SYSTEMS—NO REPAIR OR REPLACEMENT OF FAILED COMPONENTS

Consider a system consisting of exactly n components, $n = 1, 2, \ldots$. The first component starts operating at time origin, while all other components stand by idle; as soon as the first component fails, the second component takes over immediately and starts operating while the remaining $n - 2$ components stand by idle, and so on until all components are sequentially used up. The components are functionally equivalent; that is, they perform identical tasks and a failed component is not repaired or replaced when it fails. The system is considered to be operative when a component is still operative; the system fails when the last component fails. Such a system is said to have a *standby configuration* or to be a *standby system*, or the components of the system are said to be in *standby* or the system is said to have $n - 1$ *standby redundancies*. A standby system with n components is schematically represented as in Figure 3.11.

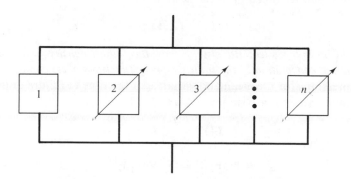

Figure 3.11. Standby system with n components.

Consider a standby system consisting of exactly n components. Let $X_1, X_2, \ldots, X_n$ be random variables denoting, respectively, the time to failure of components $1, 2, \ldots, n$. Denote, respectively, by $f_{X_1, X_2, \ldots, X_n}(x_1, x_2, \ldots, x_n)$ and $F_{X_1, X_2, \ldots, X_n}(x_1, x_2, \ldots, x_n)$ the joint PDF and joint distribution function of $X_1, X_2, \ldots, X_n$. Let Y denote the time to failure of the system. Since the system fails when the last standby component fails, it follows that

$$Y = X_1 + X_2 + \cdots + X_n$$

From this relation it is clear that the time to failure of the system will not be affected by the order of operation of the standby units. The random variable Y being the sum of n random variables, it follows that the distribution function of Y, $F_Y(y)$, is given by

$$F_Y(y) = \int \cdots \int_{0 \le x_1 + x_2 + \cdots + x_n \le y} f_{X_1, X_2, \ldots, X_n}(x_1, x_2, \ldots, x_n) \, dx_1 \, dx_2 \ldots dx_n$$

In many practical problems it is reasonable to assume that the random variables $X_1, X_2, \ldots, X_n$ are mutually independent, in which case we have an *independent standby system*. Let $f_{X_i}(x_i)$ and $F_{X_i}(x_i)$ be, respectively, the marginal PDF and marginal CDF of the random variable X_i, $i = 1, 2, , \ldots,$ n. Let also $R_{X_i}(t)$ be the reliability function of component i. Then, we have (Section 1.4-vi-a)

$$F_Y(y) = \int_0^y f_{X_1}(x) * f_{X_2}(x) * \cdots * f_{X_n}(x) \, dx, \quad y \ge 0$$

The system reliability function is

$$R_Y(t) = 1 - F_Y(t) = \int_t^\infty f_{X_1}(x) * f_{X_2}(x) * \cdots * f_{X_n}(x) \, dx, \quad t \ge 0$$

The failure density function of the system is

$$f_Y(y) = f_{X_1}(y) * f_{X_2}(y) * \cdots * f_{X_n}(y)$$

Hence, *in an independent standby system, the failure density function of the system is the convolution of the failure density functions of its components.*

Expressions for the mean and variance of the time to failure of independent standby systems are simply

$$E[Y] = \sum_{i=1}^{i=n} E[X_i]$$

$$\text{Var}\,[Y] = \sum_{i=1}^{i=n} \text{Var}\,[X_i]$$

When all n components are identical, we obtain by dropping the subscript i, $i = 1, 2, \ldots, n$ from the previous relations

$$F_Y(y) = \int_0^y f *^{(n)} (x) \, dx$$

$$R_Y(t) = \int_t^\infty f *^{(n)} (x) \, dx$$

$$f_Y(y) = f *^{(n)} (y)$$

$$E[Y] = nE[X]$$

$$\text{Var}\,[Y] = n \, \text{Var}\,[X]$$

Example 3.6

An independent standby system consists of n identical components having an exponential failure density function with rate λ. Find the reliability function of the system.

The failure density function of each component is

$$f(x) = \lambda e^{-\lambda x}, \quad \lambda > 0, x > 0$$

The failure density function of the system has therefore the gamma distribution

$$f_Y(y) = \frac{(\lambda y)^{n-1}}{\Gamma(n)} \lambda e^{-\lambda y}, \quad y \geq 0$$

The system reliability function is

$$R_Y(t) = \int_t^\infty \frac{(\lambda y)^{n-1}}{(n-1)!} \lambda e^{-\lambda y} \, dy$$

$$= \sum_{r=0}^{n-1} e^{-\lambda t} \frac{(\lambda t)^r}{r!}, \quad t \geq 0$$

which is the incomplete gamma function. The mean time to failure of the system may be found directly from this above relation.

Example 3.7

For a system consisting of n components, $n = 1, 2, \ldots$, show that a standby configuration is always more reliable than a parallel configuration.

Let $f_{X_1, X_2, \ldots, X_n}(x_1, x_2, \ldots, x_n)$ be the joint probability density function of the time to failure of the n components. For a parallel configuration, the reliability function $R_{Y_1}(t)$ of the system is

$$R_{Y_1}(t) = 1 - \int_0^t \cdots \int_0^t \int_0^t f_{X_1, X_2, \ldots, X_n}(x_1, x_2, \ldots, x_n) \, dx_1 \, dx_2 \ldots dx_n$$

For a standby configuration, the reliability function $R_{Y_2}(t)$ of the system is

$$R_{Y_2}(t) = 1 - \int \cdots \int\int_{0 \leq x_1 + x_2 + \cdots + x_n \leq t} f_{X_1, X_2, \ldots, X_n}(x_1, x_2, \ldots, x_n) \, dx_1 \, dx_2 \ldots dx_n$$

Comparing the region of integration in the two previous relations, it is evident that

$$R_{Y_2}(t) > R_{Y_1}(t), \quad \text{for all } t > 0$$

Thus a standby configuration is always more reliable than a parallel configuration. Note that independence is not assumed.

Example 3.8

Consider a system whose time-to-failure distribution is exponential with mean $1/\lambda = 1000$ hours. If one identical standby system is incorporated, determine the maximum number of hours of failure-free operation if a reliability level of .999 is to be maintained.

The reliability function of the total system is

$$R_Y(t) = e^{-\lambda t}(1 + \lambda t), \quad t \geq 0$$

To determine the maximum number of hours of failure-free operation, we need to solve the equation

$$e^{-\lambda t}(1 + \lambda t) = .999$$

or, if we let $\lambda t = T$, we can write

$$1 + T = .999 e^{T}$$

This equation may be solved graphically or numerically. We note, however, that for small T we obtain as a first approximation

$$1 + T \approx .999\left(1 + T + \frac{T^2}{2}\right)$$

The acceptable solution of this quadratic equation is $T = .0455$; hence the required value of t is approximately 45.5 hours.

iii. Supplementing Redundancy with Repair (or Replacement)

A natural question that arises when dealing with any redundant system is the possibility of still improving the system reliability by supplementing redundancy with repair (or replacement) of accessible failed components. The range of problems generated by this question has considerable practical significance, and it has merited special favor by theoretical investigators. A number of situations have been studied in the literature (e.g., see Ref. [13]) and solutions have been obtained for the time-to-failure distribution of particular systems. Among the factors to be considered in analyzing a redundant system with repair, the following should be identified in particular.

(1) The redundancy configuration: parallel, cold standby, warm standby, or mixed.

(2) The time-to-failure distribution of each component in the system: negative exponential or general.
(3) The distribution of the repair time for each component in the system: negative exponential, constant, or general.
(4) The size of the repair facility: one or more.
(5) The priority rule governing the repair.
(6) The significance of practical situations: imperfect failure detection, noninstantaneous switching time, others.

It is evident that whichever situation is considered, repair (or replacement) is feasible only if at least one component in the system is operative and that the system is considered to be in a failed state if all components are inoperative. Thus the time to failure of the system is the time elapsed from start of operation till the moment when all components are in a failed state or repair state.

We illustrate the type of analysis that may be involved by considering two special situations related to parallel and standby systems. We shall use the notation $M/M/m$ to refer to a system where (1) the time to failure of components are identically and independently distributed random variables each with a negative exponential distribution; (2) the repair time of failed components are identically and independently distributed random variables with a negative exponential distribution; and (3) the total number of repair facilities or the maximum number of failed components that can be repaired simultaneously is m, $m = 1, 2, \ldots$.

a. An $M/M/1$ Two-Unit Parallel System

We consider a parallel system consisting of two components such that the times to failure of the components are mutually independent random variables, each with a negative exponential distribution with mean $1/\lambda$. Repair time for each component has a negative exponential distribution with mean $1/\mu$. It is assumed that the times to failure of components and their repair times are mutually independent random variables. Initially, both components are operating until one of the components fails, at which time repair starts immediately, and so on. The system fails when both components are nonoperative, that is, failed or unrepaired. A typical realization of the process is given in Figure 3.12, which shows the number of operating components as a function of time.

The hazard rate function of each component is λ. We then have

$\lambda \, dt + o(dt) =$ probability that a particular component which is operating at time t will fail in the time interval $(t, t + dt]$
$\mu \, dt + o(dt) =$ probability that a failed component which is in repair state at time t will be operating in the time interval $(t, t + dt]$

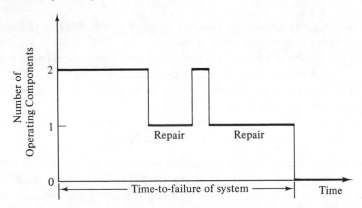

Figure 3.12. A sample function for a 2-unit parallel system with repair of failed components.

Let now $P(r, t)$ define the probability that exactly r components, $r = 0, 1, 2$, are *operating* at time t, $t \geq 0$. The initial conditions for this problem are

$$P(2, 0) = 1, \quad r = 2; t = 0$$
$$P(r, 0) = 0, \quad r = 0, 1; t = 0$$

For $t \geq 0$, we have the relations:

$$P(2, t + dt) = P(2, t)(1 - 2\lambda\, dt) + P(1, t)(1 - \lambda\, dt)\mu\, dt + o(dt)$$
$$P(1, t + dt) = P(2, t)(2\lambda\, dt)(1 - \mu\, dt) + P(1, t)(1 - \lambda\, dt)(1 - \mu\, dt) + o(dt)$$
$$P(0, t + dt) = P(1, t)\lambda\, dt(1 - \mu\, dt) + P(0, t) + o(dt)$$

These relations may be written as $(t \geq 0)$

$$\frac{P(2, t + dt) - P(2, t)}{dt} = -2\lambda P(2, t) + \mu P(1, t) + \frac{o(dt)}{dt}$$

$$\frac{P(1, t + dt) - P(1, t)}{dt} = 2\lambda P(2, t) - (\lambda + \mu)P(1, t) + \frac{o(dt)}{dt}$$

$$\frac{P(0, t + dt) - P(0, t)}{dt} = \lambda P(1, t) + \frac{o(dt)}{dt}$$

If we let $dt \to 0$, we obtain the following system of linear first-order differential equations with constant coefficients:

$$\frac{dP(2, t)}{dt} = -2\lambda P(2, t) + \mu P(1, t)$$

$$\frac{dP(1, t)}{dt} = 2\lambda P(2, t) - (\lambda + u)P(1, t)$$

$$\frac{dP(0, t)}{dt} = \lambda P(1, t)$$

subject to the given initial conditions. The solution to these equations will yield expressions for $P(0, t)$, $P(1, t)$, and $P(2, t)$ for any $t \geq 0$. Let

$$\bar{P}_r(s) = \int_0^\infty e^{-st} P(r, t)\, dt, \quad r = 0, 1, 2$$

define the Laplace transform of the function $P(r, t)$. Taking the Laplace transforms of the system of differential equations and using the initial conditions, we obtain (see Appendix)

$$s\bar{P}_2(s) - 1 = -2\lambda\bar{P}_2(s) + \mu\bar{P}_1(s)$$
$$s\bar{P}_1(s) = 2\lambda\bar{P}_2(s) - (\lambda + \mu)\bar{P}_1(s)$$
$$s\bar{P}_0(s) = \lambda\bar{P}_1(s)$$

These form a system of simultaneous linear algebraic equations, which may be written as

$$(s + 2\lambda)\bar{P}_2(s) - \mu\bar{P}_1(s) = 1$$
$$-2\lambda\bar{P}_2(s) + (s + \lambda + \mu)\bar{P}_1(s) = 0$$
$$-\lambda\bar{P}_1(s) + s\bar{P}_0(s) = 0$$

Each of the quantities $\bar{P}_0(s)$, $\bar{P}_1(s)$, and $\bar{P}_2(s)$ can thus be obtained explicitly. However, $\bar{P}_0(s)$ is the quantity of interest here. Noting that $\bar{P}_1(s)$ and $\bar{P}_2(s)$ can be solved using the first two equations only, we finally obtain for the solved value of $\bar{P}_0(s)$

$$\bar{P}_0(s) = \frac{2\lambda^2}{s[s^2 + (3\lambda + \mu)s + 2\lambda^2]}$$

Now $P_0(t)$ is the probability that no components are operating at time t, $t \geq 0$. If X is the time to failure of the system, then $P_0(t)$ is also the probability that the system has failed on or before t; thus it represents the distribution function of the random variable X,

$$P_0(t) = P\{X \leq t\} = F_X(t), \quad t \geq 0$$

The failure density function of the system is

$$f_X(t) = \frac{dP_0(t)}{dt}, \quad t > 0$$

whose Laplace transform is

$$\bar{f}_X(s) = s\bar{P}_0(s) = \frac{2\lambda^2}{s^2 + (3\lambda + \mu)s + 2\lambda^2}$$

The expression for $\bar{f}_X(s)$ is a rational function of s whose denominator has two negative real roots α_1 and α_2, where

$$\alpha_1 = \frac{-(3\lambda + \mu) + \sqrt{\lambda^2 + 6\lambda\mu + \mu^2}}{2}$$

and

$$\alpha_2 = \frac{-(3\lambda + \mu) - \sqrt{\lambda^2 + 6\lambda\mu + \mu^2}}{2}$$

Reducing the right-hand side of $\bar{f}_X(s)$ into partial fraction, we obtain

$$\bar{f}_X(s) = \frac{2\lambda^2}{\alpha_1 - \alpha_2}\left(\frac{1}{s - \alpha_1} - \frac{1}{s - \alpha_2}\right)$$

Inverting both sides yields

$$f_X(t) = \frac{2\lambda^2}{\alpha_1 - \alpha_2}(e^{\alpha_1 t} - e^{\alpha_2 t})$$

Substituting for the values of α_1 and α_2, one finally obtains

$$f_X(t) = \frac{4\lambda^2}{\sqrt{\lambda^2 + 6\lambda\mu + \mu^2}}e^{-(1/2)(3\lambda+\mu)t}\sinh\frac{\sqrt{\lambda^2 + 6\lambda\mu + \mu^2}}{2}t$$

That $f_X(t)$ is a proper density function can be verified directly from $\bar{f}_X(s)$ by noting that $\lim_{s\to 0}\bar{f}_X(s) = 1$. The mean time to failure of the system may be obtained from the relation

$$E[X] = -\lim_{s\to 0}\frac{d\bar{f}_X(s)}{ds}$$

$$= -\lim_{s\to 0}\frac{-2\lambda^2[2s + (3\lambda + \mu)]}{[s^2 + (3\lambda + \mu)s + 2\lambda^2]^2}$$

$$= \frac{2\lambda^2(3\lambda + \mu)}{4\lambda^4}$$

$$= \frac{3}{2\lambda} + \frac{\mu}{2\lambda^2}$$

In the study of a parallel system with no repair and consisting of n identical components, each with the constant hazard function λ, the mean time to failure of the system was found to be

$$\frac{1}{\lambda}\left(1 + \frac{1}{2} + \cdots + \frac{1}{n}\right)$$

Thus, for $n = 2$, the mean time to failure would be $3/2\lambda$. Consequently, it can be seen that the effect of repairing failed components with an average repair time of $1/\mu$ is to increase the mean time to failure of the system by $\mu/2\lambda^2$, or by a factor of

$$\frac{\mu/2\lambda^2}{3/2\lambda} = \frac{\mu}{3\lambda}$$

For example, if the mean time to failure of each component is equal to the mean repair time, the mean time to failure of the system would be increased by $\frac{1}{3}$.

b. AN $M/M/1$ TWO-UNIT STANDBY SYSTEM

This situation is similar to the previous problem except that one of the components, if operative, is in cold standby. Thus we assume that the idle component is not subject to deterioration or failure. The failure and repair characteristics of each component are the same as in Section 3.6-iii-a. Initially, both components are operative, and one of the components is operating while the other is in standby. When the operating unit fails, the standby unit takes over immediately, at which time repair starts on the failed component, and so on. The system fails when no components are operative, thus implying that one component failed and the other was in a repair state. Let $P(r, t)$ define the probability that exactly r components, $r = 0, 1, 2$, are *operative* (not operating as in Section 3.6-iii-a) at time t, $t \geq 0$. The initial conditions for the problem are

$$P(2, 0) = 1, \quad r = 2; t = 0$$
$$P(r, 0) = 0, \quad r = 0, 1; t = 0$$

For $t \geq 0$, we have

$$P(2, t + dt) = P(2, t)(1 - \lambda \, dt) + P(1, t)(1 - \lambda \, dt)\mu \, dt + o(dt)$$
$$P(1, t + dt) = P(2, t)\lambda \, dt(1 - \mu \, dt) + P(1, t)(1 - \lambda \, dt)(1 - \mu \, dt) + o(dt)$$
$$P(0, t + dt) = P(1, t)\lambda \, dt(1 - \mu \, dt) + P(0, t) + o(dt)$$

Transposing and passing to the limit, we obtain, for $t \geq 0$,

$$\frac{dP(2, t)}{dt} = -\lambda P(2, t) + \mu P(1, t)$$

$$\frac{dP(1, t)}{dt} = \lambda P(2, t) - (\lambda + \mu)P(1, t)$$

$$\frac{dP(0, t)}{dt} = \lambda P(1, t)$$

subject to the given initial conditions. This system of equations is identical to the one obtained in Section 3.6-iii-a except for the coefficients of $P(2, t)$, which are now λ's rather than 2λ's. Imbedding the initial conditions in the system of differential equations and taking the Laplace transforms, we obtain

$$(s + \lambda)\bar{P}_2(s) - \mu \bar{P}_1(s) = 1$$
$$-\lambda \bar{P}_2(s) + (s + \lambda + \mu)\bar{P}_1(s) = 0$$
$$-\lambda \bar{P}_1(s) + s\bar{P}_0(s) = 0$$

The solution for $\bar{P}_0(s)$ is

$$\bar{P}_0(s) = \frac{\lambda^2}{s[s^2 + (2\lambda + \mu)s + \lambda^2]}$$

The Laplace transform of the failure density function is

$$\bar{f}_X(s) = s\bar{P}_0(s) = \frac{\lambda^2}{s^2 + (2\lambda + \mu)s + \lambda^2}$$

The roots of the denominator of the right-hand side are

$$\alpha_1 = \frac{-(2\lambda + \mu) + \sqrt{4\lambda\mu + \mu^2}}{2}$$

$$\alpha_2 = \frac{-(2\lambda + \mu) - \sqrt{4\lambda\mu + \mu^2}}{2}$$

Thus the partial fraction expression for $\bar{f}_X(s)$ is

$$\bar{f}_X(s) = \frac{\lambda^2}{\alpha_1 - \alpha_2}\left(\frac{1}{s - \alpha_1} - \frac{1}{s - \alpha_2}\right)$$

Inversion yields

$$f_X(t) = \frac{\lambda^2}{\alpha_1 - \alpha_2}(e^{\alpha_1 t} - e^{\alpha_2 t})$$

$$= \frac{2\lambda^2}{\sqrt{4\lambda\mu + \mu^2}}e^{-[\lambda + (\mu/2)]t} \sinh\left(\frac{\sqrt{4\lambda\mu + \mu^2}}{2}\right)t$$

The mean time to failure of the system is

$$E[X] = -\lim_{s \to 0} \frac{d\bar{f}_X(s)}{ds}$$

$$= \frac{2\lambda + \mu}{\lambda^2} = \frac{2}{\lambda} + \frac{\mu}{\lambda^2}$$

The mean time to failure of the two-unit standby system with no repair is $2/\lambda$. Thus repair increases the mean time to failure of the system by μ/λ^2 time units, or by a factor of

$$\frac{\mu/\lambda^2}{2/\lambda} = \frac{\mu}{2\lambda}$$

For $\lambda = \mu$, this corresponds to a factor of 50%.

iv. *Design and Economic Considerations*

In general, the design of an operating system incorporating redundancies should consider the factors that invariably will be present in practice.

For example, a standby system requires the use of switching devices in each working and standby circuit. Each of the switching devices may be subject to failure; thus each component in the system may be considered to be in series with a switching device, with a resultant net decrease in the reliability of each standby unit. Since the operation of a parallel system does not require any added devices, a standby system may not, in practice, be always superior to a parallel system with the same number of elements. This, then, demands a reevaluation of the reliability of each configuration, and sometimes of mixed configuration (parallel and standby) to determine which configuration will yield the highest reliability for a given number of components.

Another problem that has received considerable attention by researchers is the design of a redundant system under given constraints. Theoretically, it is possible to increase the reliability of a given system to any level by incorporating the appropriate number of redundancies. In practice, one is often faced with the problem of limitations in the weight, size, or cost of the system. The problem then consists in the determination of the number of redundancies and their allocation to maximize the system reliability while maintaining the permissible values of weight, size, and cost. Consider, for example, an independent serial system consisting of N subsystems, N 1, 2, ... The ith subsystem, $i = 1, 2, \ldots, N$, consists of n_i components in parallel, $n_i = 1, 2, \ldots$. Assume that the system has to operate for a given number of hours. For the ith subsystem, let

R_i = reliability of each component
w_i = weight of each component

The system reliability is

$$\prod_{i=1}^{i=N} [1 - (1 - R_i)^{n_i}]$$

If the total weight of the system is not to exceed a given quantity W, then a problem of interest consists in determining the quantities $n_1, n_2, \ldots, n_N$ so as to maximize the reliability of the system. Mathematically, the problem can be formulated as

$$\max \prod_{i=1}^{i=N} [1 - (1 - R_i)^{n_i}]$$

subject to

$$\sum_{i=1}^{i=N} w_i n_i \leq W$$
$$n_i = 1, 2, \ldots, \quad i = 1, 2, \ldots, N$$

Problem 10 provides an alternative formulation involving a cost objective function subject to a reliability constraint. An equivalent problem with a cost constraint is presented in Chapter 8 of *Optimization Techniques in Operations Research* also by B. D. Sivazlian and L. E. Stanfel, Prentice-Hall,

Inc., Englewood Cliffs. N. J., 1975. The reader is also referred to Section 5.3-i of the present volume for the economic analysis and design of certain classes of replacement models in redundant type systems (see Problem 11).

SELECTED REFERENCES

[1] BARLOW, R. E., and F. PROSCHAN, *Mathematical Theory of Reliability,* John Wiley and Sons, Inc., New York, 1965.

[2] EPSTEIN, B., "The Exponential Distribution and Its Role in Life Testing," Technical Report 2, Department of Mathematics, Wayne State University, Detroit, Mich., 1958.

[3] FEDERWICZ, A. J., and M. MAZUMDAR, "Use of Geometric Programming to Maximize Reliability Achieved by Redundancy," *Operations Research*, Vol. 16, pp. 948–954, 1968.

[4] GHARE, P. M., and R. E. TAYLOR, "Optimal Redundancy for Reliability in Series System," *Operations Research*, Vol. 17, pp. 838–867, 1969.

[5] GUMBEL, E. J., *Statistics of Extremes*, Columbia University Press, New York, 1958.

[6] KAO, J. H. K., "Graphical Estimation of Mixed Weibull Parameters in Life Testing Electron Tubes," *Technometrics*, Vol. 1, No. 4, 1959.

[7] LLOYD, D. and M. LIPOW, *Reliability: Management, Methods and Mathematics,* Prentice-Hall, Inc., Englewood Cliffs, N. J., 1962.

[8] MIZUKAMI, K., "Optimum Redundancy for Maximum System Reliability by the Method of Convex and Integer Programming," *Operations Research*, Vol. 16, pp. 392–406, 1968.

[9] MUTH, E. J., "Expected Value and Variance of Failure Time in Redundant Systems," *IEEE Transactions on Reliability,* Vol. R-22, No. 2, pp. 103–105, 1973.

[10] PATTERSON, R. L., "Stochastic Failure Models Based upon Distribution of Stress Peaks," Technical Report 3, Project THEMIS, Industrial and Systems Engineering Department, University of Florida, Gainesville, Fla., 1967.

[11] POLOVKO, A. M., *Fundamentals of Reliability Theory*, Academic Press, Inc., New York, 1968.

[12] PROSCHAN, F., and T. A. BRAY, "Optimal Redundancy Under Multiple Constraints," *Operations Research*, Vol. 13, pp. 800–814, 1965.

[13] RAO, S. S., and R. NATARAJAN, "Reliability with Standbys," *Opsearch*, Vol. 7, No. 1, March 1970.

[14] RAU, J. G., *Optimization and Probability in Systems Engineering,* Van Nostrand-Reinhold, New York, 1970.

[15] SANDLER, G. H., *System Reliability Engineering*, Prentice-Hall, Inc., Englewood Cliffs, N.J., 1963.

PROBLEMS

1. Suppose that the hazard function of a system has the form

$$h_X(x) = \begin{cases} 0, & 0 \le x < a \\ \lambda, & a \le x < \infty \end{cases}$$

Show that the time to failure X of the system is described by the two-parameter exponential distribution with PDF

$$f_X(x) = \begin{cases} 0, & 0 \le x \le a \\ \lambda e^{-\lambda(x-a)}, & a < x < \infty \end{cases}$$

Find $E[X]$ and Var $[X]$.

2. Consider an independent parallel system consisting of n identical components each with a constant hazard rate function λ. Assume no repair or replacement of failed components. Show that the variance of the time to failure Y of the system is

$$\text{Var}[Y] = \frac{1}{\lambda^2} \sum_{i=1}^{i=n} \frac{1}{i^2}$$

$$\left[\textit{Hint:}\ \text{Use the relation } E[Y^2] = 2 \int_0^\infty t R_Y(t)\, dt. \right]$$

3. An independent serial system consists of n identical components each having a Weibull time-to-failure distribution. Show that the system has also a time to-failure distribution of the Weibull type, and hence determine the mean time to failure of the system.

4. The schematic configuration of a system consisting of nine identical components each with constant hazard rate function λ is as illustrated. Find the reliability function of the system.

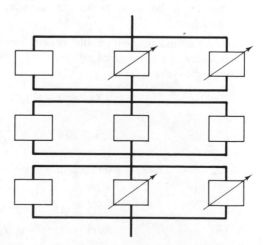

Figure P 4.

5. Suppose that a particular type of airplane engine fails with probability $1 - p = q$. We shall assume that the failure of one engine is not affected by the operation of the other engines. For a plane to make a successful flight at least half of its engines must be operating. Show that if $0 < q < \frac{1}{3}$ a four-engine plane is preferred to a two-engine plane, while if $\frac{1}{3} < q < 1$, a two-engine plane is preferred to a four-engine plane (see [15]).

6. An electrical circuit consists of four identical elements a, b, c, and d that are operating independently and with constant hazard rate functions. The configuration of the circuit is illustrated. The arrows indicate the flow of current. The circuit fails if no current is received at the output end. Calculate the reliability of the system.

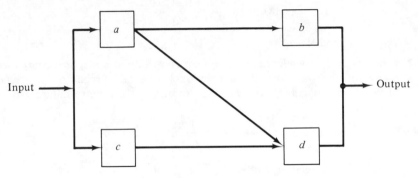

Figure P 6.

7. Consider an independent standby system consisting of two identical elements with imperfect switching. Each element has a constant hazard rate function λ and the switch has also a constant hazard rate function μ.
   (a) Obtain an expression for the reliability function of the system.
   (b) Suppose that the elements were operating in parallel, thus requiring no switching device. Under which condition would such a redundant configuration be superior to the standby configuration with imperfect switching?

8. Consider an independent parallel system consisting of n identical components each with a constant hazard rate function λ. If T_n is the time to failure of the system, show that

$$E[T_n] = E[T_{n-1}] + \frac{1}{n\lambda}$$

and $$E[T_n^2] = E[T_{n-1}^2] + \frac{2}{n\lambda}E[T_n]$$

Hence obtain an expression for $E[T_n]$ and Var $[T_n]$ (see [9]).

[*Hint:* Let $R_n(t) = 1 - (1 - e^{-\lambda t})^n$

and $H_n(t) = (1 - e^{-\lambda t})^{n-1}e^{-\lambda t}$

Then $R_n(t) = R_{n-1}(t) + H_n(t)$]

9. A system is made up of $N = 3$ subsystems in series, each subsystem consisting of a number of identical components operating in independent parallel fashion. The following characteristics are given:

Subsystem i	1	2	3
Reliability per component R_i	.60	.50	.70
Weight per component w_i (lbs)	3	2	2

If the total weight of the system is not to exceed $W = 11$ lbs, determine the number of parallel redundancies in each subsystem.

10. An independent serial system consists of N subsystems, $N = 1, 2, \ldots$. The ith subsystem, $i = 1, 2, \ldots, N$, consists of n_i components in parallel, $n_i = 1, 2, \ldots$ For the ith subsystem let

$\quad R_i$ = reliability of each component
$\quad c_i$ = cost of each component

Consider the problem of determining $n_1, n_2, \ldots, n_N$ so as to minimize the total cost of the system subject to a fixed reliability level R, that is

$$\min_{n_1, \cdots, n_N} \sum_{i=1}^{i=N} c_i n_i$$

subject to

$$\prod_{i=1}^{i=N} [1 - (1 - R_i)^{n_i}] = R$$

(a) Let for $i = 1, 2, \ldots, N$

$$1 - (1 - R_i)^{n_i} = R^{\alpha_i}, \quad 0 \leq \alpha_i \leq 1$$

Show that the optimization problem reduces to

$$\min_{\alpha_1, \cdots, \alpha_N} \sum_{i=1}^{i=N} c_i \frac{\ln(1 - R^{\alpha_i})}{\ln(1 - R_i)}$$

subject to

$$\sum_{i=1}^{i=N} \alpha_i = 1$$

(b) Assume R to be close to unity so that $R = 1 - \delta$, where δ is a small quantity. Then, $\ln R \approx -\delta$. Use the technique of Lagrange multipliers to show that the optimum value of α_i is approximately

$$\alpha_i = \frac{c_i / \ln(1 - R_i)}{\sum_{k=1}^{k=N} c_k / \ln(1 - R_k)} \quad i = 1, 2, \ldots, N$$

(c) Determine the approximate value of each n_i. Discuss.

11. An independent standby system consists of N identical components subject to failure. Following the failure of the last component, all N failed components are immediately replaced by N new identical components. Show the equivalence between this problem and the inventory problem discussed in Section 5.3-i-a. Develop an economic criterion to determine the optimum value of N.

QUEUING THEORY

4.1. Introduction and Historical Background

The theory of queues is the study of the phenomenon of waiting lines created when customers or units arrive in a generally random fashion at one or more counters or service centers offering certain facilities, and demand service. The amount of service provided to the customers is in general a random variable.

The mathematical study of queuing systems originated with the work of A.K. Erlang [1] between the years 1909 to 1920 dealing with waiting line and trunking problems for telephone exchanges. Fry (1928) [6] independently brought some major contributions to the field. Their work stimulated the research of such early contributors as Molina (1927) [11], Pollaczek (1930) [14], Khintchine (1932) [9], Palm (1943) [13], and Kendall (1951) [8]. The impetus provided by these researchers has motivated a great interest in queuing theory, particularly as a topic in applied stochastic process and as a useful subject for modeling the stochastic behavior of a large number of natural and man-created phenomena.

Examples of queuing phenomena are numerous and arise in such diverse areas as telephone traffic, airports, seaports, computer centers, supermarket counters, machine shops, shipping docks, hospitals, gas stations, and so on.

4

4.2. Characteristics of Queuing Systems

Customers or *units* demanding service originate over time from a given *input source* and join the queue of a system providing such service. The facilities in the system offering service or *servers* select members of the queue for service following some predefined rule known as *queue discipline*. The unit selected is then provided the required service by the service mechanism. Upon completion of service, the unit departs from the system and may or may not join again the population of the input source.

From this description it is clear that a queuing system is completely characterized by the following elements: (i) the source, (ii) the input process, (iii) the service mechanism, and (iv) the queue discipline. In what follows, we discuss each of these descriptive elements and simultaneously introduce some fundamental terminologies proper to queuing systems.

i. *Source*

A source is a device or group of devices from which units emanate and call for service. The total number of potential units requesting service may be either infinite or finite; the source is then said to be either *infinite* or *finite*, respectively. In all actual problems, the source is finite; however, if the population size is sufficiently large then, for all practical purposes, the source may be assumed to be infinite. This basic assumption is useful in theory since the analysis of an infinite-source model is substantially easier than a finite-source model. The fundamental difference between an infinite-source model and a finite-source model is that in the case of infinite source, departing units following service may or may not join again the units in the source without altering the input to the system. In the finite-source case the route taken by the departing units alters the rate at which the input source generates calling units. In most finite-source problems it is generally assumed that units return to the population source once service is completed.

ii. *Input Process*

In an infinite-source problem, units emanating from the source and joining the service facility can in general be assumed to form a counting process. In most theoretical problems, the interarrival times $\{T_m\}$, $m = 1, 2, \ldots$, (where the index m refers to the mth customer) between successive customers emanating from the source are, in addition, assumed to be independently and identically distributed random variables, thus forming a renewal process. A common form of this process, which is extensively used in queuing models and which has some empirical justification in its use, is the well-known Poisson process. Thus the probability that a customer will emanate from the source in the time interval $(t, t + dt]$ is $\lambda\, dt + o(dt)$, where λ is the rate of customer arrival, and $E[T] = 1/\lambda$.

For a finite-source model, a general description of the input process will not be attempted here. It is customary however, to assume that if i, $i = 0, 1, 2, \ldots, \tilde{N}$, is the population level at the source at time t, then the probability that a unit will emanate from the source in the time interval $(t, t + dt]$ is $i\lambda\, dt + o(dt)$.

iii. *Service Mechanism*

It is necessary to distinguish among three aspects of the service mechanism: the *availability*, the *capacity*, and the *duration of service*. These are three independent variables that may be known either deterministically or probabilistically. All three must be specified in order to completely define the service mechanism.

(1) The service mechanism may be *available* at certain times but not available at others. Such is the case if the server is performing ancillary duties. For example, in large computer installations routine maintenance is scheduled the first hour of each morning, during which time the computer is not available for service. Failure or

breakdown of the computer system would result in the same condition, except that the event is random. Absence of employees and closure of certain routes are examples of unavailable service mechanisms.

(2) The *capacity* of the service mechanism is usually measured in terms of the number of customers that can be served simultaneously. The capacity may be fixed in time, as, for example, the number of machines available in a shop, or the capacity may be varying over time, as, for example, the number of toll booths along turnpikes. Thus, the service facility can have one or more *channels*. A queuing system having a single channel is called a *single-channel* or *single-server* system; a queuing system with more than one channel is called a *multichannel* or *multiserver system*.

(3) Finally, the *duration of service* or *service time* must be specified. The service time can be a constant for all customers or be in general a random variable. Let S_m be the amount of service provided to the mth customer, $m = 1, 2, \ldots$. Then we shall assume that $\{S_m\}$ form a sequence of independently and identically distributed random variables. We shall also assume that the interarrival times and service times are mutually independent.

An important case arises when all S_m have the negative exponential distribution

$$\varphi_S(x) = \mu e^{-\mu x}, \quad \mu > 0, 0 < x < \infty$$

Then, for a customer in service, the probability that service will be completed in the time interval $(t, t + dt]$ is $\mu \, dt + o(dt)$, and $E[S] = 1/\mu$.

iv. *Queue Discipline*

By *discipline* is meant the manner in which customers or units are to be selected for service *once they are in the queue*.

Here we must distinguish between two cases: the single-server system and the multiserver system.

In the *single-server system* (unit capacity) the units are assigned *priorities* and they are serviced in the order of their priorities. Among the several types of priority rules, we mention the following:

(1) First-come first-served priority.
(2) Last-come first-served priority.
(3) Higher priority assigned to the shortest service time.
(4) Higher priority assigned to the longest service time.
(5) Head-of-the-line priority.

The service mechanism, upon the appearance of a unit with a higher priority, continues working on the unit it had been servicing and services the newly arrived and any other higher-priority arrivals before undertaking the task of servicing the lower-priority units.

(6) Preemptive priorities.

The service mechanism, upon the appearance of a unit with a higher priority, ceases working on the unit it had been servicing and services the newly arrived unit. In the *preemptive resume rule* the unit whose service is interrupted resumes its service, when due, subject to the preemptive rules. Thus a unit may have its service interrupted on several occasions (the shop servicing of a car by a gas attendant). In the *preemptive repeat rule* the unit whose service is interrupted has its service repeated all over again, when due, subject to the preemptive rules.

In the *discretionary priority discipline* a combination of head-of-the-line and preemption is used, depending on the amount of service provided to the low-priority unit being serviced. A cutoff service time is specified. Thus, when a low-priority unit is being serviced and a high-priority unit arrives, then if more than $e\%$ of the service has already been provided, the low-priority unit is not preempted, and if less than $e\%$ of the service has already been provided, the low-priority unit is preempted.

In the *multiserver system* (multiunit capacity) the units are again assigned priorities; however, somewhat different problems may arise in addition to the ones stated. Some units in the facility may specialize in certain services, and customers requiring that service are forced to go there. If more than one server is capable of performing the required service, *jockeying*, that is, moving from a long queue to a shorter one, becomes a factor to be considered. Such jockeying may be intentional and worked into the queue discipline, or may be random.

Two important queuing phenomena that may be present in a queuing system are

(1) *Balking*, which occurs when a customer decides not to join a queue longer than a certain length.

(2) *Reneging*, which occurs when a customer leaves the queue after having joined it and having waited for a certain length of time (impatient customer).

Unless otherwise stated, we shall assume in what follows a first-come first-served priority discipline.

v. *Configuration of Queuing Systems*

We distinguish among the following configurations in queuing systems

(1) Unlimited capacity of the waiting space.

(2) Limited capacity of the waiting space.

(3) Single-channel systems.
(4) Mutichannel systems or parallel systems (Figure 4.1).
(5) Serial systems (Figure 4.2) generating queues in tandem.
(6) Other systems.

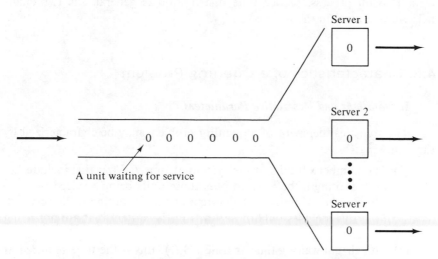

Figure 4.1. A multichannel or parallel system with *r* servers.

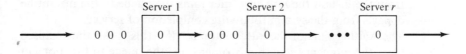

Figure 4.2. A serial system with *r* servers.

vi. *Kendall's Notation*

In 1953, Kendall proposed a symbolic classification of queues that has been extensively utilized. The following symbols will be defined:

(1) *M* (for Markov) indicates interarrival times or service times with negative exponential distribution.
(2) *GI* (for general independent) indicates an input of the general independent type; no assumptions about the distribution of the interarrival times is made.
(3) *G* (for general) indicates service time distributions of the general type.
(4) *D* (for deterministic) indicates interarrival times or service times that are constants.
(5) *C* (for channels) indicates the number of channels.

The queuing system is then characterized by the following arrangements:

input process/service process/number of channels

Thus $M/G/2$ indicates a queuing system in which arrivals are described by a Poisson process, service time distribution is general, and two channels are operating in the service facility.

4.3. Characteristics of a Queuing Problem

i. *Notations and Descriptive Parameters*

The statistical behavior of a queuing problem may be characterized by the following:

(1) The number of units in the system at time t, $N(t)$; this includes all units waiting to be serviced plus those units being serviced.
(2) The number of units in the system at time t waiting to be serviced, $L(t)$; this includes only those units in the system that are not being serviced.
(3) The virtual waiting time at time t, $V(t)$; this is the time required at time t by the server to service all the customers present in the system.
(4) The time in the system of the mth customer Θ_m; this is the time elapsed from the instant the mth customer joins the queue to the instant he departs from the system following completion of service.
(5) The waiting time of the mth customer W_m; this is the time elapsed from the moment the mth customer joins the queue to the moment he enters one of the channels for service.

In general, the study of queuing phenomena as a stochastic process attempts to characterize the statistical properties of the random processes $N(t)$, $L(t)$, $V(t)$, Θ_m, and W_m. The analysis is carried either as a time-dependent process, thus yielding the transient solutions to the queuing problem or, in much simpler means, as a steady-state process, assuming its existence.

In this chapter we will mainly study the steady-state characteristics in a queuing problem. For this limiting situation, the entities $N(t)$, $L(t)$, $V(t)$, Θ_m, and W_m will be denoted by N, L, V, Θ, and W, respectively. Note that

$$\Theta_m = W_m + S_m$$

where S_m is the service time of the mth customer.

ii. *Objective Function—Decision Variables*

In a queuing system the set of decision variables and the choice of an appropriate objective function will depend very much upon the particular queuing problem analyzed. We provide the following illustrative examples, and the reader is asked to develop additional examples of his own:

(1) In multichannel queues, the objective could be balancing the cost associated with the idle time of the servers with the cost associated in the waiting time of the customer so as to optimize the number of channels.

(2) In a single-channel problem, the objective could be to minimize the expected waiting time of the customers by the selection of an appropriate discipline.

(3) In a single-channel problem, with the server performing ancillary duties, the problem could be to determine the optimum number of units to be accumulated following a non-servicing period to minimize some cost function.

(4) In a single-channel problem with a discretionary priority rule, the problem could be to determine the optimum cutoff point, again to minimize some cost function.

iii. *Little's Formula*

Under steady-state conditions, Little [10] has established the following relations which hold under very general conditions:

$$E[N] = \frac{E[\Theta]}{E[T]}$$

$$E[L] = \frac{E[W]}{E[T]}$$

The importance of these relationships stems from the fact that important statistics of the queuing process can be determined without an elaborate mathematical analysis. The first of the relations states, for example, that the expected time spent by a customer in the system equals the expected number of customers in the system times the expected interarrival time.

4.4. The *M/M/*1 Queuing System

i. *Description of the System*

Consider a queuing system involving a single service center. The interarrival times between successive customers joining the system form a sequence of identically and independently distributed random variables $\{T_m\}$, $m = 1$, $2, \ldots$, having the exponential distribution

$$\varphi_T(x) = \lambda e^{-\lambda x}, \quad \lambda > 0, 0 < x < \infty$$

Here the index m refers to the mth arriving customer to the queuing system. Because of the memoryless property of the exponential distribution, then, given some time epoch, the time of arrival of the first customer from this time epoch is exponentially distributed.

Arriving customers form a single line, and the order at which service is provided is the order of their arrival; that is, the first customer is first in service, the second customer is next, and so on. It is evident that no service can occur when no customers are in the system.

We shall denote the amount of service provided to the mth customer by S_m. This is also called the service time or length of service of the mth customer. We shall assume that $\{S_m\}$, $m = 1, 2, \ldots$, form a sequence of independently and identically distributed random variables, each having the exponential distribution

$$\varphi_S(x) = \mu e^{-\mu x}, \quad \mu > 0, 0 < x < \infty$$

We shall also assume that $\{T_m\}$ and $\{S_m\}$ are mutually independent random variables; that is, the service times and interarrival times are independently distributed. The parameter $\rho = \lambda/\mu$ is defined as the *traffic intensity*.

Suppose now that in analyzing this queuing system as a stochastic process, we select as our index set the clock time t measured over a continuous scale from a given origin, $t \geq 0$. The arrival and service process may be alternatively characterized by saying that (see Section 2.4-i)

(1) $\lambda \, dt + o(dt)$ is the probability that a customer will join the system in the time interval $(t, t + dt]$. The probability that no customers will join the system in the time interval $(t, t + dt]$ is $[1 - \lambda \, dt + o(dt)]$; and the probability that more than one customer will join the system in the time interval $(t, t + dt]$ is $o(dt)$. Hence arrivals are generated by a Poisson process with intensity λ.

(2) $\mu \, dt + o(dt)$ is the probability that a customer in service at time t will complete service in the time interval $(t, t + dt]$. The probability that the customer in service at time t will not terminate service in the time interval $(t, t + dt]$ is $[1 - \mu \, dt + o(dt)]$.

Define the random variable $N(t)$ to denote the total number of customers in the system (i.e., being serviced and in line) at time t, $t \geq 0$; the range of $N(t)$ is discrete, and, in general, it may be defined over the set of nonnegative integers. The collection of random variables $\{N(t), 0 \leq t < \infty\}$ is a continuous-parameter discrete-valued stochastic process. We shall be interested in particular in determining the unconditional probability distribution of $N(t)$ at time t. Let then $P\{N(t) = n\} = P(n, t)$ denote the probability that exactly $N(t) = n$ customers are in the system at time t, $t \geq 0$; $n = 0, 1, 2, \ldots$. This is sometimes written as $P_n(t)$.

Because of the characteristics of the arrival and service processes, the probability distribution of $N(t + dt)$ will depend only on the probability distribution of $N(t)$. The resultant process is a time homogeneous birth and death process (Section 2.7-iii). The transition probabilities are

$$p_{ij}(dt) = \begin{cases} [1 - \lambda\,dt + o(dt)], & j = i = 0 \\ [1 - \lambda\,dt + o(dt)][\mu\,dt + o(dt)], & j = i - 1; i = 1, 2, \ldots \\ [1 - \lambda\,dt + o(dt)][1 - \mu\,dt + o(dt)], & j = i; i = 1, 2, \ldots \\ [\lambda\,dt + o(dt)][1 - \mu\,dt + o(dt)], & j = i + 1; i = 0, 1, \ldots \\ o(dt), & \text{otherwise} \end{cases}$$

Before we proceed to modeling the general $M/M/1$ system, it may be of interest to consider the following special cases.

ii. Special Case 1: A Queuing System with Poisson Arrivals and No Service

A frequent situation that may arise in practice is when, prior to the start of service at a scheduled time, customers arrive and join the queue, perhaps hoping that if they come earlier they will have to wait less. One may imagine, for example, customers arriving at the main entrance of a small post office building consisting of a single window, before the scheduled opening hour of the window (it may be the Christmas season, and they may be trying to beat the mail rush). We may be interested in finding out the number of customers waiting in line when service begins. Starting at some time origin $t = 0$ when there are no customers, the number of customers $N(t)$ that are in line at a time t, $t \geq 0$, prior to the start of service is the same as the total number of customers that have arrived up to time t. Clearly, the following initial conditions hold:

$$\begin{aligned} P(0, 0) &= 1, & n = 0; t = 0 \\ P(n, 0) &= 0, & n \geq 1; t = 0 \end{aligned} \tag{4.1}$$

Now for $n = 0$ and $t \geq 0$,

$$P(0, t + dt) = P(0, t)[1 - \lambda\,dt + o(dt)]$$

and for $n \geq 1$ and $t \geq 0$,

$$P(n, t + dt) = P(n - 1, t)[\lambda\,dt + o(dt)] + P(n, t)[1 - \lambda\,dt + o(dt)] + o(dt)$$

These equations may be written respectively as ($t \geq 0$)

$$\frac{P(0, t + dt) - P(0, t)}{dt} = -\lambda P(0, t) + \frac{o(dt)}{dt}, \quad n = 0$$

$$\frac{P(n, t + dt) - P(n, t)}{dt} = \lambda P(n - 1, t) - \lambda P(n, t) + \frac{o(dt)}{dt}, \quad n \geq 1$$

Let $dt \rightarrow 0$; we then obtain the following system of differential difference equations for $t \geq 0$:

$$\frac{dP(0, t)}{dt} = -\lambda P(0, t), \quad n = 0$$

$$\frac{dP(n, t)}{dt} = \lambda P(n - 1, t) - \lambda P(n, t), \quad n \geq 1 \tag{4.2}$$

The system of equations (4.2) subject to the initial conditions (4.1) can be identified as those of a homogeneous Poisson process. The general solution for $P(n, t)$ was found in Section 2.7-iii-a to be

$$P(n, t) = e^{-\lambda t} \frac{(\lambda t)^n}{n!}, \quad n = 0, 1, 2, \ldots ; t \geq 0 \tag{4.3}$$

Thus the number of customers in line at a time t prior to service availability is Poisson distributed with parameter λt. The expected number of customers in the system at time t is then

$$E[N(t)] = \lambda t, \quad t \geq 0$$

Since $N(t)$ does not depend on service time, the previous result will hold true irrespective of the nature of service that will be provided to the waiting customers. In general, if customer arrival forms a renewal process, and no service is in progress, the number of customers in line at a particular time t corresponds to the total number of renewals up to time t.

iii. *Special Case 2: A Queuing System with No Arrivals and Exponential Service Time Distribution*

Another special situation of interest is when no additional arrivals join the system, but service proceeds for those customers who are in line. For example, at the closing hour of the post office building, no more customers are allowed in; however, service is still provided for those customers who have previously arrived and who are in line. Suppose that time origin ($t = 0$) is selected as the closing hour, and let i be the number of customers in the system at time $t = 0$, $i \geq 1$. It is evident that service will be in progress for the first customer in the system. Because of the memoryless property of the exponential distribution, the amount of service still to be provided to the customer in service, given that the system is observed at $t = 0$, is exponentially distributed. Also the number of customers $N(t)$ in the system at time t, $t \geq 0$, is equal to i minus the total number of customers serviced up to time t. Although the range of $N(t)$ is $n = 0, 1, 2, \ldots, i$, one can easily extend this range to include all n, $n = 0, 1, \ldots$, by defining $P\{N(t) = n\} = 0$ for all $n \geq i + 1$

and $t \geq 0$. The initial conditions of this problem can be written as

$$P(n, 0) = 0, \quad 0 \leq n \leq i - 1; \, t = 0$$
$$P(i, 0) = 1, \quad n = i; \, t = 0 \qquad \qquad (4.4)$$
$$P(n, 0) = 0, \quad i + 1 \leq n < \infty; \, t = 0$$

Now for $i + 1 \leq n < \infty$ and $t \geq 0$,

$$P(n, t) = 0$$

For $n = i$ and $t \geq 0$,

$$P(i, t + dt) = P(i, t)[1 - \mu \, dt + o(dt)]$$

For $1 \leq n \leq i - 1$ and $t \geq 0$,

$$P(n, t + dt) = P(n, t)[1 - \mu \, dt + o(dt)] + P(n + 1, t)[\mu \, dt + o(dt)]$$

For $n = 0$ and $t \geq 0$,

$$P(0, t + dt) = P(0, t) + P(1, t)[\mu \, dt + o(dt)]$$

These equations may be written for $t \geq 0$ as:

$$P(n, t) = 0, \quad i + 1 \leq n < \infty$$

$$\frac{P(i, t + dt) - P(i, t)}{dt} = -\mu P(i, t) + \frac{o(dt)}{dt}, \quad n = i$$

$$\frac{P(n, t + dt) - P(n, t)}{dt} = -\mu P(n, t) + \mu P(n + 1, t) + \frac{o(dt)}{dt},$$
$$1 \leq n \leq i - 1$$

$$\frac{P(0, t + dt) - P(0, t)}{dt} = \mu P(1, t) + \frac{o(dt)}{dt}, \quad n = 0$$

If we let $dt \rightarrow 0$, we obtain for $t \geq 0$ the following system of differential difference equations:

$$P(n, t) = 0, \quad i + 1 \leq n < \infty$$

$$\frac{dP(i, t)}{dt} = -\mu P(i, t), \quad n = i$$

$$\frac{dP(n, t)}{dt} = -\mu P(n, t) + \mu P(n + 1, t), \quad 1 \leq n \leq i - 1$$

$$\frac{dP(0, t)}{dt} = \mu P(1, t), \quad n = 0$$

This system of equations together with the initial conditions (4.4) are the equations of the homogeneous death process with constant death rate μ (Section 2.7-iii-b), and we can write for the solution

$$P(n, t) = 0, \qquad\qquad n \geq i + 1; t \geq 0$$

$$P(n, t) = e^{-\mu t} \frac{(\mu t)^{i-n}}{(i - n)!}, \qquad 1 \leq n \leq i; \quad t \geq 0 \qquad\qquad (4.5)$$

$$P(0, t) = 1 - \sum_{n=1}^{n=i} P(n, t), \quad n = 0; t \geq 0$$

The expected number of customers in the system is

$$E[N(t)] = \sum_{n=1}^{n=i} n e^{-\mu t} \frac{(\mu t)^{i-n}}{(i - n)!}, \quad t \geq 0$$

iv. *General Case: Poisson Arrival and Exponential Service Time*

a. DIFFERENTIAL DIFFERENCE EQUATION

Suppose now the opening hour of the post office (when service becomes available) is selected as the time origin and assume that i customers, $i \geq 0$, are already in line. We then have as initial conditions

$$P(i, 0) = 1, \quad n = i; t = 0$$
$$P(n, 0) = 0, \quad n \neq i; t = 0 \qquad\qquad (4.6)$$

For $n = 0$ and $t \geq 0$, we have

$$P(0, t + dt) = P(0, t)[1 - \lambda \, dt + o(dt)]$$
$$+ P(1, t)[1 - \lambda \, dt + o(dt)][\mu \, dt + o(dt)]$$

For $n \geq 1$ and $t \geq 0$, we have

$$P(n, t + dt) = P(n - 1, t)[\lambda \, dt + o(dt)][1 - \mu \, dt + o(dt)]$$
$$+ P(n, t)[1 - \lambda \, dt + o(dt)][1 - \mu \, dt + o(dt)]$$
$$+ P(n + 1, t)[1 - \lambda dt + o(dt)][\mu \, dt + o(dt)]$$

These equations may be written for $t \geq 0$ as

$$\frac{P(0, t + dt) - P(0, t)}{dt} = -\lambda P(0, t) + \mu P(1, t) + \frac{o(dt)}{dt}, \quad n = 0$$

$$\frac{P(n, t + dt) - P(n, t)}{dt} = \lambda P(n - 1, t) - (\lambda + \mu) P(n, t)$$

$$+ \mu P(n + 1, t) + \frac{o(dt)}{dt}, \quad n \geq 1$$

If we let $dt \rightarrow 0$, we obtain for $t \geq 0$ the following system of differential difference equations:

$$\frac{dP(0, t)}{dt} = -\lambda P(0, t) + \mu P(1, t), \quad n = 0$$

$$\frac{dP(n, t)}{dt} = \lambda P(n - 1, t) - (\lambda + \mu)P(n, t) + \mu P(n + 1, t), \, n \geq 1$$

(4.7)

The system of equations (4.7) together with the initial conditions (4.6) can be identified as the equations of the homogeneous birth and death process with constant birth and death rate [see equation (2.101)]. The general solution for $P(n, t)$ provides an expression for the probability of exactly n customers in the system at time t, and of course it is a function of the initial number of customers in the system i. However, the method of solution requires a knowledge of analysis that goes beyond the scope of this book. With a knowledge of $P(n, t)$ it can be shown that if $\lambda/\mu < 1$, $\lim_{t \to \infty} P(n, t)$ exists; that is, the number of units in the system $N(t)$ possesses a steady-state distribution.

b. STEADY-STATE DISTRIBUTION OF THE NUMBER IN THE SYSTEM

Suppose that $\lambda/\mu < 1$ and let N be the number of customers in the system during steady state. If $\varphi_N(n)$ is the PMF of the random variable N, then

$$\varphi_N(n) = \lim_{t \to \infty} P\{N(t) = n\} = \lim_{t \to \infty} P(n, t), \quad n = 0, 1, 2, \ldots$$

This PMF is usually denoted by the symbol P_n. Since the steady-state probability distribution of $N(t)$ will be independent of i and t, we may write the system of equations (4.7) as

$$0 = -\lambda P_0 + \mu P_1, \quad n = 0 \tag{4.8}$$

$$0 = \lambda P_{n-1} - (\lambda + \mu)P_n + \mu P_{n+1}, \quad n \geq 1 \tag{4.9}$$

In addition, we have the normalizing condition

$$\sum_{n=0}^{\infty} P_n = 1 \tag{4.10}$$

Equations (4.8) and (4.9) may be solved recursively; from equation (4.8) we obtain

$$P_1 = \frac{\lambda}{\mu} P_0$$

If we set $n = 1$ in equation (4.9), we obtain

$$0 = \lambda P_0 - (\lambda + \mu)P_1 + \mu P_2$$

Thus
$$P_2 = \frac{\lambda + \mu}{\mu} P_1 - \frac{\lambda}{\mu} P_0$$

$$= \frac{\lambda + \mu}{\mu} \frac{\lambda}{\mu} P_0 - \frac{\lambda}{\mu} P_0 = \left(\frac{\lambda}{\mu}\right)^2 P_0$$

Using a simple inductive argument, it can be shown that

$$P_n = \left(\frac{\lambda}{\mu}\right)^n P_0, \quad n = 1, 2, \ldots$$

Substituting the values of P_n as a function of P_0 in, relation (4.10), we obtain

$$P_0 + \sum_{n=1}^{\infty} \left(\frac{\lambda}{\mu}\right)^n P_0 = 1$$

or
$$P_0 = \frac{1}{\sum_{n=0}^{\infty} (\lambda/\mu)^n}$$

Since $\lambda/\mu < 1$, the geometric series in the denominator converges to $1/[1 - (\lambda/\mu)]$; hence

$$P_0 = 1 - \frac{\lambda}{\mu}$$

Thus we can write in general for P_n

$$P_n = \left(1 - \frac{\lambda}{\mu}\right)\left(\frac{\lambda}{\mu}\right)^n, \quad n = 0, 1, 2, \ldots$$

which is identified as the geometric distribution. These results should be compared with the expressions obtained for the birth and death process in Chapter 2.

In terms of the traffic intensity $\rho = \lambda/\mu$, the expression for P_n, the steady-state probabilities of exactly $N = n$ customers in the system, can be written as

$$P\{N = n\} = P_n = (1 - \rho)\rho^n, \quad n = 0, 1, 2, 3, \ldots \tag{4.11}$$

c. ALTERNATIVE DERIVATION OF P_n

The system of equations (4.8) and (4.9) may be written as

$$0 = -\rho P_0 + P_1, \quad n = 0 \tag{4.12}$$

$$0 = \rho P_{n-1} - (1 + \rho)P_n + P_{n+1}, \quad n \geq 1 \tag{4.13}$$

Define the PGF of the random variable N as

$$G_N(s) = \sum_{n=0}^{\infty} s^n P_n, \quad |s| \leq 1$$

Multiplying equation (4.12) by s^0 and equation (4.13) by s^n, $n \geq 1$, and summing over all possible values of n, we obtain

$$0 = \rho \sum_{n=1}^{\infty} s^n P_{n-1} - \sum_{n=1}^{\infty} s^n P_n - \rho \sum_{n=0}^{\infty} s^n P_n + \sum_{n=0}^{\infty} s^n P_{n+1}$$

or $\quad 0 = \rho s G_N(s) - [G_N(s) - P_0] - \rho G_N(s) + \dfrac{1}{s}[G_N(s) - P_0]$

Solving for $G_N(s)$, we obtain

$$G_N(s) = \frac{P_0[(1/s) - 1]}{\rho s - 1 - \rho + (1/s)} = \frac{P_0}{1 - \rho s}$$

To obtain P_0, we note that at $s = 1$, $G_N(1) = 1$; hence $P_0 = 1 - \rho$, and

$$G_N(s) = \frac{1 - \rho}{1 - \rho s}$$

which can be recognized as the PGF of a random variable having the geometric distribution

$$P_n = (1 - \rho)\rho^n, \quad n = 0, 1, 2, \ldots$$

The expected number of customers in the system is

$$E[N] = \sum_{n=0}^{\infty} n P_n = \frac{\rho}{1 - \rho}, \quad 0 < \rho < 1$$

Also $\quad \text{Var}[N] = \sum_{n=0}^{\infty} [n - E[N]]^2 = \dfrac{\rho}{(1 - \rho)^2}, \quad 0 < \rho < 1$

The plot of $E[N]$ as a function of ρ, $0 < \rho < 1$, is shown in Figure 4.3.

Example 4.1

Determine the maximum allowable traffic intensity in the *M/M/1* queuing system so that the equilibrium probability of having customers in excess of some specified number a in the system will be less than or equal to a given level α.

Symbolically, we need to determine a value of ρ so that

$$P\{N > a\} \leq \alpha$$

or $\qquad\qquad\qquad 1 - P\{N > a\} \geq 1 - \alpha$

or $\qquad\qquad\qquad\qquad P\{N \leq a\} \geq 1 - \alpha$

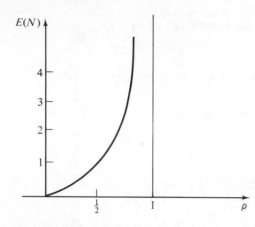

Figure 4.3. The function $E[N] = \dfrac{\rho}{1-\rho}$.

Using relation (4.11), the previous expression becomes

$$\sum_{n=0}^{n=a}(1-\rho)\rho^n \geq 1-\alpha$$

or

$$(1-\rho)\frac{1-\rho^{a+1}}{1-\rho} \geq 1-\alpha$$

or

$$\rho^{a+1} \leq \alpha$$

Hence, the maximum allowable traffic intensity is $\rho = \alpha^{1/(a+1)}$. On the basis of this expression, the chances are less than one percent that a traffic intensity of $\rho = .525$ will generate more than 8 customers in the system. The corresponding average number of customers in the system is

$$\frac{\rho}{1-\rho} = \frac{.525}{1-.525} = \frac{.525}{.475} \approx 1.1$$

d. STEADY-STATE DISTRIBUTION OF THE NUMBER WAITING IN THE SYSTEM

Let L be a random variable denoting the number of units waiting in the system in the steady state. L is also referred to as the queue length (it excludes the customer being serviced if any).

The queue length L is zero when there are no customers in the system or when exactly one customer is in the system (being serviced and not waiting). When the number of customers N in the system exceeds 1, that is, $N \geq 2$, then $L = N - 1$. We can express L as a function of the random variable N as

$$L = \begin{cases} 0, & N = 0, 1 \\ N - 1, & N \geq 2 \end{cases}$$

Let

$$\varphi_L(l) = P\{L = l\}, \quad l = 0, 1, 2, \ldots$$

For $l = 0$,

$$\varphi_L(0) = P_0 + P_1$$
$$= (1 - \rho) + (1 - \rho)\rho = 1 - \rho^2$$

For $l \geq 1$,

$$\varphi_L(l) = P\{L = l\} = P\{N - 1 = l\}$$
$$= P\{N = l + 1\}$$
$$= P_{l+1} = (1 - \rho)\rho^{l+1}$$

Hence

$$\varphi_L(l) = \begin{cases} 1 - \rho^2 & l = 0 \\ (1 - \rho)\rho^{l+1} & l \geq 1 \end{cases}$$

is the steady-state distribution of the number of customers waiting in the system.

The expected queue length or expected number of customers waiting in the system is

$$E[L] = \sum_{l=0}^{\infty} l\varphi_L(l)$$
$$= \sum_{l=1}^{\infty} l(1 - \rho)\rho^{l+1}$$
$$= (1 - \rho)\rho^2 \sum_{l=1}^{\infty} l\rho^{l-1}$$
$$= (1 - \rho)\rho^2 \cdot \frac{1}{(1 - \rho)^2}$$
$$= \frac{\rho^2}{1 - \rho}$$

Exercise 4.1

Obtain the PGF of L; hence derive $E[L]$ and $\text{Var}[L]$.

e. STEADY-STATE DISTRIBUTION OF THE VIRTUAL WAITING TIME

Let the random variable V denote the time required by the server to serve all the customers present in the system at a particular instant of time in the steady state. Let $\Phi_V(v)$ be the distribution function of V; that is,

$$\Phi_V(v) = P\{V \leq v\}, \quad 0 \leq v < \infty$$

When no customers are present in the system ($N = 0$), the server is idle; on the other hand, if $N \geq 1$ customers are present in the system, the server must terminate serving the first customer in the system and in addition he must provide service to the remaining $N - 1$ customers in the system. Let $S_1, S_2, \ldots, S_N$ denote, respectively, the service times of the customer in service, the next customer in line, . . . , the Nth customer in the system. Thus

$$V = \begin{cases} 0, & N = 0 \\ S_1' + S_2 + \cdots + S_N, & N \geq 1 \end{cases}$$

where S_1' denotes the residual service time of the customer in service. Since service time has a negative exponential distribution, it follows that it is memoryless; hence the distribution of S_1' is identical to the distribution of S_1. We can therefore write

$$P\{V \leq v, N = n\} = \begin{cases} 0, & v = 0; 1 \leq n < \infty \\ P\{N = 0\}, & v = 0; n = 0 \\ P\{S_1 + S_2 + \cdots + S_N \leq v, N = n\}, \\ & 0 < v < \infty; 1 \leq n < \infty \end{cases}$$

The (marginal) distribution function of V is

$$\Phi_V(v) = P\{V \leq v\} = \sum_{n=0}^{\infty} P\{V \leq v, N = n\} = \sum_{n=0}^{\infty} P\{V \leq v \mid N = n\} \cdot P\{N = n\}$$

or

$$\Phi_V(v) = \begin{cases} P\{N = 0\}, & v = 0 \\ \sum_{n=1}^{\infty} P\{S_1 + S_2 + \cdots + S_N \leq v \mid N = n\} \cdot P\{N = n\}, & 0 < v < \infty \end{cases}$$

Since the service times are independently and identically distributed, with PDF $\varphi_S(x) = \mu e^{-\mu x}$, it follows that the distribution of their sum $S_1 + S_2 + \cdots + S_{N=n}$ is the nth-fold convolution of $\varphi_S(x)$ with itself, that is, a gamma distribution of order n and parameter μ; hence for $n \geq 1$

$$P\{S_1 + S_2 + \cdots + S_N \leq v \mid N = n\} = \int_0^v \frac{(\mu x)^{n-1}}{\Gamma(n)} \mu e^{-\mu x} \, dx$$

Also, from (4.11),

$$P\{N = n\} = P_n = (1 - \rho)\rho^n, \quad 0 < \rho < 1; 0 \leq n < \infty$$

The expression for $\Phi_V(v)$ may be written

$$\Phi_V(v) = \begin{cases} 1 - \rho, & v = 0 \\ \sum_{n=1}^{\infty} (1 - \rho)\rho^n \int_0^v \frac{(\mu x)^{n-1}}{\Gamma(n)} \mu e^{-\mu x} \, dx, & 0 < v < \infty \end{cases}$$

$$= \begin{cases} 1 - \rho, & v = 0 \\ \int_0^v \rho(1 - \rho)\mu e^{-\mu x} \sum_{n=1}^{\infty} \frac{(\mu\rho x)^{n-1}}{(n-1)!} \, dx, & 0 < v < \infty \end{cases}$$

The sum expression can be recognized as the quantity $e^{\mu\rho x}$; hence we obtain

$$\Phi_V(v) = \begin{cases} 1 - \rho, & v = 0 \\ \lambda(1 - \rho) \int_0^v e^{-\mu(1-\rho)x} \, dx, & 0 < v < \infty \end{cases}$$

Thus the distribution of the virtual waiting time V has a discontinuity at the origin and is continuous in the range $0 < v < \infty$. Alternatively, we can write

$$\Phi_V(0) = 1 - \rho, \qquad\qquad v = 0$$
$$d\Phi_V(v) = \lambda(1 - \rho)e^{-\mu(1-\rho)v} \, dv, \quad 0 < v < \infty \tag{4.14}$$

This distribution resembles the negative exponential distribution, although it is not identical to it.

Exercise 4.2

Obtain the Laplace transform of expression (4.14). Hence or otherwise derive an expression for $E[V]$ and Var $[V]$.

For the $M/M/1$ system, it can be shown that the steady-state distribution of the time Θ a customer will spend in the system is the same as the distribution of the virtual waiting time V. This property does not hold for any queuing system, and the reader should be careful in distinguishing between the two quantities Θ and V.

f. Use of Little's Formula

Since

$$\lambda = \frac{1}{E[T]} \quad \text{and} \quad \frac{1}{\mu} = E[S]$$

we have

$$E[N] = \lambda E[\Theta]$$

and

$$E[L] = \lambda E[W]$$

Thus

$$E[\Theta] = \frac{E[N]}{\lambda} = \frac{1}{\lambda} \frac{\rho}{1 - \rho}$$

Now

$$E[\Theta] = E[W] + E[S]$$

Hence

$$E[W] = E[\Theta] - E[S]$$

$$= \frac{1}{\lambda} \frac{\rho}{1 - \rho} - \frac{1}{\mu}$$

$$= \frac{\rho^2}{\lambda(1 - \rho)}$$

g. Analyses of the Busy and Idle Cycles

We have seen that in the steady state the probability that there are no units in the system is $P_0 = 1 - \rho$. Hence P_0 is also the probability that the server is *free* or *idle* or *unoccupied*. The probability that the server is *busy* is the probability that at least one unit is present in the system and that is $1 - \rho$.

In the steady state, when $\rho < 1$, ρ would correspond to the proportion of time the server is idle while $1 - \rho$ would correspond to the proportion of time the server is busy. The server thus goes through a succession of busy and idle periods as shown:

As ρ, $0 < \rho < 1$, increases from zero to one, the proportion of time the server is busy increases proportionately; in other words, as the traffic intensity increases, the server becomes more occupied. When ρ tends to unity, the queuing system tends to a saturation state, and the server will practically be busy all the time.

A steady-state condition does not exist when $\rho \geq 1$, that is, when the arrival rate is equal to or exceeds the service rate. For such cases, as time elapses, the queue size grows up indefinitely, a condition corresponding in practice to traffic congestions or bottlenecks.

Assume now that a steady state regime has been reached ($\rho < 1$) and let

B = random variable denoting the busy cycle

I = random variable denoting the idle cycle

It is possible to obtain expressions for the expected time the server is idle and the expected time the server is busy. Since arrivals are Poisson, the distribution of the idle cycle I is negative exponential with parameter λ. This follows from the memoryless property of the interarrival time distribution: the time elapsed till the appearance of the first customer initiating a busy cycle given that the previous busy cycle has terminated has the negative exponential distribution $\lambda e^{-\lambda x}$. Hence

$$E[I] = \frac{1}{\lambda}$$

Since ρ is the proportion of time the server is busy, then

$$\rho = \frac{E[B]}{E[B] + E[I]}$$

$$= \frac{E[B]}{E[B] + (1/\lambda)}$$

It immediately follows that

$$E[B] = \frac{\rho}{\lambda(1 - \rho)}$$

4.5. The *M/M/r* Queuing System

The description of this queuing system is very similar to the $M/M/1$ system except for the following:

(1) The system consists of r servers, $r = 1, 2, \ldots$, in parallel, and arriving customers form a single line. Customers enter the service facilities when they become vacant on a first-come first-served basis.

(2) The service times provided by the stations to any of the arriving customers are identically and independently distributed, each having the exponential distribution $\varphi_S(x) = \mu e^{-\mu x}$, $\mu > 0$, $0 < x < \infty$.

The arrival and service process can thus be characterized by saying that

(1) $\lambda \, dt + o(dt)$ is the probability that one customer will join the system in the time interval $(t, t + dt]$. The probability that no customers will join the queue in the same time interval is $[1 - \lambda \, dt + o(dt)]$. Finally, the probability that more than one customer will join the queue in the interval $(t, t + dt]$ is $o(dt)$.

(2) When no customers are present in the system, no service is provided. If the number of customers n in the system is less than the number of servers, r, $1 \le n < r$, the probability that exactly one customer will terminate service in the time interval $(t, t + dt]$ is $n\mu \, dt + o(dt)$. If the number of customers in the system is greater than or equal to the number of servers r, $r \le n < \infty$, the probability that exactly one customer will terminate service in the time interval $(t, t + dt]$ is $r\mu \, dt + o(dt)$. The probability that more than one customer will terminate service in the same time interval is $o(dt)$.

The reader should set up the transition probabilities for this time homogeneous birth and death process. In what follows, $\{N(t), t \ge 0\}$ will again describe the number of customers in the system at time t, and we shall define

$$P(n, t) = P\{N(t) = n\}, \quad n = 0, 1, 2, \ldots$$

i. *Steady-State Probability Distribution of the Number of Customers in the System*

It is easy to verify that for $t \ge 0$, $P(n, t)$ satisfies the following system of equations:

$$P(0, t + dt) = P(0, t)(1 - \lambda \, dt) + P(1, t)(1 - \lambda \, dt)\mu \, dt + o(dt), \quad n = 0$$

$$\begin{aligned} P(n, t + dt) = {}& P(n, t)(1 - \lambda \, dt)(1 - n\mu \, dt) \\ & + P(n - 1, t)\lambda \, dt[1 - (n - 1)\mu \, dt] \\ & + P(n + 1, t)(1 - \lambda \, dt)(n + 1)\mu \, dt + o(dt), \quad 1 \le n \le r - 1 \end{aligned}$$

$$\begin{aligned} P(n, t + dt) = {}& P(n, t)(1 - \lambda \, dt)(1 - r\mu \, dt) + P(n - 1, t)\lambda \, dt(1 - r\mu \, dt) \\ & + P(n + 1, t)(1 - \lambda \, dt)r\mu \, dt + o(dt), \quad n \ge r \end{aligned}$$

This reduces to the following set of differential difference equations:

$$\frac{dP(0, t)}{dt} = -\lambda P(0, t) + \mu P(1, t), \quad n = 0$$

$$\frac{dP(n, t)}{dt} = \lambda P(n - 1, t) - (\lambda + n\mu)P(n, t) + (n + 1)\mu P(n + 1, t),$$
$$1 \leq n \leq r - 1$$

$$\frac{dP(n, t)}{dt} = \lambda P(n - 1, t)(\lambda + r\mu)P(n, t) + r\mu P(n + 1, t), \quad n \geq r$$

Assume steady-state conditions, and let $P_n = \lim_{t \to \infty} P(n, t)$; then

$$0 = -\lambda P_0 + \mu P_1, \quad n = 0$$
$$0 = \lambda P_{n-1} - (\lambda + n\mu)P_n + (n + 1)\mu P_{n+1}, \quad 1 \leq n \leq r - 1 \quad (4.15)$$
$$0 = \lambda P_{n-1} - (\lambda + r\mu)P_n + r\mu P_{n+1}, \quad n \geq r$$

and
$$\sum_{n=0}^{\infty} P_n = 1$$

It may be shown that existence of a steady state implies $\lambda/\mu r < 1$. The above system of equations will be solved using three different approaches.

a. Use of Probability Generating Function

Let
$$G_N(z) = \sum_{n=0}^{\infty} z^n P_n, \quad |z| \leq 1$$

denote the PGF of the random variable N for the number of customers in the system in the steady state. Then multiplying equations (4.15) by z^{n+1} and summing over the appropriate ranges of n, we obtain

$$0 = \sum_{n=1}^{\infty} \lambda z^{n+1} P_{n-1} - \sum_{n=0}^{\infty} \lambda z^{n+1} P_n - \sum_{n=1}^{n=r-1} n\mu z^{n+1} P_n - \sum_{n=r}^{\infty} r\mu z^{n+1} P_n$$
$$+ \sum_{n=0}^{n=r-1} (n + 1)\mu z^{n+1} P_{n+1} + \sum_{n=r}^{\infty} r\mu z^{n+1} P_{n+1}$$

or

$$0 = \lambda z^2 \sum_{n=1}^{\infty} z^{n-1} P_{n-1} - \lambda z \sum_{n=0}^{\infty} z^n P_n - \mu z \sum_{n=1}^{n=r} nz^n P_n - r\mu z \sum_{n=r+1}^{\infty} z^n P_n$$
$$+ \mu \sum_{n=1}^{n=r} nz^n P_n + r\mu \sum_{n=r+1}^{\infty} z^n P_n$$

or

$$0 = \lambda z^2 \sum_{n=0}^{\infty} z^n P_n - \lambda z \sum_{n=0}^{\infty} z^n P_n + \mu(1 - z) \sum_{n=1}^{n=r} nz^n P_n + r\mu(1 - z) \sum_{n=r+1}^{\infty} z^n P_n$$
$$= \lambda z^2 G_N(z) - \lambda z G_N(z) + \mu(1 - z) \sum_{n=1}^{n=r} nz^n P_n$$
$$+ r\mu(1 - z) \left[G_n(z) - \sum_{n=0}^{n=r} z^n P_n \right]$$

Transposing, we obtain

$$\lambda z(1-z)G_N(z) - r\mu(1-z)G_N(z) = \mu(1-z)\sum_{n=1}^{n=r} nz^n P_n - r\mu(1-z)\sum_{n=0}^{n=r} z^n P_n$$

Solving for $G_N(z)$ yields

$$G_N(z) = \frac{\mu\left[\sum\limits_{n=1}^{n=r} nz^n P_n - \sum\limits_{n=0}^{n=r} rz^n P_n\right]}{\lambda z - \mu r}$$

Dividing numerator and denominator by μ and setting $\rho = \lambda/\mu$, we obtain

$$G_N(z) = \frac{\sum\limits_{n=0}^{n=r}(n-r)z^n P_n}{\rho z - r}$$

or
$$G_N(z) = \frac{\sum\limits_{n=0}^{n=r}(r-n)z^n P_n}{r - \rho z} \qquad (4.16)$$

To determine the unknowns $P_0, P_1, \ldots, P_{r-1}$, we write

$$G_N(z) = \frac{1}{r} \frac{(r-0)P_0 + (r-1)zP_1 + (r-2)z^2 P_2 + \cdots + z^{r-1}P_{r-1}}{1 - \frac{\rho}{r}z}$$

Since $\rho/r < 1$, then expanding $G_N(z)$ and $[1 - (\rho/r)z]^{-1}$ we get the following identity in z:

$$P_0 + P_1 z + P_2 z^2 + P_3 z^3 + \cdots$$

$$\equiv \frac{1}{r}\left(1 + \frac{\rho}{r}z + \frac{\rho^2}{r^2}z^2 + \frac{\rho^3}{r^3}z^3 + \cdots\right)$$

$$\times [rP_0 + (r-1)zP_1 + (r-2)z^2 P_2 + \cdots + z^{r-1}P_{r-1}]$$

Equating the coefficients of equal power of z, we obtain

$$P_0 = P_0$$

$$P_1 = \frac{1}{r}\left[\frac{\rho}{r}rP_0 + (r-1)P_1\right]$$

$$P_2 = \frac{1}{r}\left[\frac{\rho^2}{r^2}rP_0 + \frac{\rho}{r}(r-1)P_1 + (r-2)P_2\right]$$

$$P_3 = \frac{1}{r}\left[\frac{\rho^3}{r^3}rP_0 + \frac{\rho^2}{r^2}(r-1)P_1 + \frac{\rho}{r}(r-2)P_2 + (r-3)P_3\right]$$

$$\cdots$$

$$P_{r-1} = \frac{1}{r}\left[\frac{\rho^{r-1}}{r^{r-1}}rP_0 + \frac{\rho^{r-2}}{r^{r-2}}(r-1)P_1 + \cdots + P_{r-1}\right]$$

Solving recursively, we can obtain $P_1, P_2, \ldots, P_{r-1}$ in terms of P_0; thus

$$rP_1 = \rho P_0 + rP_1 - P_1$$

or $$P_1 = \rho P_0$$

Also, $$rP_2 = \frac{\rho^2}{r} P_0 + \left(\rho - \frac{\rho}{r}\right) P_1 + rP_2 - 2P_2$$

or $$2P_2 = \frac{\rho^2}{r} P_0 + \rho^2 P_0 - \frac{\rho^2}{r} P_0$$

or $$P_2 = \frac{\rho^2}{2!} P_0$$

Similarly, $$P_3 = \frac{\rho^3}{3!} P_0$$

$$\ldots$$

and $$P_{r-1} = \frac{\rho^{r-1}}{(r-1)!} P_0$$

Substituting for $P_1, P_2, \ldots, P_{r-1}$ in expression (4.16) for $G_N(z)$ yields

$$G_N(z) = \frac{P_0 \sum_{n=0}^{n=r} (r-n) z^n \dfrac{\rho^n}{n!}}{r - \rho z} \tag{4.17}$$

To obtain P_0, let $z = 1$ in (4.17), then since $G_N(1) = 1$, we obtain

$$1 = \frac{P_0 \sum_{n=0}^{n=r} (r-n) \dfrac{\rho^n}{n!}}{r - \rho}$$

Solving for P_0, we obtain

$$P_0 = \frac{r - \rho}{\sum_{n=0}^{n=r} (r-n) \dfrac{\rho^n}{n!}} \tag{4.18}$$

Substituting (4.18) in expression (4.17), we finally obtain

$$G_N(z) = \frac{r - \rho}{\sum_{n=0}^{n=r} (r-n) \dfrac{\rho^n}{n!}} \frac{\sum_{n=0}^{n=r} (r-n) \dfrac{(\rho z)^n}{n!}}{r - \rho z}$$

An expression for P_n can then be obtained for all n, $n = 0, 1, 2, \ldots$

b. RECURSIVE APPROACH

It is possible to obtain the values of P_n, $n = 0, 1, 2, \ldots$, recursively using the steady-state equations. Equations (4.15) can be written as

$$0 = -\rho P_0 + P_1, \quad n = 0 \tag{4.19}$$

$$0 = \rho P_{n-1} - (\rho + n)P_n + (n + 1)P_{n+1}, \quad 1 \leq n \leq r - 1 \tag{4.20}$$

$$0 = \rho P_{n-1} - (\rho + r)P_n + rP_{n+1}, \quad r \leq n \leq \infty \tag{4.21}$$

Thus, from equation (4.19),

$$P_1 = \rho P_0$$

From equation (4.20), if we let $n = 1$, we obtain

$$2P_2 = (\rho + 1)P_1 - \rho P_0$$

$$= (\rho + 1)\rho P_0 - \rho P_0$$

hence

$$P_2 = \frac{\rho^2}{2} P_0$$

Setting $n = 2$ yields

$$3P_3 = (\rho + 2)P_2 - \rho P_1$$

$$= (\rho + 2)\frac{\rho^2}{2} P_0 - \rho^2 P_0$$

hence

$$P_3 = \frac{\rho^3}{3!} P_0$$

$$\cdots$$

$$P_{r-1} = \frac{\rho^{r-1}}{(r - 1)!} P_0$$

and

$$P_r = \frac{\rho^r}{r!} P_0$$

From equation (4.21), we obtain by setting $n = r$,

$$rP_{r+1} = (\rho + r)P_r - \rho P_{r-1}$$

$$= (\rho + r)\frac{\rho^r}{r!} P_0 - \rho \frac{\rho^{r-1}}{(r - 1)!} P_0$$

Thus

$$P_{r+1} = \frac{\rho^{r+1}}{r \cdot r!} P_0$$

and in general

$$P_n = \frac{\rho^n}{r^{n-r} r!} P_0, \quad n = r + 1, r + 2, \ldots$$

Therefore,

$$P_n = \begin{cases} \dfrac{\rho^n}{n!} P_0, & n = 0, 1, 2, \ldots, r \\[2ex] \dfrac{\rho^n}{r^{n-r} r!} P_0, & n = r + 1, r + 2, \ldots \end{cases}$$

To obtain P_0, we use the normalizing condition $\sum_{n=0}^{\infty} P_n = 1$, or

$$1 = P_0 \sum_{n=0}^{n=r} \frac{\rho^n}{n!} + P_0 \sum_{n=r+1}^{\infty} \frac{\rho^n}{r^{n-r} r!}$$

Solving for P_0, we obtain

$$P_0 = \frac{1}{\sum\limits_{n=0}^{n=r} \dfrac{\rho^n}{n!} + \sum\limits_{n=r+1}^{\infty} \dfrac{\rho^n}{r^{n-r} r!}}$$

$$= \frac{1}{\sum\limits_{n=0}^{n=r-1} \dfrac{\rho^n}{n!} + \sum\limits_{n=r}^{\infty} \rho^r \dfrac{\rho^{n-r}}{r^{n-r}} \dfrac{1}{r!}}$$

$$= \frac{1}{\sum\limits_{n=0}^{n=r-1} \dfrac{\rho^n}{n!} + \dfrac{\rho^r}{r!} \sum\limits_{i=0}^{\infty} \left(\dfrac{\rho}{r}\right)^i}$$

$$= \frac{1}{\sum\limits_{n=0}^{n=r-1} \dfrac{\rho^n}{n!} + \dfrac{\rho^r}{r!} \dfrac{r}{r - \rho}}$$

This expression can be shown to be equivalent to expression (4.18). Multiplying numerator and denominator by $e^{-\rho}$ yields

$$P_0 = \frac{e^{-\rho}}{\sum\limits_{n=0}^{n=r-1} e^{-\rho} \dfrac{\rho^n}{n!} + e^{-\rho} \dfrac{\rho^r}{r!} \dfrac{r}{r - \rho}}$$

Consider the Poisson distribution with parameter ρ. For $r = 0, 1, 2, \ldots$, let

$$\hat{p}(r; \rho) = e^{-\rho} \frac{\rho^r}{r!}$$

$$\hat{P}(r; \rho) = \sum_{i=0}^{i=r} e^{-\rho} \frac{\rho^i}{i!}$$

Then
$$P_0 = \frac{e^{-\rho}}{\hat{P}(r - 1; \rho) + \dfrac{r}{r - \rho} \hat{p}(r; \rho)}$$

Thus the probability of having exactly n customers in the system is

$$P_n = \begin{cases} \dfrac{\hat{p}(n; \rho)}{\hat{P}(r - 1; \rho) + \dfrac{r}{r - \rho} \hat{p}(r; \rho)}, & n = 0, 1, 2, \ldots, r \\[4ex] \dfrac{e^{-\rho} \dfrac{\rho^n}{r^{n-r} r!}}{\hat{P}(r - 1; \rho) + \dfrac{r}{r - \rho} \hat{p}(r; \rho)}, & n = r + 1, r + 2, \ldots \end{cases}$$

Since
$$\hat{p}(r; \rho) = \hat{P}(r; \rho) - \hat{P}(r - 1; \rho)$$

one need only have a table for the cumulative Poisson distribution to compute all values of P_n, $n = 0, 1, 2, \ldots$. Alternatively, one may use a table of incomplete gamma functions to compute the P_n by noting that (Problem 7 in Chapter 1)

$$\hat{P}(r; \rho) = 1 - \frac{1}{r!} \int_0^\rho x^r e^{-x} \, dx$$

c. Flow-Diagram Approach

The flow diagram for this problem is as shown in Figure 4.4. The steady-state equations may be obtained by analyzing either the flow at the nodes

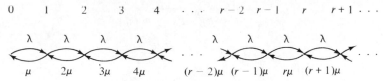

Figure 4.4. Flow diagram for the steady state $M/M/r$ queuing system

or the flow through cuts at the arcs. The reader is asked to set up the equations for the second case. We consider here the flow at the nodes and obtain the following system of equations, corresponding to the conservation of flow at each of the nodes $0, 1, 2, \ldots$ respectively:

$$\mu P_1 = \lambda P_0$$
$$\lambda P_0 + 2\mu P_2 = (\lambda + \mu)P_1$$
$$\lambda P_1 + 3\mu P_3 = (\lambda + 2\mu)P_2$$
$$\cdots$$
$$\lambda P_{r-2} + r\mu P_r = [\lambda + (r-1)\mu]P_{r-1}$$
$$\lambda P_{r-1} + r\mu P_{r+1} = (\lambda + r\mu)P_r$$
$$\lambda P_r + r\mu P_{r+2} = (\lambda + r\mu)P_{r+1}$$
$$\cdots$$

This may be written as

$$P_1 = \rho P_0$$
$$\rho P_0 + 2P_2 = (1 + \rho)P_1$$
$$\rho P_1 + 3P_3 = (2 + \rho)P_2$$
$$\cdots$$
$$\rho P_{r-2} + rP_r = [(r-1) + \rho]P_{r-1}$$
$$\rho P_{r-1} + rP_{r+1} = (r + \rho)P_r$$
$$\rho P_r + rP_{r+2} = (r + \rho)P_{r+1}$$
$$\cdots$$

Solving recursively, we obtain

$$P_1 = \rho P_0$$

$$P_2 = \frac{1}{2}[(1 + \rho)P_1 - \rho P_0]$$

$$= \frac{1}{2}[(1 + \rho)\rho P_0 - \rho P_0]$$

$$= \frac{\rho^2}{2}P_0$$

$$P_3 = \frac{1}{3}[(2 + \rho)P_2 - \rho P_1]$$

$$= \frac{1}{3}\left[(2 + \rho)\frac{\rho^2}{2}P_0 - \rho^2 P_0\right]$$

$$= \frac{\rho^3}{3!}P_0$$

$$\cdots$$

$$P_{r-1} = \frac{\rho^{r-1}}{(r-1)!}P_0$$

and
$$P_r = \frac{\rho^r}{r!}P_0$$

Also
$$P_{r+1} = \frac{1}{r}\left[(r + \rho)\frac{\rho^r}{r!}P_0 - \rho\frac{\rho^{r-1}}{(r-1)!}P_0\right]$$

$$= \frac{\rho^{r+1}}{r \cdot r!}P_0$$

and
$$P_{r+2} = \frac{1}{r}[(r + \rho)P_{r+1} - \rho P_r]$$

$$= \frac{1}{r}\left[(r + \rho)\frac{\rho^{r+1}}{r \cdot r!}P_0 - \rho\frac{\rho^r}{r!}P_0\right]$$

$$= \frac{\rho^{r+2}}{r^2 \cdot r!}P_0$$

Finally,
$$P_n = \frac{\rho^n}{r^{n-r} \cdot r!}P_0, \quad n = r + 1, r + 2, \ldots$$

Thus,
$$P_n = \begin{cases} \dfrac{\rho^n}{n!}P_0, & n = 0, 1, 2, \ldots, r \\ \dfrac{\rho^n}{r^{n-r} \cdot r!}P_0, & n = r + 1, r + 2, \ldots \end{cases}$$

In general, given the numerical value of P_0 for a particular value of $\rho = \lambda/\mu$ and r, the various values of P_n may be computed. Table 4.1 exhibits the values of P_0 for different values of r and ρ. Since steady-state conditions

Table 4.1. STEADY-STATE PROBABILITY P_0 OF AN EMPTY SYSTEM IN THE $M/M/r$ QUEUE (r = NUMBER OF CHANNELS; λ = ARRIVAL RATE; μ = SERVICE RATE PER CHANNEL)

				r				
$\rho = \lambda/\mu$	*1*	*2*	*3*	*4*	*5*	*6*	...	∞
0.250	0.75000	0.77778	0.77876	0.77880	0.77880	0.77880	...	0.77880
0.500	0.50000	0.60000	0.60606	0.60650	0.60653	0.60653	...	0.60653
0.750	0.25000	0.45455	0.47059	0.47219	0.47235	0.47237	...	0.47237
1.000		0.33333	0.36364	0.36735	0.36782	0.36787	...	0.36788
1.250		0.23077	0.27861	0.28533	0.28634	0.28648	...	0.28650
1.500		0.14286	0.21053	0.22099	0.22277	0.22307	...	0.22313
1.750		0.06667	0.15564	0.17038	0.17314	0.17366	...	0.17377
2.000			0.11111	0.13043	0.13433	0.13514	...	0.13533
2.250			0.07477	0.09881	0.10394	0.10508	...	0.10540
2.500			0.04494	0.07369	0.08010	0.08162	...	0.08209
2.750			0.02036	0.05370	0.06138	0.06329	...	0.06393
3.000				0.03774	0.04665	0.04896	...	0.04979
3.250				0.02497	0.03505	0.03774	...	0.03877
3.500				0.01475	0.02590	0.02896	...	0.03020
3.750				0.00656	0.01868	0.02208	...	0.02352
4.000					0.01299	0.01669	...	0.01832
4.250					0.00850	0.01246	...	0.01426
4.500					0.00496	0.00914	...	0.01111
4.750					0.00218	0.00654	...	0.00865
5.000						0.00451	...	0.00674
5.250						0.00293	...	0.00525
5.500						0.00169	...	0.00409
5.750						0.00074	...	0.00318

are assumed, $\rho/r < 1$. The case of $r = 1$ corresponds to the single channel system, while the case $r = \infty$ corresponds to the case of a queuing system with an infinite number of channels, namely the $M/M/\infty$ system. For small values of ρ/r, it may be noted that the $M/M/\infty$ system provides an excellent approximation for computing the value of P_0.

Exercise 4.3

Find $E[N]$ and Var $[N]$.

ii. *Probability That Exactly m Servers are Busy,* $0 \leq m \leq r$

Let the random variable $\mathfrak{M}$ denote the number of busy servers at a particular time in the steady state; then

$$P\{\mathfrak{M} = m\} = \begin{cases} P_m, & \text{for } 0 \leq m \leq r-1 \\ \sum_{n=r}^{\infty} P_n, & \text{for } m = r \end{cases}$$

This can readily be seen since if there are exactly m customers, $m < r$, in the system, exactly m servers will be busy and exactly $r - m$ servers will be idle. If, on the other hand, there are r or more customers in the system, then all servers will be busy. Now for $m = r$

$$P\{\mathfrak{M} = m\} = \sum_{n=r}^{\infty} P_n$$

$$= \sum_{n=r}^{\infty} \frac{\rho^n}{r^{n-r} r!} P_0$$

$$= \frac{\rho^r}{r!} \sum_{n=r}^{\infty} \frac{\rho^{n-r}}{r^{n-r}} P_0$$

$$= \frac{\rho^r}{r!} \frac{r}{r - \rho} P_0$$

Hence
$$P\{\mathfrak{M} = m\} = \begin{cases} \dfrac{\rho^m}{m!} P_0, & 0 \leq m \leq r - 1 \\[2ex] \dfrac{\rho^r}{r!} \dfrac{r}{r - \rho} P_0, & m = r \end{cases} \tag{4.22}$$

The expected number of busy servers is given by

$$E[\mathfrak{M}] = \sum_{m=0}^{m=r} m P\{\mathfrak{M} = m\}$$

$$= \sum_{m=0}^{m=r-1} m \frac{\rho^m}{m!} P_0 + r \frac{\rho^r}{r!} \frac{r}{r - \rho} P_0$$

$$= \rho \left[\sum_{m=1}^{m=r-1} \frac{\rho^{m-1}}{(m-1)!} + \frac{\rho^{r-1}}{(r-1)!} \frac{r}{r - \rho} \right] P_0$$

$$= \rho P_0 \left[\sum_{m=0}^{m=r-1} \frac{\rho^m}{m!} - \frac{\rho^{r-1}}{(r-1)!} + \frac{\rho^{r-1}}{(r-1)!} \frac{r}{r - \rho} \right]$$

$$= \rho P_0 \left[\sum_{m=0}^{m=r-1} \frac{\rho^m}{m!} + \frac{\rho}{r - \rho} \frac{\rho^{r-1}}{(r-1)!} \right]$$

$$= \rho P_0 \left[\sum_{m=0}^{m=r-1} \frac{\rho^m}{m!} + \frac{\rho^r}{r!} \frac{r}{r - \rho} \right]$$

Using the expression for P_0, we find

$$E[\mathfrak{M}] = \rho \tag{4.23}$$

Note that the expected number of busy servers is independent of the number of servers in the system. The expected number of idle servers is thus $r - \rho$.

It is possible to use the results of expression (4.22) to compute such quantities as the probability that a customer will not have to wait prior to being serviced or the probability that at least a given number of servers will be busy or other statistics. As an illustration, we determine the probability that a customer will not have to wait in line, that is, the probability that immediate service will be provided to an arriving customer. This is the same as the probability of having at least one server idle, which is the same as

$$1 - \text{(probability that all servers are busy)}$$

The probability that all servers are busy is

$$\sum_{n=r}^{\infty} P_n = \frac{\rho^r}{r!} \frac{r}{r - \rho} P_0$$

Hence the probability that a customer will not have to wait is

$$1 - \frac{\rho^r}{r!} \frac{r}{r - \rho} P_0$$

iii. Steady-State Distribution of the Number Waiting in the System

Let L be a random variable defining the number of customers waiting in line in the steady state. There will be no customers in line if $n \leq r$, and there will be $n - r$ customers in line if $n \geq r + 1$; thus

$$L = \begin{cases} 0, & 0 \leq N \leq r \\ N - r, & r + 1 \leq N < \infty \end{cases}$$

Therefore,

$$P\{L = l\} = \begin{cases} \sum_{n=0}^{n=r} P_n, & l = 0 \\ P\{N - r = l\}, & l \geq 1 \end{cases}$$

These quantities may be computed from the expression for P_n. The expected queue length is

$$E[L] = \sum_{l=0}^{\infty} l P\{L = l\}$$

$$= \sum_{l=0}^{\infty} l P\{N = l + r\}$$

$$= \sum_{n=r}^{\infty} (n - r) P_n$$

$$= \sum_{n=r}^{\infty} (n - r) \frac{\rho^n}{r^{n-r} r!} P_0$$

$$= \frac{\rho^r}{r!} P_0 \sum_{n=r}^{\infty} (n - r) \frac{\rho^{n-r}}{r^{n-r}}$$

$$= \frac{\rho^r}{r!} P_0 \sum_{i=0}^{\infty} i \left(\frac{\rho}{r}\right)^i$$

but
$$\sum_{i=0}^{\infty} i\left(\frac{\rho}{r}\right)^{i} = \frac{\rho}{r}\sum_{i=1}^{\infty} i\left(\frac{\rho}{r}\right)^{i-1}$$

$$= \frac{\rho}{r}\frac{\partial}{\partial\left(\frac{\rho}{r}\right)}\left\{\sum_{i=0}^{\infty}\left(\frac{\rho}{r}\right)^{i}\right\}$$

$$= \frac{\rho}{r}\frac{\partial}{\partial\left(\frac{\rho}{r}\right)}\left\{\frac{1}{1-\frac{\rho}{r}}\right\}$$

$$= \frac{\rho}{r}\frac{1}{\left(1-\frac{\rho}{r}\right)^{2}}$$

$$= \frac{\rho r}{(r-\rho)^{2}}$$

Hence
$$E[L] = \frac{\rho^{r}}{r!}\frac{\rho r}{(r-\rho)^{2}}P_{0}$$

It is possible to obtain another expression for $E[L]$. The reader should verify the following results:

For $r = 1$,
$$E[L] = \frac{\rho^{2}}{1-\rho}$$

For $r = 2$,
$$E[L] = \frac{\rho^{3}}{4-\rho^{2}}$$

For $r = 3$,
$$E[L] = \frac{\rho^{4}}{(3-\rho)(6+4\rho+\rho^{2})}$$

and, in general, for arbitrary r, we have

$$E[L] = \frac{\rho^{r+1}}{(r-\rho)\left[r! + \rho\left(\frac{r!}{1!} - \frac{(r-1)!}{0!}\right) + \cdots + \rho^{r}\left(\frac{r!}{r!} - \frac{(r-1)!}{(r-1)!}\right)\right]}$$

$$= \frac{\rho^{r+1}}{(r-\rho)\sum_{i=0}^{i=r}\frac{r!}{i!}\left(1-\frac{i}{r}\right)\rho^{i}} \tag{4.24}$$

Example 4.2

In a manufacturing center, mechanics arrive at a tool crib at the rate of 0.80 per minute according to a Poisson process. The crib is serviced by r

clerks ($r = 1, 2, \ldots$). The service time provided by each clerk to mechanics is assumed to have a negative exponential distribution with an average of 1 minute. If all clerks are busy, the arriving mechanics wait and are subsequently serviced on a first-come first-served basis by the clerks as soon as they become free. No more than one clerk may service a mechanic and it is assumed that the population of mechanics is large.

The mechanics are paid at the rate of $C_1 = \$10.00$ per hour each, while the clerks at the crib are paid at the rate of $C_2 = \$4.00$ per hour. Determine the number of clerks r to be assigned to the crib so as to minimize the sum total of expected costs per hour, consisting of the time loss of the mechanics while waiting and being serviced and the idle time of the clerks while free.

In this $M/M/r$ queuing system, $\lambda = .80$ per minute, while $\mu = 1$ per minute. Thus, $\rho = \lambda/\mu = .80/1 = .80$. For the system to be nonsaturated and reach equilibrium $\rho/r < 1$, or $r > .80$; hence, a single clerk at the crib would satisfy the condition of nonsaturation. However, this need not be the most economical configuration.

If the expected waiting time of a mechanic is $E[W]$, the expected time spent by each mechanic in the system is:

$$E[\Theta] = E[W] + \frac{1}{\mu}$$

Using Little's formula, the expected number of mechanics in the system is

$$
\begin{aligned}
E[N] &= \lambda E[\Theta] \\
&= \lambda \left[E[W] + \frac{1}{\mu} \right] \\
&= \lambda E[W] + \rho
\end{aligned}
$$

Now the expected number of mechanics in the queue waiting to be serviced is given by Little's formula as

$$E[L] = \lambda E[W]$$

Thus we may write,

$$E[N] = E[L] + \rho$$

Using relation (4.24), we obtain as an expression for $E[N]$

$$E[N] = \frac{\rho^{r+1}}{(r - \rho) \sum_{i=0}^{i=r} \frac{r!}{i!} \left(1 - \frac{i}{r} \right) \rho^i} + \rho$$

The expected cost per hour of the lost time by the mechanics while waiting and being serviced is $C_1 E[N]$. Now, from expression (4.23), the expected number of idle servers is $r - \rho$. Thus the expected cost per hour that clerks are idle in the tool crib is $C_2(r - \rho)$.

The expression for the total expected cost per hour is

$$C_1 E[N] + C_2(r - \rho)$$

It is evident that the expression for $E[N]$ is a function of r, the total number of clerks serving at the tool crib. To determine the optimum value of r, $r = 1$, $2, \ldots$, which minimizes total expected cost per hour, it is necessary to obtain the variation of $C_1 E[N] + C_2(r - \rho)$ as a function of r. The following table may be formed.

r	$E[L]$	$E[N] = E[L] + \rho$	$C_1 E[N]$	$(r - \rho)$	$C_2(r - \rho)$	$C_1 E[N] + C_2(r - \rho)$
1	3.200	4.800	48.00	0.200	0.80	48.80
2	0.153	0.953	9.53	1.200	4.80	14.33 ←
3	0.019	0.819	8.19	2.200	8.80	16.99
4	0.005	0.805	8.05	3.200	12.80	20.85
5	0.000	0.800	8.00	4.200	16.80	24.80
6	0.000	0.800	8.00	5.200	20.80	28.80

The optimum number of clerks to assign to the tool crib is seen to be 2 at a minimum total expected cost per hour of $14.33. It is interesting to note from the table that, as r gets larger, the corresponding value of $E[L]$ becomes negligible; thus, as an approximation for large r, the expression for the total expected cost per unit time is

$$C_1 E[N] + C_2(r - \rho) = C_1[E[L] + \rho] + C_2(r - \rho)$$
$$\approx C_1 \rho + C_2(r - \rho)$$

4.6. The Modified *M/G/*1 Queuing System

The modified $M/G/1$ queuing system is a variation of the $M/G/1$ queuing system originally studied by Kendall [8]. Its theoretical importance stems from its extensive utilization in analyzing queuing systems with several classes of priorities. The assumptions underlying the arrival process is the same as the $M/M/1$ system, that is, it is Poisson with intensity λ. The service times provided to each customer are such that every customer that initiates a busy period has service time S_0 with distribution function $\Phi_{S_0}(\cdot)$; all other customers have service time S with distribution function $\Phi_S(\cdot)$. We assume that service times are independently distributed from each other as well as the interarrival times.

To motivate discussion for the modified $M/G/1$ system, consider a manufacturing system where items arrive at a machine center for processing. The arrival process is Poisson, while the distribution of the processing time S for each and every item is known to be $\Phi_S(\cdot)$. At the termination of an idle cycle, at the time when the first item appears at the center, a certain amount of time U is spent resetting the machine. In general, U is a random variable. Within this context, the problem is equivalent to assuming that the processing time of that first item initiating a busy period is another random variable $S_0 = S + U$.

The $M/G/1$ queuing system in which S and S_0 have identical distributions encompasses several interesting special cases. For example, when S has a negative exponential distribution, the $M/M/1$ system results. The case of a constant service time equivalent to the $M/D/1$ system would correspond to expressions for $\Phi_S(\cdot)$ of the form

$$\Phi_S(x) = \begin{cases} 0, & -\infty < x \le a \\ 1, & a < x < \infty \end{cases} \qquad a > 0$$

where the service time length is equal to a, a constant. In general, in as much as the distribution function of the service time S is left unspecified in the analysis, the final expression obtained for the expected number of units in the system in the steady state will be seen to depend only on the first two moments of the service times, thus necessitating only the estimation of the mean and variance of the service time.

When studying the $M/M/1$ system, a statistical knowledge of the total number of customers in the system at any particular instant of time t was sufficient information to analyze the system as a birth and death process and describe its evolution at time $t + dt$. This method of analysis fails when service time is nonexponential since the state of the system at time $t + dt$ would depend not only on the number of customers in the system at time t, but as well on other variables such as the elapsed time since the last customer joined the system.

The difficulty may be overcome by using the concept of *imbedded Markov chain*, an ingenious approach first introduced by Kendall in 1951. Under this approach, the state of the queuing system is represented by the total number of customers in the system at points of departure of customers immediately following the termination of their service. These epochs of departure are known as *regeneration points*, and they are used to specify the index set of the resulting stochastic process. As will be shown next, this stochastic process is a discrete parameter Markov chain whose transition probabilities can be completely determined.

Two important remarks should be made concerning this method of analysis:

(1) The use of the imbedded Markov chain technique simplifies the problem analysis, since it reduces a problem which is in general non-Markovian to one which is Markovian and therefore simpler to handle.

(2) Since the imbedded Markov chain analysis is concerned with specifying the statistical properties of the total number of units in the system at *points of customers' departure*, no inference may be made on the distribution of the number of customers in the system at some arbitrarily selected instant of time t, $t \geq 0$. It may therefore be conjectured that the steady-state limiting distributions of the two processes, namely, the number of units in the system at departure epochs and number of units in the system at some time, would not be the same. This conjecture is, in general, true except when the input is Poisson, in which case the equilibrium distribution of the two processes are identical. Thus, for an outside observer, under steady-state conditions, the process realization describing on the state space the fluctuations of the total number of units in the system over time is the same as the one describing on the state space the fluctuation of the total number of units at epochs of departure.

In what follows, we shall define

$$\rho = \lambda E[S] \quad \text{and} \quad \rho_0 = \lambda E[S_0]$$

Note here that the service times need to be specified only by a distribution function; if corresponding density functions exist, then we shall denote the PDF's of S and S_0 by $\varphi_S(x)$ and $\varphi_{S_0}(x)$ respectively, so that

$$d\Phi_S(x) = \varphi_S(x)\,dx \quad \text{and} \quad d\Phi_{S_0} = \varphi_{S_0}(x)\,dx$$

We shall define at this stage the following Laplace transforms, which we shall utilize in the sequel:

$$\bar{\varphi}_S(s) = \int_0^\infty e^{-sx}\,d\Phi_S(x)$$

$$\bar{\varphi}_{S_0}(s) = \int_0^\infty e^{-sx}\,d\Phi_{S_0}(x)$$

i. *Use of the Imbedded Markov Chain*

For $n = 1, 2, \ldots,$ let

X_n = number in the system immediately following the departure of the nth customer

Y_n = number of arrivals during the service time of the nth customer

Then $\{X_n, n \in \mathbf{T}\}$ is a discrete-parameter Markov chain since,

$$X_{n+1} = X_n - 1 + Y_{n+1}, \quad \text{if } X_n > 0$$
$$X_{n+1} = Y_{n+1}, \qquad\qquad \text{if } X_n = 0$$

If $X_n = 0$, then when the $(n + 1)$st customer arrives, he immediately enters into service, thus initiating a busy period. In this case,

$$Y_{n+1} = Y_0$$

where Y_0 is the number of customers arriving during the service time S_0 of the $(n + 1)$st customer. If $X_n > 0$, then when the $(n + 1)$st customer arrives he joins a waiting line; that is, he joins a queue during a busy period. In this case,

$$Y_{n+1} = Y$$

where Y is the number of customer arriving during the service time S of the $(n + 1)$st customer. Thus

$$X_{n+1} = X_n - 1 + Y, \quad \text{if } X_n > 0$$
$$X_{n+1} = Y_0, \qquad\qquad \text{if } X_n = 0$$

We next develop an expression for the transition probabilities.

ii. *Transition Probabilities for the Imbedded Markov Chain*

Let $\qquad\qquad\qquad p_{ij} = P\{X_{n+1} = j \,|\, X_n = i\}$

Then $\qquad p_{ij} = \begin{cases} P\{Y = j - i + 1\}, & \text{if } i \neq 0;\, j \geq i - 1 \\ P\{Y_0 = j\}, & \text{if } i = 0;\, j \geq 0 \\ 0, & \text{if } j < i - 1 \end{cases}$

Now $\qquad\qquad P\{Y = y\} = \displaystyle\int_0^\infty \frac{(\lambda u)^y}{y!} e^{-\lambda u} \, d\Phi_s(u)$

$$P\{Y_0 = y\} = \int_0^\infty \frac{(\lambda u)^y}{y!} e^{-\lambda u} \, d\Phi_{s_0}(u)$$

$$\text{Hence} \quad p_{ij} = \begin{cases} p_{ij}^{(1)} = \int_0^\infty \frac{(\lambda u)^{j-i+1}}{(j-i+1)!} e^{-\lambda u} \, d\Phi_S(u), & \text{if } i \neq 0; j \geq i-1 \\ p_{ij}^{(2)} = \int_0^\infty \frac{(\lambda u)^j}{j!} e^{-\lambda u} \, d\Phi_{S_0}(u), & \text{if } i = 0; j \geq 0 \\ 0, & \text{if } j < i-1 \end{cases}$$

The transition probability matrix is exhibited in Table 4.2. Note, for example, that if one is in state $i = 4$, there is a positive probability to pass to a state $j = 3$, but it is impossible to pass to state $j = 0, 1, 2$, the reason being that customers are removed from the system only by serving them, and

Table 4.2. TRANSITION PROBABILITY MATRIX FOR THE IMBEDDED MARKOV CHAIN IN THE MODIFIED $M/G/1$ QUEUING SYSTEM

From i	*To j* 0	1	2	3	4	5	$\cdots$
0	$p_{00}^{(2)}$	$p_{01}^{(2)}$	$p_{02}^{(2)}$	$p_{03}^{(2)}$	$p_{04}^{(2)}$	$p_{05}^{(2)}$	$\cdots$
1	$p_{10}^{(1)}$	$p_{11}^{(1)}$	$p_{12}^{(1)}$	$p_{13}^{(1)}$	$p_{14}^{(1)}$	$p_{15}^{(1)}$	$\cdots$
2	0	$p_{21}^{(1)}$	$p_{22}^{(1)}$	$p_{23}^{(1)}$	$p_{24}^{(1)}$	$p_{25}^{(1)}$	$\cdots$
3	0	0	$p_{32}^{(1)}$	$p_{33}^{(1)}$	$p_{34}^{(1)}$	$p_{35}^{(1)}$	$\cdots$
4	0	0	0	$p_{43}^{(1)}$	$p_{44}^{(1)}$	$p_{45}^{(1)}$	$\cdots$
5	0	0	0	0	$p_{54}^{(1)}$	$p_{55}^{(1)}$	$\cdots$
.	.	.	.	.	.	.	
.	.	.	.	.	.	.	
.	.	.	.	.	.	.	

this takes place one at a time. Note also, for example, that it is possible to pass from state $i = 4$ to any state $j = 4, 5, 6, \ldots$, the reason being that an arbitrary number of customers may be added to the system during the service time of a customer.

Our next objective is to obtain a closed-form expression for the PGF of the number of customers in the system at points of departure of customers, under steady-state condition.

iii. *Steady-State Probabilities for the Number in the System X_n at Points of Customer Departure*

To determine the steady-state equations, let

$$\Pi_j = \lim_{n \to \infty} P\{X_n = j\}$$

that is, Π_j is the steady-state probability that when a customer leaves the system there will be j customers present in the system.

Using the Chapman-Kolmogorov equations, we have

$$\Pi_0 = \Pi_0 p_{00}^{(2)} + \Pi_1 p_{10}^{(1)}$$

$$\Pi_1 = \Pi_0 p_{01}^{(2)} + \Pi_1 p_{11}^{(1)} + \Pi_2 p_{21}^{(1)}$$

$$\Pi_2 = \Pi_0 p_{02}^{(2)} + \Pi_1 p_{12}^{(1)} + \Pi_2 p_{22}^{(1)} + \Pi_3 p_{32}^{(1)}$$

$$\cdots$$

$$\Pi_j = \Pi_0 p_{0j}^{(2)} + \Pi_1 p_{1j}^{(1)} + \Pi_2 p_{2j}^{(1)} + \cdots + \Pi_{j+1} p_{j+1,j}^{(1)}, \quad j = 0, 1, 2, 3, \ldots$$

or

$$\Pi_j = \Pi_0 p_{0j}^{(2)} + \sum_{i=1}^{i=j+1} \Pi_i p_{ij}^{(1)}, \quad j = 0, 1, 2, \ldots \tag{4.25}$$

It can be shown that for $\rho < 1$, Π_j exists and is finite, and that Π_j is also the probability that there will be exactly $N = j$ customers in the system in the steady state that is, $\Pi_j = P\{N = j\}$, $j = 0, 1, \ldots$

Let
$$G_N(z) = \sum_{j=0}^{\infty} z^j \Pi_j = \sum_{j=0}^{\infty} z^j P\{N = j\}, \quad |z| \le 1$$

define the PGF of N. Multiplying both sides of equation (4.25) by z^j and summing from $j = 0$ to $j = \infty$, we obtain

$$\sum_{j=0}^{\infty} z^j \Pi_j = \sum_{j=0}^{\infty} z^j \Pi_0 p_{0j}^{(2)} + \sum_{j=0}^{\infty} \sum_{i=1}^{i=j+1} z^j \Pi_i p_{ij}^{(1)} \tag{4.26}$$

Substituting in equation (4.26) for the values of $p_{0j}^{(2)}$ and $p_{ij}^{(1)}$, we obtain

$$G_N(z) = \Pi_0 \sum_{j=0}^{\infty} z^j \int_0^{\infty} \frac{(\lambda u)^j}{j!} e^{-\lambda u} \, d\Phi_{S_0}(u)$$

$$+ \sum_{j=0}^{\infty} \sum_{i=1}^{i=j+1} z^j \Pi_i \int_0^{\infty} \frac{(\lambda u)^{j-i+1}}{(j-i+1)!} e^{-\lambda u} \, d\Phi_S(u)$$

Interchanging the order of integration and summations in the right terms yields

$$G_N(z) = \Pi_0 \int_0^{\infty} e^{-\lambda u} \sum_{j=0}^{\infty} \frac{(\lambda u z)^j}{j!} \, d\Phi_{S_0}(u)$$

$$+ \int_0^{\infty} e^{-\lambda u} \sum_{i=1}^{\infty} \Pi_i \sum_{j=i-1}^{\infty} z^j \frac{(\lambda u)^{j-i+1}}{(j-i+1)!} \, d\Phi_S(u)$$

$$= \Pi_0 \int_0^{\infty} e^{-\lambda u} \sum_{j=0}^{\infty} \frac{(\lambda u z)^j}{j!} \, d\Phi_{S_0}(u)$$

$$+ \int_0^{\infty} e^{-\lambda u} \sum_{i=1}^{\infty} z^{i-1} \Pi_i \sum_{k=0}^{\infty} \frac{(\lambda u z)^k}{k!} \, d\Phi_S(u)$$

$$= \Pi_0 \int_0^{\infty} e^{-\lambda u} e^{\lambda u z} \, d\Phi_{S_0}(u) + \int_0^{\infty} e^{-\lambda u} e^{\lambda u z} \sum_{i=1}^{\infty} z^{i-1} \Pi_i \, d\Phi_S(u)$$

$$= \Pi_0 \int_0^{\infty} e^{-\lambda u(1-z)} \, d\Phi_{S_0}(u)$$

$$+ \frac{1}{z} \int_0^{\infty} e^{-\lambda u(1-z)} (\Pi_1 z + \Pi_2 z^2 + \cdots) \, d\Phi_S(u)$$

Hence, in terms of the Laplace transforms $\bar{\varphi}_S(s)$ and $\bar{\varphi}_{S_0}(s)$, we obtain

$$G_N(z) = \Pi_0 \bar{\varphi}_{S_0}[\lambda(1-z)] + \frac{G_N(z) - \Pi_0}{z} \bar{\varphi}_S[\lambda(1-z)]$$

Solving for $G_N(z)$ yields

$$G_N(z) = \Pi_0 \frac{\bar{\varphi}_S[\lambda(1-z)] - z\bar{\varphi}_{S_0}[\lambda(1-z)]}{\bar{\varphi}_S[\lambda(1-z)] - z} \tag{4.26a}$$

The quantity Π_0, which is the probability that no units are in the system, may be computed from $\sum_{n=0}^{\infty} \Pi_n = G_N(1) = 1$. Now, for $z = 1$, the right hand side of the above expression is an indeterminate form. Applying l'Hospital's rule once we obtain

$$G_N(1) = 1 = \lim_{z \to 1} \Pi_0 \frac{-\lambda\bar{\varphi}'_S[\lambda(1-z)] - \lambda z\bar{\varphi}'_{S_0}[\lambda(1-z)] - \bar{\varphi}_{S_0}[\lambda(1-z)]}{-\lambda\bar{\varphi}'_S[\lambda(1-z)] - 1}$$

$$= \Pi_0 \frac{\lambda E[S] - \lambda E[S_0] - 1}{\lambda E[S] - 1}$$

$$= \Pi_0 \frac{\rho - \rho_0 - 1}{\rho - 1}$$

Solving for Π_0 yields

$$\Pi_0 = \frac{1 - \rho}{1 - \rho + \rho_0}, \quad \rho < 1$$

The expression for $G_N(z)$ is thus

$$G_N(z) = \frac{1 - \rho}{1 - \rho + \rho_0} \frac{\bar{\varphi}_S[\lambda(1-z)] - z\bar{\varphi}_{S_0}[\lambda(1-z)]}{\bar{\varphi}_S[\lambda(1-z)] - z} \tag{4.27}$$

In the special case where S and S_0 are identically distributed, $\rho = \rho_0$ and

$$G_N(z) = \frac{(1 - \rho)(1 - z)\bar{\varphi}_S[\lambda(1-z)]}{\bar{\varphi}_S[\lambda(1-z)] - z} \tag{4.28}$$

Example 4.3

Derive an expression for the steady state distribution of the number of units in the $M/G/1$ queuing system assuming that the service time distribution is identical for each customer and is given by

(a) The negative exponential distribution $\varphi_S(x) = \mu e^{-\mu x}$, $0 < x < \infty$.
(b) The digamma distribution $\varphi_S(x) = (\mu x)\mu e^{-\mu x}$, $0 < x < \infty$.

To obtain the required expression, we shall make use of (4.28), since in the steady state, the probability of having exactly n units in the system, $P\{N = n\}$, $n = 0, 1, 2, \ldots$, is equal to Π_n, the probability that when a customer departs, there are n customers left in the system.

(a) Service time with PDF $\varphi_S(x) = \mu e^{-\mu x}, 0 < x < \infty$:

We have

$$E[S] = \frac{1}{\mu}; \quad \rho = \lambda E[S] = \frac{\lambda}{\mu} < 1$$

Also

$$\bar{\varphi}_S(z) = \frac{\mu}{\mu + z}$$

Using expression (4.28) we obtain for the probability generating function of N

$$G_N(z) = \frac{(1 - \rho)(1 - z)\dfrac{\mu}{\mu + \lambda(1 - z)}}{\dfrac{\mu}{\mu + \lambda(1 - z)} - z}$$

$$= \frac{(1 - \rho)(1 - z)\dfrac{1}{1 + \rho(1 - z)}}{\dfrac{1}{1 + \rho(1 - z)} - z}$$

$$= \frac{(1 - \rho)(1 - z)}{1 - z[1 + \rho(1 - z)]}$$

$$= \frac{(1 - \rho)(1 - z)}{(1 - z)(1 - \rho z)}$$

$$= \frac{1 - \rho}{1 - \rho z}, \quad |z| \leq 1$$

$$= (1 - \rho) \sum_{n=0}^{\infty} (\rho z)^n$$

The coefficient of z^n specifies $P\{N = n\}$; hence

$$P\{N = n\} = (1 - \rho)\rho^n, \quad n = 0, 1, 2, \ldots$$

This result should be compared with (4.11).

(b) Service time with PDF $\varphi_S(x) = (\mu x)\mu e^{-\mu x}, 0 < x < \infty$:

We have

$$E[S] = \frac{2}{\mu}; \quad \rho = \lambda E[S] = \frac{2\lambda}{\mu} < 1$$

Also

$$\bar{\varphi}_S(z) = \left(\frac{\mu}{\mu + z}\right)^2$$

From relation (4.28), the expression for the probability generating function of N is

$$G_N(z) = \frac{(1-\rho)(1-z)\left[\dfrac{\mu}{\mu + \lambda(1-z)}\right]^2}{\left[\dfrac{\mu}{\mu + \lambda(1-z)}\right]^2 - z}, \quad |z| \le 1, \rho < 1$$

$$= \frac{(1-\rho)(1-z)}{1 - z\left[1 + \dfrac{\rho}{2}(1-z)\right]^2}$$

$$= \frac{(1-\rho)(1-z)}{1 - z\left[1 + \rho(1-z) + \dfrac{\rho^2}{4}(1-z)^2\right]}$$

$$= \frac{1-\rho}{1 - \rho z - \dfrac{\rho^2}{4}z(1-z)}$$

$$= \frac{1-\rho}{1 - \left(\rho + \dfrac{\rho^2}{4}\right)z + \dfrac{\rho^2}{4}z^2}$$

$$= \frac{1-\rho}{(1 - \rho\alpha_1 z)(1 - \rho\alpha_2 z)}$$

where

$$\alpha_1 = \frac{2}{(4+\rho) + \sqrt{8\rho + \rho^2}} < 1$$

$$\alpha_2 = \frac{2}{(4+\rho) - \sqrt{8\rho + \rho^2}} < 1$$

We now expand the expression for $G_N(z)$ using partial fractions:

$$G_N(z) = \frac{1-\rho}{\alpha_2 - \alpha_1}\left[\frac{\alpha_2}{1 - \rho\alpha_2 z} - \frac{\alpha_1}{1 - \rho\alpha_1 z}\right]$$

$$= \frac{1-\rho}{\alpha_2 - \alpha_1}\left[\alpha_2 \sum_{n=0}^{\infty} (\rho\alpha_2 z)^n - \alpha_1 \sum_{n=0}^{\infty} (\rho\alpha_1 z)^n\right]$$

Identifying the coefficient of z^n, we obtain for $P\{N = n\}$

$$P\{N = n\} = \frac{1-\rho}{\alpha_2 - \alpha_1}[\alpha_2(\rho\alpha_2)^n - \alpha_1(\rho\alpha_1)^n]$$

$$= \frac{\rho^n(1-\rho)}{\alpha_2 - \alpha_1}(\alpha_2^{n+1} - \alpha_1^{n+1}), \quad n = 0, 1, 2, \ldots$$

iv. Expected Number in the System

To obtain the expected number in the system, we calulate $(\partial/\partial z)G_N(z)|_{z=1}$. We first compute a number of quantities that will be used in the sequel. Let

$$\hat{\mu} = \mu(z) = \bar{\varphi}_S[\lambda(1-z)] = \int_0^\infty e^{-\lambda(1-z)u} \, d\Phi_S(u)$$

$$\hat{v} = v(z) = \bar{\varphi}_{S_0}[\lambda(1-z)] = \int_0^\infty e^{-\lambda(1-z)u} \, d\Phi_{S_0}(u)$$

Then

$$\mu'(z) = \int_0^\infty \lambda u e^{-\lambda(1-z)u} \, d\Phi_S(u)$$

$$v'(z) = \int_0^\infty \lambda u e^{-\lambda(1-z)u} \, d\Phi_{S_0}(u)$$

$$\mu''(z) = \int_0^\infty \lambda^2 u^2 e^{-\lambda(1-z)u} \, d\Phi_S(u)$$

$$v''(z) = \int_0^\infty \lambda^2 u^2 e^{-\lambda(1-z)u} \, d\Phi_{S_0}(u)$$

Thus

$$\mu(1) = \int_0^\infty d\Phi_S(u) = 1, \qquad v(1) = \int_0^\infty d\Phi_{S_0}(u) = 1$$

$$\mu'(1) = \int_0^\infty \lambda u \, d\Phi_S(u) \qquad v'(1) = \int_0^\infty \lambda u \, d\Phi_{S_0}(u)$$

$$= \lambda E[S] = \rho, \qquad\qquad = \lambda E[S_0] = \rho_0 \qquad\qquad (4.28a)$$

$$\mu''(1) = \int_0^\infty \lambda^2 u^2 \, d\Phi_S(u) \qquad v''(1) = \int_0^\infty \lambda^2 u^2 \, d\Phi_{S_0}(u)$$

$$= \lambda^2 E[S^2], \qquad\qquad = \lambda^2 E[S_0^2]$$

Now from expression (4.26a)

$$G_N(z) = \Pi_0 \frac{\hat{\mu} - z\hat{v}}{\hat{\mu} - z}$$

Hence

$$G_N'(z) = \Pi_0 \frac{(\hat{\mu} - z)(\hat{\mu}' - z\hat{v}' - \hat{v}) - (\hat{\mu} - z\hat{v})(\hat{\mu}' - 1)}{(\hat{\mu} - z)^2} = \Pi_0 \frac{N(z)}{D(z)}$$

where $\quad N(z) = (\hat{\mu} - z)(\hat{\mu}' - z\hat{v}' - \hat{v}) - (\hat{\mu} - z\hat{v})(\hat{\mu}' - 1)$

and $\quad D(z) = (\hat{\mu} - z)^2$

To obtain $\lim_{z\to 1} G_N'(z)$, we apply l'Hospital's rule twice to remove indeterminacy. Thus

$$E[N] = \lim_{z \to 1} G'_N(z) = \Pi_0 \frac{\lim\limits_{z \to 1} N(z)}{\lim\limits_{z \to 1} D(z)} = \Pi_0 \frac{\lim\limits_{z \to 1} N'(z)}{\lim\limits_{z \to 1} D'(z)} = \Pi_0 \frac{N''(1)}{D''(1)} \quad (4.28b)$$

Now
$$\begin{aligned} N'(z) &= (\hat{\mu} - z)(\hat{\mu}'' - z\hat{v}'' - \hat{v}' - \hat{v}') + (\hat{\mu}' - 1)(\hat{\mu}' - z\hat{v}' - \hat{v}) \\ &\quad - (\hat{\mu} - z\hat{v})\hat{\mu}'' - (\hat{\mu}' - z\hat{v}' - \hat{v})(\hat{\mu}' - 1) \\ &= (\hat{\mu} - z)(\hat{\mu}'' - z\hat{v}'' - 2\hat{v}') - (\hat{\mu} - z\hat{v})\hat{\mu}'' \end{aligned}$$

and
$$\begin{aligned} N''(z) &= (\hat{\mu} - z)(\hat{\mu}''' - z\hat{v}''' - 3\hat{v}'') + (\hat{\mu}' - 1)(\hat{\mu}'' - z\hat{v}'' - 2\hat{v}') \\ &\quad - (\hat{\mu} - z\hat{v})\hat{\mu}''' - (\hat{\mu}' - z\hat{v}' - \hat{v})\hat{\mu}'' \end{aligned}$$

Substituting for $\hat{\mu}, \hat{\mu}', \hat{\mu}'', \hat{v}, \hat{v}'$, and $\hat{v}''$, evaluated at $z = 1$ as computed in expressions (4.28a), we obtain

$$N''(1) = (\rho - 1)\{\lambda^2 E[S^2] - \lambda^2 E[S_0^2] - 2\rho_0\} - (\rho - \rho_0 - 1)\lambda^2 E[S^2]$$

Also
$$D'(z) = 2(\hat{\mu}' - 1)(\hat{\mu} - z)$$

and
$$D''(z) = 2(\hat{\mu}' - 1)(\hat{\mu}' - 1) + 2\hat{\mu}''(\hat{\mu} - z)$$

Hence, using the results of (4.28a) we obtain

$$D''(1) = 2(1 - \rho)^2$$

Finally, substituting for the values of Π_0, $N''(1)$, and $D''(1)$ in equation (4.28b) yields

$$\begin{aligned} E[N] &= \frac{\partial}{\partial z} G_N(z)\bigg|_{z=1} = \frac{1 - \rho}{1 - \rho + \rho_0} \\ &\quad \times \frac{(\rho - 1)\{\lambda^2 E[S^2] - \lambda^2 E[S_0^2] - 2\rho_0\} + (1 - \rho + \rho_0)\lambda^2 E[S^2]}{2(1 - \rho)^2} \end{aligned}$$

Thus
$$E[N] = \frac{\lambda^2\{E[S_0^2] - E[S^2]\} + 2\rho_0}{2(1 - \rho + \rho_0)} + \frac{\lambda^2 E[S^2]}{2(1 - \rho)} \quad (4.29)$$

We consider now the important case when S and S_0 are identically distributed. Then $E[S_0^2] = E[S^2]$ and $\rho_0 = \rho$. Formula (4.29) then reduces to

$$E[N] = \rho + \frac{\lambda^2 E[S^2]}{2(1 - \rho)} \quad (4.30)$$

Using Little's formulas and (4.30) an expression for the expected waiting time $E[W]$ may then be obtained. Now

$$\begin{aligned} E[N] &= \lambda E[\Theta] \\ &= \lambda\{E[W] + E[S]\} \end{aligned}$$

Thus
$$E[W] = \frac{1}{\lambda} E[N] - E[S]$$

Using formula (4.30) we may write

$$E[W] = \frac{\rho}{\lambda} + \frac{\lambda E[S^2]}{2(1-\rho)} - E[S]$$

$$= \frac{\lambda E[S^2]}{2(1-\rho)}$$

$$= \frac{\lambda\{\text{Var}\,[S] + \{E[S]\}^2\}}{2(1-\rho)}$$

$$= \frac{\lambda\{E[S]\}^2\left\{1 + \frac{\text{Var}\,[S]}{\{E[S]\}^2}\right\}}{2(1-\rho)}$$

$$= \frac{\rho E[S]\left\{1 + \frac{\text{Var}\,[S]}{\{E[S]\}^2}\right\}}{2(1-\rho)}$$

When the service time has the negative exponential distribution, then Var $[S] = \{E[S]\}^2$, yielding for the $M/M/1$ queue

$$E[W] = \frac{\rho E[S]}{1-\rho}$$

For constant service time, Var $[S] = 0$; thus, for the $M/D/1$ queue

$$E[W] = \frac{\rho E[S]}{2(1-\rho)}$$

For a given traffic intensity ρ and a fixed mean service time $E[S]$, the expected customer's delay is twice as much in the $M/M/1$ system as in the $M/D/1$ system.

The relations obtained so far have shown that for Poisson arrival, the mean characteristics of the queuing system depend not only on the mean arrival rate and on the mean service rate, but as well on the variance of the service time. Such characteristics, however, are independent of any other moments of the service-time distribution.

Example 4.4

In a manufacturing center units arrive for processing in a Poisson fashion at the rate of λ units per unit time. Each unit is identical and the amount of service provided to each item is a constant equal to a time units. At the appearance of the first unit initiating a busy cycle, an amount of time b is spent setting up the equipment. It is possible to speed up or slow down the processing time on each item by adjusting the processing speed or service rate $v = 1/a$; v is expressed in terms of number of units processed per unit time. Let $C_1(v)$ be the cost of processing one item as a function of the processing speed and suppose that $C_1(v)$ is a strictly increasing function of v. For each item being held in the system per unit time, a cost C_2 is charged. It is required to determine the optimum processing speed to minimize the total expected cost per unit time of processing and of holding units in the system.

Since the average arrival rate is λ, the expected number of units processed per unit time is also λ. Hence, $\lambda C_1(v)$ is the expected processing cost per unit time, which is an increasing function of v. Now, in the steady state, for $\lambda/v < 1$, the expected number of units in the system is $E[N]$. As would be expected, $E[N]$ is a strictly decreasing function of v for $v > \lambda$, since fewer items would be in the system with an increase in service rate. It follows that the expected cost of holding units in the system per unit time, $C_2 E[N]$, is a decreasing function of v. Naturally, one would wonder whether an optimum value of $v > \lambda$ exists which minimizes the sum of these costs, namely

$$\text{total cost (TC)} = \lambda C_1(v) + C_2 E[N]$$

To obtain an explicit expression for $E[N]$, we refer to formula (4.29) and note that

$$E[S] = \frac{1}{v}; \qquad\qquad E[S_0] = \frac{1}{v} + b$$

$$E[S^2] = \frac{1}{v^2}; \qquad\qquad E[S_0^2] = \left(\frac{1}{v} + b\right)^2$$

$$\rho = \lambda E[S] = \frac{\lambda}{v}; \qquad \rho_0 = \lambda E[S_0] = \lambda\left(\frac{1}{v} + b\right)$$

Substituting these relations in expression (4.29) yields:

$$E[N] = \frac{\lambda^2\left[\left(\frac{1}{v} + b\right)^2 - \frac{1}{v^2}\right] + 2\lambda\left(\frac{1}{v} + b\right)}{2\left[1 - \frac{\lambda}{v} + \lambda\left(\frac{1}{v} + b\right)\right]} + \frac{\frac{\lambda^2}{v^2}}{2\left(1 - \frac{\lambda}{v}\right)}$$

$$= \frac{\lambda^2(2b + b^2 v) + 2\lambda(1 + bv)}{2v(1 + \lambda b)} + \frac{\lambda^2}{2v(v - \lambda)}$$

This then provides explicitly an expression for $E[N]$ as a function of v, the processing speed. Mathematically, the optimization problem may be formulated as

$$\min \text{TC} = \lambda C_1(v) + C_2\left[\frac{\lambda^2(2b + b^2 v) + 2\lambda(1 + bv)}{2v(1 + \lambda b)} + \frac{\lambda^2}{2v(v - \lambda)}\right]$$

subject to $v > \lambda$

Explicit derivation of the optimum value of v will depend on the form of the function $C_1(v)$. Suppose that $C_1(v)$ is a linear function of v, that is, $C_1(v) = C_1 v$. Further let $b = 0$. Then the total expected cost per unit time is

$$\text{TC} = \lambda C_1 v + C_2 \left[\frac{\lambda}{v} + \frac{\lambda^2}{2v(v - \lambda)} \right]$$

Differentiating with respect to v and setting the result equal to zero yields

$$\frac{d(\text{TC})}{dv} = 0 = \lambda C_1 - \frac{\lambda C_2}{v^2} - \frac{\lambda^2 C_2}{2} \frac{(2v - \lambda)}{v^2(v - \lambda)^2}$$

Thus, if an optimum value of v exists, it will be solution to the equation

$$2 \frac{C_1}{C_2} v^2 (v - \lambda)^2 - 2(v - \lambda)^2 - \lambda(2v - \lambda) = 0$$

This is a quartic equation in v. For given numerical values of λ and C_1/C_2, this equation may be solved numerically or otherwise. It is evident that the positive root whose value exceeds λ determines the optimum value of v.

v. Steady-State Distribution of the Time Θ Spent in the System from Arrival to Departure

When an arriving customer departs after waiting in the system a time Θ, he will leave behind him N customers. Thus N is the number of customers in the system at points of departure, or equivalently, it is the number of customers joining the system in the time interval Θ. Since arrivals are Poisson, we have

$$P\{N = n | \Theta = \theta\} = e^{-\lambda\theta} \frac{(\lambda\theta)^n}{n!}, \quad n = 0, 1, 2, \ldots$$

Hence, if $\Phi_\Theta(x)$ denotes the distribution function of Θ, we may write

$$P\{N = n\} = \int_0^\infty P\{N = n | \Theta = \theta\} d\Phi_\Theta(\theta)$$

$$= \int_0^\infty e^{-\lambda\theta} \frac{(\lambda\theta)^n}{n!} d\Phi_\Theta(\theta), \quad n = 0, 1, 2, \ldots$$

Multiplying both sides by z^n and summing over all possible values of n we get the following expression for the PGF of N:

$$G_N(z) = \sum_{n=0}^\infty z^n P\{N = n\} = \sum_{n=0}^\infty z^n \int_0^\infty e^{-\lambda\theta} \frac{(\lambda\theta)^n}{n!} d\Phi_\Theta(\theta)$$

Interchanging the order of summation and integration in the right hand side term yields

$$G_N(z) = \int_0^\infty \left[\sum_{n=0}^\infty e^{-\lambda\theta} \frac{(\lambda\theta z)^n}{n!} \right] d\Phi_\Theta(\theta)$$

$$= \int_0^\infty e^{-\lambda(1-z)\theta} d\Phi_\Theta(\theta) \tag{4.31}$$

Define now the Laplace transform

$$\bar{\varphi}_\Theta(s) = \int_0^\infty e^{-sx} d\Phi_\Theta(x)$$

Expression (4.31) may then be written as

$$G_N(z) = \bar{\varphi}_\Theta[\lambda(1 - z)]$$

If we now set $s = \lambda(1 - z)$, we obtain

$$\bar{\varphi}_\Theta(s) = G_N\left(1 - \frac{s}{\lambda}\right) \tag{4.32}$$

Finally using expression (4.27) for $G_N(z)$ gives

$$\bar{\varphi}_\Theta(s) = \frac{1 - \rho}{1 - \rho + \rho_0} \frac{\bar{\varphi}_S(s) - \left(1 - \frac{s}{\lambda}\right)\bar{\varphi}_{S_0}(s)}{\bar{\varphi}_S(s) - \left(1 - \frac{s}{\lambda}\right)}$$

$$= \frac{1 - \rho}{1 - \rho + \rho_0} \frac{\lambda\bar{\varphi}_S(s) - (\lambda - s)\bar{\varphi}_{S_0}(s)}{\lambda\bar{\varphi}_S(s) - \lambda + s} \tag{4.33}$$

The expression for the distribution function of Θ can then be obtained by inverting expression (4.33) given the distribution functions of S and S_0.

We now use expression (4.32) to obtain the well known relation $E[N] = \lambda E[\Theta]$. We have

$$-E[\Theta] = \lim_{s \to 0} \bar{\varphi}'_\Theta(s) = \lim_{s \to 0} G'_N\left(1 - \frac{s}{\lambda}\right)$$

$$= -\frac{1}{\lambda} G'_N(1)$$

But $$E[N] = G'_N(1)$$

Thus $$E[N] = \lambda E[\Theta]$$

An explicit expression for $E[\Theta]$ may then be obtained using relation (4.29).

We consider now the important special case when S and S_0 have identical distributions; then from (4.33), an expression for $\bar{\varphi}_\Theta(s)$ is

$$\bar{\varphi}_\Theta(s) = \frac{(1 - p)s\bar{\varphi}_S(s)}{\lambda\bar{\varphi}_S(s) - \lambda + s} \tag{4.34}$$

Since $p = \lambda E[S]$, we may write expression (4.34) alternatively as

$$\bar{\varphi}_\Theta(s) = \frac{(1 - p)\bar{\varphi}_S(s)}{1 - \dfrac{p}{E[S]}\left[\dfrac{1}{s} - \dfrac{\bar{\varphi}_S(s)}{s}\right]} \tag{4.35}$$

Let now $H_S(x)$ be the complementary cumulative function of the service time S, that is

$$H_S(x) = 1 - \int_0^x d\Phi_S(u)$$

We know that

$$E[S] = \int_0^\infty H_S(x)\,dx$$

We now define arbitrarily a new random variable U with the known PDF

$$\varphi_U(x) = \frac{H_S(x)}{E[S]}, \quad 0 < x < \infty$$

$$= \frac{1}{E[S]}\left[1 - \int_0^x d\Phi_S(u)\right]$$

Defining by $\bar{\varphi}_U(s)$ the Laplace transform of $\varphi_U(x)$, we have

$$\bar{\varphi}_U(s) = \frac{1}{E[S]}\left[\frac{1}{s} - \frac{1}{s}\bar{\varphi}_S(s)\right] \tag{4.36}$$

Using expression (4.36) in expression (4.35) yields

$$\bar{\varphi}_\Theta(s) = \frac{(1 - p)\bar{\varphi}_S(s)}{1 - p\bar{\varphi}_U(s)} \tag{4.37}$$

Since for $p < 1, |p\bar{\varphi}_U(s)| < 1$, we have using the binomial expansion

$$\bar{\varphi}_\Theta(s) = (1 - p)\bar{\varphi}_S(s) \sum_{n=0}^\infty [p\bar{\varphi}_U(s)]^n$$

$$= (1 - p)\left\{\bar{\varphi}_S(s) + \sum_{n=1}^\infty p^n\bar{\varphi}_S(s)[\bar{\varphi}_U(s)]^n\right\}$$

Inverting term by term and using $*$ to denote the convolution operation, we obtain as an expression for the PDF of the time spent in the system Θ

$$\varphi_\Theta(x) = (1 - \rho)\left\{\varphi_S(x) + \sum_{n=1}^{\infty} \rho^n \varphi_S(x) * [\varphi *_U^{(n)}(x)]\right\}$$

In Section 2.8-iii, it was shown that the expression for the Laplace transform of the expected number of renewals occurring up to a specified time t was similar to expression (4.37). Through this analogy, we may infer that $\varphi_\Theta(x)$ is solution to the integral equation

$$\varphi_\Theta(x) = (1 - \rho)\varphi_S(x) + \rho \int_0^x \varphi_\Theta(u)\varphi_U(x - u) \, du$$

Example 4.5

Obtain the steady-state distribution of the time spent by a customer in the $M/G/1$ queuing system if service time has the negative exponential distribution

$$\varphi_S(x) = \mu e^{-\mu x}, \quad \mu > 0, \quad 0 < x < \infty$$

The Laplace transform of $\varphi_S(x)$ is

$$\bar{\varphi}_S(s) = \frac{\mu}{\mu + s}$$

Substituting this expression in (4.34) yields

$$\bar{\varphi}_\Theta(s) = \frac{(1 - \rho)s \dfrac{\mu}{\mu + s}}{\dfrac{\lambda\mu}{\mu + s} - \lambda + s}$$

$$= \frac{(1 - \rho)\mu s}{\lambda\mu - (\lambda - s)(\mu + s)}$$

$$= \frac{(1 - \rho)\mu s}{-\lambda s + \mu s + s^2}$$

$$= \frac{(1 - \rho)\mu}{(1 - \rho)\mu + s}$$

which may be recognized as the Laplace transform of the negative exponential distribution

$$\varphi_\Theta(x) = (1 - \rho)\mu e^{-(1-\rho)\mu x}, \quad 0 < x < \infty$$

vi. *Steady-State Distribution of The Waiting Time W from Arrival to Service*

In order to obtain the distribution of the waiting time W of a customer prior to entering into service, we note that Θ, the time spent in the system, equals the waiting time W plus the service time S, i.e., $\Theta = W + S$. Since W and S are independently distributed, we must have

$$\varphi_\Theta(x) = \varphi_W(x) * \varphi_S(x)$$

In terms of Laplace transforms, this relation is equivalent to

$$\bar{\varphi}_\Theta(s) = \bar{\varphi}_W(s) \cdot \bar{\varphi}_S(s) \tag{4.38}$$

Using relation (4.34) in relation (4.38), we obtain the following expression for the Laplace transform of the PDF of the waiting time W:

$$\bar{\varphi}_W(s) = \frac{(1 - \rho)s}{\lambda \bar{\varphi}_S(s) - \lambda + s} \tag{4.39}$$

Using the same reasoning as for Θ in the previous section, the reader should verify that the explicit expression for $\varphi_W(x)$ is

$$\varphi_W(x) = (1 - \rho)\left\{1 + \sum_{n=1}^{\infty} \rho^n \varphi*_U^{(n)}(x)\right\} \tag{4.40}$$

where
$$\varphi_U(x) = \frac{1}{E[S]}\left[1 - \int_0^x d\Phi_S(u)\right]$$

The method of derivation used here is mostly due to Kendall [8]. Expression (4.39) is usually referred to as the *Pollaczek-Khintchine formula*, and it gives an expression for the Laplace transform $\bar{\varphi}_W(s)$ of the steady-state distribution of the waiting time in terms of the Laplace transform $\bar{\varphi}_S(s)$ of the service time distribution. Formula (4.40) is due to Benes [2].

An expression for $E[W]$ may be obtained from relation (4.39) by using the formula

$$E[W] = \lim_{s \to 0} - \bar{\varphi}'_W(s)$$

The reader may verify by differentiation that

$$E[W] = \frac{\lambda E[S^2]}{2(1 - \rho)}, \quad \rho < 1$$

$$= \frac{\lambda\{\text{Var}\,[S] + \{E[S]\}^2\}}{2(1 - \rho)}$$

Example 4.6

Obtain the steady-state distribution of the waiting time of a customer in the $M/G/1$ queuing system if service time has the negative exponential distribution

$$\varphi_S(x) = \mu e^{-\mu x}, \quad \mu > 0, \ 0 < x < \infty$$

We shall make use of relation (4.39) with $\bar{\varphi}_S(x) = \mu/(\mu + s)$. We have for the Laplace transform expression $\bar{\varphi}_W(s)$

$$\bar{\varphi}_W(s) = \frac{(1 - \rho)s}{\dfrac{\lambda \mu}{\mu + s} - \lambda + s}$$

$$= \frac{(1 - \rho)(\mu + s)}{-\lambda + \mu + s}$$

$$= (1 - \rho) + \frac{\lambda(1 - \rho)}{\mu(1 - \rho) + s}$$

If $\Phi_W(x)$ is the distribution function of W, then formal inversion of $\bar{\varphi}_W(s)$ yields

$$\Phi_W(0) = (1 - \rho), \qquad\qquad x = 0$$
$$d\Phi_W(x) = \lambda(1 - \rho)e^{-\mu(1-\rho)x} \, dx, \quad 0 < x < \infty$$

Comparing this result with expression (4.14) we conclude that for the $M/M/1$ system the steady-state distributions of the waiting time W and the virtual waiting time V are identical.

4.7. The $M/G/\infty$ Queuing System

When the number of servers or channels is infinite, all incoming units enter immediately into service. Thus no arriving customers will have to wait, and at any instant of time t, $t \geq 0$, the total number of customers $N(t)$ in the system is the same as the total number of busy servers. Our objective will be to study the stochastic process $\{N(t), \ t \geq 0\}$ and in particular to obtain a closed form expression for $P\{N(t) = n\}$, $n = 0, 1, 2, \ldots$, when arrival of customers is a Poisson process with intensity λ and service time S of each customer has the identical general distribution function $\Phi_S(\cdot)$. We shall assume that initially, at time $t = 0$, the system is empty; thus

$$P\{N(0) = n\} = \begin{cases} 1, & n = 0 \\ 0, & n = 1, 2, 3, \ldots \end{cases} \tag{4.41}$$

Prior to carrying the formal proof, we need to establish a property of the Poisson process which will be used later on in the analysis.

i. *Property of the Poisson Process*

We consider the Poisson process $\{X(t), t \geq 0\}$ with intensity λ such that

$$P\{X(t) = m\} = e^{-\lambda t}\frac{(\lambda t)^m}{m!}, \quad m = 0, 1, 2, \ldots \tag{4.42}$$

The following property will be established.

Theorem 4.1

Let Θ be a random variable denoting the time of occurence of an event in a Poisson process over $(0, t]$ given that at least one event has occurred in $(0, t]$, that is $m \geq 1$. Then the probability of the occurrence of any of the m events in the interval $(\theta, \theta + d\theta]$, $0 < \theta < t$, is

$$P\{\theta < \Theta \leq \theta + d\theta\} = \varphi_\Theta(\theta)\,d\theta = \frac{d\theta}{t} \tag{4.43}$$

and is independent of the frequency of occurrences, that is the number m, or the time of occurrences in $(0, t]$.

To show this property, suppose that in the Poisson process, the waiting times for the occurrences of the first, second, $\ldots$, mth events are respectively $W_1, W_2, \ldots, W_m$. We have shown (Section 2.4-c) that the interarrival times between successive events form a sequence of identically and independently distributed random variables each having a negative exponential distribution with parameter λ. Thus, the joint probability that $X(t) = m \geq 1$ events have occurred in the time $(0, t]$ and that the first event occurs in $(t_1, t_1 + dt_1]$, the second occurs in $(t_2, t_2 + dt_2]$, $\ldots$, and the mth event occurs in $(t_m, t_m + dt_m]$, $0 < t_1 < t_2 < \cdots < t_m < t$, is

$$\begin{aligned}
P\{X(t) &= m \geq 1; t_1 < W_1 \leq t_1 + dt_1, \\
&\quad t_2 < W_2 \leq t_2 + dt_2, \ldots, t_m < W_m \leq t_m + dt_m\} \\
&= (e^{-\lambda t_1}\lambda dt_1)(e^{-\lambda(t_2 - t_1)}\lambda dt_2)\ldots(e^{-\lambda(t_m - t_{m-1})}\lambda dt_m)e^{-\lambda(t - t_m)} \\
&= \lambda^m e^{-\lambda t}\,dt_1\,dt_2\ldots dt_m
\end{aligned} \tag{4.44}$$

Hence the conditional probability that the first event occurs in $(t_1, t_1 + dt_1]$, the second occurs in $(t_2, t_2 + dt_2]$, $\ldots$, and the mth event occurs in

$(t_m, t_m + dt_m]$, given that $X(t) = m \geq 1$ events have occurred in $(0, t]$, is

$$P\{t_1 < W_1 \leq t_1 + dt_1, t_2 < W_2 \leq t_2 + dt_2, \ldots, t_m < W_m \leq t_m + dt_m | X(t) = m \geq 1\}$$

$$= \frac{P\{X(t) = m \geq 1; t_1 < W_1 \leq dt_1, t_2 < W_2 \leq t_2 + dt_2, \ldots, t_m < W_m \leq t_m + dt_m\}}{P\{X(t) = m \geq 1\}}$$

$$= \frac{\lambda^m e^{-\lambda t} dt_1 \, dt_2 \ldots dt_m}{e^{-\lambda t} \frac{(\lambda t)^m}{m!}} \tag{4.45}$$

$$= \frac{m!}{t^m} dt_1 \, dt_2 \ldots dt_m, \quad m = 1, 2, \ldots$$

Relation (4.45) follows from relations (4.42) and (4.44). It represents the joint probability that given at least one Poisson event has occured in $(0, t]$, the first event will happen in $(t_1, t_1 + dt_1]$, the second in $(t_2, t_2 + dt_2]$, and so forth. From relation (4.45), it may be noted that the joint PDF $m!/t^m$ is independent of $t_1, t_2, \ldots, t_m$, thus implying that the occurrence of the events as specified are independent random variables.

We now establish the equivalence of (4.45) and Theorem 4.1. Assuming that the theorem is true, then given that exactly one event has occurred in $(0, t]$, i.e., $m = 1$, the probability that it will occur in $(t_1, t_1 + dt_1]$, $0 < t_1 < t$ is, from expression (4.43), dt_1/t; that is, it is equally likely to occur in any small interval dt_1 in $(0, t]$. This corresponds in relation (4.45) to the case, when $m = 1$. Suppose now that two events are known to have occurred in $(0, t]$. Based on the assumption of the Poisson process, no more than one single event may occur in an infinitesimal time interval $d\theta$. Hence, from Theorem 4.1, the probability that any one of the two events occurs in $(t_1, t_1 + dt_1]$ is $2dt_1/t$, while the probability of the other event occurring in $(t_2, t_2 + dt_2]$ is dt_2/t. Thus, $(2dt_1/t)(dt_2/t) = (2!/t^2) dt_1 \, dt_2$ is the probability that one of the events occurs in $(t_1, t_1 + dt_1]$ while the other occurs in $(t_2, t_2 + dt_2]$, and this expression corresponds to (4.45) when $m = 2$. In general, the probability that one of the $m \geq 1$ events occurs in each of the m intervals $dt_1, dt_2, \ldots, dt_m$ is given by formula (4.45).

Thus, given that at least one Poisson event has occurred in $(0, t]$, the probability of it occurring in some interval $d\theta$ in $(0, t]$ is given by expression (4.43).

ii. *Probability Distribution of the Number of Customers* *N(t) in the System*

Since $\Phi_S(\cdot)$ is the service time distribution of a customer, then the probability that a customer arriving at some time $\Theta = \theta$ will have his service completed on or before time t, $0 < \theta \leq t$, is

$$P\{S \leq t - \theta | \Theta = \theta\} = \Phi_S(t - \theta)$$

Using Theorem 4.1, the probability of an arrival in $(\theta, \theta + d\theta]$ and a service completion on or before t, given that at least one customer has arrived in $(0, t]$, is

$$P\{S \le t - \theta \,|\, \Theta = \theta\} \cdot P\{\theta < \Theta \le \theta + d\theta\} = \Phi_s(t - \theta) \cdot \frac{d\theta}{t}, \quad 0 < \theta < t$$

Finally, the probability $q(t)$ that a customer arriving in $(0, t]$ will have his service completed on or before t is

$$
\begin{aligned}
q(t) &= \int_0^t P\{S \le t - \theta \,|\, \Theta = \theta\} \cdot P\{\theta < \Theta \le \theta + d\theta\} \\
&= \int_0^t \Phi_s(t - \theta) \cdot \frac{d\theta}{t}
\end{aligned}
$$

Changing the variable of integration to $u = t - \theta$, we obtain

$$q(t) = \frac{1}{t} \int_0^t \Phi_s(u)\, du \tag{4.46}$$

It is clear that the probability $p(t)$ that an arriving customer in $(0, t]$ will not have his service completed on or before t is

$$p(t) = 1 - q(t)$$

It must be noted that the probabilities $q(t)$ and $p(t)$ do not depend on the number of customers arriving in $(0, t]$ or their order of arrivals; further, the events associated with these probabilities are independent.

Suppose now that exactly $m \ge 1$ customers arrive in $(0, t]$; then the probability that exactly n out of these m, $0 \le n \le m$, do not complete service has the binomial distribution

$$\binom{m}{n} [p(t)]^n [1 - p(t)]^{m-n}, \quad 0 \le n \le m, \quad m \ge 1 \tag{4.47}$$

Since, in our $M/G/\infty$ system we have assumed an initially empty system, dictated by condition (4.41), expression (4.47) gives also the probability that exactly n customers are in service at time t, given that exactly m customers arrived in $(0, t]$. Thus

$$P\{N(t) = n \,|\, X(t) = m\} = \binom{m}{n} [p(t)]^n [1 - p(t)]^{m-n}, \quad 0 \le n \le m, \quad m \ge 1 \tag{4.48}$$

Finally, using (4.42) in (4.48) we have

$$P\{N(t) = n\} = \sum_{m=n}^{\infty} P\{N(t) = n \mid X(t) = m\} \cdot P\{X(t) = m\}$$

$$= \sum_{m=n}^{\infty} \binom{m}{n} [p(t)]^n [(1 - p(t)]^{m-n} \cdot e^{-\lambda t} \frac{(\lambda t)^m}{m!}$$

$$= \sum_{m=n}^{\infty} \frac{m!}{n!(m-n)!} [p(t)]^n [1 - p(t)]^{m-n} \cdot e^{-\lambda t} \frac{(\lambda t)^m}{m!}$$

Making the change in variable $i = m - n$ yields

$$P\{N(t) = n\} = e^{-\lambda t} \frac{[p(t)]^n}{n!} \sum_{i=0}^{\infty} \frac{1}{i!} [1 - p(t)]^i (\lambda t)^{i+n}$$

$$= e^{-\lambda t} \frac{[p(t)]^n}{n!} (\lambda t)^n e^{\lambda t[1 - p(t)]}$$

$$= e^{-\lambda t p(t)} \frac{[\lambda t p(t)]^n}{n!} \tag{4.49}$$

$$= e^{-\lambda t \left[1 - \frac{1}{t} \int_0^t \Phi_S(u)\, du\right]} \frac{\left\{\lambda t \left[1 - \frac{1}{t} \int_0^t \Phi_S(u)\, du\right]\right\}^n}{n!}$$

$$= e^{-\lambda \int_0^t [1 - \Phi_S(u)]\, du} \frac{\left\{\lambda \int_0^t [1 - \Phi_S(u)]\, du\right\}^n}{n!}, \quad n = 0, 1, 2, \ldots$$

Expression (4.49) can be identified as a nonhomogeneous Poisson process with parameter $\lambda \int_0^t [1 - \Phi_S(u)]\, du$ (Section 2.7-iii-a-2). It immediately follows that the expected number of customers in the system at time t is

$$E[N(t)] = \lambda \int_0^t [1 - \Phi_S(u)]\, du \tag{4.50}$$

iii. *Steady-State Distribution of the Number of Customers in the System*

If we denote by $P\{N = n\}$ the steady-state distribution of the number of customers $N(t)$ in the system as $t \rightarrow \infty$, and use the relation

$$\int_0^{\infty} [1 - \Phi_S(u)]\, du = E[S]$$

we obtain from (4.49)

$$P\{N = n\} = \lim_{t \to \infty} P\{N(t) = n\}$$

$$= e^{-\lambda \int_0^\infty [1 - \Phi_S(u)] du} \frac{\left\{ \lambda \int_0^\infty [1 - \Phi_S(u)] du \right\}^n}{n!}$$

$$= e^{-\lambda E[S]} \frac{\{\lambda E[S]\}^n}{n!}, \quad n = 0, 1, 2, \dots$$

Example 4.7

Passengers arrive at a bus terminal according to a Poisson process with intensity λ per minute. Buses are scheduled to leave the terminal every t minutes. A bus does not leave the terminal unless at least one passenger is present to board it at time of departure. It is further assumed that a bus has sufficient capacity to accomodate all passengers present at departure.

Passengers have a tendency to renege, that is to become impatient and leave the terminal after having waited a certain time following their arrival, to opt for other modes of transportation. Let S be a random variable denoting the time in minutes elapsed from customer arrival to his reneging assuming that the bus has not yet departed. Suppose that the distribution function of S is

$$P\{S \leq u\} = \Phi_S(u) = 1 - e^{-\alpha u}, \quad \alpha > 0, \quad u \geq 0$$

The ticket price per passenger is a dollars and it costs b dollars per bus to perform a trip. No other passengers may board the bus during the trip.

(a) Show that if $(b/a) \leq 1$ the operation of the buses is always profitable, while if $(b/a) > 1$ the operation is profitable only if a value of t exists such that $t > (1/\alpha) \ln [1 - (\alpha/\lambda)x^*]^{-1}$ where x^* is a root of the equation $e^{-x} = 1 - (a/b)x$ such that $0 < x^* < \lambda/\alpha$.

(b) Determine the optimum value of t that will maximize the expected profit per unit time when $\lambda = 1$ passenger per minute, $1/\alpha = 20$ minutes and $b/a = 10$.

If we select as time origin the instant just following a scheduled departure time, then, over the interval $0 < \theta \leq t$, the system may be viewed as a $M/G/\infty$ queuing system, in which the arrival process is dictated by the incoming passengers and the duration of service for each passenger is his or her waiting time S from arrival until reneging. Let $N(\theta)$ denote the number of passengers at the terminal at time θ, then, from relation (4.49), we have

$$P\{N(\theta) = n\} = e^{-\lambda \int_0^\theta e^{-\alpha u} du} \frac{\left[\lambda \int_0^\theta e^{-\alpha u} du \right]^n}{n!}$$

$$= e^{-(\lambda/\alpha)(1 - e^{-\alpha\theta})} \frac{\left[\frac{\lambda}{\alpha} (1 - e^{-\alpha\theta}) \right]^n}{n!}, \quad n = 0, 1, 2, \dots$$

At the time of the scheduled bus departure t, the probability that no passengers are present to board the bus is

$$p_0(t) = P\{N(t) = 0\} = e^{-(\lambda/\alpha)(1-e^{-\alpha t})}$$

which is also the probability that a bus will not leave as scheduled.

On the other hand, the probability that a bus will leave as scheduled is the probability that at least one passenger is present at the terminal at time t, and that is $[1 - p_0(t)]$. In this case, the expected number of passengers boarding the bus is

$$\sum_{n=1}^{\infty} nP\{N(t) = n\} = \sum_{n=0}^{\infty} nP\{N(t) = n\}$$

$$= E[N(t)] = \frac{\lambda}{\alpha}(1 - e^{-\alpha t})$$

The expected profit associated with a scheduled bus departure is equal to the expected revenue minus the expected cost given that a bus departs. Now

$$\text{expected revenue over } t = a \sum_{n=1}^{\infty} nP\{N(t) = n\}$$

$$= a\frac{\lambda}{\alpha}(1 - e^{-\alpha t})$$

$$\text{expected cost over } t = b[1 - p_0(t)]$$

$$= b[1 - e^{-(\lambda/\alpha)(1-e^{\alpha t})}]$$

$$\text{expected profit over } t = a\frac{\lambda}{\alpha}(1 - e^{-\alpha t}) - b[1 - e^{-(\lambda/\alpha)(1-e^{-\alpha t})}]$$

Denoting by $P(t)$ the expected profit per unit time, we have

$$P(t) = \frac{1}{t}\left\{a\frac{\lambda}{\alpha}(1 - e^{-\alpha t}) - b[1 - e^{-(\lambda/\alpha)(1-e^{-\alpha t})}]\right\}$$

$$= \frac{a}{t}\left\{\frac{\lambda}{\alpha}(1 - e^{-\alpha t}) - \frac{b}{a}[1 - e^{-(\lambda/\alpha)(1-e^{-\alpha t})}]\right\} \qquad (4.51)$$

(a) The operation of the buses is profitable if $P(t) > 0$, that is for $t > 0$,

$$\frac{\lambda}{\alpha}(1 - e^{-\alpha t}) - \frac{b}{a}[1 - e^{-(\lambda/\alpha)(1-e^{-\alpha t})}] > 0 \qquad (4.52)$$

Let
$$x = \frac{\lambda}{\alpha}(1 - e^{-\alpha t}) \qquad (4.53)$$

Then, inequality (4.52) becomes

$$x - \frac{b}{a}(1 - e^{-x}) > 0$$

or
$$e^{-x} > 1 - \frac{a}{b}x \qquad (4.54)$$

The graphs of the functions $y = e^{-x}$ and $y = 1 - (a/b)x$ are displayed in Figure 4.5. Since at $x = 0$ the slope of the curve $y = e^{-x}$ is -1, it follows that

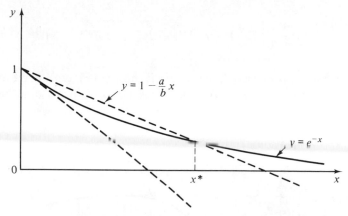

Figure 4.5. The functions $y = e^{-x}$ and $y = 1 - \frac{a}{b}x$

as long as the slope of the line $y = 1 - (a/b)x$ is less than -1, i.e., whenever $(a/b) > 1$, inequality (4.54) will always be satisfied and the operation of the buses will always be profitable.

If $(a/b) < 1$, then the equation $e^{-x} = 1 - (a/b)x$ has a second root x^* different from 1 such that for all $x > x^*$ inequality (4.54) is satisfied. Thus, the operation is profitable if

$$x = \frac{\lambda}{\alpha}(1 - e^{-\alpha t}) > x^*$$

or if
$$(1 - e^{-\alpha t}) > \frac{\alpha x^*}{\lambda} \qquad (4.55)$$

Since $0 < e^{-\alpha t} < 1$, then the value of x^* must also satisfy the condition

$$\frac{\alpha x^*}{\lambda} < 1$$

or
$$0 < x^* < \frac{\lambda}{\alpha}$$

and from inequality (4.55), t must satisfy the condition

$$t > \frac{1}{\alpha} \ln \left(1 - \frac{\alpha}{\lambda} x^*\right)^{-1} \qquad (4.56)$$

(b) For $\lambda = 1$, $\alpha = .05$, and $(b/a) = 10$, the reader may verify that the condition for profitability is satisfied when $t \geq 14.0$. Substituting the numerical values of λ, α and b/a, we obtain from relation (4.51) the following expression for the total expected profit per minute:

$$P(t) = \frac{a}{t} \{20(1 - e^{-.05t}) - 10[1 - e^{-20(1 - e^{-.05t})}]\}$$

The variation of $(1/a)P(t)$ as a function of t is shown in Figure 4.6. We note that profitability starts at $t = 14.0$. As t increases, $(1/a)P(t)$ increases at first until

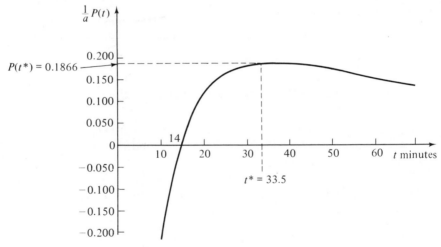

Figure 4.6. The function $\frac{1}{a}P(t) = \frac{1}{t} \{20(1 - e^{-.05t})$
$- 10[1 - e^{-20(1 - e^{-.05t})}]\}$

it achieves a maximum at $t^* = 33.5$, and then it tends asymptotically to zero as t tends to infinity. Thus, for maximum expected profit per unit time, the buses should be scheduled for departure every 33.5 minutes. Table 4.3 exhibits the various characteristics of the system as a function of the variable t when $a = 1$. The probability that a bus will not leave as scheduled under optimal operating conditions is negligible.

Table 4.3. CHARACTERISTICS OF THE BUS-PASSENGER SYSTEM ANALYZED IN EXAMPLE 4.7 WHEN $\lambda = 1$ PASSENGER PER MINUTE, $1/\alpha = 20$ MINUTES, $b = \$20.00$ AND $a = \$1.00$ PER PASSENGER

Time Interval t (minutes)	Expected Number of Passengers Boarding a Bus $E[N(t)] = 20(1 - e^{-.05t})$	Probability of a Bus Not Leaving the Terminal $p_0(t) = e^{-20(1 - e^{-.05t})}$	(1) Expected Revenue per Minute (\$) $\frac{1}{t}E[N(t)]$	(2) Expected Cost per Minute (\$) $\frac{1}{t}\frac{a}{b}[1 - p_0(t)]$	(3) Expected Profit per Minute (\$) (1) − (2)
3.00	2.786	0.062	0.929	3.128	−2.199
6.00	5.184	0.006	0.864	1.657	−0.793
9.00	7.247	0.001	0.805	1.110	−0.305
12.00	9.024	0.000	0.752	0.833	−0.081
15.00	10.553	0.000	0.704	0.667	0.037
18.00	11.869	0.000	0.659	0.556	0.104
21.00	13.001	0.000	0.619	0.476	0.143
24.00	13.976	0.000	0.582	0.417	0.166
27.00	14.815	0.000	0.549	0.370	0.178
30.00	15.537	0.000	0.518	0.333	0.185
33.00	16.159	0.000	0.490	0.303	0.187
36.00	16.694	0.000	0.464	0.278	0.186
39.00	17.155	0.000	0.440	0.256	0.183
42.00	17.551	0.000	0.418	0.238	0.180
45.00	17.892	0.000	0.398	0.222	0.175
48.00	18.186	0.000	0.379	0.208	0.171
51.00	18.438	0.000	0.362	0.196	0.165
54.00	18.656	0.000	0.345	0.185	0.160
57.00	18.843	0.000	0.331	0.175	0.155
60.00	19.004	0.000	0.317	0.167	0.150
63.00	19.143	0.000	0.304	0.159	0.145
66.00	19.262	0.000	0.292	0.152	0.140
69.00	19.365	0.000	0.281	0.145	0.136
72.00	19.454	0.000	0.270	0.139	0.131
75.00	19.530	0.000	0.260	0.133	0.127
78.00	19.595	0.000	0.251	0.128	0.123
81.00	19.652	0.000	0.243	0.123	0.119
84.00	19.700	0.000	0.235	0.119	0.115
87.00	19.742	0.000	0.227	0.115	0.112
90.00	19.778	0.000	0.220	0.111	0.109
93.00	19.809	0.000	0.213	0.108	0.105
96.00	19.835	0.000	0.207	0.104	0.102
99.00	19.858	0.000	0.201	0.101	0.100

4.8. Control of Single Server Queuing System

In many of the example problems solved so far in this chapter, some decision variable was usually identified and, based on a selected economic criterion, an optimum operating doctrine was derived. In a single server queuing system, one may sometimes have control of the operation of the station by having the option of varying the idle fraction through removal of the server for a portion of the time. Here, the server, if inactive, remains inactive until Q customers are present in the system, at which time operation starts and the station remains open until the system is empty, that is when zero customers are in the system. Service is reinstated when the queue size has accumulated to a level of Q, $Q = 1, 2, \ldots$. The quantity Q thus becomes a decision variable whose value would be dictated according to some criterion.

It is reasonable to assume a cost structure whereby whenever the server starts his busy cycle a fixed cost K is incurred, while a holding cost h is associated when a customer is in residence in the system for one unit time. As Q gets larger, the average frequency of setups gets smaller while the customers are being held longer on the average, in the system. It is therefore reasonable to conjecture that an optimum value of Q exists that would balance the contribution of the expected setup cost and the expected holding cost per unit time.

The example that follows analyzes this class of problem in the simple case of a $M/M/1$ queuing system and simultaneously illustrates the use of the flow diagram in setting up the balance equations. The original work of Yadin and Naor [17] should be consulted for other extensions.

Example 4.8

Items arrive for processing at a manufacturing center according to a Poisson process with intensity λ. The service time necessary to process an item is negative exponential with mean $1/\mu$. The center does not operate when no items are present. It starts its operation at the appearance of the Qth item following a shut down. All items in the system are then processed, and the working cycle is terminated when no more items are in the system; the center then closes down and the cycle is repeated. Assume $\rho = \lambda/\mu < 1$ and steady-state condition has been reached.

If a fixed cost of K dollars is associated in setting up the processor every time an operation starts, and a holding cost of h dollars is charged for holding one item in the center for one unit time, determine the optimum value of Q so as to minimize the total expected cost per unit time.

If the states $i = 0$ and $i = 1$ correspond respectively to the processor being idle and busy, while the state n, $n = 0, 1, 2, \ldots$, corresponds to the number of units in the system, then the pair (i, n) completely specifies the state of the system in the steady state. The flow diagram describing this steady-state regime is as shown in Figure 4.7. Let

$P(0, n) =$ probability that the processor is idle and there are n units in
the system $(n = 0, 1, \ldots, Q - 1)$

$P(1, n) =$ probability that the processor is busy and there are n units in
the system $(n = 1, 2, \ldots)$

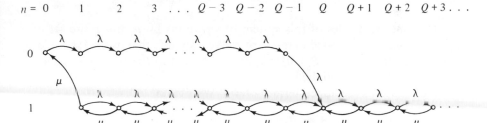

Figure 4.7. Flow diagram for the steady state $M/M/1$ queuing system with removable server (Example 4.8).

Applying the principle of conservation of flow at the nodes corresponding to the states $(0, n)$, $n = 1, \ldots, Q - 1$, we obtain the following equations

$$\lambda P(0, 0) = \lambda P(0, 1) = \cdots = \lambda P(0, Q - 1)$$

or

$$P(0, 0) = P(0, 1) = \cdots = P(0, Q - 1)$$

Define now the generating function

$$G_0(z) = \sum_{n=0}^{n=Q-1} z^n P(0, n), \quad |z| \leq 1$$

In terms of $P(0, 0)$ we may write

$$G_0(z) = P(0, 0) \sum_{n=0}^{n=Q-1} z^n$$

$$= P(0, 0) \frac{1 - z^Q}{1 - z}$$

(4.57)

The balance equations corresponding to the nodes $(0, 0)$ and $(1, n)$, $n =$

$1, 2, \ldots$, are respectively

$$\mu P(1, 1) = \lambda P(0, 0)$$
$$\mu P(1, 2) = (\lambda + \mu) P(1, 1)$$
$$\lambda P(1, 1) + \mu P(1, 3) = (\lambda + \mu) P(1, 2)$$
$$\lambda P(1, 2) + \mu P(1, 4) = (\lambda + \mu) P(1, 3)$$
$$\cdots$$
$$\lambda P(1, Q - 2) + \mu P(1, Q) = (\lambda + \mu) P(1, Q - 1)$$
$$\lambda P(0, Q - 1) + \lambda P(1, Q - 1) + \mu P(1, Q + 1) = (\lambda + \mu) P(1, Q)$$
$$\lambda P(1, Q) + \mu P(1, Q + 2) = (\lambda + \mu) P(1, Q + 1)$$
$$\cdots$$

Dividing both sides of this system of equations by μ, the entire set of equations may be expressed in terms of ρ. We also note that in the $(Q + 1)$st equation, $P(0, Q - 1) = P(0, 0)$. Suppose now that in the resultant system, the first equation is multiplied by z, the second equation by z^2, and so forth, and the equations are added term by term. We obtain

$$\rho z^{Q+1} P(0, 0) + \rho \sum_{n=3}^{\infty} z^n P(1, n - 2) + \sum_{n=1}^{\infty} z^n P(1, n)$$
$$= \rho z P(0, 0) + (1 + \rho) \sum_{n=2}^{\infty} z^n P(1, n - 1)$$

Alternatively, we have

$$\rho z^{Q+1} P(0, 0) + \rho z^2 \sum_{n=1}^{\infty} z^n P(1, n) + \sum_{n=1}^{\infty} z^n P(1, n)$$
$$= \rho z P(0, 0) + (1 + \rho) z \sum_{n=1}^{\infty} z^n P(1, n) \tag{4.58}$$

We now introduce the generating function

$$G_1(z) = \sum_{n=1}^{\infty} z^n P(1, n), \quad |z| \leq 1$$

Expression (4.58) then becomes

$$\rho z^{Q+1} P(0, 0) + \rho z^2 G_1(z) + G_1(z) = \rho z P(0, 0) + (1 + \rho) z G_1(z)$$

Solving for $G_1(z)$, we obtain

$$G_1(z) = P(0, 0) \frac{\rho z - \rho z^{Q+1}}{\rho z^2 + 1 - (1 - \rho) z}$$
$$= P(0, 0) \frac{\rho z (1 - z^Q)}{(1 - \rho z)(1 - z)} \tag{4.59}$$

Let now

> $N =$ random variable denoting the number of units in the system under steady state conditions
>
> $P_n = P\{N = n\} =$ probability that exactly n units are in the system, $n = 0, 1, 2, \ldots$

Let $G_N(z)$ denote the probability generating function of N; thus

$$G_N(z) = \sum_{n=0}^{\infty} z^n P_n$$

It is evident that

$$P_n = P(0, n) + P(1, n)$$

and hence that

$$G_N(z) = G_0(z) + G_1(z)$$

Thus, using relations (4.57) and (4.59), we obtain for $G_N(z)$ the expression

$$G_N(z) = P(0, 0)\left[\frac{1 - z^Q}{1 - z} + \frac{pz(1 - z^Q)}{(1 - pz)(1 - z)}\right]$$

$$= P(0, 0)\frac{1 - z^Q}{(1 - z)(1 - pz)} \tag{4.60}$$

To determine $P(0, 0)$, which corresponds to the probability that the system is empty, we make use of the normalizing condition

$$G_N(1) = \sum_{n=0}^{\infty} P_n = 1$$

Setting $z = 1$ in expression (4.60) yields an indeterminate form in the right side. Applying l'Hospital's rule once, we obtain from (4.60)

$$\lim_{z \to 1} G_N(z) = 1 = \lim_{z \to 1} P(0, 0)\left\{\frac{-Qz^{Q-1}}{-p(1 - z) - (1 - pz)}\right\}$$

$$= P(0, 0)\frac{Q}{1 - p}$$

Solving for $P(0, 0)$ results in

$$P(0, 0) = \frac{1 - p}{Q} \tag{4.61}$$

Substituting (4.61) in expression (4.60) yields

$$G_N(z) = \frac{1 - p}{Q}\frac{1 - z^Q}{(1 - z)(1 - pz)} \tag{4.62}$$

Consider now a full cycle consisting of an empty/non-empty system. Let

T_1 = random variable denoting the length of time the system is empty per cycle

T_2 = random variable denoting the length of time the system is not empty per cycle

Since the processor starts operating once per cycle, the expected number of setups per unit time is $1/(E[T_1] + E[T_2])$. Now

$$E[T_1] = \frac{1}{\lambda}$$

Using (4.61), we obtain for the proportion of time the system is empty

$$\frac{E[T_1]}{E[T_1] + E[T_2]} = P(0, 0) = \frac{1 - \rho}{Q}$$

Hence, an expression for the expected number of setups per unit time is

$$\frac{1}{E[T_1] + E[T_2]} = \frac{1}{E[T_1]} \cdot \frac{1 - \rho}{Q} = \frac{\lambda(1 - \rho)}{Q}$$

The expected cost of setups per unit time is clearly $K/(E[T_1] + E[T_2])$.

We now wish to determine an expression for the expected number of items in the system, $E[N]$, in order to compute the expected holding cost per unit time, which is $hE[N]$. Now, expression (4.62) for $G_N(z)$ may be written as

$$G_N(z) = \frac{1 - \rho}{Q(1 - \rho z)} \sum_{n=0}^{n=Q-1} z^n$$

Then
$$G'_N(z) = \frac{(1 - \rho)}{Q(1 - \rho z)^2} \left[(1 - \rho z) \sum_{n=1}^{n=Q-1} n z^{n-1} + \rho \sum_{n=0}^{n=Q-1} z^n \right]$$

and
$$E[N] = G'_N(1) = \frac{1 - \rho}{Q(1 - \rho)^2} \left[(1 - \rho) \sum_{n=1}^{n=Q-1} n + \rho \sum_{n=0}^{n=Q-1} 1 \right]$$

$$= \frac{1 - \rho}{Q(1 - \rho)^2} \left[(1 - \rho) \frac{Q(Q - 1)}{2} + \rho Q \right]$$

$$= \frac{Q - 1}{2} + \frac{\rho}{1 - \rho}$$

The total expected cost per unit time, $F(Q)$, expressed as a function of Q, is the sum total of the expected cost of setups per unit time plus the expected cost of holding items in the system per unit time. Thus

$$F(Q) = \frac{K}{E[T_1] + E[T_2]} + hE[N]$$

$$= \frac{K\lambda(1 - \rho)}{Q} + h\left(\frac{Q - 1}{2} + \frac{\rho}{1 - \rho} \right)$$

(4.63)

This cost expression is very similar to many cost expressions obtained in inventory theory (see Chapter 5). Since Q is a positive integer $Q = 1, 2, \ldots$, the optimum value of Q, Q^* that minimizes the function $F(Q)$ must satisfy the conditions

$$F(Q^*) \leq F(Q^* - 1)$$
$$F(Q^*) \leq F(Q^* + 1)$$

Using relation (4.63) in the first of these conditions yields the inequality

$$\frac{K\lambda(1-p)}{Q^*} + h\left(\frac{Q^*-1}{2} + \frac{p}{1-p}\right) \leq \frac{K\lambda(1-p)}{Q^*-1} + h\left(\frac{Q^*-2}{2} + \frac{p}{1-p}\right)$$

or

$$K\lambda(1-p)\left(\frac{1}{Q^*-1} - \frac{1}{Q^*}\right) \geq \frac{h}{2}$$

or

$$(Q^*-1)Q^* \leq \frac{2K\lambda(1-p)}{h}$$

Similarly, the second condition yields the inequality

$$Q^*(Q^*+1) \geq \frac{2K\lambda(1-p)}{h}$$

These last two inequalities may be combined to yield as a condition for an optimal value of Q^*

$$(Q^*-1)Q^* \leq \frac{2K\lambda(1-p)}{h} \leq Q^*(Q^*+1)$$

We now form the following table:

Q^*	1	2	3	4	5	6	7	8	9	10	$\cdots$
$(Q^*-1)Q^*$	0	2	6	12	20	30	42	56	72	90	$\cdots$
$Q^*(Q^*+1)$	2	6	12	20	30	42	56	72	90	110	$\cdots$

Given a particular value of $2K\lambda(1-p)/h$, the optimum value Q^* may be determined. Thus, if $2K\lambda(1-p)/h = 45$, then $Q^* = 7$. We note that a double solution is possible.

As an approximation, if Q is treated as a continuous variable, then differentiating the expression for $F(Q)$ in (4.63) with respect to Q and setting the result equal to zero yields

$$\frac{-K\lambda(1-p)}{Q^2} + \frac{h}{2} = 0$$

Thus Q^* is approximately given by

$$Q^* \approx \sqrt{\frac{2K\lambda(1-\rho)}{h}}$$

a square root formula frequently appearing in inventory theory.

SELECTED REFERENCES

[1] Brochmeyer, E., H. L. Halstrom, and A. Jensen, "The Life and Works of A. K. Erlang," *Transactions of the Danish Academy of Technical Sciences*, Vol. 2, 1948.

[2] Cohen, J. W., *The Single Server Queue*, North-Holland Publishing Company, Amsterdam, 1969.

[3] Conway, R. W., W. L. Maxwell, and L. W. Miller, *Theory of Scheduling*, Addison-Wesley Publishing Co., Reading, Mass., 1967.

[4] Cooper, R. B., *Introduction to Queuing Theory*, The Macmillan Company, New York, 1972.

[5] Cox, D. R., and W. L. Smith, *Queues*, John Wiley & Sons, Inc., New York, 1961.

[6] Fry, T. C., *Probability and Its Engineering Uses*, Van Nostrand Reinhold Company, New York, 1928.

[7] Jaiswal, N. Y., *Priority Queues*, Academic Press, Inc., New York, 1968.

[8] Kendall, D. G., "Some Problems in the Theory of Queues," *Journal of the Royal Statistical Society*, Vol. B-13, pp. 151–185, 1951.

[9] Khintchine, A. Y., "Studied Poisson-Input Arbitrary-Holding Time, Single-Channel Problems," *Mathematicheskii Sbornik*, Vol. 39, p. 37, 1953.

[10] Little, J. D. C., "A Proof for the Queuing Formula: $L = \lambda W$," *Operations Research*, Vol. 9, pp. 383–387, 1967.

[11] Molina, E. C., "Application of the Theory of Probability to Telephone Trunking Problems," *Bell System Technical Journal*, Vol. 6, pp. 461–494, 1927.

[12] Morse, P. M., *Queues, Inventories and Maintenance*, John Wiley & Sons, Inc., 1958.

[13] Palm, C., "Intensitätsschwankungen in Fernsprechverkehr," *Ericsson Technics*, Vol. 44, pp. 1–189, 1943.

[14] Pollaczek, F., "Über eine Aufgabe der Wahrscheinlichkeitsstheorie," *Mathematische Zeitschrift*, Vol. 32, pp. 64–100 and 729–750, 1930.

[15] Prabhu, N. U., *Queues and Inventories*, John Wiley & Sons, Inc., 1965.

[16] Saaty, T. L., *Queuing Theory*, McGraw-Hill Book Company, New York, 1961.

[17] YADIN, M, and P. NAOR, "Queuing Systems with a Removable Service Station," *Operational Research Quarterly*, Vol. 14, pp. 393–405, 1963.

PROBLEMS

1. Items arrive at a machine center, which can service them one at a time. The arrivals are Poisson at a rate λ and service time has a negative exponential distribution with mean $1/\mu$. The last step in the service is a quality-control inspection, and a fraction p of the items are rejected, in which case they go back into the queue for repeated service. Determine the steady-state distribution of the number of units in the system.

2. *Bulk Arrival.* Items arrive by pair at a service center, which can service them one at a time. The arrivals of the pair are Poisson at a rate λ, and service time has a negative exponential distribution with mean $1/\mu$. Assuming the system will reach steady state, determine the distribution of the number of units in the system. What is the condition on λ and μ for a steady state?

3. *Balking with Constant Value. A System with Finite Queue.* Suppose that customers arrive at a service counter in a Poisson fashion at a constant rate λ, until the number in the system is M, $M \geq 1$. When M units are present, further arriving customers leave immediately without waiting for service; this is called balking with constant balking value. Let the service time have a negative exponential distribution with mean $1/\mu$. Let $N(t)$ be the number of customers in the system at time t, and let $\rho = \lambda/\mu$.
   (a) Show that

$$\lim_{t \to \infty} P\{N(t) = n\} = \rho^n \frac{1 - \rho}{1 - \rho^{M+1}}, \quad n = 0, 1, \ldots, M$$

   (b) Assume that at time $t = 0$ the system is empty and that $M = 2$. For $n = 0, 1, 2$ and $t \geq 0$, show that $P\{N(t) = n\}$ has the form

$$P\{N(t) = n\} = A_n + B_n e^{-\mu(1 + \rho + \sqrt{\rho})t} + C_n e^{-\mu(1 + \rho - \sqrt{\rho})t}$$

   Determine A_n, B_n, and C_n.

4. *Balking with Random Value.* In a single-server queue with Poisson arrivals at rate λ and service time with a negative exponential distribution with mean $1/\mu$, the following phenomenon was observed.

   An arriving customer decides to join the queue if the number in the system at time t, $N(t)$, is less than or equal to M, $M \geq 1$. However, the quantity M is a random variable independently distributed from customer to customer with PMF $\varphi_M(x) = P\{M = x\}$.
   (a) Write down the steady-state equations for P_n, the probability of n customers in the system, $n = 0, 1, 2, \ldots$.
   (b) Determine P_n in terms of P_0.

(c) Given that

$$\varphi_M(x) = \begin{cases} 0, & x = 0 \\ \dfrac{1}{A}, & x = 1, 2, \ldots, A \\ 0, & \text{otherwise} \end{cases}$$

where $A > 0$, find P_n.

5. *The $M/M/\infty$ Queuing System.* Let $N(t)$ denote the number of customers in the $M/M/\infty$ queuing system at time t, and let

$$P(n, t) = P\{N(t) = n\}$$

Show that $P(n, t)$ satisfies the following system of differential-difference equations

$$\frac{dP(0, t)}{dt} = -\lambda P(0, t) + \mu P(1, t), \quad n = 0$$

$$\frac{dP(n, t)}{dt} = \lambda P(n - 1, t) - (\lambda + \mu)P(n, t) + (n + 1)\mu P(n + 1, t), \quad n \geq 1$$

Derive the equations for $P_n = \lim_{t \to \infty} P(n, t)$. Show that the steady state solution always exists and is given by

$$P_n = e^{-\lambda/\mu} \frac{(\lambda/\mu)^n}{n!}, \quad n = 0, 1, 2, \ldots$$

Compare this result with expression (4.49).

6. *Machine-Interference Problem with Several Repairmen.* Suppose that there are M machines, each of which when running has a probability of $\lambda\, dt + o(dt)$ of breaking down in the time interval $(t, t + dt]$. There are r repairmen, and only one man can work on a machine at a time. If $n > r$ machines are out of action, then $n - r$ of them wait until a repairman is free, at which time repair starts on a machine on a first-come first-served basis. The time to complete a repair has a negative exponential distribution with parameter μ.
   (a) Show that when n machines, $n = 0, 1, 2, \ldots, M$, are broken down, the probability of a further breakdown in the time interval $(t, t + dt]$ is $(M - n)\lambda\, dt + o(dt)$, and that the probability of a service being completed for $n = 1, 2, \ldots, M$ is $n\mu\, dt + o(dt)$ if $1 \leq n \leq r$ and $r\mu\, dt + o(dt)$ if $r + 1 \leq n \leq M$. Find the steady-state distribution of the number of machines out of operation.
   (b) Suppose that the cost of a repairman is C dollars per hour and the gross profit on a machine which is running is p dollars per hour. Formulate an appropriate objective function to determine the optimum number of repairmen r, $r = 1, 2, \ldots$.

7. *A Preemptive Resume Queue Discipline.* Suppose that units arrive at a service center following a Poisson law with intensity λ. Service time has a negative exponential distribution with mean $1/\mu$. Assume that the service station experi-

ences breakdowns, which occur following a Poisson law with intensity ζ, in a manner which is independent of any of the other system characteristics. Thus a breakdown may occur even if no customers are present in the system. Following a breakdown, the service of any customer is interrupted, and repair of the service station starts immediately. The time to repair is assumed to follow a negative exponential distribution with mean $1/\nu$. Following the termination of repair, the interrupted customer resumes his servicing. The server thus may be considered to be in the two states 0 or 1, depending on whether it is operative or in a state of repair. Determine the expected number of customers in the system in the steady state.

8. Using the expression for $\bar{\varphi}_\Theta(z)$, the Laplace transform of the PDF of the steady-state distribution of the time Θ spent in the system, find an expression for $E[\Theta]$. Verify your results using Little's formula.

9. Obtain the steady state distribution of the time spent in the $M/G/1$ queuing system where service time S has PDF

$$\varphi_S(x) = (\mu x)\mu e^{-\mu x}, \quad \mu > 0, \quad 0 < x < \infty$$

10. Passengers arrive at a bus terminal according to a Poisson process with intensity λ. Assume no reneging takes place. Each bus departs with Q passengers as soon as the Qth passenger arrives following the depature of the previous bus. Let $P(n, t)$, $n = 0, 1, \ldots, Q - 1$, $t \geq 0$, denote the probability that exactly n passengers are waiting at the bus terminal at time t, and assume that initially the bus terminal is empty.

   (a) Show that $P(n, t)$ satisfies the following system of differential-difference equations

   $$\frac{dP(1, t)}{dt} = \lambda P(Q - 1, t) - \lambda P(0, t), \quad n = 0$$

   $$\frac{dP(n, t)}{dt} = \lambda P(n - 1, t) - \lambda P(n, t), \quad n = 1, 2, \ldots, Q - 1$$

   with the initial conditions

   $$P(n, 0) = \begin{cases} 1, & \text{if } n = 0 \\ 0, & \text{otherwise} \end{cases}$$

   (b) Show that in the steady state, $P_n = \lim_{t \to \infty} P(n, t) = \dfrac{1}{Q}$. Determine the expected number of passengers waiting at the bus terminal and the expected frequency of bus departure.

INVENTORY THEORY
FOR SINGLE-COMMODITY
SINGLE-INSTALLATION SYSTEMS

5.1. Introduction

Inventory management of physical goods or other products or elements is an integral part of logistic systems common to all sectors of the economy, such as business, industry, agriculture, and defense. In an economy that is perfectly predictable, inventory may be needed to take advantage of the economic features of a particular technology, or to synchronize human tasks, or to regulate the production process to meet the changing trends in demand. When uncertainty is present, inventories are used as a protection against risk of stockout.

The existence of inventory in a system generally implies the existence of an organized complex system involving inflow, accumulation, and outflow of some commodities or goods or items or products. For example, in business the inflow of goods is generated through procurement, purchase, or production. The outflow is generated through demand for the goods. Finally, the difference between the rate of outflow and the rate of inflow generates an inventory for the goods.

The regulation and control of inventory must proceed within the context of this organized system. Thus inventories, rather than being interpreted

5

as idle resources, should be regarded as a very essential element, the study of which may provide insight in the aggregate operation of the system. The scientific analysis of inventory systems defines the degree of interrelationship between inflow, accumulation, and outflow and identifies economic control methods for operating such systems.

i. *Historical Background and Examples of Inventory Systems*

The first quantitative analysis in inventory studies applied to business goes back to the work of Harris [11] in 1915, who formulated and optimized a simple inventory situation. Wilson rediscovered the same formula in 1918, and apparently was more successful in popularizing its use. The formula derived by Harris is an expression for an optimal production lot size given as a square root function of a fixed cost, an investment or holding cost, and the demand. It is often referred to as the *simple lot-size formula* or the *economic order quantity* (*EOQ*) *formula*, or the *Wilson formula*. Raymond's book [15], which appeared in 1931, is the first treatise to have attempted to explain extensions and applications in the practice of the EOQ formula in production and inventory operations.

Several variations of Wilson's deterministic model have been studied; however, the stochastic nature of inventory problems was mainly investigated in the post World War II period. In this respect, the works of Massé [13] in 1946, Arrow, Harris, and Marschak [2] in 1951, Dvoretsky, Kiefer, and Wolfowitz [8] in 1952, and Whitin [23] in 1953 should be noted. These authors take into consideration the effect of randomness in the demand, and apply the theory of stochastic processes and/or dynamic programming to formulate and solve complex inventory problems.

Since then much interest has been directed in studying the theory of single product single-installation systems with the successful utilization of several well-known optimization techniques to arrive at optimal solutions (see, for example, references at end of chapter). Present emphasis in research is mainly related to investigating the theory and economic control of multicommodity, multiinstallation systems, a subject that is introduced in the next chapter.

Inventory systems can arise in several different ways, and the type of commodity involved may be extremely varied. The following are some examples of inventory systems:

1. A retailer has to satisfy his customers' demand for a particular product. In general, such demand is stochastic in nature. To this end the retailer stores a certain amount of the commodity in his store. At a particular instant of time, he must decide whether to replenish his stock through purchase from the wholesaler, or not to replenish it. If he overstocks, he will tie up capital in his stock, and may run the risk of having on hand obsolete or perished goods. If, on the other hand, he runs out of stock, he will not be able to satisfy the demand from his customers; thus he may lose profit or customer good will. His policy for replenishing his stock should account for the demand from his customers, and should incorporate the economic factors associated with procurement, capital in stock, and loss of customers' good will.

2. This next example is a classic inventory problem, although at first it may not appear to be of the type: An airline company runs a school for training air hostesses. Classes start at the beginning of each month, and it takes 3 months to train the candidates. New hostesses are required as a result of turnover and improved carrier services; it is assumed that the demand for trained hostesses in a particular month is a random variable. If, following her training, a hostess is not required for duties, the airline company has still to pay her salary. On the other hand, if insufficient hostesses are available to meet the demand in a particular month, the airline company has to hire temporarily standby substitute hostesses at a premium. The problem is to determine the optimum class size.

3. This final example is drawn from the area of production management. A manufacturer produces on one of his production facilities a commodity to meet the demand of a class of customers. He may satisfy customers' orders by producing as many units as necessitated by each separate order, and thus

avoid any inventories. However, this will induce an excessive number of production setups. On the other hand, should he reduce the number of production setups, he will have to produce lots of increasing larger sizes and consequently increase his stocking capacity to meet present and future demands. This last alternative will reduce his production setup costs, but will increase inventory carrying and storage charges. Between these two extreme alternatives there exists an optimum number of setups that minimizes the combined contribution of the fixed production cost and the inventory costs. The Harris and Wilson models are idealized descriptions of this class of problems which will be analyzed in detail later in the chapter.

ii. *Motives for Carrying Inventory*

The three basic motives for carrying inventory are the (a) transaction motive, (b) precautionary motive, and (c) speculative motive.

a. TRANSACTION MOTIVE

Consider a situation in which the requirements for the demand are known with total certainty and all factors and parameters affecting the management and control of an inventory system in the future can be completely determined with certainty. Even with these considerations, a certain amount of inventory may still be carried to synchronize the inflow and outflow of goods. Examples of this nature can be drawn from mass-production systems. It is possible that a production system manufactures goods in runs, each run involving an expensive setup; it then becomes necessary to take advantage of the economic features of the technology of the process and carry a certain amount of inventory (Figure 5.1). Other production systems produce goods in a continuous fashion; to meet excess demand under peak seasonal conditions without offsetting unduly the rate of production, or when operat-

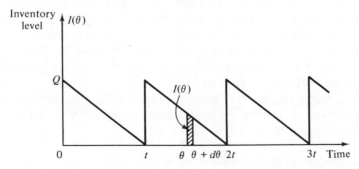

Figure 5.1. Inventory movement for a lot size production problem under a fixed production lot and constant demand rate.

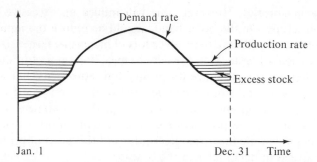

Figure 5.2. Inventory fluctuation in the constant rate production of a seasonal commodity.

ing under limited production capacity, an overproduction during the slack seasonal periods would naturally generate inventory (Figure 5.2).

b. PRECAUTIONARY MOTIVE

In many practical situations there exists an inability to predict exactly demands and/or the amount of time necessary for the delivery of goods (known as lead time). More specifically, demand and lead time are in general random variables with known distributions. The uncertainty element thus introduced may generate an unpredictable shortage with a high penalty cost. A consequent need for a safety allowance in stock should then be made on and above any amount of stock carried within the stipulation of the transaction motive. This extra stock allowance, sometimes known as safety stock, is essentially a precautionary move toward future uncertainty.

c. SPECULATIVE MOTIVE

If a rise in the product price is anticipated in the near future, a manufacturer would normally overproduce and a retailer would normally overpurchase because of the contemplated extra profit resulting from disposing of the future goods, which are presently inventoried. This type of speculation generates a third motive for holding inventories. Other future uncertain factors that may motivate a speculative move toward holding additional inventory are strikes, national or foreign trade policies, per capita income, defense expenditures, and so forth.

The present chapter will not elaborate on the speculative aspect for carrying inventory. Rather, we shall discuss topics related to the transaction and precautionary motive for holding stock.

iii. *Analysis of the Inventory System*

Inventory systems may be broadly classified as continuous review systems or periodic review systems. In *continuous review systems*, the system is monitored continuously over time. In *periodic review systems,* the system is monitored at discrete, equally spaced instants of time. Processes generated in each of these two systems correspond, respectively, to continuous-parameter processes and discrete-parameter processes.

The simplest type of inventory system is characterized by the movement or flow of a single commodity into and out of a single storage point (Figure 5.3).

Figure 5.3. Schematic representation in the flow of commodity in a simple inventory system.

The inflow of the commodity is governed by some type of *replenishment* from production or from purchase acquisition, generally referred to as procurement. The outflow of the commodity is induced by *demand* associated with either customer orders or production orders. Because, in general, it is either impossible or uneconomical to balance exactly the inflow with the outflow, inventory is created at the storage point.

To broaden the basis of future discussion, we interpret inventory in the most general sense, and assume that it can be measured both at the positive and negative level. A positive inventory level implies the existence of a *stock on hand* in the amount equal to the inventory level. A zero or negative inventory level implies *shortage.* We shall assume throughout this chapter that *back orders* are allowed; that is, if the commodity is not available in stock to satisfy any demands when they presently occur, then all such demands are backlogged and can be satisfied at future times as soon as the commodity becomes available. The amount of back order at some instant of time may thus be measured by the absolute value of the negative inventory level at that time. Let, for $\theta \geq 0$,

$I(\theta)$ = inventory level at time θ
$H(\theta)$ = stock on hand at time θ
$B(\theta)$ = back order at time θ

Then

$$H(\theta) = \begin{cases} I(\theta), & \text{if } I(\theta) > 0 \\ 0, & \text{if } I(\theta) \leq 0 \end{cases}$$

$$B(\theta) = \begin{cases} 0, & \text{if } I(\theta) > 0 \\ -I(\theta), & \text{if } I(\theta) \leq 0 \end{cases}$$

It follows that

$$I(\theta) = H(\theta) - B(\theta)$$

Suppose now that we consider a continuous review system, and let

$p(\theta)$ = rate of inflow or procurement rate at time θ
$d(\theta)$ = rate of outflow or demand rate at time θ

The balance equation dictating the change in inventory level over the infinitesimal time interval $(\theta, \theta + d\theta)$ is

$$I(\theta + d\theta) = I(\theta) + p(\theta)\, d\theta - d(\theta)\, d\theta$$

which may be rewritten as

$$\frac{I(\theta + d\theta) - I(\theta)}{d\theta} = p(\theta) - d(\theta)$$

If then the limit of the left-hand quantity exists as $d\theta \rightarrow 0$, we obtain

$$\frac{dI(\theta)}{d\theta} = p(\theta) - d(\theta) \tag{5.1}$$

which states that the rate of change of the inventory level equals the rate of inflow minus the rate of outflow.

Suppose now that, at $\theta = 0$, $I(0)$ is given; then clearly

$$I(\theta) = I(0) + \int_0^\theta p(u)\, du - \int_0^\theta d(u)\, du \tag{5.2}$$

Hence the inventory level at any particular instant of time θ equals the inventory level originally available at the stocking point plus the cumulative contribution of all procurements since time origin minus the cumulative contribution of all demands since origin (see Figures 5.1 and 5.4). It is then apparent that in order to measure the inventory level at any particular time it is necessary to know the characteristics of procurement and demand.

In a periodic review system when time θ is measured over the discrete scale, $\theta = 0, 1, 2, \ldots$, let

$p(\theta)$ = amount of inflow during the time interval $[\theta - 1, \theta)$
$d(\theta)$ = amount of outflow during the time interval $[\theta - 1, \theta)$
$I(\theta)$ = inventory level at end of time θ

The discrete counterparts of equations (5.1) and (5.2) are, respectively,

$$I(\theta) = I(\theta - 1) + p(\theta) - d(\theta)$$

and

$$I(\theta) = I(0) + \sum_{u=1}^{u=\theta} p(u) - \sum_{u=1}^{u=\theta} d(u) \tag{5.3}$$

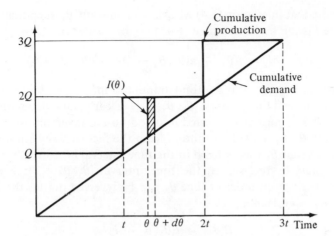

Figure 5.4. Alternative representation of the inventory movement of Figure 5.1.

In general, if we consider two distinct instants of time θ_1 and θ_2 such that $\theta_1 < \theta_2$, and let

$\pi(\theta_1, \theta_2)$ = total inflow or addition to inventory in the interval $[\theta_1, \theta_2)$
$D(\theta_1, \theta_2)$ = total outflow or demand or depletion from inventory in the interval $[\theta_1, \theta_2)$

then, from either relations (5.2) or (5.3), we obtain

$$I(\theta_2) = I(\theta_1) + \pi(\theta_1, \theta_2) - D(\theta_1, \theta_2) \qquad (5.4)$$

The regulation of inflow in an inventory system through procurement orders constitutes the main avenue for controlling the inventory level. Usually, there exists a time lapse from the moment a directive or order is placed to replenish inventory to the moment the actual addition to the inventory takes place. This leads us to introduce the notion of *lead time l*, which is defined as the time elapsed from the moment an order is initiated to the moment the order is received in stock.

We now define two additional concepts in inventory systems: outstanding orders and position inventory.

The *on order* or *outstanding orders* at a particular instant θ, denoted by $R(\theta)$, is the total quantity of orders that have been placed on or before θ, but that have not been received as yet in inventory.

The *position inventory* or *systems stock* at a particular instant θ, denoted by $\mathcal{P}(\theta)$, is the sum total of the inventory level $I(\theta)$ at time θ plus the outstanding orders $R(\theta)$ at time θ, or

$$I(\theta) = \mathcal{P}(\theta) - R(\theta) \qquad (5.5)$$

Suppose that in relation (5.4) we select the instant θ_2 such that $\theta_2 = \theta_1 + l$, where l is the lead time; then (5.4) may be written as

$$I(\theta_1 + l) = I(\theta_1) + \pi(\theta_1, \theta_1 + l) - D(\theta_1, \theta_1 + l) \qquad (5.6)$$

The quantity $\pi(\theta_1, \theta_1 + l)$ is the total amount of inflow or addition to inventory during the lead time interval $[\theta_1, \theta_1 + l)$. Clearly, all outstanding orders at time θ_1, $R(\theta_1)$, must be converted into addition to inventory over the lead time interval $[\theta_1, \theta_1 + l)$. This follows from the fact that any amount which is on order at time θ_1 was ordered in the time interval $[\theta_1 - l, \theta_1)$, and, consequently, must be received in the time interval $[\theta_1, \theta_1 + l)$; conversely, nothing that is not on order at time θ_1 can be received during the time interval $[\theta_1, \theta_1 + l)$. Hence

$$R(\theta_1) = \pi(\theta_1, \theta_1 + l)$$

Using relations (5.5) and (5.6), we have

$$\begin{aligned} I(\theta_1 + l) &= I(\theta_1) + R(\theta_1) - D(\theta_1, \theta_1 + l) \\ &= \mathcal{P}(\theta_1) - D(\theta_1, \theta_1 + l) \end{aligned} \qquad (5.7)$$

This important relation states that at any particular instant of time, the level of inventory equals the position inventory l units of time ago minus the demand during this lead time period.

In analyzing an inventory system with lead time, it is often more convenient to analyze first the process generated by the position inventory, and then use either relation (5.6) or (5.7) to obtain the characteristics of the inventory level and hence the expression for the stock on hand and the quantity back-ordered. In the simplest inventory models we shall be discussing, instantaneous delivery of orders or zero lead time is assumed. Then, at any particular instant of time, the quantity on order is zero and the position inventory equals the inventory level. If, in addition, no shortages are allowed, then the position inventory measures also the stock on hand.

To the operation of an inventory system are usually associated some costs. It is reasonable to consider three cost components related to procurement, positive inventory, and negative inventory, and known respectively as procurement cost, holding cost, and shortage cost.

The *procurement cost* has usually two basic components: a fixed component that is independent of the quantity ordered, incurred only if an order is placed for replenishment but not otherwise, and a variable component that is a function of the quantity ordered. The *fixed cost* may include the cost of processing an order, the setup cost for a production run, the minimum freight charge in the delivery of orders, the cost of inspection and testing, and so forth. The *variable cost* expressed in dollars per unit procurement usually accounts for the cost of the units and the cost of delivery, excluding the fixed component. We shall henceforth assume that the variable component is proportional to the quantity procured. Thus, if

$P(z)$ = cost of procuring z units of the commodity
 K = fixed order cost in dollars
  c = variable unit procurement cost in dollars per unit

then

$$P(z) = \begin{cases} 0, & z = 0 \\ K + cz, & z > 0 \end{cases}$$

The *holding cost* expresses the cost of carrying in stock the units of the commodity over time. This cost component includes insurance, taxes, cost of storage, and the cost of capital invested in stocking goods. The holding cost will be assumed to be proportional to the quantity of stock on hand, and the length of time over which such stock is carried. In what follows we let

h = unit holding cost per unit time in dollars/[(unit)(unit time)]

If i denotes the interest rate in dollars/[(dollar)(unit time)], then ic represents the portion of the unit holding cost h associated with capital tied up in inventoried goods. The difference $\hat{h} = h - ic$ reflects the holding cost associated with other factors.

Finally, the *shortage or penalty cost* is the cost incurred when demands occur when the system is out of stock. The cost of back order is usually taken as the penalty associated with the loss of customers' good will and other related costs, and, in general, will depend on the number of units back-ordered and the length of time over which the units are back-ordered. In a way similar to the holding cost, we shall assume that the back-order cost is proportional to the quantity back-ordered and the length of time of back orders. We shall let

p = unit back-order cost per unit time in dollars/[(unit)(unit time)]

As a final remark, the occurrence of shortage may result in *lost sales*; that is, present unsatisfied demands are never fulfilled. Although this type of shortage will not be analyzed, the modeling of the lost-sales case is identical to the backlog case when lead time is zero.

iv. *Inventory Policies—Decision Variables*

An *inventory policy* is a set of decision rules that dictate the *when* and the *how much* to order. It is most convenient to use policies based on the movement of the position inventory incorporating the actual inventory and the quantity on order. Several policies may be used to control an inventory system; of these, the single most important policy is the (s, S) *policy*. Under this particular policy, whenever the position inventory is equal to or less than a value s, a procurement is made to bring its level to S. Under a continuous review system, the (s, S) policy will usually imply the procurement of a fixed quantity $Q = S - s$ of the commodity while in periodic review systems the procurement quantity will vary, but will always be greater than or equal to $Q = S - s$ (Figure 5.5). The (s, S) policy incorporates two *decision variables*

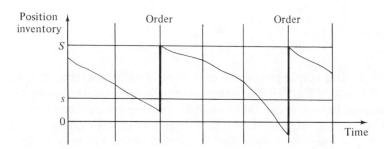

Figure 5.5. Sample function for the inventory movement in a periodic review system operating under the (s, S) policy.

s and S. The variable s, known sometimes as the reorder level, identifies the "when to order," while both variables s and S identify the "how much to order." The inventory problems we shall be studying in this chapter will be assumed to operate under the (s, S) policy, and we shall address ourselves to the determination of the optimum value of s and S so as to satisfy some objective function.

At any instant of time, it is possible to represent the position inventory level of the commodity by a point on a directed line; the movement of this point on the line simulates the movement of the position inventory. The strategies for ordering, as dictated by a particular policy, may be represented in terms of decision regions on the line. These regions define for all possible inventory states when an order should be placed and how much to order. If we consider in particular policies of the (s, S) type, the strategies for ordering could be represented as in Figure 5.6. Thus, whenever the level of the position inventory falls on or below s, an order is initiated to raise the level to S. If, on the other hand, the position inventory exceeds s, then no action is taken.

v. *Objective Functions*

In an inventory problem, the *objective function* may take several forms, and these usually involve the *minimization of a cost function* or the *maximization of a profit function*. The *planning period* or *horizon period*, which is the length of time over which the system is assumed to operate, may be either *finite* or *infinite*. For a finite horizon period, the total cost (profit) experienced

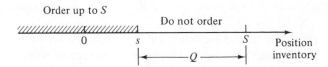

Figure 5.6. Strategies for ordering in the (s, S) policy.

over the entire horizon may be the criterion, and this could be either undiscounted or discounted; alternatively, the criterion may be selected as the average of the total cost (profit) per unit time. If, on the other hand, the horizon period is infinite, the undiscounted long run average cost (profit) per unit time, or the total discounted cost (profit) experienced over the infinite horizon, is selected as the criterion. In stochastic models the *expected* values of costs are measured, and the criterion consists in the minimization of the total expected costs per period or of the total expected discounted cost over a finite or infinite horizon. The cost function will, in general, consist of the additive contribution of the procurement cost, the holding cost, and the shortage cost. A similar set of criteria may be devised for the case of a profit function. The available alternatives for selecting a particular criterion are summarized in the following table.

Objective	Horizon Period	Undiscounted Total Cost	Discounted Total Cost	Per Unit Time Cost
Minimize (expected) cost	Finite	✓	✓	✓
	Infinite		✓	✓
Maximize (expected) profit	Finite	✓	✓	✓
	Infinite		✓	✓

Under the (s, S) policy, the objective function will, in general, be expressible as a function of the two decision variables s and S. The ensuing optimization problem consists then in determining the values of s and S to achieve the selected criterion. In a number of situations, it will be more convenient to determine the optimal values of the pair (s, Q) or (S, Q). Often it will be advantageous to use other decision variables with appropriate relations tying them to the original decision variables.

We have assumed that the reader possesses some familiarity with such basic concepts as compound interest, annuities, and discounting. To provide a brief review, suppose that at some time origin, a *principal* in the amount of A_0 dollars is invested at a *compound interest* of i dollars/[(dollar)(year)]. It is required to determine the net value of the investment $A(t)$ at some future time t.

If time is measured on the discrete scale, $t = 0, 1, 2, \ldots$, then we note that between two successive times $t - 1$ and t we have the following relation:

$$A(t) = A(t - 1) + iA(t - 1), \quad t = 1, 2, \ldots$$

with $\qquad A(0) = A_0$

This difference equation may be solved recursively or otherwise to yield

$$A(t) = A_0(1 + i)^t, \quad t = 0, 1, 2, \ldots$$

If now time is measured over the continuous scale $t \geq 0$, then

$$A(t + dt) = A(t) + iA(t)$$

which may be written as

$$\frac{A(t + dt) - A(t)}{dt} = iA(t)$$

If we let $dt \to 0$, we obtain

$$\frac{dA(t)}{dt} - iA(t) = 0$$

The solution of this differential equation with the initial condition $A(0) = A_0$ is

$$A(t) = A_0 e^{it}, \quad t \geq 0$$

Conversely, if at some time t an investment in the amount $A(t)$ is contemplated, the *present worth* of this investment or the *discounted value* of this investment is

$$A_0 = A(t)(1 + i)^{-t}, \quad \text{if time is discrete}$$
$$A_0 = A(t)e^{-it}, \qquad \text{if time is continuous}$$

The discount factor is $1/(1 + i)$ for discrete time. For continuous time, the discount factor is e^{-i}, and the value of a unit dollar investment discounted over an infinitesimal time interval dt is

$$e^{-i\,dt} = 1 - i\,dt + \frac{i^2}{2!}(dt)^2 + \cdots$$

$$\approx 1 - i\,dt$$

5.2. Deterministic Inventory Models

Deterministic inventory models are characterized by the fact that all input parameters are known in value for all future times. Thus, for example, demand is perfectly predictable and may entirely be expressed as a mathematical function of time. With the absence of uncertainty, the only motive to carry inventory would be of the transaction type. We shall only investigate situations in which inventory is used to synchronize the outflow of goods with their inflow.

Deterministic inventory models have been investigated for many situations. In particular, solution methods have been developed for continuous review systems involving continuous demand and for periodic review systems involving discrete demand. Most of these models qualify as applicable to many practical systems. An explicit elaboration on each of these models would certainly be desirable both at the theoretical and practical level; nevertheless, their development would require considerable exposition, which would be beyond the scope of the present text. Consequently, the present section is restricted to the discussion of selected numbers of continuous review and periodic review inventory systems.

i. *Continuous Review System*

The first model discussed derives the well-known Wilson formula. It assumes infinite production rate, constant demand rate, immediate delivery of orders, disallowance of shortage, an infinite horizon period, and finally, an objective function in which cost per unit time is minimized.

The remaining set of models in this section investigates variations of this first model in which one or more of the aforementioned assumptions are relaxed or altered. These models provide excellent illustrations in modeling systems and in the application of optimization techniques.

a. MODEL 1. THE BASIC MODEL AND THE WILSON FORMULA

The following assumptions underlie this model:

(1) Immediate delivery of orders occur and replenishment rate is infinite.
(2) Demand rate is a constant D; that is, $d(t) = D$.
(3) The system operates over an infinite horizon period.
(4) No shortages are allowed.
(5) Replenishment of stock is accomplished using the (s, S) policy; that is, whenever the inventory level reaches s, a quantity is ordered to bring its level up to S. It immediately follows that each order quantity is fixed in the amount of $Q = S - s$.

(6) Initially, an order has just been placed, and, consequently, immediately received.

(7) There are two competing cost elements, which are the cost of procurement associated with placing an order and the cost of carrying stock.

(8) The objective function is selected as the long-run total cost per unit time or the average cost per unit time.

Referring to Figure 5.7, let t be the time elapsed between two successive orders; t is known as the *length of a cycle*. Consider an arbitrary cycle, and

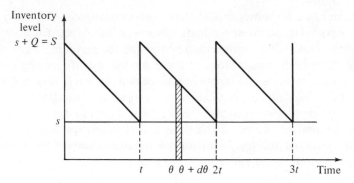

Figure 5.7. Inventory movement in the basic model.

assume time origin is measured at the start of a cycle coinciding with the receipt of an order or procurement. Let

$I(\theta)$ = inventory level at instant θ *following* a procurement, $0 \leq \theta < t$

Since no procurements occur over the time interval $0 < \theta < t$, it follows from equation (5.1) that

$$\frac{dI(\theta)}{d\theta} = -D, \quad 0 < \theta < t \tag{5.8}$$

with

$$I(0) = S = s + Q \tag{5.9}$$

Integrating equation (5.8) and using (5.9) as initial conditions, we find

$$I(\theta) = s + Q - D\theta, \quad 0 \leq \theta < t \tag{5.10}$$

Now, at the end of the cycle the inventory level is s; that is,

$$I(t) = s \tag{5.11}$$

But from (5.10)

$$I(t) = s + Q - Dt \tag{5.12}$$

From expressions (5.11) and (5.12), it follows that

$$Q = Dt \tag{5.13}$$

Since exactly one procurement in the amount of Q is made over one cycle, the total procurement cost per cycle is $K + cQ$. Therefore,

$$\text{average procurement cost per unit time} = \frac{K + cQ}{t} \tag{5.14}$$

Since no shortage is allowed, the inventory level $I(\theta)$ equals the stock level $H(\theta)$; thus the average stock level over one cycle is

$$\frac{1}{t} \int_0^t H(\theta) \, d\theta = \frac{1}{t} \int_0^t I(\theta) \, d\theta$$

and

$$\text{average holding cost per unit time} = \frac{h}{t} \int_0^t I(\theta) \, d\theta \tag{5.15}$$

The average total cost per unit time $\tilde{F}(Q, s, t)$ is the sum total of the average procurement cost per unit time and the average holding cost per unit time, or, from expressions (5.14) and (5.15),

$$\tilde{F}(Q, s, t) = \frac{K + cQ}{t} + \frac{h}{t} \int_0^t I(\theta) \, d\theta$$

Using relation (5.10), we write

$$\tilde{F}(Q, s, t) = \frac{K + cQ}{t} + \frac{h}{t} \int_0^t (s + Q - D\theta) \, d\theta$$

$$= \frac{K + cQ}{t} + \frac{h}{t} \left[(s + Q)t - \frac{Dt^2}{2} \right]$$

$$= \frac{K + cQ}{t} + h(s + Q) - h\frac{Dt}{2}$$

Since a replenishment presumes the addition of a positive quantity, we must have $Q > 0$. On the other hand, the disallowance of shortages implies that $s \geq 0$. In addition $Q = Dt$. The problem now is to determine the optimum values of the two decision variables s and Q so as to minimize the average total cost per unit time, taking into account the operating constraints on the system. Mathematically, this problem is formulated as

$$\min_{Q, s, t} \left\{ \tilde{F}(Q, s, t) = \frac{K + cQ}{t} + h(s + Q) - h\frac{Dt}{2} \right\}$$

subject to $\qquad\qquad Q > 0, \qquad s \geq 0, \qquad Q = Dt$

We may use the constraint $Q = Dt$, and eliminate the variable t in the objective function. The new objective function is

$$\min_{s, Q}\left\{ F(s, Q) = cD + \frac{KD}{Q} + hs + h\frac{Q}{2} \right\} \tag{5.16}$$

subject to $\qquad\qquad\qquad Q > 0, \qquad s \geq 0$

We note that $F(s, Q)$ is a separable and convex function of s and Q, namely, $cD + hs$, and $\hat{F}(Q) = (KD/Q) + hQ/2$ respectively. Since for $s \geq 0$ the minimum value of the linear function $CD + hs$ occurs at $s = 0$, it follows that the optimal value s^* of s is $s^* = 0$. This implies that the most economic reorder level is selected when the stock on hand reaches a zero level. To determine the optimal value Q^* of Q, we consider the problem

$$\min_{Q}\left\{ \hat{F}(Q) = \frac{KD}{Q} + h\frac{Q}{2} \right\} \tag{5.17}$$

subject to $\qquad\qquad\qquad Q > 0$

Now $\hat{F}(Q)$ is the sum of a strictly convex and a convex function (Figure 5.8);

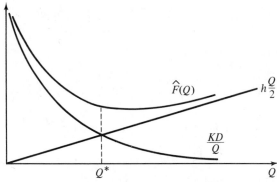

Figure 5.8. The function $\hat{F}(Q) = \dfrac{KD}{Q} + h\dfrac{Q}{2}$.

thus $\hat{F}(Q)$ is a strictly convex function of Q (e.g., see Chapter 2 of *Optimization Techniques in Operations Research*, also by B. D. Sivazlian and L. E. Stanfel, Prentice-Hall, Inc., Englewood Cliffs, N.J., 1975). Differentiating $\hat{F}(Q)$ with respect to Q and setting the result equal to zero, we obtain

$$\frac{d\hat{F}(Q)}{dQ} = 0 = -\frac{KD}{Q^2} + \frac{h}{2}$$

which yields

$$Q^2 - \frac{2KD}{h} = 0$$

or

$$\left(Q + \sqrt{\frac{2KD}{h}}\right)\left(Q - \sqrt{\frac{2KD}{h}}\right) = 0$$

The only valid solution being for $Q > 0$, it follows that $Q = \sqrt{2KD/h}$. The optimum values of s and Q, respectively, are therefore

$$\boxed{\begin{array}{c} s^* = 0 \\ \hline Q^* = \sqrt{\frac{2KD}{h}} \end{array}}$$

(5.18)

These values are those that minimize the value of the objective function $F(s, Q)$, while simultaneously satisfying the constraints. The corresponding value of the minimum cost function is

$$
\begin{aligned}
F(s^*, Q^*) &= cD + KD\sqrt{\frac{h}{2KD}} + \frac{h}{2}\sqrt{\frac{2KD}{h}} \\
&= cD + \sqrt{\frac{1}{2}KDh} + \sqrt{\frac{1}{2}KDh} \\
&= cD + \sqrt{2KDh}
\end{aligned}
$$

(5.19)

Because the function $F(s, Q)$ is convex, the existence of the unique relative minimum (5.18) ensures that at (s^*, Q^*) the function $F(s, Q)$ achieves its absolute minimum value, which is given by formula (5.19). The optimum value of S, S^* is

$$S^* = s^* + Q^* = \sqrt{\frac{2KD}{h}}$$

The reader should verify that at optimality the contribution of the average fixed cost per unit time is equal to the contribution of the average holding cost per unit time. The variable unit procurement cost c does affect the values of the optimal decision variables s^* and Q^*, through the holding cost h.

The optimum value of Q^* involving the square root expression is often referred to as the *Wilson formula* or the *economic order quantity (EOQ) formula*. It is seen that the optimal order quantity is proportional to the square root of the fixed setup cost and the demand, and inversely proportional to the square root of the holding cost. Readers familiar with dimensional analysis should attempt to determine the expression for Q^* using dimensional expressions. We shall often refer to the present model as the *basic model* or Model 1.

Example 5.1

A subcontractor has to supply an automobile manufacturer with roller bearings of a given type and size at the rate of 48,000 units a year. The subcontractor must ship a week's supply of the bearing each week to the manufacturer. The cost of producing a single bearing is estimated at $5 and the setup cost per run is $500. The subcontractor estimates that stocked items cost him 20 cents on the dollar per year. What should be the production-run size if no shortages are allowed? Determine the average number of production runs per year.

Using the conventional symbols, we have

$$D = 48,000 \text{ units/year}$$
$$K = \$500$$
$$c = \$5.00/\text{unit}$$
$$h = .20 \times 5.00 = 1.00 \text{ dollar/[(unit)(year)]}$$

From formula (5.18), the optimal production-run size is

$$Q^* = \sqrt{\frac{2 \times 500 \times 48,000}{1.00}} = 6920 \text{ units}$$

The average number of production runs per year is

$$\frac{D}{Q^*} = \frac{48,000}{6920} \approx 7$$

b. MODEL 2. EFFECT OF LEAD TIME

The assumptions in this model are the same as the basic model except that we assume a finite lead time l elapses from the moment an order is placed to the moment an order is received. If we consider the movement of the position inventory, then, under the (s, S) policy, an order will be placed in the amount of Q once the position inventory reaches a level of s. Thus the position inventory will fluctuate between s and $S = s + Q$. Since no shortage is allowed, the stock movement will vary between 0 and Q (Figure 5.9). It is again optimal to order when the stock level is 0. From (5.7), the optimal reorder level s^* is obtained from the relation

$$0 = s^* - Dl$$

or
$$s^* = Dl$$

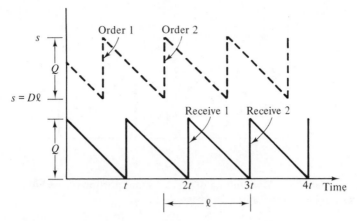

Figure 5.9. Effect of lead time *l*: position inventory movement (. . .) and inventory movement (—).

while again we have

$$Q^* = \sqrt{\frac{2KD}{h}}$$

The reorder level s^* based on the position inventory is thus equal to the demand during the lead time.

c. MODEL 3. EFFECT OF DISCOUNTING

The assumptions in this model are similar to those of Model 1, except
that the objective function is selected as the minimization of the total discounted cost. From the outset we shall let $s = 0$, since intuitively it is optimal
to reorder whenever the inventory level reaches zero. The reader is encouraged, however, to reformulate the problem incorporating the variable *s* and
showing that it is optimal to have $s = 0$.

Consider one particular cycle (Figure 5.10) of length *t*, and let $Z(Q, t)$ be
the total discounted cost over this particular cycle. The holding cost excluding investment, associated in keeping in stock a quantity $Q - D\tau$ of the
product over a length of time $d\tau$ is $\hat{h}(Q - D\tau)\, d\tau$. This quantity if discounted
to the start of the cycle is $e^{-i\tau}\hat{h}(Q - D\tau)\, d\tau$, where

i = interest rate in dollars/[(dollar)(unit time)]
e^{-i} = discount factor over continuous time
$\hat{h}$ = holding cost in dollars/[(unit)(unit time)] excluding investment
   cost; $\hat{h} = h - ic$

The total value of all costs incurred during this particular cycle discounted
to the start of the cycle consists of the procurement cost $K + cQ$ incurred at

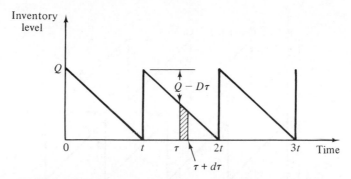

Figure 5.10. Discounting of an elemental portion of the holding cost to the start of a cycle.

the beginning of the cycle plus the discounted holding costs over the entire cycle. Thus

$$Z(Q, t) = K + cQ + \int_0^t e^{-i\tau}\hat{h}(Q - D\tau)\, d\tau$$

Evaluating the integral expression by parts, we obtain

$$Z(Q, t) = K + cQ - \hat{h}(Q - D\tau)\frac{e^{-i\tau}}{i}\Big|_0^t - \int_0^t \hat{h}D\frac{e^{-i\tau}}{i}\, d\tau$$

$$= K + cQ - \left[\hat{h}(Q - Dt)\frac{e^{-it}}{i} - \frac{\hat{h}Q}{i}\right] + \frac{\hat{h}D}{i^2}(e^{-it} - 1)$$

Since $Q = Dt$, we have

$$Z(Q, t) = K + cQ + \frac{\hat{h}Q}{i} - (1 - e^{-it})\frac{\hat{h}D}{i^2} \tag{5.20}$$

The total of all costs incurred over successive cycles discounted to time origin is

$$F(Q, t) = Z(Q, t) + Z(Q, t)e^{-it} + Z(Q, t)e^{-2it} + \cdots$$

$$= Z(Q, t) \sum_{n=0}^{\infty} e^{-int}$$

$$= \frac{Z(Q, t)}{1 - e^{-it}}$$

and, using relation (5.20), we obtain for the expression of the total discounted costs

$$F(Q, t) = \frac{1}{1 - e^{-it}}\left[K + cQ + \frac{\hat{h}Q}{i} - (1 - e^{-it})\frac{\hat{h}D}{i^2}\right] \tag{5.21}$$

This expression is to be minimized subject to

$$Q = Dt$$
$$Q > 0$$

Letting $Q = Dt$, we may write the optimization problem corresponding to (5.21) as

$$\min_t \left\{ \tilde{F}(t) = \frac{K + cDt + (\hat{h}Dt/i)}{1 - e^{-it}} - \frac{\hat{h}D}{i^2} \right\} \tag{5.22}$$

subject to $t > 0$. To obtain the optimal value of t, we differentiate $\tilde{F}(t)$ with respect to t and set the result equal to zero, yielding

$$\frac{d\tilde{F}(t)}{dt} = 0 = \frac{(1 - e^{-it})[cD + (\hat{h}D/i)] - te^{-it}[K + cDt + (\hat{h}Dt/i)]}{(1 - e^{-it})^2}$$

From this we obtain as a necessary condition for an optimum t,

$$(1 - e^{-it})\left(cD + \frac{\hat{h}D}{i}\right) = ie^{-it}\left(K + cDt + \frac{\hat{h}Dt}{i}\right)$$

Dividing both sides by $[cD + (hD/i)]e^{-it}$ yields

$$\frac{1 - e^{-it}}{e^{-it}} = i\frac{K + t\left(cD + \frac{\hat{h}D}{i}\right)}{cD + \frac{\hat{h}D}{i}}$$

or, since $\hat{h} + ic = h$, the total unit holding cost per unit time, then

$$e^{it} = 1 + \frac{i^2K}{Dh} + it \tag{5.23}$$

Using the second-derivative test on (5.22), the reader should verify that the optimum value of t obtained from equation (5.23) actually minimizes the function $\tilde{F}(t)$.

The optimum value of Q is thus the solution to the equation

$$e^{iQ/D} = 1 + \frac{i^2K}{Dh} + \frac{iQ}{D} \tag{5.24}$$

A graphical solution of equation (5.24) is shown in Figure 5.11. It is evident that (5.24) possesses a unique root $iQ/D > 0$, obtained as the positive intercept of the two functions

$$y(Q) = e^{i(Q/D)}$$

and

$$y(Q) = 1 + \frac{i^2K}{Dh} + i\frac{Q}{D}$$

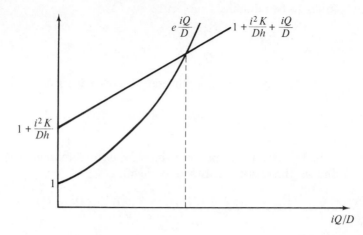

Figure 5.11. Solution of the equation $e^{iQ/D} = 1 + \dfrac{i^2 K}{Dh} = \dfrac{iQ}{D}$.

For small values of $iQ/D < 1$, we have upon expanding the exponential function on the left-hand side of equation (5.24) as a power series and retaining the first three terms

$$1 + \frac{iQ}{D} + \frac{1}{2!}\left(\frac{iQ}{D}\right)^2 + \cdots = 1 + \frac{i^2 K}{Dh} + \frac{iQ}{D}$$

Thus an approximate expression for the optimal value of Q is obtained by solving the quadratic equation

$$\frac{1}{2}\left(\frac{iQ}{D}\right)^2 \approx \frac{i^2 K}{Dh}$$

yielding

$$Q^* \approx \sqrt{\frac{2KD}{h}}$$

which is the Wilson formula. The reader should also verify that the expression for the minimum total discounted cost is

$$K + \frac{cD}{i} + \frac{h}{i}Q^*$$

d. Model 4. Effect of Shortage

In this particular model we assume that shortages are allowed in the form of backlog; that is, all unfilled demands as a result of a stockout are satisfied in the future following receipt of an order. All other assumptions

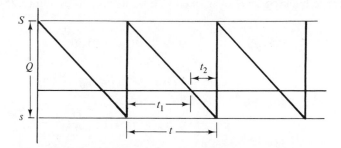

Figure 5.12. Effect of a backlogged shortage.

are identical to the basic model. We are again interested in determining the optimal values of s and S to minimize the total cost per unit time, consisting of the procurement cost, the holding cost, and the shortage cost.

We consider again one complete cycle of length t (Figure 5.12). Following the same reasoning as in the basic model, we obtain for $I(\theta)$, the inventory level at instant θ following a procurement $0 \leq \theta < t$ [see relation (5.10)],

$$I(\theta) = s + Q - D\theta, \quad 0 \leq \theta < t \tag{5.25}$$

The inventory level in this interval can either be positive or negative; the positive inventory level measures the stock on hand $H(\theta)$ and the negative inventory level measures the backlog $B(\theta)$. Since

$$H(\theta) = \begin{cases} I(\theta), & 0 < \theta < t_1 \\ 0, & t_1 \leq 0 < t \end{cases}$$

the average stock level over one cycle is, using expression (5.25),

$$\frac{1}{t} \int_0^t H(\theta)\, d\theta = \frac{1}{t}\left[\int_0^{t_1} I(\theta)\, d(\theta) + \int_{t_1}^t 0\, d\theta \right]$$

$$= \frac{1}{t} \int_0^{t_1} (s + Q - D\theta)\, d\theta \tag{5.26}$$

$$= \frac{1}{t}\left[(s + Q)t_1 - \frac{Dt_1^2}{2} \right]$$

Now since

$$B(\theta) = \begin{cases} 0, & 0 < \theta < t_1 \\ -I(\theta), & t_1 \leq \theta < t \end{cases}$$

the average backlog level over one cycle is, using expression (5.25),

$$\frac{1}{t}\int_0^t B(\theta)\, d\theta = \frac{1}{t}\left[\int_0^{t_1} 0\, d\theta + \int_{t_1}^t -I(\theta)\, d\theta\right]$$

$$= -\frac{1}{t}\int_{t_1}^t (s + Q - D\theta)\, d\theta \qquad (5.27)$$

$$= -\frac{1}{t}\left[(s + Q)(t - t_1) - \frac{D}{2}(t^2 - t_1^2)\right]$$

In addition, we note that the following conditions prevail:

$$S = Q + s = Dt_1$$
$$Q = Dt > 0$$

The average total cost per unit time is the sum total of

(1) The average procurement cost per unit time, which is the same as (5.14), or

$$\frac{K + cQ}{t}$$

(2) The average holding cost per unit time, which from (5.26) is

$$\frac{h}{t}\left[(s + Q)t_1 - \frac{Dt_1^2}{2}\right]$$

(3) The average shortage cost per unit time, which from (5.27) is

$$\frac{p}{t}\left[\frac{D}{2}(t^2 - t_1^2) - (s + Q)(t - t_1)\right]$$

The average total cost per unit time is thus

$$\frac{K + cQ}{t} + \frac{h}{t}\left[(s + Q)t_1 - \frac{Dt_1^2}{2}\right]$$

$$+ \frac{p}{t}\left[\frac{D}{2}(t^2 - t_1^2) - (s + Q)(t - t_1)\right] \qquad (5.28)$$

This expression is to be minimized with respect to s, Q, t and t_1 subject to

$$S = Q + s = Dt_1$$
$$Q = Dt$$
$$Q > 0, \qquad S > 0, \qquad 0 < t_1 < t$$

In expression (5.28), eliminate t_1, t, and s and express the objective function in terms of Q and S, obtaining

$$F(Q, S) = \frac{KD}{Q} + cD + \frac{hD}{Q}\left(S\frac{S}{D} - \frac{D}{2}\frac{S^2}{D^2}\right)$$

$$+ \frac{pD}{Q}\left[\frac{D}{2}\left(\frac{Q+S}{D}\right)\left(\frac{Q-S}{D}\right) - S\frac{Q-S}{D}\right]$$

$$= \frac{KD}{Q} + cD + \frac{hS^2}{2Q} + \frac{p(Q-S)^2}{2Q}$$

By forming the Hessian of the function $F(Q, S)$ (e.g., see Chapter 2 of *Optimization Techniques in Operations Research*, also by B. D. Sivazlian and L. E. Stanfel, Prentice-Hall, Inc., Englewood Cliffs, N.J., 1975), the reader may show that such Hessian is positive definite and that the function $F(Q, S)$ is a strictly convex function of Q and S, for all $Q > 0$ and $S > 0$.

    To obtain the optimum value of Q and S, we differentiate $F(Q, S)$ partially with respect to Q and S and set the result equal to zero.

Thus

$$\frac{\partial F(Q, S)}{\partial S} = 0 = \frac{2hS}{2Q} - \frac{2p(Q - S)}{2Q}$$

or

$$\frac{S}{Q} = \frac{p}{h + p} \tag{5.29}$$

Also

$$\frac{\partial F(Q, S)}{\partial Q} = 0 = -\frac{KD}{Q^2} - \frac{hS^2}{2Q^2} + \frac{p(Q^2 - S^2)}{2Q^2} \tag{5.30}$$

Using (5.29) in (5.30), we obtain

$$0 = -\frac{KD}{Q^2} - \frac{h}{2}\left(\frac{p}{h + p}\right)^2 + \frac{p}{2}\left(1 - \frac{p^2}{(h + p)^2}\right)$$

which yields after simplification

$$Q^2 - \frac{2KD}{p}\frac{h + p}{h} = 0$$

or

$$\left(Q - \sqrt{\frac{2KD}{h}}\sqrt{1 + \frac{h}{p}}\right)\left(Q + \sqrt{\frac{2KD}{h}}\sqrt{1 + \frac{h}{p}}\right) = 0$$

Since $Q > 0$, the optimal value of Q, Q^*, is

$$Q^* = \sqrt{\frac{2KD}{h}}\sqrt{1 + \frac{h}{p}} \tag{5.31}$$

Finally, from relation (5.29), we obtain for the optimum value of S,

$$S^* = \sqrt{\frac{2KD}{h}}\sqrt{\frac{p}{p + h}} \tag{5.32}$$

The optimum value of s, is

$$s^* = S^* - Q^* = -\sqrt{\frac{2KDh}{p(p+h)}} \tag{5.33}$$

Several important remarks can be noted at this point. First, since t_1/t is the proportion of time stock is on hand, $1 - (t_1/t)$ is the proportion of time α the system is out of stock. Since $t_1/t = S/Q$, it can be noted from (5.29) that at optimality the proportion of time the system is out of stock is

$$\alpha = 1 - \frac{S^*}{Q^*} = 1 - \frac{p}{h+p}$$

$$= \frac{h}{h+p}$$

$$= \frac{1}{1+(p/h)}$$

Thus, if the ratio of the penalty cost to the holding cost is $p/h = 50$, the proportion of time the system is out of stock is $\alpha = \frac{1}{51} \approx .02$. In practice, backorder costs are extremely difficult to measure, and very often the decision maker is more prepared to provide an intelligent measurement of the desired proportion of time the system should be out of stock. Thus, with a knowledge of α, one can obtain a corresponding imputed shortage cost p. Second, we note from the formula of Q^* that if we let $p = \infty$, that is, if we attach an infinite penalty cost, the Wilson formula is recovered. Thus, allowing no shortage is equivalent to a situation where shortage cost is infinitely large. Finally, we note that the optimal reorder level s^* is negative.

e. MODEL 5. EFFECT OF FINITE PRODUCTION RATE AND SHORTAGE

This model is similar to Model 4 except that we assume the replenishment of stock is done at the finite rate of R units per unit time, where $R > D$. The replenishment policy is of the (s, S) type, and again it is required to determine the optimal values of s and S so as to minimize the long run total cost per unit time of procurement, holding, and shortage.

Figure 5.13 displays a typical inventory movement for this particular problem. If we consider one complete cycle of length t between two successive production terminations, we note that over the interval t_1, inventory is depleted through demand at a rate of D units per unit time. Over the remaining interval t_2, production and demand occur simultaneously, and inventory is built up at the rate of $(R - D)$ units per unit time. Although a total of Q units is produced over one cycle, the maximum inventory level is $S < Q + s$ units. Note that the value of s is negative.

Using a simple geometric argument, we would like to show that this problem is reducible to the one encountered in Model 4. It is evident that over

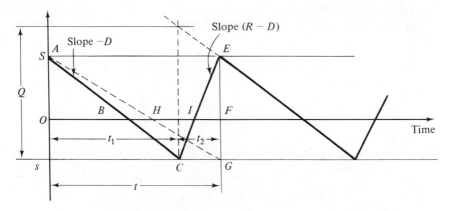

Figure 5.13. Effect of finite production rate and shortage.

one complete cycle t, the stock carrying charge is proportional to the areas of the triangles OAB and EIF; while the penalty cost is proportional to the area of the triangle BCI. If the line AG is drawn, then, from the geometry of the figure, $BH = IF$. Hence area $EIF =$ area ADH. Therefore, the stock carrying charge is proportional to the area of the triangle OAH. Is is easy to also verify that area $BCI =$ area HFG. Thus, the penalty cost is proportional to the area HFG.

As a consequence, each cycle in Figure 5.13 may be made to correspond to a cycle in Figure 5.12, in which the demand rate would be equal to the slope of the line AG. Let D_1 denote the absolute value of the slope of AG. Then

$$D_1 = \frac{S-s}{t} = \frac{S-s}{t_1 + t_2}$$

But the slopes of the lines AC and CE are respectively given in absolute values by:

$$D = \frac{S-s}{t_1} \quad \text{and} \quad (R-D) = \frac{S-s}{t_2}$$

From which we obtain

$$t_1 = \frac{S-s}{D} \quad \text{and} \quad t_2 = \frac{S-s}{R-D}$$

Hence

$$D_1 = \frac{S-s}{\dfrac{S-s}{D} + \dfrac{S-s}{R-D}}$$

$$= \frac{D(R-D)}{R}$$

The optimal values of S and s, S^* and s^*, are immediately given by (5.32) and (5.33) respectively, by replacing in these formulas the value of D by that of D_1. Thus

$$S^* = \sqrt{\frac{2KD}{h}\left(\frac{R-D}{R}\right)}\sqrt{\frac{p}{p+h}}$$

and

$$s^* = -\sqrt{\frac{2KD}{h}\left(\frac{R-D}{R}\right)}\sqrt{\frac{h}{p+h}}$$

The reader should verify that the optimal order quantity $Q^* = Dt$ is

$$Q^* = \sqrt{\frac{R}{R-D}}\sqrt{\frac{2KD}{h}}\sqrt{\frac{p+h}{p}}$$

f. Model 6. An Inventory Model with Decay

In a number of practical situations, a certain amount of decay or waste is experienced on the stocked items. For example, this may arise in certain food products subject to deterioration, or radioactive materials where decay is present, or volatile fluids under evaporation. The amount of loss will in general depend on many factors, including the nature of the particular product under study. In what follows, we restrict our analysis to a case where the rate of loss is proportional to the level of undecayed material. We shall denote by μ this proportional constant and term it the decay rate. The inventory system will be assumed to operate under the same set of assumptions as in Model 1. Holding cost will be charged only to the amount of undecayed stock.

Assuming an (s, S) policy, the inventory movement may appear as in Figure 5.14. Let t be the length of a cycle and let

$I(\theta)$ = inventory level of undecayed product at instant θ following a procurement, $0 \leq \theta < t$.

During the time interval $(\theta, \theta + d\theta)$, the inventory level will be depleted by an amount $\mu I(\theta)$ due to decay and by an amount $D d\theta$ due to demand. Thus

$$I(\theta + d\theta) = I(\theta) - \mu I(\theta)\, d\theta - D d\theta$$

Transposing and dividing by $d\theta$, we obtain

$$\frac{I(\theta + d\theta) - I(\theta)}{d\theta} = -\mu I(\theta) - D$$

Taking limits as $d\theta \to 0$ yields

$$\frac{dI(\theta)}{d\theta} = -\mu I(\theta) - D, \quad 0 \leq \theta < t \tag{5.34}$$

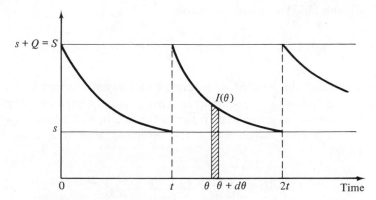

Figure 5.14. Inventory movement of a product subject to decay.

At time $\theta = 0$, a procurement is initiated in the amount of Q and the inventory level is brought up to S, thus

$$I(0) = S = s + Q \qquad (5.35)$$

The linear differential equation (5.34) may be solved by using the integrating factor $e^{\mu\theta}$; we then have

$$\frac{d}{d\theta}[e^{\mu\theta}I(\theta)] = -De^{\mu\theta}$$

Integrating both sides yields

$$e^{\mu\theta}I(\theta) = -\int De^{\mu\theta}\, d\theta + A$$

where A is an arbitrary constant. Finally, we obtain for $0 \leq \theta < t$:

$$I(\theta) = -\frac{D}{\mu} + Ae^{-\mu\theta} \qquad (5.36)$$

The value of A is determined using condition (5.35). From relation (5.36) we have at $\theta = 0$

$$I(0) = -\frac{D}{\mu} + A$$

And using equation (5.35) yields

$$A = s + Q + \frac{D}{\mu} \qquad (5.37)$$

Using (5.37) in (5.36), we obtain as the full solution of the differential equation (5.34) with (5.35) as initial condition

$$I(\theta) = -\frac{D}{\mu} + \left(s + Q + \frac{D}{\mu}\right)e^{-\mu\theta}, \quad 0 \leq \theta < t \qquad (5.38)$$

This then specifies the level of inventory of undecayed product in a given cycle, at time θ following a procurement. This level is dictated by a negative exponential function as compared to the linear function of Model 1. Now, at the end of a cycle, when $\theta = t$, the inventory level is s, thus, from relation (5.38)

$$s = I(t) = -\frac{D}{\mu} + \left(s + Q + \frac{D}{\mu}\right)e^{-\mu t}$$

or
$$Q = \left(s + \frac{D}{\mu}\right)(e^{\mu t} - 1) \qquad (5.39)$$

Since a single procurement in the amount of Q units is made over one cycle, the average procurement cost per unit time is

$$\frac{K + cQ}{t} \qquad (5.40)$$

Also, the average stock level of undecayed material is:

$$\frac{1}{t}\int_0^t I(\theta)\,d\theta = \frac{1}{t}\int_0^t \left[-\frac{D}{\mu} + \left(s + Q + \frac{D}{\mu}\right)e^{-\mu\theta}\right]d\theta$$

$$= \frac{1}{t}\left|-\frac{D}{\mu}\theta - \frac{1}{\mu}\left(s + Q + \frac{D}{\mu}\right)e^{-\mu\theta}\right|_0^t$$

$$= -\frac{D}{\mu} + \frac{1}{\mu t}\left(s + Q + \frac{D}{\mu}\right)(1 - e^{-\mu t})$$

The average holding cost per unit time is thus

$$\frac{h}{t}\int_0^t I(\theta)\,d\theta = h\left[-\frac{D}{\mu} + \frac{1}{\mu t}\left(s + Q + \frac{D}{\mu}\right)(1 - e^{-\mu t})\right] \qquad (5.41)$$

The average total cost per unit time, $\tilde{F}(s, Q, t)$ is the sum of (5.40) and (5.41) or

$$\tilde{F}(s, Q, t) = \frac{K + cQ}{t} + h\left[-\frac{D}{\mu} + \frac{1}{\mu t}\left(s + Q + \frac{D}{\mu}\right)(1 - e^{-\mu t})\right] \qquad (5.42)$$

Using (5.39) and eliminating Q from (5.42) yields the following cost function in s and t

$$F(s, t) = \frac{K}{t} + \frac{c}{t}\left(s + \frac{D}{\mu}\right)(e^{\mu t} - 1)$$

$$- \frac{hD}{\mu} + \frac{h}{\mu t}\left[s + \left(s + \frac{D}{\mu}\right)(e^{\mu t} - 1) + \frac{D}{\mu}\right](1 - e^{-\mu t})$$

$$= \frac{K}{t} + \frac{c}{t}\left(s + \frac{D}{\mu}\right)(e^{\mu t} - 1) - \frac{hD}{\mu} + \frac{h}{\mu t}\left(s + \frac{D}{\mu}\right)(e^{\mu t} - 1)$$

$$= -\frac{hD}{\mu} + \frac{K}{t} + \frac{1}{t}\left(c + \frac{h}{\mu}\right)\left(s + \frac{D}{\mu}\right)(e^{\mu t} - 1)$$

From this relation, it is evident that the optimal value of s is zero. To determine the optimal value t^* of t, we set

$$\frac{\partial F(s, t)}{\partial t} = 0 = -\frac{K}{t^2} + \left(c + \frac{h}{\mu}\right)\left(s + \frac{D}{\mu}\right)\frac{t(\mu e^{\mu t}) - (e^{\mu t} - 1)}{t^2}$$

Thus, after setting $s = 0$, t^* is solution to the equation

$$0 = -K + \left(c + \frac{h}{\mu}\right)\frac{D}{\mu}\left(\mu t e^{\mu t} - e^{\mu t} + 1\right)$$

or

$$\frac{K}{\left(c + \frac{h}{\mu}\right)\left(\frac{D}{\mu}\right)} = \mu t e^{\mu t} - e^{\mu t} + 1$$

or

$$\mu t - 1 = \left[-1 + \frac{K}{\left(c + \frac{h}{\mu}\right)\left(\frac{D}{\mu}\right)}\right]e^{-\mu t}$$

A graphical procedure to solve this equation is shown in Figure 5.15. The equation has a unique root t^* determined as the intersection of the straight line

$$\mu t - 1$$

with the curve

$$\left[-1 + \frac{K}{\left(c + \frac{h}{\mu}\right)\left(\frac{D}{\mu}\right)}\right]e^{-\mu t}$$

g. MODEL 7. A FINITE HORIZON MODEL WITH VARIABLE DEMAND RATE

We consider now a continuous review inventory system operating under the following set of assumptions:

(1) Delivery of orders is instantaneous;

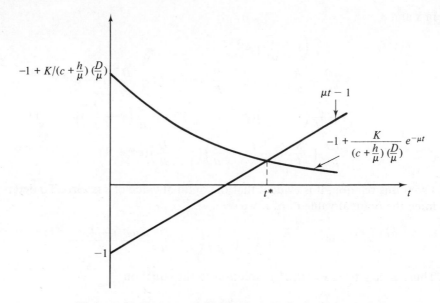

Figure 5.15. On the graphical solution of the equation.

$$\mu t - 1 = \left[-1 + \frac{K}{\left(C + \frac{h}{\mu}\right)\left(\frac{D}{\mu}\right)} \right] e^{-\mu t}$$

(2) The system is to operate over a finite horizon T;

(3) At time $t, 0 \leq t < T$, demand rate $d(t)$ is a function of time t. We shall let $D(t)$ be the cumulative demand at time t, so that

$$D(t) = \int_0^t d(\theta) \, d\theta$$

or

$$d(t) = \frac{d}{dt} D(t);$$

(4) No shortages are allowed;

(5) Replenishment of stock is made by placing n orders, $n = 1, 2, \ldots$, during the time interval $[0, T)$. Let

Q_i = size of the ith order, $i = 1, 2, \ldots, n$

t_i = time at which the $(i + 1)^{st}$ order is placed and received,

$0 = t_0 < t_1 < t_2 < \cdots < t_{n-1} < t_n = T$;

(6) The initial inventory level is zero; thus, at time $t_0 = 0$, the first order is immediately placed;

(7) There are two costs associated with the operation of the system

(a) The procurement cost incurred when placing an order. For an order of size Q_i, the procurement cost is assumed to be

$$P(Q_i) = \begin{cases} 0, & Q_i = 0 \\ K + cQ_i, & Q_i > 0 \end{cases}$$

(b) The holding cost assumed to be proportional to the stock level. We shall denote by h the holding cost of one unit for one unit time.

(8) The objective is to determine the sequence of pairs $\{Q_i, t_{i-1}\}$, $i = 1, 2, \ldots, n$ and the value of n, $n = 1, 2, \ldots$, so as to minimize the total cost of operation over the entire horizon period T.

As may be expected, the stock level at the end of the horizon period T must be equal to zero, since otherwise excess stock would be carried at T implying unnecessary procurement. Further, since procurement and holding costs do not depend on time, it is always more economical to order only when the level of inventory is exactly zero and thus avoid extra stock carrying charges. A particular realization of the movement of inventory is shown in Figure 5.16. Note that the first order in the amount of Q_1 is placed at $t = 0$ and is depleted at time $t = t_1$, while the last order Q_n is placed at time t_{n-1} and is completely depleted by time $t = T$. The following relations hold true

$$\begin{aligned} Q_1 &= D(t_1) \\ Q_2 &= D(t_2) - D(t_1) \\ &\quad \cdot \quad \cdot \quad \cdot \\ Q_{n-1} &= D(t_{n-1}) - D(t_{n-2}) \\ Q_n &= D(T) - D(t_{n-1}) \end{aligned} \tag{5.43}$$

Therefore, the sequence of numbers $\{t_{i-1}\}$ completely specifies the sequence of numbers $\{Q_i\}$ and vice versa. Thus, it is sufficient to consider as decision variables the sequence $\{t_{i-1}\}$, $i = 1, 2, \ldots, n$ and the number n, $n = 1, 2, \ldots$

Let now $P(t)$ denote the cumulative level of procurement at time t. Thus, using relations (5.43) we may write

$$P(t) = \begin{cases} Q_1 = D(t_1), & 0 \leq t < t_1 \\ Q_1 + Q_2 = D(t_2), & t_1 \leq t < t_2 \\ \quad \cdot & \quad \cdot \\ \quad \cdot & \quad \cdot \\ \quad \cdot & \quad \cdot \\ Q_1 + Q_2 + \cdots + Q_n = D(T), & t_{n-1} \leq t < T \end{cases} \tag{5.44}$$

The total procurement cost over the horizon period T is

$$\sum_{i=1}^{i=n} (K + cQ_i) = nK + c \sum_{i=1}^{i=n} Q_i$$
$$= nK + cD(T)$$

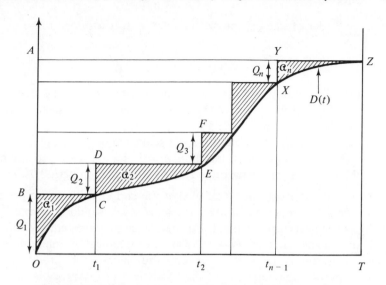

Figure 5.16. Finite horizon model with varying demand rate. Cumulative demand $D(t)$ and ordering sequences.

The stock carrying cost over T is: $h \int_0^T [P(t) - D(t)]\, dt$. Thus, the total cost incurred over the entire horizon period T is:

$$(TC) = nK + cD(T) + h \int_0^T [P(t) - D(t)]\, dt \qquad (5.45)$$

For a fixed number of orders placed, the minimization of (5.45) would be achieved by minimizing the quantity $\int_0^T [P(t) - D(t)]\, dt$ which is the sum of the areas $\mathcal{C}_1, \mathcal{C}_2, \ldots, \mathcal{C}_n$ in Figure 5.16. Now

$$\sum_{i=1}^{i=n} \mathcal{C}_i = (\text{area } AOCE \ldots XZ) - (\text{area } ABCDEF \ldots XYZ)$$

Therefore, since area $AOCE \ldots XZ$ is fixed, minimizing (TC) for a fixed n, corresponds to maximizing the area $ABCDEF \ldots XYZ$. This area is given by the expression

$$\mathcal{C} = Q_2 t_1 + Q_3 t_2 + \cdots + Q_n t_{n-1}$$

Using relations (5.43), we may write

$$\mathcal{C} = t_1 [D(t_2) - D(t_1)] + t_2 [D(t_3) - D(t_2)] + \cdots$$
$$+ t_{n-2}[D(t_{n-1}) - D(t_{n-2})] + t_{n-1}[D(T) - D(t_{n-1})] \qquad (5.46)$$

The problem then consists of determining the optimal values of $t_1, t_2, \ldots, t_{n-1}$ so as to maximize $\mathcal{C}$. Taking in (5.46) the partial derivatives of $\mathcal{C}$ with respect to each of the variables $t_1, t_2, \ldots, t_{n-1}$ and setting the result equal to zero yields the following system of equations:

$$\frac{\partial \alpha}{\partial t_1} = 0 = -t_1 d(t_1) + D(t_2) - D(t_1)$$

$$\frac{\partial \alpha}{\partial t_2} = 0 = -(t_2 - t_1) d(t_2) + D(t_3) - D(t_2)$$

$$\cdots \qquad\qquad\qquad\qquad (5.47)$$

$$\frac{\partial \alpha}{\partial t_{n-2}} = 0 = -(t_{n-2} - t_{n-3}) d(t_{n-2}) + D(t_{n-1}) - D(t_{n-2})$$

$$\frac{\partial \alpha}{\partial t_{n-1}} = 0 = -(t_{n-1} - t_{n-2}) d(t_{n-1}) + D(T) - D(t_{n-1})$$

The solution of this system of $(n-1)$ equations in the $(n-1)$ unknowns $t_1, t_2, \ldots, t_{n-1}$ will yield the optimal values of $t_1, t_2, \ldots, t_{n-1}$ for a given n. Since for each n, $n = 1, 2, \ldots$, the optimal value of t_i, $i = 1, 2, \ldots, n-1$ are known, expression (5.45) for the total cost can be computed for each n, and the optimal value of n which minimizes (TC) may be determined.

Alternatively, the system of equations (5.47) could have been obtained by using (5.44) in expression (5.45) and setting

$$\frac{\partial (TC)}{\partial t_i} = 0, \quad i = 1, 2, \ldots, n-1$$

Example 5.2

$$\text{Let } d(t) = D = \text{constant}, \ 0 < t < T$$
$$\text{Then } D(t) = Dt$$

The system of equations (5.47) may be written:

$$0 = -Dt_1 + Dt_2 - Dt_1$$
$$0 = -D(t_2 - t_1) + Dt_3 - Dt_2$$
$$\cdots \qquad\qquad\qquad\qquad (5.48)$$
$$0 = -D(t_{n-2} - t_{n-3}) + Dt_{n-1} - Dt_{n-2}$$
$$0 = -D(t_{n-1} - t_{n-2}) + DT - Dt_{n-1}$$

Solving the system of equations (5.48) is equivalent to solving the difference equation

$$t_{i+1} - 2t_i + t_{i-1} = 0, \quad i = 1, 2, \ldots, n-1$$

where
$$t_0 = 0 \quad \text{and} \quad t_n = T$$

We find recursively, or otherwise,

$$t_i = \frac{iT}{n} \quad i = 1, 2, \ldots, n-1 \tag{5.49}$$

It is easily verified that this solution corresponds to placing orders at equally spaced intervals.

We are now in a position to write down expression (5.45) explicity as a function of n, $n = 1, 2, \ldots,$ only. Using (5.44) in (5.45) we obtain

$$F(n) = nK + cDT + h \sum_{i=1}^{i=n} \int_{t_{i-1}}^{t_i} (Dt_i - Dt) \, dt$$
$$= nK + cDT + \frac{hD}{2} \sum_{i=1}^{i=n} (t_i - t_{i-1})^2 \tag{5.50}$$

Substituting in expression (5.50) the values of t_i as obtained in (5.49), we obtain:

$$F(n) = nK + cDT + \frac{hD}{2} \sum_{i=1}^{i=n} \left[\frac{iT}{n} - \frac{(i-1)T}{n} \right]^2$$
$$= nK + cDT + \frac{hDT^2}{2n^2} \sum_{i=1}^{i=n} [i - (i-1)]^2 \tag{5.51}$$
$$= nK + cDT + \frac{hDT^2}{2n}, \quad n = 1, 2, \ldots$$

The optimum value of n, n^*, is obtained by minimizing $F(n)$ with respect to n over the set of positive integers n. n^* must satisfy the following conditions

$$F(n^*) - F(n^* + 1) \leq 0 \tag{5.52}$$
$$F(n^*) - F(n^* - 1) \leq 0 \tag{5.53}$$

Using (5.51), inequality (5.52) may be written:

$$n^*K + cDT + \frac{hDT^2}{2n^*} - \left[(n^* + 1)K + cDT + \frac{hDT^2}{2(n^* + 1)} \right] \leq 0$$

or

$$\frac{hDT^2}{2} \left(\frac{1}{n^*} - \frac{1}{n^* + 1} \right) \leq K$$

or

$$n^*(n^* + 1) \geq \frac{hDT^2}{2K} \tag{5.54}$$

Similarly, using expression (5.51) in inequality (5.53) we obtain

$$n^*(n^* - 1) \leq \frac{hDT^2}{2K} \tag{5.55}$$

Combining inequalities (5.54) and (5.55) results in the conditions

$$n^*(n^* - 1) \leq \frac{hDT^2}{2K} \leq n^*(n^* + 1)$$

These are necessary and sufficient conditions to be satisfied by n^* for $F(n)$ to be a minimum. If we now form the table

n^*	1	2	3	4	5	6	7	8	9	10 ...
$n^*(n^* - 1)$	0	2	6	12	20	30	42	56	72	90 ...
$n^*(n^* + 1)$	2	6	12	20	30	42	56	72	90	110 ...

then, given the known quantity $hDT^2/2K$, the optimum n^* may be determined. For example, if $hDT^2/2K = 60$, then $n^* = 8$.

A Geometric Interpretation of the Optimal Solution

The system of equations (5.47) may be written as:

$$\begin{aligned}
(t_1 - 0)\, d(t_1) &= D(t_2) - D(t_1) \\
(t_2 - t_1)\, d(t_2) &= D(t_3) - D(t_2) \\
&\cdots \\
(t_{n-2} - t_{n-3})\, d(t_{n-2}) &= D(t_{n-1}) - D(t_{n-2}) \\
(t_{n-1} - t_{n-2})\, d(t_{n-1}) &= D(T) - D(t_{n-1})
\end{aligned} \tag{5.56}$$

We shall provide a geometric interpretation to this system of equations and show how one can generate graphically the optimal values of $t_2, t_3, \ldots, t_{n-1}$ given the optimal value of t_1.

Suppose that the optimal value t_1 at which the second order is placed is known. Referring to Figure 5.17, the following geometric construction is performed:

Let P_1 be the point on $D(t)$ corresponding to abcissa t_1. From P_1 draw a line parallel to the time axis; this intersects the ordinate axis at U. From U draw a line parallel to the tangent to $D(t)$ at point P_1; this intersects the ordinate at t_1 at the point V. Finally, from V draw a line parallel to the time axis, intersecting $D(t)$ at the point P_2. The abcissa of the point P_2 determines the optimal value of t_2. To see this, we note that

$$UP_1 = t_1 - 0$$

$$\tan \alpha = \left. \frac{d}{dt} D(t) \right|_{t=t_1} = d(t_1)$$

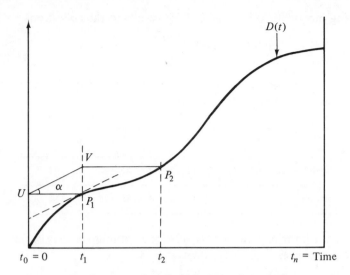

Figure 5.17. A geometric solution to the finite horizon problem for a given number of orders.

Thus
$$P_1 V = U P_1 \tan \alpha$$
$$= (t_1 - 0)\, d(t_1)$$

But
$$P_1 V = D(t_2) - D(t_1)$$

Hence
$$(t_1 - 0)\, d(t_1) = D(t_2) - D(t_1)$$

and this generates the first equation in (5.56). Having generated t_2, the same geometric construction is applied to determine t_3 and so on. Since a priori the optimal value of t_1 is not known, a trial and error procedure will have to be used. For example, for n orders placed, one may initialize a value of t_1 equal to T/n. Applying the graphical procedure, the value of t_{n-1} generated is used to determine if $t_n = T$. If $t_n \gtrless T$, one selects a next value of t_1, and so on.

A Dynamic Programming Formulation

Suppose that for the problem with horizon period T an optimal solution is found involving n orders, and suppose that the *last order* is placed at time x, $0 \leq x < T$ (see Figure 5.18). The total cost incurred over the time interval $[x, T)$ consists of a procurement cost in the amount of $K + c[D(T) - D(x)]$, and a holding cost equal to $h \int_x^T [D(T) - D(t)]\, dt$. Thus the total cost experienced over the time interval $[x, T)$ is

$$K + c[D(T) - D(x)] + h \int_x^T [D(T) - D(t)]\, dt$$

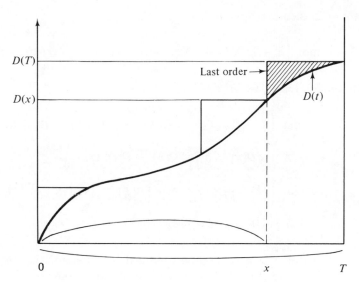

Figure 5.18. If solution is optimal over [0, T), it must be optimal over [0, x).

Since the ending inventory at time x, before an order is placed, is equal to zero, then, if the solution is optimal for horizon period T, it must be optimal to consider the x horizon period problem by itself (e.g., see Chapter 7 of *Optimization Techniques in Operations Research*, also by B. D. Sivazlian and L. E. Stanfel, Prentice-Hall, Inc., Englewood Cliffs, N. J., 1975). Let

$f_n(T)$ = the minimum total cost over a T horizon period, given that n orders are placed over the interval [0, T).

Then, we have the following functional equation, $n = 1, 2, \ldots$

$$f_n(T) = \min_{0 \le x < T} \left\{ K + c[D(T) - D(x)] + h \int_x^T [D(T) - D(t)]\, dt + f_{n-1}(x) \right\}$$

$$f_0(T) = 0$$

(5.57)

And if $f(T)$ is the minimum total cost over a horizon time T, then

$$f(T) = \min_{n=1,2,\ldots} \{f_n(T)\}$$

The functional equation (5.57) would have to be solved recursively, by first determining $f_1(T)$, then $f_2(T)$, and so forth.

Let $n = 1$; then necessarily $x = 0$ and $f_0(T) = 0$ for all T. From (5.57)

$$f_1(T) = \min_{x=0} \left\{ K + c[D(T) - D(x)] + h \int_x^T [D(T) - D(t)]\, dt \right\}$$

$$= K + cD(T) + h \int_0^T [D(T) - D(t)]\, dt$$

(5.58)

Let $n = 2$, then

$$f_2(T) = = \min_{0 \leq x < T} \left\{ K + c[D(T) - D(x)] + h \int_x^T [D(T) - D(t)] \, dt + f_1(x) \right\}$$

Using expression (5.58), we have

$$\begin{aligned}
f_2(T) &= \min_{0 \leq x < T} \left\{ K + c[D(T) - D(x)] + h \int_x^T [D(T) - D(t)] \, dt \right. \\
&\quad \left. + K + cD(x) + h \int_0^x [D(x) - D(t)] \, dt \right\} \\
&= \min_{0 \leq x < T} \left\{ 2K + cD(T) + h \int_x^T [D(T) - D(t)] \, dt \right. \\
&\quad \left. + h \int_0^x [D(x) - D(t)] \, dt \right\}
\end{aligned} \tag{5.59}$$

The optimum value of x may then be obtained by differentiating the right hand expression of (5.59) with respect to x and setting the result equal to zero, or

$$\frac{d}{dx} \left\{ 2K + cD(T) + h \int_x^T [D(T) - D(t)] \, dt + h \int_0^x [D(x) - D(t)] \, dt \right\} = 0$$

or

$$-[D(T) - D(x)] + \int_0^x D'(x) \, dt = 0$$

and since $D'(x) = d(x)$, we obtain the relation

$$(x - 0) \, d(x) = D(T) - D(x) \tag{5.60}$$

This corresponds exactly to the first equation of (5.47) or (5.56). The reader may verify that as the functional equation is solved recursively, each of the equations in (5.47) is generated. Finally, we ask the reader to apply the functional equation technique to solve the special case when $d(t) = D = \text{constant}$.

ii. *Periodic Review System*

In the operation of a periodic review inventory system, the inventory level is reviewed periodically at the beginning of equally spaced time intervals, and a decision to order or not is made according to the inventory status. In this section, we analyze a periodic review inventory system operating under the following set of assumptions:

(1) Delivery of orders is instantaneous.
(2) The inventory system operates over a finite horizon consisting of T periods, $T = 1, 2, \ldots$.

(3) The sequence of known demand $\{d(t)\}$ over the successive periods $t, t = 1, 2, \ldots, T$ forms a set of non-negative quantities which may vary from period to period.

(4) No shortages are allowed.

(5) Replenishment of stock is made at the beginning of a period; the quantity procured at the beginning of period t will be denoted by $p(t), p(t) \geq 0, t = 1, 2, \ldots, T$.

(6) The initial inventory level is zero; thus, at the beginning of the first period, an order is immediately placed and received.

(7) There are two competing cost elements:

(a) The procurement cost associated with placing an order. If $p(t)$ is the quantity procured at the beginning of period $t, t = 1, 2, \ldots, T$, then a procurement cost in the amount of

$$P[p(t)] = \begin{cases} 0, & p(t) = 0 \\ K + cp(t), & p(t) > 0 \end{cases}$$

is incurred at the beginning of period t.

(b) The holding or stock carrying cost. Let $I(t)$ measure the inventory level at the end of period $t, t = 1, 2, \ldots, T$; then from (5.3) we have the balance equation

$$I(t) = I(t - 1) + p(t) - d(t), \qquad t = 1, 2, \ldots, T$$
$$I(0) = 0$$

The stock carrying cost for period t may be measured at the end of period t, in which case it will be a function of $I(t)$. It may be measured at the beginning of period t, following a decision to order, in which case it will be a function of $[I(t - 1) + p(t)]$. Finally, it may be measured as a function of both $I(t)$ and $[I(t - 1) + p(t)]$. For simplicity, we shall assume in what follows that the carrying charges are made at the end of a period and are proportional to the stock levels. Thus, the holding cost incurred during period $t, t = 1, 2, \ldots, T$, is

$$\mathcal{K}[I(t)] = hI(t)$$

(8) The objective is to determine the sequence of non-negative quantities $\{p(t)\}, t = 1, 2, \ldots, T$ so as to minimize the total cost of procurement and holding over the entire horizon period. Such a sequence will be called an *optimal program*.

A realization of the stock movement is shown in Figure 5.19. It is evident that this model is the discrete time counterpart of Model 7. An

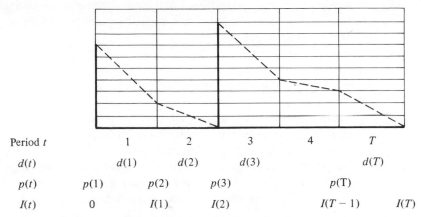

Period t		1	2	3	4	T	
$d(t)$		$d(1)$	$d(2)$	$d(3)$		$d(T)$	
$p(t)$	$p(1)$		$p(2)$	$p(3)$		$p(T)$	
$I(t)$	0		$I(1)$	$I(2)$		$I(T-1)$	$I(T)$

Figure 5.19. A deterministic periodic review inventory system with varying demand.

optimal program possesses the property that $I(t-1) \cdot p(t) = 0$, $t = 1, 2, \ldots,$ T. This means that it is optimal to order only when at the end of a period the stock level equals zero, and not to order otherwise. Suppose to the contrary that an optimal program implies $I(t-1) \cdot p(t) > 0$. Then, it would be no more expensive to reschedule the procurement in the future and order an amount $p(t) + I(t-1)$, thus saving on the holding cost on $I(t-1)$. The reader is referred to [22] for further properties of an optimal program.

Suppose that in the problem involving a horizon of T periods, $T = 1$, $2, \ldots,$ an optimal program is found and that the *last order* is placed at the beginning of period x, $x = 1, 2, \ldots, T$. Thus, a procurement $p(x) > 0$ would meet the total demand of periods $x, x+1, \ldots, T$. The total cost experienced over the periods $x, x+1, \ldots, T$ is

$$K + cp(x) + \sum_{t=x}^{t=T} hI(t)$$

Since the ending inventory at the end of period $x-1$, $I(x-1) = 0$, then, if the program is optimal for the T period horizon problem, it is optimal to consider the $(x-1)$ period horizon problem by itself. Let

$f(T)$ = the minimum total cost over a horizon of T periods, $T = 1, 2, \ldots$

We have the following functional equation: for $T = 1, 2, \ldots$

$$f(T) = \min_{1 \le x \le T} \left\{ K + cp(x) + \sum_{t=x}^{t=T} hI(t) + f(x-1) \right\}$$

and $f(0) = 0$

This equation may be solved recursively, by computing first $f(1)$, next $f(2)$, and so forth, until an optimal program for the T horizon period problem is found. The reader should note the similarity between this equation and equation (5.57).

We illustrate the procedure by solving the following example

Example 5.3

We are given a $T = 3$ period problem with $K = \$5.00$, $c = \$3.00$ per unit, $h = \$.30$ per unit per period. The demands over the three successive periods are: $d(1) = 4$, $d(2) = 1$, and $d(3) = 2$.

We consider first the problem involving the first period as horizon, then, $T = 1$ and (see Figure 5.20).

$$f(1) = \min_{x=1} \left\{ K + cp(x) + \sum_{t=x}^{t=1} hI(t) + f(x-1) \right\}$$

$$= K + cp(1) + hI(1) + f(0)$$

$$= (5.00) + (3.00)(4) + (.30)(0) + 0$$

$$= 17.00$$

Thus, the corresponding optimal program is $p(1) = 4$.

Next, we consider the problem involving the first two periods as horizon, then, $T = 2$, and (see Figure 5.21).

$$f(2) = \min_{1 \le x \le 2} \left\{ K + cp(x) + \sum_{t=x}^{t=2} hI(t) + f(x-1) \right\}$$

$$= \min \begin{cases} K + cp(1) + h[I(1) + I(2)] + f(0), & x = 1 \\ K + cp(2) + hI(2) + f(1), & x = 2 \end{cases}$$

$$= \min \begin{cases} (5.00) + (3.00)(4+1) + (.30)(1+0) + 0, & x = 1 \\ (5.00) + (3.00)(1) + (.30)(0) + 17.00, & x = 2 \end{cases}$$

$$= \min \begin{cases} 20.30 \\ 25.00 \end{cases} = 20.30 \quad \text{with } x = 1$$

The corresponding optimal program is $p(1) = 5$, $p(2) = 0$.

We finally consider the problem involving the three periods as horizon; then $T = 3$, and (see Figure 5.22).

$$f(3) = \min_{1 \le x \le 3} \left\{ K + cp(x) + \sum_{t=x}^{t=3} hI(t) + f(x-1) \right\}$$

$$= \min \begin{cases} K + cp(1) + h[I(1) + I(2) + I(3)] + f(0), & x = 1 \\ K + cp(2) + h[I(2) + I(3)] + f(1), & x = 2 \\ K + cp(3) + hI(3) + f(2), & x = 3 \end{cases}$$

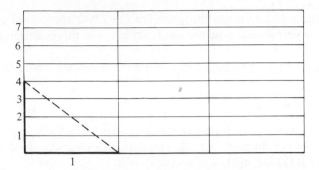

$x = 1$ Last order placed at the beginning of period 1, Cost $\$17.00$

Figure 5.20. The one-period case of Example 5.3. The optimal program is when $x = 1$; $f(1) = \$17.00$.

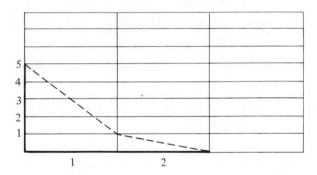

$x = 1$;   Last order placed at the beginning of period 1, Cost $\$20.30$

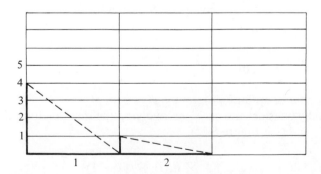

$x = 2$;   Last order placed at the beginning of period 2 and the optimal program for period 1 is dictated by the 1-period horizon problem, Cost $\$25.00$

Figure 5.21. The 2-period case for Example 5.3. The optimal program is when $x = 1$; $f(2) = \$20.30$.

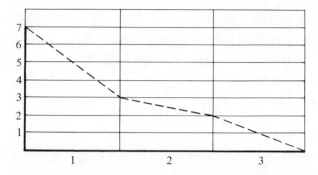

$x = 1$; Last order is placed at the beginning of period 1, Cost $\$27.50$

a.

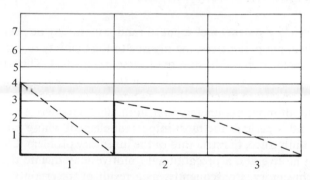

$x = 2$; Last order is placed at the beginning of period 2 and the optimal program for period 1 is dictated by the 1-period horizon problem, Cost $\$31.60$

b.

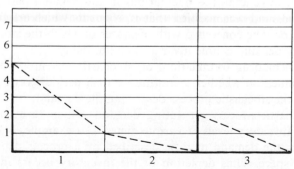

$x = 3$; Last order is placed at the beginning of period 3 and the optimal program for periods 1 and 2 is dictated by the 2-period horizon problem, Cost $\$31.30$

c.

Figure 5.22. The 3-period horizon case for Example 5.3. The optimal program is when $x = 1$; $f(3) = \$27.50$.

$$= \min \begin{cases} (5.00) + (3.00)(4 + 1 + 2) + .30(3 + 2) + 0, & x = 1 \\ (5.00) + (3.00)(1 + 2) + .30(2 + 0) + 17.00, & x = 2 \\ (5.00) + (3.00)(2) + .30(0) + 20.30, & x = 3 \end{cases}$$

$$= \min \begin{Bmatrix} 27.50 \\ 31.60 \\ 31.30 \end{Bmatrix} = 27.50 \quad \text{with } x = 1$$

The corresponding optimal program is $p(1) = 7$, $p(2) = 0$, $p(3) = 0$.

5.3. Stochastic Inventory Models

In an inventory system an element of randomness may be present in one or more of the input parameters, mainly due to the inability to provide exact estimation or prediction on the nature of these parameters. Thus, in general, to account for the uncertainty over time in the values taken by these parameters, it is necessary to describe them as random variables.

Of the set of input parameters that normally characterize an inventory system, demand is perhaps the most important element whose predictability could significantly affect the structure of the inventory problem. Clearly then, a transaction motive and a precautionary motive may dictate the necessity for carrying inventory. Consequently, as a result of uncertainty one would normally expect the presence of additional inventory. In this section we restrict our analysis to inventory situations in which demand is the only element incorporating randomness.

Although, in general, the fluctuations in the demand pattern may incorporate trends and seasonalities, that is, elements which are functions of time, we shall only be concerned with instances in which the stochastic process generating demand is time invariant, or stationary.

The mathematical formulation of stochastic inventory models will depend very much on whether a continuous or a periodic review system is analyzed and the stochastic process describing the demand.

The stochastic process generating demand will in turn structure the movement of inventory levels at various times. For example, in continuous review systems the demand may be described by a process defined over a discrete state space. Thus depletion of the inventory occurs in a discrete fashion, and the resultant stochastic process for the inventory level is a discrete-valued continuous-parameter process.

Two distinct approaches may be taken in formulating and solving the stochastic inventory problem both in theory and practice. In the first approach, the system is viewed as a multistage decision process, and the technique of dynamic programming is employed in finding the optimal policy that minimizes the total expected cost over the duration of the process. To ensure that the optimal policy is of a given type, say (s, S), certain restrictions are imposed on the cost factors that affect the decision rules. For example, when the ordering cost is linear with a fixed setup, and when holding and shortage costs are both proportional to the amount on hand and backlogged, respectively, the optimal policy is of the (s, S) type. We shall illustrate the use of this approach when analyzing the single-period problem in periodic review systems, as a one-stage dynamic-programming problem.

When the duration of the process is infinite, a second approach is often used: an ordering policy of a given type is chosen and the stationary behavior of the inventory levels is analyzed without reference to the cost structure of the problem. Such entities as the expected frequency of orders and the expected quantity on hand and back-ordered at particular instants of time are computed. A cost structure is then imposed on the system, and the stationary total expected cost for operating the inventory system per unit time is minimized. As a result, the optimal values of the prevailing decision variables are obtained. The use of the steady-state or stationary solutions to the inventory problem will be illustrated both for continuous and periodic review systems.

Some of the advantages of the stationary approach over the dynamic-programming approach are

(1) The stationary approach provides us with information about the statistical characteristics of the system. In general, this is not provided by following the dynamic-programming approach, since the latter imbeds from the very outset the cost elements in the formulation of the problem, thus making difficult the statistical analysis of the inventory system under cost-free conditions.

(2) The stationary solution may be used to analyze the relationship between the optimal policy variables and the various parameters involved in the model, as well as to test the sensitivity of costs as functions of the particular policy used in controlling the inventory system.

(3) Whereas the dynamic-programming formulation is the correct approach to structure the optimal policy of an inventory system, the solution to the stationary problem lends itself more easily to computation and approximation analysis, and, as such, is more amenable to practical implementation.

i. *Continuous Review System*

The set of models analyzed in this section assumes that demand is defined as a discrete-valued process such that units are demanded or withdrawn from stock one at a time. This type of demand pattern may prevail in dealers of appliances or cars. Other realistic examples would be the demand for a particular book in a bookstore or the demand generated by the failure of a component in a system incorporating several such components.

We assume that the (s, S) policy is in effect; thus the quantity ordered to replenish stock is assumed to be a constant $Q = S - s$. Although the initial formulation of the stochastic models will invariably assume that the system is operating over a finite horizon, nevertheless, the economic criterion will measure the expected cost for operating the system per unit time under steady-state conditions.

As usual, we assume the presence of three cost factors in the operation of the inventory system. These are

    (1) The procurement cost, consisting of a fixed cost K and a variable cost per unit c.

    (2) The holding cost h per unit per unit time.

    (3) The penalty cost p per unit backlogged per unit time.

The first two models studied involve instantaneous delivery of orders with no shortage allowed. Thus the procurement cost and the holding cost are balanced in determining the optimal value of s and S. The third model considers a situation involving a finite nonzero lead time in which demand for the item forms a Poisson process, and unfilled demand is completely backlogged, thus incurring an additional penalty cost.

Figure 5.23 is a realization of the stochastic process generated by the movement of stock or position inventory over time.

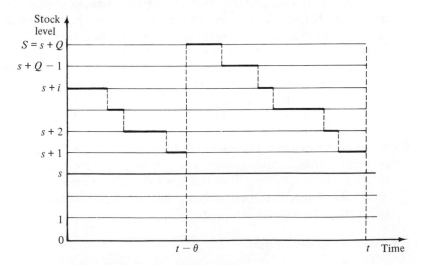

Figure 5.23. Sample function of the inventory pattern in a continuous review system with immediate delivery of orders operating under the (s, S) policy.

a. Model Involving No Lead Time and Unit Demand

(1) Problem

In this first model we assume instantaneous delivery of orders. It is evident that since the inventory levels are continuously monitored, no shortages will occur: as soon as the reorder level s is reached, the order in the amount of Q units is received in stock. Initially, at time $t = 0$, we assume that a demand has just taken place, resulting in a stock level of $s + i$, $i = 1$, $2, \ldots, Q$.

We assume that the interarrival times between successive demands form a sequence of random variables $\{X_j\}$, where the random variables X_j are independently and identically distributed, possessing a common distribution function $\Phi(\cdot)$ and PDF $\varphi(\cdot)$. Thus $\{X_j\}$ forms an ordinary renewal process.

(2) Analysis of the Stochastic Process

The stock level $H(t)$ at a particular time t, $t \geq 0$, is a discrete-valued continuous-parameter stochastic process $\{H(t), t \geq 0\}$ taking on values of $s + 1, s + 2, \ldots, s + Q$. Let $P\{\tilde{H}(t) = n\}$, $n = 1, 2, \ldots, Q$, be the probability that at time t the stock level is $s + n$.

Suppose now that we consider the sequence of events consisting of the times at which an order in the amount of Q is placed and immediately received. Let Y_1 be the time elapsed from origin until the first order is placed, Y_2 be the time elapsed between the first and second order, and so on. The sequence of random variables $\{Y_k\}$, $k = 1, 2, \ldots$, forms a modified renewal process in which the distribution function of Y_1 is given by

$$P\{Y_1 \leq y\} = \int_0^y \varphi *^{(i)} (u) \, du$$

where $\varphi *^{(i)} (u)$ denotes the ith-fold convolution of $\varphi(u)$ with itself, and the distribution function of the random variables $Y_2, Y_3, \ldots$ is given by

$$P\{Y \leq y\} = \int_0^y \varphi *^{(Q)} (u) \, du$$

The probability that the first order will be placed between times t and $t + dt$ is

$$P\{t < Y_1 \leq t + dt\} = \varphi *^{(i)} (t) \, dt, \quad i = 1, 2, \ldots, Q$$

The probability that the mth order, $m = 2, 3, \ldots$, will be placed in the time interval t and $t + dt$ is

$$P\{t < Y_1 + Y_2 + \cdots + Y_m \leq t + dt\} = [\varphi *^{(i)} (t)] * [\varphi *^{[(m-1)Q]}(t)] \, dt$$
$$= \varphi *^{[i+(m-1)Q]}(t) \, dt, \quad m = 2, 3, \ldots$$

Let $P\{\tilde{H}(t) = n \mid k, t - \theta\}$ be the probability that the stock level is $s + n$,

$n = 1, 2, \ldots, Q$, at time t, given that the kth order was placed at time $t - \theta$ and the $(k + 1)$st order is not as yet placed; then, for $k = 1, 2, \ldots$, we have from equation (2.42),

$$P\{\tilde{H}(t) = Q \,|\, k, t - \theta\} = 1 - P\{X_1 \leq \theta\}$$
$$P\{\tilde{H}(t) = Q - 1 \,|\, k, t - \theta\} = P\{X_1 \leq \theta\} - P\{X_1 + X_2 \leq \theta\}$$

$$\cdot \quad \cdot \quad \cdot$$

$$P\{\tilde{H}(t) = 1 \,|\, k, t - \theta\} = P\{X_1 + \cdots + X_{Q-1} \leq \theta\} - P\{X_1 + \cdots + X_Q \leq \theta\}$$

Thus, in general, for $n = 1, 2, \ldots, Q - 1$,

$$P\{\tilde{H}(t) = n \,|\, k, t - \theta\} = \int_0^\theta \varphi *^{(Q-n)} (u)\, du - \int_0^\theta \varphi *^{(Q-n+1)} (u)\, du \quad (5.61)$$

and, for $n = Q$,

$$P\{\tilde{H}(t) = Q \,|\, k, t - \theta\} = 1 - \int_0^\theta \varphi(u)\, du \quad (5.62)$$

Now we let $P\{\tilde{H}(t) = n \,|\, 0\}$ be the probability that the stock level is $s + n$ at time t between time origin and the occurrence of the first order; then for $n = i + 1, i + 2, \ldots, Q$,

$$P\{\tilde{H}(t) = n \,|\, 0\} = 0 \quad (5.63)$$

For $n = i$,

$$P\{\tilde{H}(t) = i \,|\, 0\} = 1 - \int_0^t \varphi(u)\, du \quad (5.64)$$

For $n = 1, 2, \ldots, i - 1$,

$$P\{\tilde{H}(t) = n \,|\, 0\} = \int_0^t \varphi *^{(i-n)} (u)\, du - \int_0^t \varphi *^{(i-n+1)} (u)\, du \quad (5.65)$$

An expression for $P\{\tilde{H}(t) = n\}$ may then be obtained from the relation

$$\begin{aligned}
P\{\tilde{H}(t) = n\} = {} & P\{\tilde{H}(t) = n \,|\, 0\} \\
& + \sum_{k=1}^{\infty} \int_{\theta=0}^{\theta=t} P\{\tilde{H}(t) = n \,|\, k, t - \theta\} \\
& \cdot P\{t - \theta < Y_1 + Y_2 + \cdots + Y_k \leq t - \theta + d\theta\}
\end{aligned} \quad (5.66)$$

Using the relations previously obtained, we compute this expression for various values of n, as well as expressions for the Laplace transform

$$\bar{P}(n, s) = \int_0^\infty e^{-st} P\{\tilde{H}(t) = n\}\, dt$$

Let
$$\bar{\varphi}(s) = \int_0^\infty e^{-st}\, d\Phi(t) = \int_0^\infty e^{-st} \varphi(t)\, dt$$

Now, for $n = 1, 2, \ldots, i - 1$, we have, using expressions (5.61) and (5.65) in (5.66),

$$P\{\tilde{H}(t) = n\} = \int_0^t \varphi *^{(i-n)}(u)\, du - \int_0^t \varphi *^{(i-n+1)}(u)\, du$$
$$+ \sum_{k=1}^{\infty} \int_{\theta=0}^{\theta=t} \left[\int_0^{\theta} \varphi *^{(Q-n)}(u)\, du \right.$$
$$\left. - \int_0^{\theta} \varphi *^{(Q-n+1)}(u)\, du \right] \varphi *^{[i+(k-1)Q]}(t - \theta)\, d\theta$$

The corresponding expression for $\bar{P}(n, s)$ is

$$\bar{P}(n, s) = \frac{1}{s}[\bar{\varphi}(s)]^{i-n} - \frac{1}{s}[\bar{\varphi}(s)]^{i-n+1}$$
$$+ \sum_{k=1}^{\infty} \left\{ \frac{1}{s}[\bar{\varphi}(s)]^{Q-n} - \frac{1}{s}[\bar{\varphi}(s)]^{Q-n+1} \right\} [\bar{\varphi}(s)]^{i+(k-1)Q} \quad (5.67)$$
$$= \frac{1}{s}[\bar{\varphi}(s)]^{i-n}[1 - \bar{\varphi}(s)] + \frac{1}{s}[\bar{\varphi}(s)]^{Q-n+i} \frac{1 - \bar{\varphi}(s)}{1 - [\bar{\varphi}(s)]^Q}$$

For $n = i$ we have, using expressions (5.61) and (5.64) in (5.66),

$$P\{\tilde{H}(t) = n\} = 1 - \int_0^t \varphi(u)\, du + \sum_{k=1}^{\infty} \int_{\theta=0}^{\theta=t} \left[\int_0^{\theta} \varphi *^{(Q-i)}(u)\, du \right.$$
$$\left. - \int_0^{\theta} \varphi *^{(Q-i+1)}(u)\, du \right] \varphi *^{[i+(k-1)Q]}(t - \theta)\, d\theta$$

Hence, for $n = i$,

$$\bar{P}(n, s) = \frac{1}{s}[1 - \bar{\varphi}(s)] + \frac{1}{s}[\bar{\varphi}(s)]^Q \frac{1 - \bar{\varphi}(s)}{1 - [\bar{\varphi}(s)]^Q} \quad (5.68)$$

For $n = i + 1, i + 2, \ldots, Q - 1$, we have, using expressions (5.61) and (5.63) in (5.66),

$$P\{\tilde{H}(t) = n\} = \sum_{k=1}^{\infty} \int_{\theta=0}^{\theta=t} \left[\int_0^{\theta} \varphi *^{(Q-n)}(u)\, du \right.$$
$$\left. - \int_0^{\theta} \varphi *^{(Q-n+1)}(u)\, du \right] \varphi *^{[i+(k-1)Q]}(t - \theta)\, d\theta$$

and, consequently,

$$\bar{P}(n, s) = \frac{1}{s}[\bar{\varphi}(s)]^{Q-n+i} \frac{1 - \bar{\varphi}(s)}{1 - [\bar{\varphi}(s)]^Q} \quad (5.69)$$

Finally, for $n = Q$, we have, using expressions (5.62) and (5.63) in (5.66),

$$P\{\tilde{H}(t) = n\} = \sum_{k=1}^{\infty} \int_{\theta=0}^{\theta=t} \left[1 - \int_0^{\theta} \varphi(u)\, du \right] \varphi *^{[i+(k-1)Q]}(t - \theta)\, d\theta$$

and

$$\bar{P}(n, s) = \frac{1}{s}[\bar{\varphi}(s)]^i \frac{1 - \bar{\varphi}(s)}{1 - [\bar{\varphi}(s)]^Q} \tag{5.70}$$

Combining (5.67), (5.68), (5.69), and (5.70), we may write in general

$$\bar{P}(n, s) = \begin{cases} \dfrac{1}{s}[\bar{\varphi}(s)]^{i-n}[1 - \bar{\varphi}(s)] + \dfrac{1}{s}[\bar{\varphi}(s)]^{Q-n+i}\dfrac{1 - \bar{\varphi}(s)}{1 - [\bar{\varphi}(s)]^Q}, & \\ & n = 1, 2, \dots, i \quad (5.71) \\ \dfrac{1}{s}[\bar{\varphi}(s)]^{Q-n+i}\dfrac{1 - \bar{\varphi}(s)}{1 - [\bar{\varphi}(s)]^Q}, & n = i + 1, \dots, Q \end{cases}$$

(3) Steady-State Distribution of the Number of Units in Stock

Let P_n be the probability that exactly $H = s + n$ units, $n = 1, 2, \dots, Q$, are in stock in the steady state. Then

$$P_n = P\{H = s + n\} = \lim_{t \to \infty} P\{\tilde{H}(t) = n\}$$

$$= \lim_{s \to 0} s\bar{P}(n, s)$$

This last expression follows from the final value theorem (see Appendix). It is easy to verify from (5.71) that, for all $n = 1, 2, \dots, Q$,

$$P_n = \lim_{s \to 0} \frac{1 - \bar{\varphi}(s)}{1 - [\bar{\varphi}(s)]^Q}$$

To obtain the limiting value of this indeterminate expression, we apply l'Hospital's rules once, yielding

$$P_n = \lim_{s \to 0} \frac{-\bar{\varphi}'(s)}{-Q\bar{\varphi}'(s)[\bar{\varphi}(s)]^{Q-1}} \tag{5.72}$$

$$= \frac{1}{Q}, \quad n = 1, 2, \dots, Q$$

Thus, in the steady state the number of units in stock are uniformly distributed over the interval $\{s + 1, \dots, s + Q\}$ and is independent of the distribution of the interarrival times between demands.

(4) Objective Function and Optimal Decision Rules

We select as our objective function the steady-state total expected cost per unit time; the decision variables s and Q are to be selected so as to minimize the objective function.

If $\bar{D}$ denotes the expected demand per unit time, then $\bar{D} = 1/E[X]$, where $E[X]$ denotes the expected interarrival time between two successive demands. Then the expected time elapsed between two successive orders is

$$E[Y] = QE[X] = \frac{Q}{D}$$

Thus the expected number of orders placed per unit time is

$$\frac{1}{E[Y]} = \frac{\bar{D}}{Q} \tag{5.73}$$

The expected stock level at any given instant of time is

$$E[H] = s + \sum_{n=1}^{n=Q} nP_n$$

Using (5.72), we have

$$E[H] = s + \frac{1}{Q} \sum_{n=1}^{n=Q} n$$

$$= s + \frac{1}{Q} \frac{Q(Q+1)}{2} \tag{5.74}$$

$$= s + \frac{Q+1}{2}$$

The total expected cost per unit time, expressed as a function of s and Q, is

$$F(s, Q) = \frac{K + cQ}{E[Y]} + hE[H]$$

Using expressions (5.73) and (5.74), we obtain for $s \geq 0$, $Q \geq 1$

$$F(s, Q) = K\frac{\bar{D}}{Q} + c\bar{D} + h\left[s + \frac{Q+1}{2}\right] \tag{5.75}$$

This expression is to be minimized with respect to s and Q. It is clear that $F(s, Q)$ is a separable function of s and Q and that the optimum value of s is $s^* = 0$. The optimum value of Q is obtained by minimizing the function

$$\tilde{F}(Q) = K\frac{\bar{D}}{Q} + h\frac{Q}{2}$$

over the set of positive integers Q.

It may be verified that an optimum value of Q, Q^*, exists and that, for such value(s), the following conditions prevail

$$\tilde{F}(Q^*) - \tilde{F}(Q^* + 1) \leq 0$$
$$\tilde{F}(Q^*) - \tilde{F}(Q^* - 1) \leq 0$$

From the first inequality we obtain

$$K\frac{\bar{D}}{Q^*} + h\frac{Q^*}{2} - \left(K\frac{\bar{D}}{Q^* + 1} + h\frac{Q^* + 1}{2}\right) \leq 0$$

or

$$K\bar{D}\left(\frac{1}{Q^*} - \frac{1}{Q^* + 1}\right) \leq \frac{h}{2}(Q^* + 1 - Q^*)$$

or

$$\frac{K\bar{D}}{Q^*(Q^* + 1)} \leq \frac{h}{2}$$

or

$$\frac{2K\bar{D}}{h} \leq Q^*(Q^* + 1)$$

Similarly, it may be shown that the second inequality implies

$$\frac{2K\bar{D}}{h} \geq Q^*(Q^* - 1)$$

Combining these two conditions, we obtain

$$Q^*(Q^* - 1) \leq \frac{2K\bar{D}}{h} \leq Q^*(Q^* + 1)$$

If we form the table

Q^*	1	2	3	4	5	6	7	8	9	10	...
$Q^*(Q^* - 1)$	0	2	6	12	20	30	42	56	72	90	...
$Q^*(Q^* + 1)$	2	6	12	20	30	42	56	72	90	110	...

then, for a given value of $2K\bar{D}/h$, the optimum Q^* may be selected. For example, if $2K\bar{D}/h = 10$, $Q^* = 3$, while if $2K\bar{D}/h = 42$, then $Q^* = 6$ or 7.

The reader should note that the expression for $F(s, Q)$ is only affected by the first moment of the interarrival times $\{X_j\}$ and is independent of any other statistical characteristics of the X_j's. This holds true so long as delivery of orders is instantaneous and no shortages are allowed. A case involving nonzero lead time will be discussed in another section.

As a special case, the present model covers the deterministic situation when the time interval between demand occurrence is a constant. The model then corresponds to the discrete-demand version of the basic model, and the reader should note the similarity between the two cost expressions. In fact, for large values of Q, the optimal value of Q, Q^*, is approximately given by the square root formula

$$Q^* \approx \sqrt{\frac{2K\bar{D}}{h}}$$

b. ANOTHER MODEL INVOLVING NO LEAD TIME AND UNIT DEMAND

(1) Problem of Multiple Satellite Launch

In establishing a network of communication satellites, it is necessary to maintain a minimum number of operating satellites in orbit while simultaneously taking into consideration the economy of the total program, including the launch of satellites.

In one such network, a minimum of s orbiting operating satellites is necessary to maintain operational capability of the system. However, satellites are subject to failure, and it is assumed that the probability that a satellite will be inoperative in the time interval $(t, t + dt]$ is $\mu dt + o(dt)$.

Satellites may be replaced individually as soon as they fail by launching a new satellite into orbit. However, it is also possible to initiate a multiple launch, that is, to launch simultaneously a single rocket carrying several satellites. Thus an alternative program would be to maintain a maximum number of $S = s + Q$ satellites in orbit and launch a single rocket carrying Q satellites whenever the number of orbiting satellites reaches the level s. We assume that a negligible time elapses between the failure of the last satellite and the launch.

The cost $P(Q)$ of making a multiple launch of Q satellites is assumed to be of the form

$$P(Q) = \begin{cases} 0, & Q = 0 \\ K + cQ, & Q = 1, 2, \dots \end{cases}$$

K is interpreted as a fixed cost associated with any launch. The additional cost is proportional to the number of satellites launched. The cost of having one satellite launched into orbit is assumed to represent a loss in investment opportunity of h dollars/[(unit)(unit time)].

It is required to determine the optimum number Q of satellites to be launched simultaneously so as to minimize the long-run total expected cost per unit time.

(2) Analysis of the Stochastic Process

In this problem the stock level $H(t)$ at a particular time $t, t \geq 0$, represents the number of operating satellites in orbit. Here $\{H(t), t \geq 0\}$ is a discrete-valued continuous-parameter Markov process taking on values of $s + 1, s + 2, \dots, S = s + Q$. Let

$$P(n, t) = P\{H(t) = n\}$$

$n = s + 1, \dots, S, t \geq 0$, denote the probability that at any instant of time t there are exactly n operating satellites in orbit. We may specify an initial condition of the form

$$P(n, 0) = \begin{cases} 1, & n = i \\ 0, & \text{otherwise} \end{cases}$$

That is, initially, exactly i operating satellites are in orbit. These initial conditions will not be used explicitly in the sequel, since we shall only be interested in steady-state conditions. The following equations may be written for $P(n, t)$. For $s + 1 \leq n \leq S - 1$ and $t \geq 0$,

$$P(n, t + dt) = P(n, t)[1 - n\mu \, dt + o(dt)] \\ + P(n + 1, t)[(n + 1)\mu \, dt + o(dt)] \tag{5.76}$$

For $n = S$ and $t \geq 0$,

$$P(S, t + dt) = P(S, t)[1 - S\mu \, dt + o(dt)] \\ + P(s + 1, t)[(s + 1)\mu \, dt + o(dt)] \tag{5.77}$$

In this last equation the last term in the right-hand side is the probability of the event "exactly one out of the $s + 1$ operating satellites at time t fails in the time interval $(t, t + dt]$, bringing the level of satellites to s; thus, immediately, a multiple launch is triggered, resulting in a net total of S operating satellites." Transposing, dividing by dt, and letting $dt \rightarrow 0$, equations (5.76) and (5.77) may be written as ($t \geq 0$)

$$\frac{dP(n, t)}{dt} = -n\mu P(n, t) + (n + 1)\mu P(n + 1, t), \quad s + 1 \leq n \leq S - 1$$

$$\tag{5.78}$$

$$\frac{dP(S, t)}{dt} = -S\mu P(S, t) + (s + 1)\mu P(s + 1, t), \quad n = S \tag{5.79}$$

*(3) Steady-State Distribution of the Number of
Operating Satellites in Orbit*

Let P_n be the probability that in the steady state exactly $H = n$ operating satellites are in orbit, $n = s + 1, \ldots, S$; that is,

$$P_n = P\{H = n\} = \lim_{t \to \infty} P\{H(t) = n\}$$

Noting that $S = s + Q$, the following steady-state equations are obtained from (5.78) and (5.79):

$$0 = -n\mu P_n + (n + 1)\mu P_{n+1}, \quad s + 1 \leq n \leq s + Q - 1 \\ 0 = -(s + Q)\mu P_{s+Q} + (s + 1)\mu P_{s+1}, \quad n = s + Q \tag{5.80}$$

Solving recursively the system of equations (5.80), we have

$$P_{s+2} = \frac{s+1}{s+2} P_{s+1}$$

$$P_{s+3} = \frac{s+2}{s+3} P_{s+2}$$

$$= \frac{s+2}{s+3} \cdot \frac{s+1}{s+2} P_{s+1}$$

$$= \frac{s+1}{s+3} P_{s+1}$$

In general, we find

$$P_n = \frac{s+1}{n} P_{s+1}, \quad n = s+1, \ldots, s+Q \tag{5.81}$$

To determine P_{s+1}, we use the normalizing condition

$$\sum_{n=s+1}^{n=s+Q} P_n = (s+1)P_{s+1} \sum_{n=s+1}^{n=s+Q} \frac{1}{n} = 1$$

from which it follows that

$$P_{s+1} = \frac{1}{(s+1) \sum_{i=1}^{i=Q} \frac{1}{s+i}} \tag{5.82}$$

Substituting (5.82) in (5.81), we obtain for the steady-state distribution of the number of units in the system

$$P_n = \frac{1}{n \sum_{i=1}^{i=Q} \frac{1}{s+i}}, \quad n = s+1, \ldots, s+Q \tag{5.83}$$

The expected number of operating satellite units in the steady state is

$$E[H] = \sum_{n=s+1}^{n=s+Q} nP_n$$

$$= \frac{Q}{\sum_{i=1}^{i=Q} \frac{1}{s+i}} \tag{5.84}$$

*(4) Steady-State Distribution of the Time Elapsed Between
Two Successive Launches*

Define an independent parallel system composed of S identical components. The system is considered operating if at least $s+1$ of the components are required to function; the system is considered failed if $Q = S - s$

components out of the S fail for the first time. Such a system is known as the $s + 1$ out of S system. We further assume that all components have a negative exponential time-to-failure distribution with mean $1/\mu$, and that at time origin all components are operating. Let $\tilde{P}(n, t)$ be the probability that at time $t, t \geq 0$, exactly n components, $n = s + 1, \ldots, S$ are in operation. It is easy to verify that the differential difference equations for $\tilde{P}(n, t)$ are

$$\frac{d\tilde{P}(S, t)}{dt} = -S\mu\tilde{P}(S, t), \quad n = S$$

$$\frac{d\tilde{P}(n, t)}{dt} = -n\mu\tilde{P}(n, t) + (n + 1)\mu\tilde{P}(n + 1, t), \quad s + 1 \leq n \leq S - 1 \tag{5.85}$$

with initial conditions

$$P(n, 0) = \begin{cases} 1, & n = S \\ 0, & \text{otherwise} \end{cases} \tag{5.86}$$

Solving equations (5.85) recursively and using (5.86) yields (Section 2.7-iii-b-3)

$$\tilde{P}(n, t) = \binom{S}{n}(e^{-\mu t})^n(1 - e^{-\mu t})^{S-n}, \quad s + 1 \leq n \leq S; t \geq 0$$

Let $\tilde{P}_0(t)$ denote the probability that the system is in a failed state at time t. An expression for $\tilde{P}_0(t)$ is

$$\tilde{P}_0(t) = 1 - \sum_{n=s+1}^{n=S} \tilde{P}(n, t)$$

$$= 1 - \sum_{n=s+1}^{n=S} \binom{S}{n}(e^{-\mu t})^n(1 - e^{-\mu t})^{S-n}, \quad t \geq 0 \tag{5.87}$$

Now let T be a random variable denoting the time to failure of the system. The system reliability is, from (5.87),

$$P\{T > t\} = 1 - \tilde{P}_0(t)$$

$$= \sum_{n=s+1}^{n=S} \binom{S}{n}(e^{-\mu t})^n(1 - e^{-\mu t})^{S-n}, \quad t \geq 0 \tag{5.88}$$

Returning to our launch problem, let Y be a random variable denoting the time elapsed between two successive launches in the steady state. Since the time to failure of each satellite has a negative exponential distribution, it follows that given a multiple launch replacing all failed satellites the S operating satellites in orbit are as good as new (memoryless property). It is then evident that

$$P\{Y > t\} = P\{T > t\}$$

Using (5.88), we obtain

$$E[Y] = \int_0^\infty P\{T > t\}\, dt$$

$$= \int_0^\infty \sum_{n=s+1}^{n=S} \binom{S}{n} (e^{-\mu t})^n (1 - e^{-\mu t})^{S-n}\, dt \qquad (5.89)$$

Consider the integral

$$I(n) = \int_0^\infty (e^{-\mu t})^n (1 - e^{-\mu t})^{S-n}\, dt$$

Let $z = e^{-\mu t}$; we obtain

$$I(n) = \frac{1}{\mu} \int_0^1 z^{n-1}(1 - z)^{S-n}\, dz$$

$$= \frac{1}{\mu} \frac{\Gamma(n)\Gamma(S - n + 1)}{\Gamma(S + 1)} \qquad (5.90)$$

Interchanging the order of integration and summation in (5.89) and using (5.90), we have

$$E[Y] = \sum_{n=s+1}^{n=S} \binom{S}{n} \int_0^\infty (e^{-\mu t})^n (1 - e^{-\mu t})^{S-n}\, dt$$

$$= \sum_{n=s+1}^{n=S} \frac{\Gamma(S + 1)}{\Gamma(n + 1)\Gamma(S - n + 1)} \cdot \frac{1}{\mu} \frac{\Gamma(n)\Gamma(S - n + 1)}{\Gamma(S + 1)} \qquad (5.91)$$

$$= \frac{1}{\mu} \sum_{n=s+1}^{n=S} \frac{1}{n}$$

$$= \frac{1}{\mu} \sum_{i=1}^{i=Q} \frac{1}{s + i}$$

(5) Objective Function and Optimal Decision Rules

Under steady-state conditions, the expected cost of launch per unit time is $(K + cQ)/E[Y]$, and the expected loss of investment opportunity per unit time is $hE[H]$. Using (5.84) and (5.91), we obtain the following expression for the steady-state total expected cost per unit time:

$$F(Q) = \frac{K + cQ}{E[Y]} + hE[H]$$

$$= \frac{K + cQ}{\dfrac{1}{\mu} \displaystyle\sum_{i=1}^{i=Q} \frac{1}{s + i}} + \frac{hQ}{\displaystyle\sum_{i=1}^{i=Q} \frac{1}{s + i}} \qquad (5.92)$$

$$= \frac{\mu K + (c\mu + h)Q}{\displaystyle\sum_{i=1}^{i=Q} \frac{1}{s + i}}$$

The reader should verify that the following conditions for an optimum value of Q, Q^*, which minimizes $F(Q)$ in (5.92), must be satisfied

$$(Q^* + s)(c\mu + h) \le F(Q^*) \le (Q^* + 1 + s)(c\mu + h) \qquad (5.93)$$

Using expression (5.92), the inequalities (5.93) may be written as

$$(Q^* + s)\sum_{i=1}^{i=Q^*} \frac{1}{s+i} - Q^* \le \frac{\mu K}{c\mu + h} \le (Q^* + 1 + s)\sum_{i=1}^{i=Q^*} \frac{1}{s+i} - Q^* \quad (5.94)$$

The following numerical example is considered. The communication network requires at least fifty satellites ($s = 50$) and the average life of a satellite is $1/\mu = 10$ years. Assume that $K/c = 2$ and that $h/c = .10$. Utilizing these quantities in inequalities (5.93) or (5.94), it may readily be verified that the optimum number of satellites to launch is $Q^* = 10$.

c. MODEL INVOLVING FIXED LEAD TIME

(1) Introduction

At this stage some comments on the effect of lead time are appropriate. So far we have assumed instantaneous delivery of orders in the two example models we have worked out, and no shortages were allowed. The effect of randomness in demand, combined with the presence of lead time, will invariably induce, once in a while, a shortage of the product. Such shortage cannot be avoided, and in practice one has to accept it and regulate the control of stock in such a way that the simultaneous contribution of the expected costs of procurement, holding, and shortage over a given time period is minimized.

The effect of lead time will in general add to the complexity of the problem analysis, which will also depend on the type of shortage incorporated in the model. Except for some exceptional situations, the treatment of the lost-sales case is inordinately complex. However, more often, the backlog case may be approached analytically, although the mathematical formulation still presents some inherent difficulties.

Several possible methods may be used to characterize the inventory problem with random demand, fixed lead time with backlog allowed. In one method, the joint probability distribution of the position inventory $\mathcal{P}(t)$ and the number of outstanding orders $\mathcal{R}(t)$ at a particular instant of time t are determined. The distributions of the inventory level $I(t)$, the stock level $H(t)$, and the backlog level $B(t)$ at that instant of time t are then determined from the respective relations.

$$I(t) = \mathcal{P}(t) - Q\mathfrak{N}(t)$$

$$H(t) = \begin{cases} \mathcal{P}(t) - Q\mathfrak{N}(t), & \text{if } \mathcal{P}(t) > Q\mathfrak{N}(t) \\ 0, & \text{if } \mathcal{P}(t) \leq Q\mathfrak{N}(t) \end{cases}$$

$$B(t) = \begin{cases} 0, & \text{if } Q\mathfrak{N}(t) < \mathcal{P}(t) \\ Q\mathfrak{N}(t) - \mathcal{P}(t), & \text{if } Q\mathfrak{N}(t) \geq \mathcal{P}(t) \end{cases}$$

In another method, the joint distribution of the position inventory $\mathcal{P}(t)$ and demand occurrence $D(t, t + l)$ over the lead time interval $(t, t + l)$ is determined. The expressions for the distributions of $I(t + l)$, $H(t + l)$, and $B(t + l)$ at time $t + l$ are then obtained from the relations

$$I(t + l) = \mathcal{P}(t) - D(t, t + l) \tag{5.95}$$

$$H(t + l) = \begin{cases} \mathcal{P}(t) - D(t, t + l), & \text{if } \mathcal{P}(t) > D(t, t + l) \\ 0, & \text{if } \mathcal{P}(t) \leq D(t, t + l) \end{cases} \tag{5.96}$$

$$B(t + l) = \begin{cases} 0, & \text{if } D(t, t + l) < \mathcal{P}(t) \\ D(t, t + l) - \mathcal{P}(t), & \text{if } D(t, t + l) \geq \mathcal{P}(t) \end{cases} \tag{5.97}$$

From either method the steady-state distributions of the inventory level, the stock level, and the backlog level may be derived to be ultimately used in the objective function. In the example that follows, the use of the second approach is illustrated, although the problem structure is considerably simplified because of the characteristics of the demand process.

(2) Problem

The demand characteristics of this model are a special case of the model developed in Section 5.3-i-a. Here we assume that the interarrival times between successive demands form a sequence $\{X_j\}$ of independently and identically distributed random variables with the negative exponential distribution

$$\varphi_X(x) = \lambda e^{-\lambda x}, \quad \lambda > 0, \quad 0 < x < \infty$$

Equivalently, the total number of units $D(\theta)$ in demand over an arbitrary time interval θ, $\theta > 0$, has the Poisson distribution

$$P\{D(\theta) = r\} = e^{-\lambda\theta} \frac{(\lambda\theta)^r}{r!}, \quad r = 0, 1, \ldots \tag{5.98}$$

We assume that there is a fixed time elapsed l from the moment an order is placed to replenish stock to the moment the order is received. All unsatisfied demand is backlogged, and must be satisfied in future times.

(3) *Analysis of the Stochastic Process*

If we assume that initially at time $t = 0$, a demand has just taken place, resulting in an inventory position level of $s + i$, $i = 1, 2, \ldots, Q$, then the inventory position $\mathcal{P}(t)$ at a particular time t, $t \geq 0$, will be described by the same stochastic process as the stock level $H(t)$ analyzed in the first model (Section 5.3-i-a). Thus, in the steady state, the position inventory is uniformly distributed over the interval $\{s + 1, \ldots, s + Q\}$. For this particular situation, since $\{\mathcal{P}(t), t \geq 0\}$ is Markovian, this result could have been derived as follows: Let

$$P\{\mathcal{P}(t) = s + n\} = P(n, t), \quad n = 1, 2, \ldots, Q; t \geq 0$$

be the probability that at time t the position inventory level is $s + n$; then $P(n, t)$ satisfies the following relations for $t \geq 0$:

$$P(n, t + dt) = P(n, t)(1 - \lambda\, dt) + P(n + 1, t)\lambda\, dt + o(dt),$$
$$1 \leq n \leq Q - 1 \quad (5.99)$$
$$P(Q, t + dt) = P(Q, t)(1 - \lambda\, dt) + P(1, t)\lambda\, dt + o(dt), \quad n = Q$$

with the initial conditions

$$P(n, 0) = \begin{cases} 1, & n = i \\ 0, & \text{otherwise} \end{cases} \quad (5.100)$$

Equations (5.99) reduce to the following system of differential difference equations:

$$\frac{dP(n, t)}{dt} = -\lambda P(n, t) + \lambda P(n + 1, t), \quad n = 1, \ldots, Q - 1$$
$$\frac{dP(Q, t)}{dt} = -\lambda P(Q, t) + \lambda P(1, t), \quad n = Q \quad (5.101)$$

The transient solution may be obtained by solving (5.101) using (5.100) as initial conditions.

Let P_n be the probability that in the steady state the position inventory $\mathcal{P}$ consists of exactly $s + n$ units, $n = 1, \ldots, Q$. Then

$$P_n = P\{\mathcal{P} = s + n\} = \lim_{t \to \infty} P\{\mathcal{P}(t) = s + n\}$$

From equation (5.101) one obtains for $P_n, n = 1, \ldots, Q$,

$$0 = -\lambda P_n + \lambda P_{n+1}, \quad n = 1, \ldots, Q - 1$$
$$0 = -\lambda P_Q + \lambda P_1, \quad n = Q \quad (5.102)$$

Solving (5.102) recursively, we find

$$P_1 = P_2 = \ldots = P_Q$$

Since

$$\sum_{i=1}^{i=Q} P_i = 1$$

it immediately follows that

$$P_n = P\{\mathcal{P} = s + n\} = \frac{1}{Q}, \quad n = 1, 2, \ldots, Q \tag{5.103}$$

Therefore, the position inventory is uniformly distributed on the set $\{s+1, \ldots, s+Q\}$.

In this problem it is possible to determine the joint distribution of the position inventory at any particular instant of time t and the demand occurring over the time interval $(t, t + \theta]$. Since demand is Poisson, it is a function of θ only; hence, using expressions (5.98) and (5.103), we obtain

$$P\{\mathcal{P}(t) = s + n; \ D(\theta) = r\} = P\{D(\theta) = r | \mathcal{P}(t) = s + n\} P\{\mathcal{P}(t) = s + n\}$$
$$= P\{D(\theta) = r\} P\{\mathcal{P}(t) = s + n\}$$
$$= \varrho^{-\lambda\theta} \frac{(\lambda\theta)^r}{r!} \cdot \frac{1}{Q}, \quad r = 0, 1, \ldots; 1 \leq n \leq Q \tag{5.104}$$

A knowledge of this joint distribution leads us to using relations (5.95), (5.96), and (5.97) to determine the important characteristics of the system in the steady state. Let

$I =$ random variable defining the inventory level in the steady state
$H =$ random variable defining the stock level in the steady state
$B =$ random variable defining the backlog level in the steady state

From relation (5.95), we obtain

$$E[I] = E[\mathcal{P}] - E[D(l)]$$
$$= s + \frac{Q+1}{2} - \lambda l \tag{5.105}$$

From relation (5.97) we have, using (5.104),

$$E[B] = \sum_{n=1}^{n=Q} \left\{ \sum_{r=0}^{r=s+n-1} 0 + \sum_{r=s+n}^{\infty} [r - (s + n)] \right\} e^{-\lambda l} \frac{(\lambda l)^r}{r!} \cdot \frac{1}{Q}$$
$$= \frac{1}{Q} \sum_{n=1}^{n=Q} \sum_{r=s+n}^{\infty} [r - (s + n)] e^{-\lambda l} \frac{(\lambda l)^r}{r!} \tag{5.106}$$

From relations (5.95), (5.96), and (5.97), it is evident that

$$E[H] = E[I] - E[B] \tag{5.107}$$

From (5.73), the expected number of orders placed per unit time is

$$\frac{1}{E[Y]} = \frac{\lambda}{Q} \tag{5.108}$$

The steady-state total expected cost per unit time expressed as a function of s and Q is

$$F(s, Q) = \frac{K + cQ}{E[Y]} + hE[H] + pE[B]$$

Using (5.107) and (5.108) in this last expression, we obtain

$$F(s, Q) = \frac{\lambda}{Q}(K + cQ) + hE[I] + (h + p)E[B]$$

Using (5.105) and (5.106), we get

$$\begin{aligned} F(s, Q) = c\lambda + \frac{K\lambda}{Q} + h\left(s + \frac{Q+1}{2} - \lambda l\right) \\ + (h + p)\frac{1}{Q}\sum_{n=1}^{n=Q}\sum_{r=s+n}^{\infty}[r - (s + n)]e^{-\lambda l}\frac{(\lambda l)^r}{r!} \end{aligned} \tag{5.109}$$

Expression (5.109) may be further reduced by interchanging the order of summation in the double-sum expression. The final form may be expressed in terms of the cumulative Poisson distribution or, equivalently, the incomplete gamma function. We leave this as an exercise to the reader, and further ask him to analyze from expression (5.109) the special case when lead time is zero, (see also [10]).

ii. *Periodic Review Systems*

The analysis of periodic review systems with zero lead time assumes that at the beginning of equal time intervals the inventory of the commodity is reviewed and a decision to order or not is made. Not all decisions result in an inventory replenishment; however, if they do, delivery of orders is immediate. Depletion of inventory during each period is induced by some type of consumption or customer demand. The demands $\{D(t)\}$ over the successive periods of time $t = 1, 2, \ldots$ form a sequence of independently and identically distributed nonnegative random variables with distribution function $\Phi(\xi)$, $0 \leq \xi < \infty$. Demand may either be defined as a discrete random variable or a continuous random variable. Throughout this section, we shall assume demand to be a nonnegative continuous random variable, in which case we assume the existence of a probability density function $\varphi(\xi), 0 < \xi < \infty$. The expected demand per period will be denoted by $\bar{D} = \int_0^\infty \xi\,\varphi(\xi)\,d\xi$.

The sequence of ordering decisions at the beginning of each period of time will depend on the inventory status of the commodity, the demand for the commodity during present and future periods, and, finally, the cost factors influencing the operation of the system. Consider the beginning of any given period, and let

x = inventory level of the commodity prior to making a decision; in case of shortage, x is nonpositive

z = quantity ordered and received following a decision; if a decision not to order is made, then $z = 0$, while a decision to order would result in some $z > 0$

y = inventory level of the commodity following a decision assumed to be always positive; in case of backlog, $y = x + z$

We assume that the procurement cost $P(z)$ in ordering z units of the commodity has the standard form

$$P(z) = \begin{cases} 0, & z = 0 \\ K + cz, & z > 0 \end{cases}$$

That is, a fixed cost is incurred only if a procurement is made, while the variable procurement cost is proportional to the amount procured.

We now introduce the function $L(y)$ to denote the total expected holding and shortage cost incurred during the period, given that at the beginning of the period, following a decision, the inventory level is y. From an analysis standpoint, the introduction of this function is very convenient for reasons we shall see later. The analytic expression for $L(y)$ will depend on the assumptions underlying the measurement of the holding and shortage costs. For example, the amount of stock on hand may be measured either at the beginning of the period following a decision or at the end of the period. The cost of holding stock per period may be a linear or nonlinear function of the stock on hand. In a similar fashion, the shortage cost per period may or may not be a linear function of the quantity short.

Although the explicit expression of the function $L(y)$ will exhibit different forms, in general $L(y)$ will be a convex function of y. We illustrate this property by developing the form of $L(y)$ in the very important case when the holding and shortage costs are both measured at the end of the period and are respectively proportional to the amount on hand and the amount of shortage. Let

$\hat{H}(D\,|\,y)$ = stock level at the end of the period, given that initially the inventory level is y, when the demand during the period is described by the random variable D

$\hat{B}(D\,|\,y)$ = shortage level at the end of the period, given that initially the inventory level is y, when the demand during the period is described by the random variable D

Each of $\hat{H}(D|y)$ and $\hat{B}(D|y)$ being functions of the random variable D are themselves random variables. The inventory level at the end of the period will be positive if the demand D during the period is less than the initial inventory level y, $y \geq 0$. On the other hand, the inventory level at the end of the period will be negative, thus resulting in a shortage, if the demand D during the

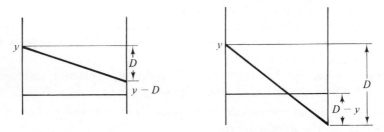

Figure 5.24. Stock level and shortage level measured at end of period.

period exceeds the initial level y, $y \geq 0$. This is illustrated in Figure 5.24. We may then write down $\hat{H}(D|y)$ and $\hat{B}(D|y)$ explicitly as

$$\hat{H}(D|y) = \begin{cases} y - D, & 0 \leq D < y \\ 0, & y \leq D < \infty \end{cases} \qquad (5.110)$$

$$\hat{B}(D|y) = \begin{cases} 0, & 0 \leq D < y \\ D - y, & y \leq D < \infty \end{cases} \qquad (5.111)$$

Now let

$h =$ holding cost per unit of stock on hand measured at the end of the period

$p =$ penalty cost per unit of shortage measured at the end of the period

$\hat{C}(D|y) =$ total cost of holding and shortage incurred during the period, given that initially the inventory level is y

By definition $\hat{C}(D|y)$ is a random variable such that

$$\hat{C}(D|y) = h\hat{H}(D|y) + p\hat{B}(D|y)$$

Using expressions (5.110) and (5.111), we may write explicitly

$$\hat{C}(D|y) = \begin{cases} h(y - D), & 0 \leq D < y \\ p(D - y), & y \leq D < \infty \end{cases} \qquad (5.112)$$

The graph of $\hat{C}(D|y)$ expressed as a function of D is as shown in Figure 5.25.

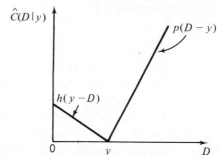

Figure 5.25. The function $\hat{C}(D\,|\,y)$ for linear holding and shortage costs.

By definition $L(y)$ is the expectation of the random variable $\hat{C}(D\,|\,y)$; thus

$$L(y) = E[\hat{C}(D\,|\,y)]$$
$$= \int_0^\infty \hat{C}(\xi\,|\,y)\varphi(\xi)\,d\xi$$

and, using (5.112), we write

$$L(y) = \int_0^y h(y - \xi)\varphi(\xi)\,d\xi + \int_y^\infty p(\xi - y)\varphi(\xi)\,d\xi$$

or

$$L(y) = h \int_0^y (y - \xi)\varphi(\xi)\,d\xi$$
$$+ p\left[\int_0^\infty (\xi - y)\varphi(\xi)\,d\xi - \int_0^y (\xi - y)\varphi(\xi)\,d\xi\right]$$
$$= (h + p) \int_0^y (y - \xi)\varphi(\xi)\,d\xi + p \int_0^\infty (\xi - y)\varphi(\xi)\,d\xi$$

Integrating by parts the first integral and using the symbol $\bar{D}$ for the expectation of D in the second integral, we obtain

$$L(y) = (h + p) \int_0^y \Phi(\xi)\,d\xi + p(\bar{D} - y), \quad y \geq 0 \qquad (5.113)$$

It is easily verified that

$$L'(y) = (h + p)\Phi(y) - p$$

and

$$L''(y) = (h + p)\varphi(y)$$

Thus $L''(y) \geq 0$ for all values of $y \geq 0$; hence the function $L(y)$ is convex over the set $y \geq 0$.

The first model we study is the one-period problem, implying the opportunity to make a single decision. The second model studies the steady-state characteristics of periodic review systems operating under the (s, S) policy.

a. ONE-PERIOD MODEL

(1) Problem

Several practical situations arise that make the study of the *one-period* or *static* problem useful. A classical example is the Christmas tree problem in which a merchant must order Christmas trees before the beginning of the season; assuming that he has only one opportunity to order before the season and that any unsold tree at the end of the season is a total loss, how much should he order to maximize his profit?

This one-shot-type decision occurs frequently in the manufacturing of highly seasonal products, such as toys or high-fashion dresses. On the one hand, excessive production may cause undesirable surplus of the product resulting in obsolete goods, while, on the other hand, a limited production may induce shortage of goods resulting in a loss of profit. It is reasonable then to wonder whether an optimum production run exists that would attain some objective function normally formulated as the minimization of the expected cost or the maximization of the expected profit. In all these problems an irrevocable decision must be made at the start of some period of time having a specific duration on whether to initiate a replenishment, and if so, how much to order.

(2) Objective Function and Optimal Ordering Strategies

Our objective will be to determine an optimal replenishment policy so as to minimize the sum total of the procurement cost and the expected holding and shortage costs. Let

$x =$ initial inventory prior to a decision
$y =$ inventory level following a decision, $y > 0$, $y \geq x$
$z = y - x =$ quantity procured, $z \geq 0$
$C(x) =$ minimum total expected cost over the period when following an optimal policy

Then we can write

$$C(x) = \min_{x \leq y} \{P(y - x) + L(y)\}$$

$$= \min \begin{cases} \min_{x=y} L(y) \\ \min_{x \leq y} \{K + c(y - x) + L(y)\} \end{cases} \tag{5.114}$$

In expression (5.114) the first alternative corresponds to a no-order decision, in which case the initial inventory level remains at x and the only cost experienced is the expected holding and shortage costs. The second alternative corresponds to a decision to order, in which case a fixed cost is incurred in the amount of K, plus a variable procurement cost proportional to the quantity ordered $(y - x)$, and finally the expected holding and shortage costs starting with an initial inventory level of y.

We now define the function

$$G(y) = cy + L(y) \tag{5.115}$$

Using (5.115), expression (5.114) may be written as

$$C(x) = \min \begin{cases} \min_{x=y} \{-cy + G(y)\}, & \text{do not order} \\ \min_{x \leq y} \{K - cx + G(y)\}, & \text{order} \end{cases} \tag{5.116}$$

For an initial inventory level x, the solution of system (5.116) will specify the *optimal ordering strategy;* that is, it will dictate which ordering strategy to select, as well as the quantities to be ordered, so as to minimize the total expected cost. We now consider each of these two alternatives separately.

Suppose that the alternative is not to order; the minimum total expected cost associated with this decision is clearly, from (5.116),

$$C_1(x) = cx + G(x) \tag{5.117}$$

Suppose now that the alternative is to order; then, from (5.116), the minimum cost associated with this decision is

$$\begin{aligned} C_2(x) &= \min_{x \leq y} \{K - cx + G(y)\} \\ &= K - cx + \min_{x \leq y} G(y) \end{aligned} \tag{5.118}$$

Now, from (5.115), if $L(y)$ is assumed to be a convex function, $G(y)$, which is the sum of two convex functions, is also convex. Let S^* be the value of y for which $G(y)$ assumes an absolute minimum over the set $y \geq x$. Then, from (5.118),

$$C_2(x) = K - cx + G(S^*) \tag{5.119}$$

The alternative "order" is preferred to the alternative "do not order" iff

$$C_2(x) < C_1(x)$$

and, using expressions (5.117) and (5.119), we have

$$K - cx + G(S^*) < -cx + G(x)$$

or
$$K + G(S^*) \leq G(x) \tag{5.120}$$

Now the equation

$$K + G(S^*) = G(x) \tag{5.121}$$

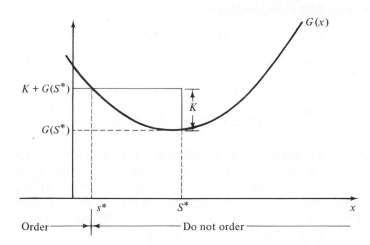

Figure 5.26. On the solution of the equation $K + G(S^*) = G(x)$.

will yield as a solution the value $x = s^*$ (Figure 5.26). It is evident that inequality (5.120) will be satisfied if and only if $x < s^*$. It immediately follows that

(1) If $x \leq s^*$, then an order should be placed to bring the inventory level up to S^*, at which point the expected cost function associated with an order decision is minimum and is equal to $C_2(x)$.
(2) If $x > s^*$, then no order should be placed, since otherwise a higher cost would be incurred with an order placed.

The preceding discussion demonstrates that under conditions involving a fixed procurement cost, a linear variable procurement cost, and an expected holding and shortage cost which is convex the optimal ordering policy is of the (s, S) type. The optimal value of S, S^*, is that value which minimizes the function $G(x) = cx + L(x)$, while the optimum value of s, s^*, is a solution to the equation $K + G(S^*) = G(s^*)$.

An important special case to this problem is a situation in which there is no fixed procurement charge K; that is, $K = 0$. If, in the previous problem, we parametrize K and let K take on smaller and smaller values, the corresponding values of s^*, which is the solution to (5.121), will get closer and closer to the value of S^* (Figure 5.27). Finally, when $K = 0$, we note that $s^* = S^*$, and the policy then consists in ordering up to S^* if $x < S^*$, and otherwise not to place any order.

We illustrate the procedure for determining S^* and s^* by considering

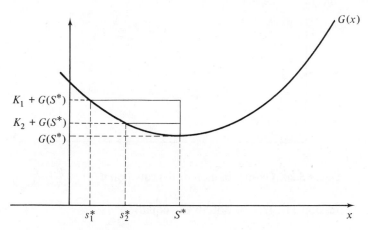

Figure 5.27. Parametric analysis of the equation $K + G(S^*)$ $= G(x)$ when K varies.

the specific functional form $L(y)$ as given in expression (5.113). We have

$$G(x) = cx + L(x)$$
$$= cx + (h + p) \int_0^x \Phi(\xi)\, d\xi + p(\bar{D} - x), \quad x \geq 0 \qquad (5.122)$$

where $\Phi(\cdot)$ is the cumulative distribution of demand during the period and $\bar{D}$ is the average demand. A relative minimum for $G(x)$ may be obtained by differentiating expression (5.122) with respect to x and setting the result equal to zero. We then have

$$\frac{dG(x)}{dx} = 0 = c + (h + p)\Phi(x) - p$$

Thus S^* is a solution to the equation

$$\Phi(S^*) = \frac{p - c}{h + p} \qquad (5.123)$$

The solution to this equation will exist provided $p > c$. A graphical procedure for obtaining the value of S^* is illustrated in Figure 5.28. Note that the cumulative function $\Phi(x)$ is such that $0 \leq \Phi(x) \leq 1$, and that, for $p - c > 0$, the quantity $(p - c)/(h + p)$ is positive and less than 1.

To determine the optimum value s^*, we have to solve equation (5.121) or

$$G(s^*) - G(S^*) = K$$

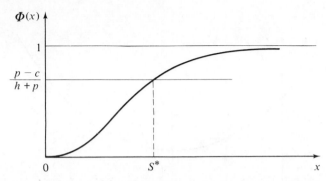

Figure 5.28. On the solution of the equation $\Phi(x) = \dfrac{p - c}{h + p}$.

Using expression (5.122), we obtain the equation

$$\left[cs^* + (h + p) \int_0^{s^*} \Phi(\xi)\,d\xi + p(\bar{D} - s^*) \right]$$
$$- \left[cS^* + (h + p) \int_0^{S^*} \Phi(\xi)\,d\xi + p(\bar{D} - S^*) \right] = K$$

or $\qquad (p - c)(S^* - s^*) - (h + p) \int_{s^*}^{S^*} \Phi(\xi)\,d\xi = K \qquad (5.124)$

If we let $Q^* = S^* - s^*$, and make the change in variable, $u = S^* - \xi$, in the integral in (5.124), we obtain

$$(p - c)Q^* - (h + p) \int_0^{Q^*} \Phi(S^* - u)\,du = K \qquad (5.125)$$

Thus, knowing the value of S^*, the value of Q^*, and hence s^*, may be determined from equation (5.125).

Example 5.4

Assume the demand D to be a random variable with the negative exponential distribution

$$\varphi(\xi) = \lambda e^{-\lambda \xi}, \quad \lambda > 0, 0 < \xi < \infty$$

Then $\qquad \Phi(\xi) = 1 - e^{-\lambda \xi} \qquad (5.126)$

Using (5.126) in equation (5.123), we obtain

$$1 - e^{-\lambda S^*} = \frac{p - c}{h + p}$$

or $\qquad e^{-\lambda S^*} = \frac{h + c}{h + p} \qquad (5.127)$

Taking logarithms, we obtain explicitly

$$S^* = \frac{1}{\lambda} \ln \frac{h + p}{h + c}$$

Using (5.126) in equation (5.125), we obtain

$$(p - c)Q^* - (h + p) \int_0^{Q^*} [1 - e^{-\lambda(S^* - u)}] \, du = K$$

or

$$(p - c)Q^* - (h + p)\left[Q^* + e^{-\lambda S^*} \frac{1}{\lambda}(e^{\lambda Q^*} - 1) \right] = K \qquad (5.128)$$

Using (5.127) in equation (5.128) and simplifying results in the expression

$$-(c + h)Q^* + \frac{1}{\lambda}(c + h)(e^{\lambda Q^*} - 1) = K$$

Finally we obtain, by dividing both sides by K,

$$\frac{c + h}{\lambda K}(e^{\lambda Q^*} - 1 - \lambda Q^*) = 1 \qquad (5.129)$$

The quantity $X = \lambda Q^*$ can be obtained as the positive intercept of the curves (Figure 5.29).

$$Y = e^X \quad \text{and} \quad Y = \frac{\lambda K}{c + h} + 1 + X$$

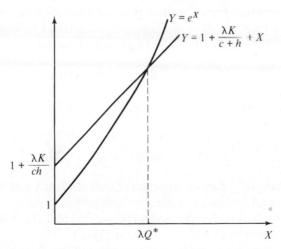

Figure 5.29. On the solution of the equation $\dfrac{(c + h)}{\lambda k}(e^X - 1 - X) = 1$.

We note from Figure 5.29 that a solution will always exist, since $\lambda K/(c + h)$ > 0. From equation (5.129), it is possible to obtain an approximate closed-form expression for Q^*, provided that λQ^* is small. If, in equation (5.129), the quantity $e^{\lambda Q^*}$ is expanded as a power series and the first three terms are retained, we obtain

$$\frac{c + h}{\lambda K}\left[1 + \lambda Q^* + \frac{1}{2!}(\lambda Q^*)^2 + \cdots -(1 + \lambda Q^*)\right] = 1$$

or

$$\frac{(c + h)\lambda}{2K}Q^{*2} \approx 1$$

Hence

$$Q^* \approx \sqrt{\frac{2K}{\lambda(c + h)}}$$

This square-root expression should be compared to the Wilson formula.

b. Steady-State Periodic Review Model Using the (s, S) Policy

(1) Problem

We consider a periodic review system operating under the (s, S) policy with immediate delivery of orders and backlog allowed (Figure 5.5). Thus, when the inventory level falls at or below s, an order is placed to raise instantaneously the level to $S > 0$; when the inventory level exceeds s, no order is placed. Let for $t = 1, 2, \ldots$

$D(t) =$ demand during period t
$X(t) =$ inventory level at end of period t before a decision
$Y(t) =$ inventory level at the beginning of period t following a decision
$Z(t) =$ quantity ordered at the beginning of period t

Then
$$Y(t + 1) = Y(t) + Z(t) - D(t) \tag{5.131}$$

Under the (s, S) policy

$$Y(t + 1) = \begin{cases} S, & \text{if } -\infty < X(t) \leq s \\ Y(t) - D(t), & \text{if } \quad s < X(t) < S \end{cases} \tag{5.132}$$

We assume that $\{D(t)\}$ form a sequence of independently and identically distributed non-negative random variables from period to period, with distribution function $\Phi(\xi)$, $0 \leq \xi < \infty$, and PDF $\varphi(\xi)$. Under the stated conditions and the given policy, the stochastic process $\{Y(t), t = 1, 2, \ldots\}$ is Markovian since the state at $t + 1$ depends only on the state at t. We shall assume that steady state exists and has been reached. Each of $X(t)$, $Y(t)$, and $Z(t)$ will be

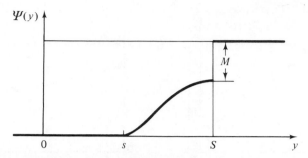

Figure 5.30. Inventory levels at beginning of period in a periodic review system.

independent of the initial conditions and of t. We shall denote them respectively by X, Y, and Z. For the resulting stationary problem, let

$\Psi(y) = $ distribution function of the inventory level Y at the beginning of a period just following a decision, $-\infty < y < \infty$

$M = $ probability of placing an order at the beginning of a period, $y = S$

$\varphi(y)\, dy = $ probability that the inventory level Y at the beginning of a period lies between y and $y + dy$, $s < y < S$.

The quantity M and the function $\psi(y)$ are unknowns to be determined. We may simply write (see Figure 5.31)

$$d\Psi(y) = \begin{cases} \psi(y)\, dy, & s < y < S \\ M, & y = S \\ 0, & \text{otherwise} \end{cases} \qquad (5.133)$$

If we now consider the transitions between two successive periods, we note from Figure 5.30 that an inventory level of y, $s < y < S$, is attained at the beginning of a period if the previous period starts with an inventory level of S and there is a demand depletion in the amount $\xi = S - y$, or if the previous period starts with an inventory level of x, $s < x < S$, and there is a depletion of demand in the amount of $\xi = x - y$. Thus, the probability that the inventory level Y at the beginning of a period will be in the interval $y < Y \leq y + dy$, $s < y < S$, is equal to the sum of the following: 1) the probability M that an order has been placed in the previous period thus raising the inventory level to S, times the probability that a demand D in the interval $S - y - dy < D \leq S - y$ has occurred; 2) the additive contribution of the probabilities that the inventory level is x, $s < x < S$, in the previous period, times the probability that a demand D in the interval $x - y - dy < D \leq$

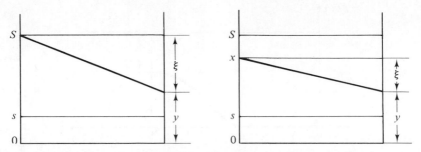

Figure 5.31. Distribution function $\Psi(y)$ of inventory level at beginning of periods following a decision in a periodic review system.

$x - y$ has occurred. Remembering that demand during a period has probability density function $\varphi(\cdot)$, we may write

$$\psi(y)\, dy = M\varphi(S - y)\, dy + \int_{x=0}^{x=y} \psi(x)\varphi(x - y)\, dx\, dy$$

or
$$\psi(y) = M\varphi(S - y) + \int_0^y \psi(x)\varphi(x - y)\, dx, \quad s < y < S \qquad (5.134)$$

In equation (5.134), let $u = S - x$, $v = S - y$; then we obtain

$$\psi(S - v) = M\varphi(v) + \int_0^{S-y} \psi(S - u)\varphi(v - u)\, du \qquad (5.135)$$

If we now set $Q = S - s$, and let

$$f(v) = \psi(S - v), \quad 0 < v < Q \qquad (5.136)$$

expression (5.135) may be written as

$$f(v) = M\varphi(v) + \int_0^v f(u)\varphi(v - u)\, du, \quad 0 < v < Q \qquad (5.137)$$

Equation (5.137) is the integral equation for the renewal density (see 2.8-ii) whose solution is

$$f(v) = M \sum_{n=1}^{\infty} \varphi *^{(n)} (v), \quad 0 < v < Q \qquad (5.138)$$

where $\varphi *^{(n)} (v)$ is the nth-fold convolution of $\varphi(\cdot)$ with itself. Let

$$\tau(v) = \sum_{n=1}^{\infty} \varphi *^{(n)} (v) \qquad (5.139)$$

Using expressions (5.138) and (5.139) in (5.136), we obtain

$$\psi(y) = M\tau(S - y), \quad s < y < S \tag{5.140}$$

To determine the constant M, we use, from (5.133), the normalizing condition

$$\int_{-\infty}^{\infty} d\Psi(y) = 1$$

or

$$M + \int_{s}^{S} \psi(y)\, dy = 1$$

or, using (5.140),

$$M + \int_{s}^{S} M\tau(S - y) = 1 \tag{5.141}$$

Let $S - y = v$ in relation (5.141); then we obtain

$$M + M \int_{0}^{Q} \tau(v)\, dv = 1 \tag{5.142}$$

Finally, if we define

$$\Delta(Q) = \int_{0}^{Q} \tau(v)\, dv \tag{5.143}$$

then, using expression (5.143) in (5.142), we obtain

$$M = \frac{1}{1 + \Delta(Q)} \tag{5.144}$$

It is to be noted that M is also the expected number of orders placed during a given period.

(2) Objective Function and Optimal Decision Rules

We select as our objective function the minimization of the total expected cost per period consisting of the sum of the expected procurement cost and the expected holding and shortage costs.

Since backlog is allowed, all demands are ultimately satisfied; therefore, the average quantity procured per period equals the expected demand per period $\bar{D} = E[D]$. Thus the expected variable procurement cost per period is $c\bar{D}$. Also, the expected fixed procurement cost per period equals K times the expected number of orders placed during a given period, or KM. The expected procurement cost per period is thus

$$c\bar{D} + KM = c\bar{D} + \frac{K}{1 + \Delta(Q)} \tag{5.145}$$

The reader will now recall that $L(y)$ was defined as the expected holding and shortage costs incurred over a period, given that the inventory level at the beginning of the period is y. Since the steady-state distribution function of the inventory level at the beginning of a period following a decision is $\Psi(y)$, the unconditional total expected holding and shortage costs per period is, using (5.133),

$$\int_{-\infty}^{\infty} L(y) \, d\Psi(y) = ML(S) + \int_{s}^{S} L(y)\psi(y) \, dy \qquad (5.146)$$

Let $S - y = v$ in the right-hand integral of expression (5.146); then

$$\int_{-\infty}^{\infty} L(y) \, d\Psi(y) = ML(S) + \int_{0}^{Q} L(S - v)\psi(S - v) \, dv \qquad (5.147)$$

Now using expressions (5.140) and (5.144) in (5.147), we obtain

$$\int_{-\infty}^{\infty} L(y) \, d\Psi(y) = \frac{L(S) + \int_{0}^{Q} L(S - v)\tau(v) \, dv}{1 + \Delta(Q)} \qquad (5.148)$$

Therefore, the total expected cost per period expressed as a function of the two decision variables Q and S is the sum total of expressions (5.145) and (5.148), or

$$F(Q, S) = c\bar{D} + KM + \int_{-\infty}^{\infty} L(y) \, d\Psi(y)$$

$$= c\bar{D} + \frac{K + L(S) + \int_{0}^{Q} L(S - v)\tau(v) \, dv}{1 + \Delta(Q)} \qquad (5.149)$$

The optimum value of Q and S may be obtained by differentiating partially $F(Q, S)$ with respect to each of Q and S and setting the result equal to zero.

Example 5.5

Suppose that the demand per period has the negative exponential distribution

$$\varphi(\xi) = \lambda e^{-\lambda \xi}, \quad \lambda > 0, 0 < \xi < \infty$$

Furthermore, suppose that holding and shortage costs are linear and are measured at the end of a period; thus the function $L(y)$ has the form of (5.113).

The nth-fold convolution of the negative exponential distribution being the gamma distribution of order n, we have

$$\varphi *^{(n)}(\xi) = \frac{(\lambda \xi)^{n-1}}{(n - 1)!} \lambda e^{-\lambda \xi}, \quad n = 1, 2, 3, \ldots$$

Hence from expression (5.139) we obtain

$$\tau(v) = \sum_{n=1}^{\infty} \frac{(\lambda v)^{n-1}}{(n-1)!} \lambda e^{-\lambda v}$$

$$= \lambda e^{-\lambda v} \sum_{i=0}^{\infty} \frac{(\lambda v)^i}{i!}, \quad 0 < v < Q$$

The infinite series can be recognized as the exponential function $e^{\lambda v}$. Hence we have

$$\tau(v) = \lambda e^{-\lambda v} \cdot e^{\lambda v}$$

$$= \lambda, \quad 0 < v < Q \tag{5.150}$$

The expression for $\Delta(Q)$ may then be obtained by substituting (5.150) in (5.143):

$$\Delta(Q) = \int_0^Q \lambda \, dv = \lambda Q \tag{5.151}$$

From (5.113), the expression for $L(y)$ is

$$L(y) = (h + p) \int_0^y \Phi(\xi) \, d\xi + p(\bar{D} - y)$$

$$= (h + p) \int_0^y (1 - e^{-\lambda \xi}) \, d\xi + p\left(\frac{1}{\lambda} - y\right) \tag{5.152}$$

$$= (h + p)\left(y + \frac{1}{\lambda} e^{-\lambda y} - \frac{1}{\lambda}\right) + p\left(\frac{1}{\lambda} - y\right)$$

$$= h\left(y + \frac{1}{\lambda} e^{-\lambda y} - \frac{1}{\lambda}\right) + \frac{p}{\lambda} e^{-\lambda y}$$

Substituting (5.150), (5.151), and (5.152) in the expression for the cost function (5.149), we obtain

$$F(Q, S) = c\bar{D} + \frac{1}{1 + \lambda Q}\left\{K + h\left(S + \frac{1}{\lambda} e^{-\lambda S} - \frac{1}{\lambda}\right) + \frac{p}{\lambda} e^{-\lambda S}\right.$$

$$+ \lambda \int_0^Q \left[h(S - v) + \frac{h}{\lambda} e^{-\lambda(S-v)} - \frac{h}{\lambda} + \frac{p}{\lambda} e^{-\lambda(S-v)}\right] dv \bigg\}$$

$$= c\bar{D} + \frac{1}{1 + \lambda Q}\left\{K + h\left(S - \frac{1}{\lambda}\right) + \frac{h+p}{\lambda} e^{-\lambda S}\right.$$

$$+ \lambda\left[h\left(SQ - \frac{Q^2}{2}\right) - hQ + \frac{h+p}{\lambda} e^{-\lambda S} \cdot \frac{1}{\lambda}(e^{\lambda Q} - 1)\right]\bigg\}$$

$$= c\bar{D} + \frac{1}{1 + \lambda Q}\left[K + h\left(S - \frac{1}{\lambda} - Q + \lambda SQ - \frac{\lambda Q^2}{2}\right)\right.$$

$$+ \frac{h+p}{\lambda} e^{-\lambda(S-Q)}\bigg]$$

If we now set $S = s + Q$, we get the following function in s and Q:

$$\tilde{F}(s, Q) = c\bar{D} + \frac{1}{1 + \lambda Q}\left[K + h\left(s - \frac{1}{\lambda} + \lambda sQ + \frac{\lambda Q^2}{2}\right) + \frac{h + p}{\lambda}e^{-\lambda s}\right]$$

(5.153)

Differentiating $\tilde{F}(s, Q)$ partially with respect to s and setting the result equal to zero yields

$$\frac{\partial \tilde{F}}{\partial s} = 0 = \frac{1}{1 + \lambda Q}[h(1 + \lambda Q) - (h + p)e^{-\lambda s}]$$

Hence

$$e^{-\lambda s} = \frac{h(1 + \lambda Q)}{h + p} \qquad (5.154)$$

Differentiating $\tilde{F}(s, Q)$ partially with respect to Q and setting the result equal to zero yields

$$\frac{\partial \tilde{F}}{\partial Q} = 0$$

$$= \frac{(1 + \lambda Q)h(\lambda s + \lambda Q) - \lambda\left[K + h\left(s - \frac{1}{\lambda} + \lambda sQ + \frac{\lambda Q^2}{2}\right) + \frac{h + p}{\lambda}e^{-\lambda s}\right]}{(1 + \lambda Q)^2}$$

Using expression (5.154) and simplifying we obtain as the optimal value of Q,

$$Q^* = \sqrt{\frac{2K}{\lambda h}} \qquad (5.155)$$

The optimal value of s^* is then obtained by substituting formula (5.155) in equation (5.154).

SELECTED REFERENCES

[1] ACKOFF, R. L. and M. W. SASIENI, *Fundamentals of Operations Research*, John Wiley & Sons, Inc., New York, 1968.

[2] ARROW, K. J., T. HARRIS, and J. MARSCHAK, "Optimal Inventory Policy," *Econometrica*, XIX, pp. 250–272, 1951.

[3] ARROW, K. J., S. KARLIN, and H. SCARF, *Studies in the Mathematical Theory of Inventory and Production*, Stanford University Press, Stanford, California, 1958.

[4] ARROW, K. J., S. KARLIN, and H. SCARF (Eds.), *Studies in Applied Probability and Management Science*, Stanford University Press, Stanford, California, 1962.

[5] BELLMAN, R., *Dynamic Programming*, Princeton University Press, Princeton, New Jersey, 1957.

[6] BELLMAN, R., I. GLICKSBERG, and O. GROSS, "On the Optimal Inventory Equation," *Management Science*, Vol. 2, pp. 83–104, 1955.

[7] CHURCHMAN, C. W., R. L. ACKOFF, and E. L. ARNOFF, *Introduction to Operations Research*, John Wiley & Sons, Inc., New York, 1957.

[8] DVORETSKY, A., KIEFER, and J. WOLFOWITZ, "The Inventory Problem: I, Case of Known Distribution of Demand; II, Case of Unknown Distribution of Demand," *Econometrica*, XX, pp. 187–222 and 450–466, 1952.

[9] GALLIHER, H. P., *Production Scheduling*, Chapter 10 in *Notes on Operations Research 1959*, The Technology Press, MIT, Cambridge, Massachusetts, 1959.

[10] HADLEY, G., and T. M. WHITIN, *Analysis of Inventory Systems*, Prentice-Hall, Inc., Englewood Cliffs, New Jersey, 1963.

[11] HARRIS, F., *Operations and Cost* (Factory Management Series), A. W. Shaw Co., Chicago, pp. 48–52, 1915.

[12] LESOURNE, J., *Technique Economique et Gestion Industrielle*, Dunod, Paris, 1958.

[13] MASSÉ, P., *Les Réserves et la Régulation de l'Avenir dans la Vie Economique*, 2 vols., Herman & Cie, Paris, 1946.

[14] MASSÉ, P., *Optimal Investment Decisions*, Prentice-Hall, Inc., Englewood Cliffs, New Jersey, 1958.

[15] RAYMOND, F. E., *Quantity and Economy in Manufacture*, McGraw-Hill Book Company, New York, 1931.

[16] RUBALSKIY, G. B. "On the Level of Supplies in a Warehouse with a Lag in Procurement", *Engineering Cybernetics*, Vol. 10, No. 1, pp. 52–57, 1972.

[17] RUBALSKIY, G. B. "Calculation of Optimum Parameters in an Inventory Control Problem", *Engineering Cybernetics*, Vol. 10, No. 2, pp. 182–187, 1972.

[18] SASIENI, M., A. YASPAN, and L. FRIEDMAN, *Operations Research, Methods and Problems*, John Wiley & Sons, Inc., New York, 1959.

[19] SCARF, H. E., D. M. GILFORD, and M. W. SHELLY (Eds.), *Multistage Inventory Models and Techniques*, Stanford University Press, Stanford, California, 1963.

[20] SIVAZLIAN, B. D., "Dimensional and Computational Analysis in Stationary (s, S) Inventory Problems with Gamma Distributed Demand", *Management Science*, Vol. 17, No. 6, pp. B-307–B-311, 1971.

[21] SIVAZLIAN, B. D., "A Continuous Review (s, S) Inventory System with Arbitrary Interarrival Distribution between Unit Demand" *Operations Research*, Vol. 22, No. 1, pp. 65–71, 1974.

[22] WAGNER, H. M., and T. M. WHITIN, "Dynamic Version of the Economic Lot Size Model," *Management Science*, Vol. 5, No. 1, pp. 89–91, 1958.

[23] WHITIN, T. M., *The Theory of Inventory Management*, Princeton University Press, Princeton, New Jersey, 1953.

PROBLEMS

1. A company uses a liquid volatile chemical stored in large cylindrical tanks of length L feet and diameter a feet. Consumption is at the rate of D cubic feet per day. Losses from evaporation are proportional to the free liquid surface and occur at the rate of α cubic feet per square foot of liquid. Determine how often to order the tanks if order cost is negligible and no shortage is allowed. Consider the two cases where the tank stands horizontally and vertically.

2. The Acme Manufacturing Co. produces a certain type of a small appliance part at a constant rate of 5000 units per month. Production cost per unit is estimated at $2. These units are delivered to the manufacturer of the small appliances who is subcontracting from Acme Manufacturing Co. the production of the parts. The delivery of these units is done by a single truck owned and operated by Acme. The truck has a total shipping capacity of 2500 units, and it costs $50 to make a shipment, irrespective of the quantity shipped. Inventory carrying charges are estimated at .20 dollar/[($)(year)]. Determine how often shipments should be made to the manufacturer of small appliances.

3. Show that the results of the basic model are not altered if initially the stock level is $x, 0 < x \leq S$.

4. Suppose that the demand for an item is deterministic, whose demand rate as a function of time is $d(t) = 2t$. Let T be the horizon period, K the fixed procurement cost, and h the unit holding cost. Assume that the objective function is to minimize the undiscounted total cost function. What is the optimum replenishment policy?

5. In a periodic review system, obtain an expression for $L(y)$, the conditional expected holding and shortage costs per period, given that the inventory level at the beginning of the period is y, for the following cases, and test for convexity of $L(y), y \geq 0$.
   (a) Holding cost is linear and is measured at the beginning of the period; shortage cost is linear and is measured at the end of the period.
   (b) Holding and shortage costs are quadratic and are measured at the end of the period.
   (c) Holding cost is linear and is measured at the beginning of the period; shortage cost is quadratic and is measured at the end of the period.

6. For the one-period problem, show how to determine the values of S and s when $L(y)$ takes on the various analytic expressions of Problem 5.

7. For the one-period problem with probabilistic demand, assume that the demand has the uniform distribution

$$\varphi(\xi) = \frac{1}{a}, \quad 0 < \xi < a, a > 0$$

Under the assumptions of a fixed setup cost, a linear variable procurement cost, and an expected holding and shortage cost $L(y)$, having the form (5.113), determine·the optimum values s^* and S^*.

8. A candy store must decide how many large chocolate eggs to order for the Easter season. The orders must be placed 6 weeks in advance and there is no possibility of placing a reorder. Each egg costs the store $3 and sells for $6, and any egg not sold at the end of the Easter season is marked down at $3. The store estimates that it can sell at least 50 eggs, but no more than 250 eggs, and that it is equally likely to sell any number of eggs between these two numbers.

   Determine how many chocolate eggs the store has to order to maximize its expected profit. What is this maximum expected profit? What would the profit have been if the store had ordered the average number of 150 eggs?

9. The weekly demand for a product is a random variable D with probability mass function $P\{D = r\} = \frac{1}{4}, r = 0, 1, 2, 3$.

   Demands in successive weeks are independently and identically distributed. At the beginning of every week the stock on hand is measured, and if its value is zero, three units are ordered. Any shortages are lost sales, and delivery of orders is immediate. Assume that the system starts with no units in stock. Let $p(n, t)$ be the probability that following a decision at the beginning of the week t there are n units in stock, $n = 1, 2, 3$.
   (a) Find $p(n, t)$.
   (b) Show that as $t \to \infty$ a steady-state condition is reached. Calculate $p(n) = \lim_{t \to \infty} p(n, t)$. What is $p(n)$?
   (c) Under steady-state conditions find the probability of an order being placed in any given period. Calculate the expected stock on hand and the expected shortage at the end of a period.

10. In a periodic review system the demand per period is a discrete random variable identically and independently distributed from period to period. The random variable takes the values of 0, 1, or 2 with probabilities of p_0, p_1, and p_2, respectively, $p_0 + p_1 + p_2 = 1$. Unsatisfied demand is *lost sales*. Orders placed at the beginning of the period are delivered *two* periods in the future. The following rule for replenishing stock is used: at the end·of a period, the stock on hand (including new deliveries if any) is reviewed, and if its value is zero, an order for two units is placed; otherwise no order is placed.
    (a) Determine the steady-state probability of placing an order in a given period.
    (b) If $p_0 = p_1 = p_2$, find the expected stock on hand at the end of a period for the steady-state problem.

11. A subcontractor must produce and deliver on the first of each month for a period of one year, the following quantities of a particular item: 30, 30, 45, 45, 50, 50, 60, 60, 45, 45, 30, 30. The item is produced in lots and the setup cost is $150. Each item costs $10 to produce, and the inventory carrying charge is .20 dollars/[(dollar)(year)]. Determine an optimal production program.

INVENTORY THEORY
FOR MULTICOMMODITY
MULTIINSTALLATION SYSTEMS

6.1. Introduction

Whereas in the previous chapter we have studied inventory systems consisting of a single commodity located at a single installation, many inventory systems are of the multicommodity multiinstallation type. That is, the system involves several distinct classes of commodities which may be stored at different locations. This may be the case of a manufacturing concern producing several types of products and shipping such products to a number of regional warehouses. Another interesting case emanates, for example, in the production of several commodities using a single facility consisting of two stages. Following the processing at the first stage, an in-process inventory of the semifinished goods is built up to supply the second stage, which ultimately results in an inventory of finished goods.

The theoretical results derived from an analysis of a single-commodity single-installation system are valid in multicommodity multiinstallation systems where each commodity at each installation can be treated as a separate entity; that is, when the complex system is decomposable and treated as separate and independent single product systems. Unfortunately, this type of independency leading to a problem decomposition is not very common in

6

practical situations, and it becomes necessary to incorporate in the model at least some of the significant interacting factors that most naturally arise in this type of problem.

One possible factor that may induce interaction in a multicommodity multiinstallation system is the imposition of one or more constraints. Thus, at a particular installation, several products may be competing for a limited storage space. Or a fixed budgetary allocation may severely constrain the aggregate level of capital tied up in inventory for the entire system. Often, the mathematical formulation of this class of problems consists in the setting up of a cost objective function made up of terms, each term corresponding to the cost contribution of a particular product at a particular location as if treated independently, subject to one or more constraints. Various optimization techniques can then be used to solve the problem. The purpose of this chapter is not to study this particular class of problems, but rather to extend the analysis of the single-commodity single-installation system to a situation where interdependency is introduced through another important factor, that is, the procurement or production of the commodities, and, more specifically, in the setup or order cost. Such study is intended to fulfill two purposes. First, it may be shown that substantial economic advantages are derived by studying inventory systems at the aggregate level. Second, the study of multicommodity multiinstallation systems may be used to establish structural equivalences between various classes of complex inventory distribution systems.

As may be expected, the study of such problems is invariably more complex and involves the manipulation of a large number of decision variables. However, once insight is gained in the problem structure, the level of abstraction necessary to formulate the mathematical models is considerably simplified.

The present chapter is broadly divided into two sections. The first section deals with continuous review inventory systems with deterministic demands. In the second section, the analog of the one-period model in single-commodity single-installation systems with stochastic demand is treated when more than one commodity is involved. The reader should consult references [1] to [8] for additional reading material on multicommodity multiinstallation inventory systems.

i. *Classification of the Multicommodity Multiinstallation Systems*

The three simplest structured multicommodity multiinstallation inventory systems are

(1) The *multicommodity single-installation system* (Figure 6.1a). In this system, each product at the installation is procured from a single exogenous source, and there is a market demand for each product.

(2) The *single-commodity multiinstallation parallel system* (Figure 6.1b). Here, a single exogenous source supplies several installations situated in different geographical location with a single product. There is a market demand for the product at each installation, and, in the simplest case transshipments between installations are not permitted.

(3) The *single-commodity multiinstallation serial system* (Figure 6.1c). The system composed of installation in series has a cascade configuration in which the installation at the highest level or the source supplies the product to the installation at the next to highest level and so on, each of the remaining installations receiving its product from the preceding installation of the system. There is a market demand for the product at the lowest installation. A typical example of such a configuration would be a production system involving the manufacturing of a product in stages with in-process inventories between stages.

In general in a multiinstallation system the various stocking points are usually so arranged as to form an echelon structure. A typical multiechelon inventory system consisting of three levels or echelons is shown in Figure 6.2. In this particular system, the highest echelon (level 1) is the source point that

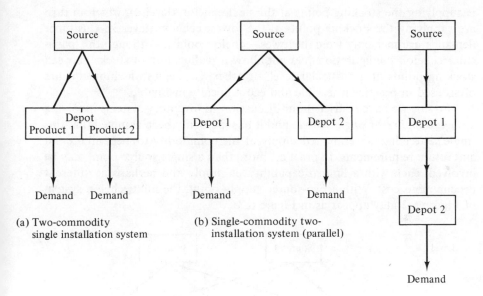

Figure 6.1. Examples of multicommodity multiinstallation inventory systems.

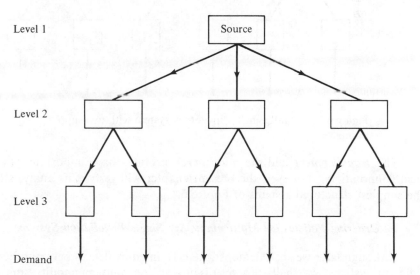

Figure 6.2. A multiechelon inventory system with a single supply point.

is supplying the stocking points at the next echelon (level 2), which in turn are supplying the stocking points at the lowest echelon (level 3). Customer demands are met only from the lowest echelon points. Although this particular echelon configuration does not allow a reallocation of stocks between stocking points at a particular level, or otherwise, such transshipments are often used in practice if feasible and economically justifiable.

So far, we have restricted our discussion to inventory systems involving a *single supplier* or *single source*, and it has implicitly been assumed that this single source had an unlimited supply of the commodity to meet all present and future requirements. In practice, more than a single source point may be involved, each with a limited capability of supply and perhaps at different procurement cost. With a two-source supply point, the multiechelon system of Figure 6.2 may appear as in Figure 6.3.

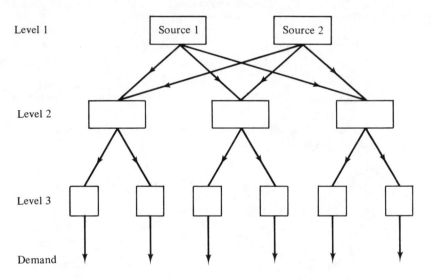

Figure 6.3. A multiechelon inventory system with two supply points.

The *transshipment* and the *multisource* problem have important practical connotations; however, the present chapter will restrict its analysis to the simplest structured systems of Figure 6.1.

ii. *Ordering Policies for Multicommodity Single-Installation Systems*

Although later we shall demonstrate certain equivalences between these different systems, we shall start by discussing the multicommodity single-installation system.

For the multicommodity system there are three types of ordering policies:

(1) Individual ordering policy.
(2) Joint ordering policy.
(3) Mixed ordering policy.

For an *individual ordering policy* each product is treated as if the other products did not exist. Thus the "when" and "how much" to order of each product is determined without consideration of possible cost saving by simultaneous ordering of several products. For a *joint ordering policy* all products are ordered simultaneously when the ordering of at least one product is required. A *mixed ordering policy* allows for ordering all the products or some of the products.

The setup costs that are the fixed portion of procurement costs greatly affect the optimal ordering policy. If there is no setup cost and the holding and shortage cost functions are separable by products, then an individual ordering policy will always be optimal. If there are setup costs, then, intuitively, there may be ordering policies that allow for ordering several products simultaneously, thus obtaining a saving by having one setup cost that is less than the sum of the individual setup costs.

The setup costs associated with each ordering strategy of a mixed ordering policy may be *invariant* or *variant*. Invariant setup costs have the same value whether ordering some or all the products simultaneously, whereas variant setup costs may have different values, depending on the product or products ordered.

When decision for replenishing stock is to be made using a mixed ordering policy, one has to decide which commodities to order, if any, and the quantity of the particular commodity. For example, there are four possible ordering strategies for a two-commodity inventory system available when making such a decision; these are

(1) Order commodity 1, but not commodity 2.
(2) Order commodity 2, but not commodity 1.
(3) Order both commodities.
(4) Do not order anything.

To each of the above ordering decisions there corresponds a fixed cost. Thus, if $k(z_1, z_2)$ represents the setup cost for ordering z_1 units of commodity 1 and z_2 units of commodity 2, then, in general

$$k(z_1, z_2) = \begin{cases} K_1, & \text{if } z_1 > 0, z_2 = 0 \\ K_2, & \text{if } z_1 = 0, z_2 > 0 \\ K, & \text{if } z_1 > 0, z_2 > 0 \\ 0, & \text{if } z_1 = 0, z_2 = 0 \end{cases}$$

For example, a fixed cost K_2 is incurred if commodity 2 is ordered but not commodity 1 ($z_1 = 0$, $z_2 > 0$). The particular situation will dictate the relative magnitudes of K_1, K_2, and K, and, in general, the following inequalities must be satisfied

$$\max(K_1, K_2) \leq K \leq K_1 + K_2$$

In words, this states that if a joint order for the two commodities is initiated the fixed order cost, K, incurred cannot exceed the sum of the fixed cost K_1 and K_2 associated with the initiation of separate individual orders; in addition, the fixed joint order cost K should at least be equal to the greater of the individual fixed order costs. In particular, under invariant setup costs, we have $K = K_1 = K_2$.

Although some economic advantage is derived by using a joint order policy, through a reduction in the fixed cost, it is not at all true that a joint order policy is always optimal. As may be expected, the cost of operating an inventory system under a particular policy will also depend on such factors as demand for the commodities and their holding and shortage costs. It is the purpose of this chapter to analyze and compare the alternative policies that may be used in the management of multicommodity multiinstallation inventory systems, and to arrive at optimal decision rules.

As in the single-commodity single-installation case, we assume that there exists a variable procurement cost which is proportional to the quantity procured for each commodity. Thus, in the two-commodity case, if c_1 and c_2 are the unit procurement costs for commodities 1 and 2, respectively, the total variable procurement cost will be of the form

$$c_1 z_1 + c_2 z_2, \quad z_1 \geq 0; z_2 \geq 0$$

In general, c_i will be used to denote the unit variable procurement cost for commodity i in an m-commodity system, $i = 1, 2, \ldots, m$. Again, there will in general be a holding cost which for commodity i will be h_i dollars/[(unit) (unit time)]. If in addition there is shortage, the associated cost for commodity i will be denoted by p_i dollars/[(unit)(unit time)].

iii. *Some Equivalence Problems*

If in Figure 6.1a representing a two-commodity single-installation system, each commodity at the installation is treated as a separate installation, one recovers the system of Figure 6.1b, that is, a single-commodity two-installation system in parallel. This equivalence can obviously be extended to an arbitrary number of commodities stocked at an arbitrary number of parallel installations. Thus, a system consisting of m_1 commodities inventoried at m_2 parallel installations may be viewed as a system consisting of $m = m_1 m_2$ distinct installations in a parallel system. The reader should satisfy himself that each of the following problems describing a different situation is equivalent to the other:

(1) m different commodities are purchased from a common vendor, and it is possible to order all these commodities simultaneously through a single purchase order, thus experiencing one order cost.

(2) A single factory supplies an arbitrary number m of major depots, which are used as storage and distribution points. A given commodity is stored at these various depots. Whenever a depot orders from the factory to replenish its stock, a fixed production setup cost is incurred whether the depot orders individually or simultaneously with the other depots.

(3) A production center can manufacture m different commodities using a unique production setup that remains invariant. The commodities are stored at a single depot for distribution.

Later in the chapter when treating the mixed ordering policy, it will be shown that the single-commodity serial two-installation system of Figure 6.1(c) is equivalent to a two-commodity single installation system. This equivalence can be generalized to a serial multiinstallation system. Thus, without any loss in generality, we shall be analyzing throughout this chapter the multicommodity single-installation system.

6.2. Deterministic Models—The Two-Commodity Problem

We start our analysis by considering the simplest inventory situation involving two commodities. The assumptions underlying this system are identical to the assumptions of the basic Wilson model. No shortages are allowed, and it may be possible to order both commodities individually or simultaneously. Three possible alternatives in the management of these two commodities are available. In the first alternative, each commodity is managed independently of the other commodity, thus giving rise to the individual ordering policy. In the second alternative, the possibility of order interaction between commodities is accounted for, and if an order is initiated, both commodities are simultaneously procured; this forms the basis of the joint ordering policy. Finally, in the third alternative, ordering of commodities may be executed either individually or jointly, resulting in the use of the mixed ordering policy.

i. *Joint Ordering Policy with No Shortages*

a. ALTERNATIVE DERIVATION OF THE WILSON FORMULA

For the single-commodity system discussed in the basic model in Chapter 5, define the following:

$$D = \text{demand in units/unit time}$$
$$h = \text{holding cost in dollars/[(unit) (unit time)]}$$
$$K = \text{fixed setup or order cost in dollars}$$
$$c = \text{variable procurement cost in dollars/unit}$$
$$t = \text{time interval between orders in unit time}$$
$$Q = \text{order quantity in units}$$
$$(TC) = \text{total cost per unit time}$$

Referring to Figure 6.4, we have

$$(TC) = cD + \frac{K}{t} + \frac{1}{2}hDt \tag{6.1}$$

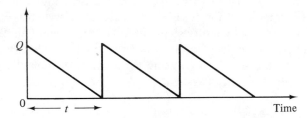

Figure 6.4. Stock level variation in a single-commodity inventory system.

To determine the optimum value of t, we form $d(TC)/dt$ and set the result equal to zero; this yields

$$\frac{d(TC)}{dt} = 0 = -\frac{K}{t^2} + \frac{h}{2}D$$

The optimum value of t, t^*, is then

$$t^* = \sqrt{\frac{2K}{hD}} \tag{6.2}$$

and the optimum value of Q, Q^*, is

$$Q^* = Dt^* = \sqrt{\frac{2DK}{h}} \tag{6.3}$$

The minimum value of (TC), say $(TC)^*$, is

$$(TC)^* = cD + \sqrt{2hKD} \tag{6.4}$$

b. INDIVIDUAL ORDERING FOR A TWO-COMMODITY SYSTEM

In a two-commodity system with no order interaction, each commodity is dealt with as if the other commodity did not exist, thus leading to an individual ordering policy. If we introduce the same symbols as for the single-commodity case, and use the subscript 1 and 2 to denote, respectively commodity 1 and commodity 2, we obtain, for $i = 1, 2$, the following:

D_i = demand in units/unit time
h_i = holding cost in dollars/[(unit)(unit time)]
K_i = fixed setup or order cost in dollars
c_i = variable procurement cost in dollars/unit
t_i = time interval between orders in unit time
Q_i = order quantity in units
$(TC)_I$ = total cost per unit time when following an individual ordering policy

The objective function can be written as

$$(TC)_I = \sum_{i=1}^{i=2} \left(c_i D_i + \frac{K_i}{t_i} + \frac{h_i}{2} D_i t_i \right) \tag{6.5}$$

This last expression is a decomposable function of the two decision variables t_1 and t_2 such that each of the two functions is of the same form as expression (6.1). We can immediately write down for the optimum values of t_i and Q_i, $i = 1, 2$,

$$t_i^* = \sqrt{\frac{2K_i}{h_i D_i}} \tag{6.6}$$

and

$$Q_i^* = \sqrt{\frac{2D_i K_i}{h_i}} \tag{6.7}$$

The minimum value of $(TC)_I$ is then given by

$$(TC)_I^* = \sum_{i=1}^{i=2} (c_i D_i + \sqrt{2h_i K_i D_i}) \tag{6.8}$$

c. JOINT ORDERING FOR A TWO-COMMODITY SYSTEM

When following a joint order policy, the ordering times of each commodity are synchronized in such a way that simultaneous ordering of the two commodities is possible. This is achieved by setting the time interval between orders for one commodity to be equal to the time interval between orders for the other commodity (see Figure 6.5). For commodity i, $i = 1, 2$, we shall make use of the previous symbols and let

K = fixed setup or order cost associated with a joint order expressed in dollars
t = common time interval between orders in unit times ($t_1 = t_2 = t$)
$(TC)_J$ = total cost per unit time when following a joint ordering policy

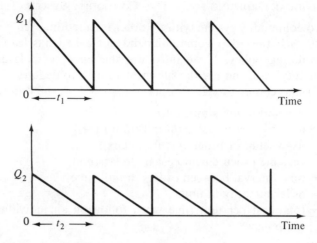

Figure 6.5. Stock level variations in a two-commodity inventory system with joint order ($t_1 = t_2$).

The expression for the objective function is

$$(TC)_J = (c_1 D_1 + c_2 D_2) + \frac{K}{t} + \frac{h_1}{2} D_1 t + \frac{h_2}{2} D_2 t$$
$$= \sum_{i=1}^{i=2} (c_i D_i) + \frac{K}{t} + \frac{1}{2}\left(\sum_{i=1}^{i=2} h_i D_i\right)t \tag{6.9}$$

Expression (6.1) would be identical to expression (6.9) if cD were replaced by $\sum_{i=1}^{i=2} c_i D_i$ and hD were replaced by $\sum_{i=1}^{i=2} h_i D_i$. Hence we can immediately write down the expressions for the optimum value of t by simply replacing in (6.2) the quantity hD by the quantity $\sum_{i=1}^{i=2} h_i D_i$; this yields

$$t^* = \sqrt{\frac{2K}{\sum_{i=1}^{i=2} h_i D_i}} \tag{6.10}$$

The optimal values of Q_1 and Q_2 are obtained by using the relations $Q_1 = D_1 t$ and $Q_2 = D_2 t$, respectively; thus

$$Q_i^* = \sqrt{\frac{2KD_i^2}{\sum_{i=1}^{i=2} h_i D_i}}, \quad i = 1, 2 \tag{6.11}$$

The minimum value of $(TC)_J$ can be derived from expression (6.4) by the appropriate substitution for hD:

$$(TC)_J^* = \sum_{i=1}^{i=2} (c_i D_i) + \sqrt{2K \sum_{i=1}^{i=2} h_i D_i} \tag{6.12}$$

d. Discussions and Comparison of the Two Ordering Policies

In discussing the results of the basic model for the single-commodity inventory system, we have seen that, at optimality, the holding cost per unit time equals the fixed cost per unit time. This same property can be verified to hold true in the case of the two-commodity system, irrespective of the ordering policy used, provided that one compares the lumped sum of the two holding costs per unit time with the setup cost(s) per unit time.

To determine the conditions under which a particular policy should be selected, it is necessary to compare the minimum cost per unit time for each policy, as given by expressions (6.8) and (6.12), and select the one with the least value. Thus, joint ordering policy is preferred to an individual ordering policy if and only if

$$(TC)_J^* < (TC)_I^*$$

or

$$c_1 D_1 + c_2 D_2 + \sqrt{2K(h_1 D_1 + h_2 D_2)} < c_1 D_1 + c_2 D_2 + \sqrt{2K_1 h_1 D_1} + \sqrt{2K_2 h_2 D_2}$$

or

$$\sqrt{2K(h_1 D_1 + h_2 D_2)} < \sqrt{2K_1 h_1 D_1} + \sqrt{2K_2 h_2 D_2}$$

$$(6.13)$$

Dividing both sides of (6.13) by $\sqrt{2K(h_1 D_1 + h_2 D_2)}$, we obtain

$$1 < \sqrt{\frac{K_1}{K} \frac{h_1 D_1}{h_1 D_1 + h_2 D_2}} + \sqrt{\frac{K_2}{K} \frac{h_2 D_2}{h_1 D_1 + h_2 D_2}} \qquad (6.14)$$

Define the dimensionless parameters λ_1 and λ_2 as follows:

$$\lambda_1 = \sqrt{\frac{K_1}{K} \frac{h_1 D_1}{h_1 D_1 + h_2 D_2}}, \qquad \lambda_2 = \sqrt{\frac{K_2}{K} \frac{h_2 D_2}{h_1 D_1 + h_2 D_2}}$$

It is easily seen that $0 < \lambda_1, \lambda_2 < 1$. Relation (6.14) becomes

$$1 < \lambda_1 + \lambda_2$$

This of course leads us to a geometric interpretation for selecting the proper policy. Each of λ_1 and λ_2 is a function of the known input parameters to the problem. The pair (λ_1, λ_2) can be represented by a point P in the square region $\{\lambda_1, \lambda_2 : 0 < \lambda_1 < 1, 0 < \lambda_2 < 1\}$ (Figure 6.6). The relation $\lambda_1 + \lambda_2 = 1$ represents the equation of the diagonal line AB. Hence, if $1 < \lambda_1 + \lambda_2$, the point P lies inside the triangle OAB and it would be best to select the individual ordering policy. If $1 > \lambda_1 + \lambda_2$, the point P would lie within the triangle ABC and joint ordering policy would be preferred. If the point P lies on the line AB, either of the policies could be used.

We now show that under invariant setup costs, that is, when $K_1 = K_2 = K$, the joint ordering policy is always preferred. Squaring both sides of ex-

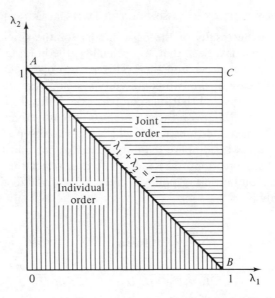

Figure 6.6. Decision regions for selecting individual ordering policy or joint ordering policy.

pression (6.13) and then dividing by $2K(h_1 D_1 + h_2 D_2)$, we obtain

$$1 < \frac{K_1 h_1 D_1 + K_2 h_2 D_2}{K(h_1 D_1 + h_2 D_2)} + \frac{2\sqrt{K_1 K_2 h_1 D_1 h_2 D_2}}{K(h_1 D_1 + h_2 D_2)}$$

When $K_1 = K_2 = K$, this inequality becomes

$$1 < 1 + 2\frac{\sqrt{h_1 D_1 h_2 D_2}}{h_1 D_1 + h_2 D_2}$$

a relation that is always satisfied.

Suppose that, in addition, both commodities have the same characteristics, and let $h_1 = h_2 = h$ and $D_1 = D_2 = D$. Then the cost savings per unit time associated with a joint ordering policy over an individual ordering policy is

$$2\sqrt{2}\sqrt{KhD} - 2\sqrt{KhD} = 2(\sqrt{2} - 1)\sqrt{KhD}$$

The relative savings based on the combined fixed order and holding costs of individual ordering policy is

$$\frac{2(\sqrt{2} - 1)\sqrt{KhD}}{2\sqrt{2}\sqrt{KhD}} = 1 - \frac{1}{\sqrt{2}}$$

or approximately 30%.

Example 6.1

Consider a two-commodity inventory system for which either an individual ordering policy or a joint ordering policy may be instituted. The characteristics of each commodity are as follows:

Commodity i	c_i, $/units	K_i, $	D_i, units/year	h_i, $/[(unit)(year)]$
1	10	10	2000	1
2	20	10	3000	2

The fixed cost associated with a joint order is $K = \$15$. Show that a joint ordering policy is superior to an individual ordering policy, and compute the annual operating cost distribution for each policy.

For each commodity, we compute the parameters λ_1 and λ_2 to obtain

$$\lambda_1 = \sqrt{\frac{K_2}{K} \frac{h_1 D_1}{h_1 D_1 + h_2 D_2}} = \sqrt{\frac{10}{15} \frac{(1)(2000)}{(1)(2000) + (2)(3000)}} = .412$$

$$\lambda_2 = \sqrt{\frac{K_2}{K} \frac{h_2 D_2}{h_1 D_1 + h_2 D_2}} = \sqrt{\frac{10}{15} \frac{(2)(3000)}{(1)(2000) + (2)(3000)}} = .702$$

Since $\lambda_1 + \lambda_2 = 1.114 > 1$, a joint ordering policy is more economical. We now consider each of these policies separately.

Joint Ordering Policy: The optimal time interval between orders is

$$t^* = \sqrt{\frac{2K}{h_1 D_1 + h_2 D_2}} = \sqrt{\frac{2(15)}{(1)(2000) + (2)(3000)}} = .06125 \text{ years}$$

The optimal order quantity for each commodity is

$$Q_1^* = D_1 t^* = (2000)(.06125) = 122 \text{ units}$$
$$Q_2^* = D_2 t^* = (3000)(.06125) = 184 \text{ units}$$

When following an optimal policy, the annual distribution of the fixed order cost and holding cost will be as given in Table 6.1.

Individual Ordering Policy: The optimal time interval between orders for commodities 1 and 2, respectively, is

$$t_1^* = \sqrt{\frac{2K_1}{h_1 D_1}} = \sqrt{\frac{2(10)}{(1)(2000)}} = .1000 \text{ years}$$

$$t_2^* = \sqrt{\frac{2K_2}{h_2 D_2}} = \sqrt{\frac{2(10)}{(1)(3000)}} = .0577 \text{ years}$$

Table 6.1. ANNUAL COST DISTRIBUTION FOR JOINT ORDERING POLICY

Cost/year, $	Commodity 1	Commodity 2	Total
Fixed order cost[a]	122.50	122.50	245.00
Holding cost	61.50	183.50	245.00
Total	184.00	306.00	490.00

[a]Assumes equal apportionment between products for the fixed order cost.

The respective optimal order quantities are

$$Q_1^* = D_1 t_1^* = (2000)(.1000) = 200 \text{ units}$$
$$Q_2^* = D_2 t_2^* = (3000)(.0577) = 173 \text{ units}$$

Under an optimal policy, the yearly distribution of the fixed order cost and holding cost is as given in Table 6.2.

Table 6.2. ANNUAL COST DISTRIBUTION FOR INDIVIDUAL ORDERING POLICY

Cost/year, $	Commodity 1	Commodity 2	Total
Fixed order cost	100.00	173.00	273.00
Holding cost	100.00	173.00	273.00
Total	200.00	346.00	546.00

Comparison of Tables 6.1 and 6.2 shows that the use of a joint ordering policy results in an economy of $56.00 per year. In particular, we note that the effect of a joint order is to reduce the total operating cost per year through a reduction in both the fixed order cost *and* the inventory carrying cost; thus, although the holding cost of commodity 2 shows an increase of $(183.50 − 173.00) = $10.50 per year, the holding cost of commodity 1 shows a reduction of $(100.00 − 61.50) = $38.50, with a net total reduction of $28.00 per year in holding cost. The simultaneous reduction in the yearly fixed order cost is also $28.00.

ii. *Mixed Ordering Policy with No Shortages*

In a mixed ordering policy, a combination of individual and joint orderings is used. Thus sometimes the commodities are ordered jointly, hence incurring a setup cost K, or commodity i, $i = 1, 2$, is ordered individually with an associated setup cost K_i. In its simplest form, a mixed ordering would involve the individual ordering of only one of the products (see Figure 6.7). The time at which an order for either commodity occurs is of course a time for

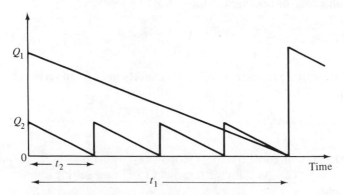

Figure 6.7. Stock level variation of a two-commodity or a parallel two-installation inventory system.

joint ordering. We shall call the product with the longer time between orders, the *mother product* or *M-product*, and the product with the shorter time between orders the *daughter product* or *D-product*.

Mixed ordering policy problems present in general some inherent difficulties and, as a consequence, result in a more complex mathematical analysis. Even in the simplest form of mixed ordering policy involving only two commodities, one is faced at the outset with the problem of identifying which is the *M*-product and which is the *D*-product. Another added difficulty relates to the larger number of decision variables appearing in the objective function and the fact that one of the variables is an integer.

In what follows, we shall restrict our analysis to a two-commodity system operating under the simplest form of mixed ordering policy. To start with, let product 1 be the *M*-product and product 2 be the *D*-product. We will use the same symbols introduced previously to identify the various parameters and variables in the problem. We immediately have the relations

$$Q_1 = D_1 t_1$$
$$Q_2 = D_2 t_2 \qquad\qquad (6.15)$$
$$t_1 = n t_2$$

where n is the number of times commodity 2 is ordered over the time interval t_1 between the ordering of commodity 1, $(n = 2, 3, \ldots)$. Now, the average stock level for commodities 1 and 2 is respectively $(Q_1 t_1)/2$ and $(Q_2 t_2)/2$. Over a cycle of length t_1, exactly one joint order is placed. Thus, $1/t_1$ is the frequency of placing a joint order. On the other hand, over the same cycle, exactly $(n - 1)$ individual orders for the *D*-product are placed; hence, $(n - 1)/t_1$ is the frequency of ordering the *D*-product. Therefore, the long run total cost per unit time of procurement and holding stock is:

$$(TC) = c_1 D_1 + c_2 D_2 + \frac{K + (n-1)K_2}{t_1} + \frac{1}{2} h_1 Q_1 t_1 + \frac{1}{2} h_2 Q_2 t_2$$

Using relations (6.15) to express all the variables in (TC) in terms of n and t_2, we obtain

$$(TC) = c_1 D_1 + c_2 D_2 + \frac{K + (n-1)K_2}{n t_2} + \frac{1}{2}(h_1 D_1 n + h_2 D_2)t_2 \quad (6.16)$$

(TC) is to be minimized with respect to n and t_2, where $n = 2, 3, \ldots$ and $t_2 > 0$. Differentiating (TC) partially with respect to t_2 and setting the result equal to zero yields

$$\frac{\partial (TC)}{\partial t_2} = 0 = -\frac{K + (n-1)K_2}{n t_2^2} + \frac{1}{2}(h_1 D_1 n + h_2 D_2)$$

Solving for t_2 we obtain

$$t_2 = \sqrt{\frac{K + (n-1)K_2}{\frac{n}{2}(h_1 D_1 n + h_2 D_2)}} \quad (6.17)$$

This expression gives the optimal value of t_2 for a given value of n. Substituting (6.17) in (6.16) yields for the total cost per unit time as a function of n

$$\tilde{F}(n) = c_1 D_1 + c_2 D_2 + 2\sqrt{\frac{1}{2n}[K + (n-1)K_2](h_1 D_1 n + h_2 D_2)}$$

$$= c_1 D_1 + c_2 D_2 + \sqrt{2(K - K_2 + nK_2)\left(h_1 D_1 + \frac{h_2 D_2}{n}\right)} \quad (6.18)$$

$$= c_1 D_1 + c_2 D_2$$
$$+ \sqrt{2\left[(K - K_2)h_1 D_1 + h_2 D_2 K_2 + \frac{h_2 D_2 (K - K_2)}{n} + h_1 D_1 K_2 n\right]}$$

Let $n = n^*$ be the optimal value of n which minimizes expression (6.18). If product 1 is to be the M-product, then a value of $n^* \geq 2$ must exist. That is, the mixed ordering policy must be at least as economical as a joint ordering policy, and this would imply that

$$\tilde{F}(n^*) \leq \tilde{F}(1)$$

or, using expression (6.18)

$$c_1 D_1 + c_2 D_2 + \sqrt{2(K - K_2 + n^* K_2)\left(h_1 D_1 + \frac{h_2 D_2}{n^*}\right)}$$
$$\leq c_1 D_1 + c_2 D_2 + \sqrt{2K(h_1 D_1 + h_2 D_2)}$$

Simplifying and squaring both sides yields

$$2(K - K_2 + n^*K_2)\left(h_1 D_1 + \frac{h_2 D_2}{n^*}\right) \leq 2K(h_1 D_1 + h_2 D_2)$$

or

$$Kh_1 D_1 + \frac{Kh_2 D_2}{n^*} + (n^* - 1)K_2 h_1 D_1 + \frac{(n^* - 1)}{n^*}K_2 h_2 D_2 \leq Kh_1 D_1 + Kh_2 D_2$$

Simplifying we obtain

$$\left(\frac{n^* - 1}{n^*}\right)(-Kh_2 D_2 + n^*K_2 h_1 D_1 + K_2 h_2 D_2) \leq 0$$

Since $n^* \geq 2$, $(n^* - 1)/n^* > 0$, and we must have

$$n^* \leq \frac{(K - K_2)h_2 D_2}{K_2 h_1 D_1} \tag{6.19}$$

To summarize, if product 1 is the M-product, then

$$2 \leq n^* \leq \frac{(K - K_2)h_2 D_2}{K_2 h_1 D_1} \tag{6.20}$$

We next show that if product 1 is the M-product, product 2 cannot be the M-product. To do this, assume to the contrary, and let $m^* \geq 2$ be the optimum number of orders placed for product 1 between two consecutive orders of product 2. Then, we should have

$$2 \leq m^* \leq \frac{(K - K_1)h_1 D_1}{K_1 h_2 D_2}$$

This last inequality, together with (6.20) would imply

$$4 \leq m^*n^* \leq \frac{(K - K_1)(K - K_2)}{K_1 K_2}$$

or

$$4K_1 K_2 \leq K^2 - K(K_1 + K_2) + K_1 K_2$$

or

$$K_1 + K_2 \leq K - \frac{3K_1 K_2}{K}$$

This last inequality contradicts the assumption that

$$\max(K_1, K_2) \leq K \leq K_1 + K_2$$

Therefore, product 1 and product 2 cannot be M-products simultaneously. Thus, inequality (6.20) can be used to determine which product is the M-product, if any. If inequality (6.20) is not satisfied, then either product 2 is the *M*-product or there is no *M*-product, in which case a joint ordering policy is to be followed.

From expression (6.18), we note that the optimum value n^* of n can be obtained by minimizing the function

$$F_2(n) = \frac{h_2 D_2 (K - K_2)}{n} + h_1 D_1 K_2 n, \quad n = 2, 3, \ldots$$

with respect to the positive integer variable n. The minimization of such a function was already investigated both in Chapters 4 and 5. The existence of an optimum value n^* implies that

$$F_2(n^*) - F_2(n^* + 1) \leq 0$$
$$F_2(n^*) - F_2(n^* - 1) \leq 0$$

These two inequalities can be shown to reduce to

$$n^*(n^* - 1) \leq \frac{(K - K_2) h_2 D_2}{K_2 h_1 D_1} \leq n^*(n^* + 1)$$

This last expression should be compared to inequalities (6.20).

Example 6.2

Consider a two-commodity inventory system with the following characsitics

Commodity i	c_i, $/unit$	K_i, $	D_i, *units/year*	h_i, $/[(unit)(year)]$
1	500	9	4	5
2	800	3	20	8

The fixed cost of a joint order is $K = \$10$. Determine
(a) The *M*-product,
(b) The optimum time elapsed between two successive orders for the *D*-product and the *M*-product,
so as to minimize the long run total cost per year.

(a) To determine which product is the *M*-product, we compute the quantities:

$$\frac{(K - K_2) h_2 D_2}{K_2 h_1 D_1} = \frac{(10 - 3)(8)(20)}{(3)(5)(4)} = \frac{56}{3} = 18.66 > 2$$

and $\dfrac{(K - K_1) h_1 D_1}{K_1 h_2 D_2} = \dfrac{(10 - 9)(5)(4)}{(9)(8)(20)} = \dfrac{1}{72} < 2$

Hence, product 1 is the M-product

(b) To determine the optimal value n^* of n, we use the inequalities

$$n^*(n^* - 1) \leq 18.66 \leq n^*(n^* + 1)$$

and it is easily verified that $n^* = 4$. From formula (6.17), we have for the optimal value of t_2:

$$t_2^* = \sqrt{\frac{K + (n^* - 1)K_2}{\frac{n^*}{2}(h_1 D_1 n^* + h_2 D_2)}}$$

$$= \sqrt{\frac{10 + (4 - 1)3}{\frac{4}{2}[(5)(4)(4) + (8)(20)]}}$$

$$= \sqrt{\frac{19}{480}} = .199 \text{ years}$$

Now the optimum value of t_1 is given by

$$t_1^* = n^* t_2^* = (4)(.199) = .796 \text{ years}$$

The optimal order quantities Q_1^* and Q_2^* are respectively given by

$$Q_1^* = D_1 t_1^* = (4)(.796) = 3.184 \text{ units}$$
$$Q_2^* = D_2 t_2^* = (20)(.199) = 3.980 \text{ units}$$

The reader may also verify that the minimum long run total cost per year excluding the variable procurement cost is \$47.75. This compares with a minimum long run total cost per year of \$49.96 when using a joint ordering policy.

iii. *Two-Echelon Single Commodity Systems*

We now show how the results obtained for the mixed ordering policy may be utilized to formulate a serial two-installation inventory system. Figure 6.8 represents the stock movement at both the high and the low echelons. The high echelon orders from the source at equal intervals t_1 a quantity Q_1, while the low echelon orders each time a quantity Q_2 from the high echelon at equal intervals t_2. Between the time of arrivals of successive orders to the high echelon, the low echelon will have placed a total of n orders, $n = 1, 2, \ldots$ It is then evident that

$$Q_1 = nQ_2$$
$$t_1 = nt_2$$

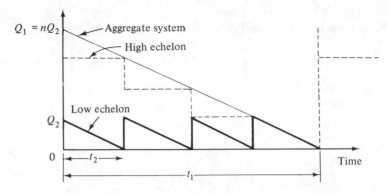

Figure 6.8. Stock level variation of a two-echelon system.

When the high echelon receives an order in the amount of Q_1 units, Q_2 of these units are immediately shipped to the low echelon to replenish its stock. Thus, the high echelon starts its cycle with a net stock level of $(Q_1 - Q_2)$ units. The reduction of stock at the high echelon level is in steps of Q_2 units. The last order quantity Q_2 that the high echelon disposes of prior to replenishing its own stock level, induces a zero stock level for a length of time t_2.

It should be noted that the movement of stock at the low echelon level is typical of the stock movement of a single commodity at a single installation. This also is the case of the stock movement of the *aggregate* system at the low and at the high echelons as shown in Figure 6.8.

If we consider now a cycle of length t_1, then the average stock level at the high echelon is

$$\frac{1}{t_1}\{(Q_1 - Q_2)t_2 + (Q_1 - 2Q_2)t_2 + \cdots + Q_2 t_2 + 0 t_2\}$$

$$= \frac{1}{nt_2}\{(n-1)Q_2 t_2 + (n-2)Q_2 t_2 + \cdots + Q_2 t_2\}$$

$$= \frac{Q_2}{n}\sum_{i=1}^{i=n-1} i$$

$$= \frac{Q_2}{2}(n-1)$$

The frequency of orders at the high echelon is $1/t_1 = 1/nt_2$. At the low echelon, the average stock level is $Q_2/2$ and the frequency of orders is $1/t_2$.

Let D be the demand rate at the low echelon. We now define the following parameters for $i = 1, 2$ in which the index 1 refers to the high echelon and the index 2 refers to the low echelon:

h_i = holding cost in \$/[(unit)(unit time)];

K_i = fixed set-up or order cost in \$;

c_i = variable procurement cost in $/(unit);
(TC) = total cost per unit time in $/(unit time).

The long run total costs per unit time associated with the operation of the high echelon and the low echelon are respectively:

$$c_1 \frac{Q_1}{t_1} + \frac{K_1}{nt_2} + h_1 \frac{Q_2}{2}(n - 1)$$

and

$$c_2 \frac{Q_2}{t_2} + \frac{K_2}{t_2} + h_2 \frac{Q_2}{2}$$

Thus, the long run total cost of operation per unit time of the entire system is

$$(TC) = c_1 \frac{Q_1}{t_1} + \frac{K_1}{nt_2} + h_1 \frac{Q_2}{2}(n - 1) + c_2 \frac{Q_2}{t_2} + \frac{K_2}{t_2} + h_2 \frac{Q_2}{2}$$

Using the relations

$$Q_1 = nQ_2$$
$$t_1 = nt_2$$
$$Q_2 = Dt_2$$

and expressing the variables in (TC) in terms of the variables n and t_2, we obtain:

$$(TC) = c_1 D + c_2 D + \frac{1}{t_2}\left(\frac{K_1}{n} + K_2\right) + \frac{t_2}{2}[h_1 D(n - 1) + h_2 D]$$

$$= c_1 D + c_2 D + \frac{(K_1 + K_2) + (n - 1)K_2}{nt_2} + \frac{1}{2}[h_1 Dn + (h_2 - h_1)D]t_2$$

This expression should be compared to expression (6.16). It could have been obtained from (6.16) if we had set

$$D_1 = D_2 = D$$
$$K = K_1 + K_2$$

and $(h_2 - h_1)$ as the holding cost for the D-product. We now can state the following equivalence between a two-commodity single-installation system and a single-commodity two-echelon serial system, for the particular situation considered:

A serial two-echelon single-commodity inventory system is equivalent to a two-commodity single-installation system provided that the low-echelon is considered as the D-product, the aggregate system (low and high echelons) is considered as the M-product and

(1) the demand rate for the M-product and the D-product is equal, that is $D_1 = D_2 = D$;

(2) the joint fixed order cost K is computed as the sum of fixed order costs at the low and high echelons, that is $K = K_1 + K_2$;

(3) a holding cost h_1 is attached to the M-product and a holding cost $(h_2 - h_1)$ is attached to the D-product.

6.3. Deterministic Models—The Multicommodity Problem

The extension of the results of a two-commodity inventory system to a multicommodity inventory system is straightforward for individual ordering policy and joint ordering policy. The problem formulation is not, however, a trivial matter for mixed ordering policies. In this section, we discuss briefly an inventory system consisting of m commodities, $m = 1, 2, \ldots$, for which either individual ordering or joint ordering is possible with no shortage allowed.

Referring to relations (6.6) and (6.8), we may immediately write down for an individual ordering policy, the expressions for t_i^*, the optimal time interval between orders for commodity i, $i = 1, 2, \ldots, m$, and for $(TC)_I^*$, the minimum value of the total cost function per unit time:

$$t_i^* = \sqrt{\frac{2K_i}{h_i D_i}}$$

$$(TC)_I^* = \sum_{i=1}^{i=m} (c_i D_i + \sqrt{2h_i K_i D_i})$$

For a joint ordering policy, let t be the common time interval between orders (Figure 6.9). Let again $(TC)_J$ be the total cost per unit time; then similar to expressions (6.10) and (6.12) we have at optimality

$$t^* = \sqrt{\frac{K}{\sum_{i=1}^{i=m} h_i D_i}}$$

$$(TC)_J^* = \sum_{i=1}^{i=m} c_i D_i + \sqrt{2K \sum_{i=1}^{i=m} h_i D_i}$$

Joint ordering is preferred to individual ordering if and only if

$$\sqrt{2K \sum_{i=1}^{i=m} h_i D_i} < \sum_{i=1}^{i=m} \sqrt{2h_i K_i D_i}$$

This inequality may be discussed in a way similar to the case of a two-commodity system. In particular, if for all i, $K_i = K$, then it can be shown that a

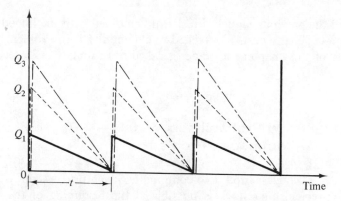

Figure 6.9. Stock level variations for a three-commodity inventory system with joint order.

joint ordering policy is always superior to an individual ordering policy. The reader should also verify that the relative savings in using a joint ordering policy based on the combined ordering and holding cost for individual ordering is

$$\frac{m\sqrt{2} - \sqrt{2m}}{m\sqrt{2}} = 1 - \frac{1}{\sqrt{m}}$$

This is plotted as a function of m in Figure 6.10.

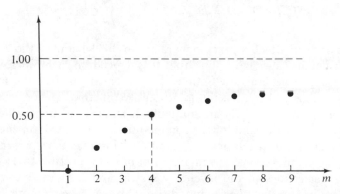

Figure 6.10. Variation of $\left(1 - \dfrac{1}{\sqrt{m}}\right)$ as a function of m.

Remark. The deterministic models discussed so far assumed immediate delivery of orders and no shortages. The case of a nonzero lead time would have been treated in the same fashion as in the single-commodity single-installation systems, resulting in an ordering rule based on the position inventory of the commodities rather than the stock on hand. When shortage is allowed, all the previous models are slightly modified with the introduction of appropriate penalty costs p_i attached to commodity i. In Problems 3 and 4 at the end of this chapter the reader is asked to formulate and solve such models.

6.4. Single-Period Probabilistic Models— The Two-Commodity Problem

Consider a single-period two-commodity inventory system where a decision for replenishing stock is possible at the beginning of the period through individual or joint orders placed at a single supply center. We shall assume that at the time when such a decision is made backlog might exist. We shall also assume that delivery of orders is instantaneous, and that immediately following the decision process and receipt of orders any amount of outstanding backorders is instantaneously filled and a positive level of stock is left over for all the commodities.

During the period, a depletion of the inventory of the two commodities is induced by customers' demand described by the random variable D_i, $i = 1, 2$, for commodity i. Let the joint density function of (D_1, D_2) be $\varphi(\xi_1, \xi_2)$. We shall define the marginal density function of D_i, $i = 1, 2$, by $\varphi_i(\xi_i)$, and let

$$\Phi_i(\xi_i) = \int_0^{\xi_i} \varphi_i(u)\, du \quad \text{and} \quad \overline{D}_i = \int_0^{\infty} \xi_i \varphi_i(\xi_i)\, d\xi_i$$

Although one can attach a general character to the demand of the commodities by assuming, as we have done, that they depend on each other, we shall later see that, in general, the entities of interest are in fact the marginal density functions rather than the joint density function. Hence even if the random variables D_1 and D_2 are mutually dependent, as, for example, in a situation where the demand for the commodities is correlated, one need only know the marginal distributions of D_1 and D_2. This, of course, presents some practical and computational advantages. For one, it is possible to decompose certain characteristics of the inventory system and treat each commodity independently from the other commodities. This will become more apparent when we discuss the functional form of the expected holding and shortage cost and on the determination of the optimal values of certain decision variables. When the number of commodities exceeds two, this decomposition feature becomes naturally handy in introducing simple computational schemes. In summary, although it might appear at first that an interdependence in the demand of the commodities might formally affect the structure of the problem in contradistinction to the case when demands are independent, that is,

$$\varphi(\xi_1, \xi_2) = \varphi_1(\xi_1) \cdot \varphi_2(\xi_2)$$

this actually is not so, for an equivalent problem can be formulated by simply dealing with the marginal distribution of demands.

When a decision for replenishing stock is to be made, one has to decide which commodity to order, if any, and the magnitude of the order for a particular commodity. An exhaustive enumeration shows that $2^2 = 4$ possible courses of action, or *ordering strategies* are available when deciding which of the commodities to order. These are

Strategy 1: order commodity 1 but not commodity 2.
Strategy 2: order commodity 2 but not commodity 1.
Strategy 3: order commodities 1 and 2.
Strategy 4: do not order anything.

These are of the mixed ordering type and, it can be noted that not all these decisions result in an inventory replenishment. We shall say that these four strategies are the elements of the *set of alternative ordering decisions*. Thus, whereas in a single-period single-commodity problem one had to decide on how much to order if any, in multicommodity problems one has to decide on which commodities to order and how much to order if any.

Consider the beginning of the period, and let for commodity i, $i = 1, 2$,

$x_i =$ inventory level prior to making a decision (x_i can be negative)
$z_i =$ quantity ordered ($z_i \geq 0$)
$y_i =$ inventory level following a decision ($y_i = x_i + z_i > 0$)

The selection of the proper ordering strategy together with the magnitudes of the quantities ordered (z_1, z_2) will be dictated by such factors as the joint stock status of the commodities at the beginning of the period (x_1, x_2), the statistical characteristics of the demands (D_1, D_2), and the various interplaying cost elements.

i. Cost Structure of the Problem

In general, three basic cost elements will be associated with the operation of the inventory system; they will be the procurement cost, holding cost, and shortage cost.

a. PROCUREMENT COST

The procurement cost will be assumed to have two components: a fixed cost incurred only if an order for replenishment is placed, but not otherwise; a variable cost that depends on the quantity of each commodity procured.

Thus, in general, with each ordering strategy one would associate a quantity equal to the fixed order cost; this cost could possibly take a zero value. The variable procurement cost will be assumed to be *linear* in form; that is, it will be taken to be directly proportional to the quantity of each commodity procured. Symbolically, let $P(z_1, z_2)$ denote the procurement cost function; then

$$P(z_1, z_2) = k(z_1, z_2) + c_1 z_1 + c_2 z_2, \quad z_1 \geq 0, z_2 \geq 0 \qquad (6.21)$$

where $k(z_1, z_2)$ represent the fixed portion of the procurement cost and c_1 and c_2 denote, respectively, the variable procurement costs per unit of commodities 1 and 2. In general, $k(z_1, z_2)$ could take on as many different values as the number of elements in the set of alternative ordering decisions; its general form will be

$$k(z_1, z_2) = \begin{cases} K_1, & \text{if } z_1 > 0, z_2 = 0 \\ K_2, & \text{if } z_1 = 0, z_2 > 0 \\ K, & \text{if } z_1 > 0, z_2 > 0 \\ 0, & \text{if } z_1 = 0, z_2 = 0 \end{cases} \qquad (6.22)$$

Thus, for example, if a decision to order both commodities 1 and 2 is made $(z_1 > 0, z_2 > 0)$, a fixed order cost K will be incurred. The relative magnitudes of the constants K_1, K_2, and K will depend on the particular problem faced; and, as we have seen, the inequalities

$$\max(K_1, K_2) \leq K \leq K_1 + K_2 \qquad (6.23)$$

are satisfied. A case of particular interest to us will be when $K = K_1 = K_2$; that is, the fixed order cost is invariant and does not depend on whether a joint order or an individual order is placed.

b. HOLDING AND SHORTAGE COSTS

In addition to the procurement cost, a holding and a shortage cost will in general exist. At the termination of the period one could end up either with an amount left over or an amount short for either of the commodities.

It is evident that the total expected holding and shortage cost incurred over the period for both commodities 1 and 2 is a function of the stock level (y_1, y_2) of the two commodities following the decision process at the start of the period; let this be denoted by $L(y_1, y_2)$. The magnitude of $L(y_1, y_2)$ is thus directly affected by the quantity of each commodity ordered; if, for example, an order for $z_1 > 0$ units is placed for commodity 1 and nothing is ordered for commodity 2 $(z_2 = 0)$, then $L(y_1, y_2) = L(x_1 + z_1, x_2)$. We shall not make any particular stipulation on the components of $L(y_1, y_2)$ or on how it is measured; the only assumption we shall make about $L(y_1, y_2)$ is that its second-order derivatives exist and that it is a strictly convex function of y_1

and y_2. The existence of the second derivatives is not absolutely essential, for the bulk of the results we shall derive hold true under the weaker assumption of $L(y_1, y_2)$ being continuous in y_1 and y_2. However, partly for mathematical convenience and partly because of its validity with many practical problems, we shall adhere to the assumption that $L(y_1, y_2)$ is twice differentiable.

In a large number of practical situations, the function $L(y_1, y_2)$ is also separable; that is, it can be expressed as the sum of a function of y_1 only and of a function of y_2 only:

$$L(y_1, y_2) = L_1(y_1) + L_2(y_2) \tag{6.24}$$

The major inherent advantage of the separability condition is the simplification introduced in the characterization of the function $L(y_1, y_2)$. For example, the convexity of $L(y_1, y_2)$ can be established by showing that each of $L_1(y_1)$ and $L_2(y_2)$ is convex. Also, the minimum of $L(y_1, y_2)$ can be located by simply identifying the minimum of each of $L_1(y_1)$ and $L_2(y_2)$ separately. Thus, using the separability condition, one can study the properties of a function of m variables by studying m functions of one variable. The order of simplification is perhaps marginal when analyzing a function of two variables; however, it becomes quite significant when dealing with functions of more than two variables.

To illustrate the discussion on the function $L(y_1, y_2)$, consider the case when for each commodity holding and shortage costs are linear and are measured at the end of the period. Let h_i and p_i, $i = 1, 2$ denote, respectively, the unit holding and shortage penalty costs; then for $y_1, y_2 > 0$

$$
\begin{aligned}
L(y_1, y_2) = h_1 &\int_0^\infty \int_0^{y_1} (y_1 - \xi_1)\varphi(\xi_1, \xi_2)\, d\xi_1\, d\xi_2 \\
+ p_1 &\int_0^\infty \int_{y_1}^\infty (\xi_1 - y_1)\varphi(\xi_1, \xi_2)\, d\xi_1\, d\xi_2 \\
+ h_2 &\int_0^{y_2} \int_0^\infty (y_2 - \xi_2)\varphi(\xi_1, \xi_2)\, d\xi_1\, d\xi_2 \\
+ p_2 &\int_{y_2}^\infty \int_0^\infty (\xi_2 - y_2)\varphi(\xi_1, \xi_2)\, d\xi_1\, d\xi_2
\end{aligned}
$$

Introducing the marginal density functions of the demands, we obtain

$$L(y_1, y_2) = \sum_{i=1}^{i=2} \left\{ h_i \int_0^{y_i} (y_i - \xi_i)\varphi_i(\xi_i)\, d\xi_i + p_i \int_{y_i}^\infty (\xi_i - y_i)\varphi_i(\xi_i)\, d\xi_i \right\} \tag{6.25}$$

The separability of $L(y_1, y_2)$ can be verified by setting (see relation 5.113)

$$
\begin{aligned}
L_i(y_i) &= h_i \int_0^{y_i} (y_i - \xi_i)\varphi_i(\xi_i)\, d\xi_i + p_i \int_{y_i}^\infty (\xi_i - y_i)\varphi_i(\xi_i)\, d\xi_i \\
&= (h_i + p_i) \int_0^{y_i} \Phi_i(\xi_i)\, d\xi_i - p_i y_i + p_i \overline{D}_i, \quad i = 1, 2
\end{aligned} \tag{6.26}
$$

We also note that $L_i''(y_i) = (h_i + p_i)\varphi_i(y_i)$, $i = 1, 2$. Hence $L(y_1, y_2)$ will be twice differentiable at all points y_1, $y_2 > 0$, for which $\varphi_i(y_i)$ is continuous; also $L(y_1, y_2)$ will be strictly convex at all points y_i at which $\varphi_i(y_i) > 0$. The role played by the marginal density functions becomes apparent in this illustration. In fact, $L_i(y_i)$ is the expected holding and shortage cost for commodity i, and what has been shown for this example is that $L(y_1, y_2)$ is expressible as the sum of these individual expected costs for the commodities involved. Thus, by treating each commodity separately using their marginal demand distribution, it becomes possible to structure the function $L(y_1, y_2)$ as a separable function. We shall have in the sequel occasions to use the function $L(y_1, y_2)$ as it appears in expression (6.25) for illustrating specific analytic results.

ii. Objective Function and Optimal Ordering Strategies

Our objective is to determine an optimal replenishment policy so as to minimize the sum total of the procurement cost $P(z_1, z_2)$ and the expected holding and shortage costs $L(y_1, y_2)$ experienced over the period. Let $C(x_1, x_2)$ denote the minimum total expected cost for the period when following an optimal policy; then, using expression (6.21),

$$C(x_1, x_2) = \min_{y_1 \geq x_1, y_2 \geq x_2} \{P(y_1 - x_1, y_2 - x_2) + L(y_1, y_2)\}$$

$$= \min_{y_1 \geq x_1, y_2 \geq x_2} \{k(y_1 - x_1, y_2 - x_2) + c_1(y_1 - x_1) \quad (6.27)$$
$$+ c_2(y_2 - x_2) + L(y_1, y_2)\}$$

Define

$$G(y_1, y_2) = c_1 y_1 + c_2 y_2 + L(y_1, y_2) \quad (6.28)$$

Then

$$C(x_1, x_2) = \min_{y_1 \geq x_1, y_2 \geq x_2} \{k(y_1 - x_1, y_2 - x_2) - c_1 x_1 - c_2 x_2 \\ + G(y_1, y_2)\} \quad (6.29)$$

Using (6.22), we can write expression (6.29) more explicitly as

$$C(x_1, x_2) = \min \begin{cases} \min_{y_1 \geq x_1, y_2 = x_2} \{K_1 - c_1 x_1 - c_2 x_2 + G(y_1, y_2)\} \\ \min_{y_1 = x_1, y_2 \geq x_2} \{K_2 - c_1 x_1 - c_2 x_2 + G(y_1, y_2)\} \\ \min_{y_1 \geq x_1, y_2 \geq x_2} \{K - c_1 x_1 - c_2 x_2 + G(y_1, y_2)\} \\ \min_{y_1 = x_1, y_2 = x_2} \{-c_1 x_1 - c_2 x_2 + G(y_1, y_2)\} \end{cases} \quad (6.30)$$

Note that the four expressions appearing on the right-hand side of (6.30) correspond to the four ordering strategies previously defined. For a particular initial stock (x_1, x_2), the solution of (6.30) will specify the *optimal ordering strategy;* that is, it will dictate which ordering strategy to select as well as the quantities to be ordered so as to minimize the total expected cost. We shall also be interested in specifying the optimal ordering strategies for the whole range of values potentially taken by the initial stock levels (x_1, x_2).

It is possible to represent simultaneously the inventory levels of the two commodities by a point in the two-dimensional Euclidean space E^2. Thus, the point $\mathbf{x} = (x_1, x_2)$ would represent the inventory levels prior to making a decision, whereas the point $\mathbf{y} = (y_1, y_2)$ would represent the levels following a decision. The vector whose origin is $\mathbf{x}$ and terminal point $\mathbf{y}$ has a magnitude and a direction defining completely the decision associated with an initial state $\mathbf{x}$. It is evident that this vector can be oriented only in certain directions since $y_1 - x_1 \geq 0$ and $y_2 - x_2 \geq 0$, and this pair of nonnegative scalars, components of the vector, asserts that a decision has been made to order a quantity $y_1 - x_1$ of commodity 1 and a quantity $y_2 - x_2$ of commodity 2. The optimal ordering strategies for the range of values of $\mathbf{x}$ can also be described on the plane E^2 in terms of decision regions. These regions define for each possible initial inventory state $\mathbf{x}$ which commodity to order and how much to order so as to minimize the total expected cost. Such a geometric illustration becomes particularly useful for presenting and interpreting results; for all practical purposes, however, its usefulness can be said to be limited only to the two-commodity problem.

For notational convenience, we shall introduce the general vector notation $\boldsymbol{\tau} = (\tau_1, \tau_2)$ for any pair of variables and define $\boldsymbol{\tau}^T = \begin{pmatrix} \tau_1 \\ \tau_2 \end{pmatrix}$ to denote the transpose of $\boldsymbol{\tau}$.

iii. *The Function G(y)*

The characterization of the optimal ordering strategies will in part depend on the structure of the function $G(\mathbf{y})$ defined by relation (6.28). It is evident that if $L(\mathbf{y})$ is strictly convex, so will be the function $G = G(\mathbf{y})$; thus $G(\mathbf{y})$ will possess an absolute minimum occurring at a point $\mathbf{S} = (S_1, S_2)$ such that at $y_1 = S_1$ and $y_2 = S_2$, $\partial G/\partial y_1 = 0 = \partial G/\partial y_2$. Although the strict convexity of $G(\mathbf{y})$ will guarantee the existence of $\mathbf{S}$, a meaningful interpretation to the problem can be attached only if $S_i > 0$, $i = 1, 2$.

Consider now the two curves $\mathbf{M}_1$ and $\mathbf{M}_2$ defined respectively as

$$\mathbf{M}_1 = \left\{ \mathbf{y} : \frac{\partial G}{\partial y_1} = 0 \right\} \qquad \mathbf{M}_2 = \left\{ \mathbf{y} : \frac{\partial G}{\partial y_2} = 0 \right\} \tag{6.31}$$

The reader should establish the following basic properties for the curves $\mathbf{M}_1$ and $\mathbf{M}_2$:

(1) $\mathbf{M}_1$ and $\mathbf{M}_2$ intersect uniquely at the point $\mathbf{S}$.
(2) $\mathbf{M}_1$ and $\mathbf{M}_2$ are monotone functions such that
  (a) If $\partial^2 G/\partial y_1\,\partial y_2 > 0$ (or < 0), $\mathbf{M}_1$ and $\mathbf{M}_2$ are strictly decreasing (or increasing) functions of y_1; and for any given y_1, the slope of $\mathbf{M}_1$ (or $\mathbf{M}_2$) is steeper than the slope of $\mathbf{M}_2$ (or $\mathbf{M}_1$).
  (b) If $\partial^2 G/\partial y_1\,\partial y_2 = 0$, $\mathbf{M}_1$ and $\mathbf{M}_2$ are lines parallel to the y_2 and y_1 axis, respectively.
  (c) As a corollary, for $\partial^2 G/\partial y_1\,\partial y_2 \lessgtr 0$, an inverse function exists and defines each of the curves $\mathbf{M}_1$ and $\mathbf{M}_2$. We can thus express the two relations (6.31) equivalently by

$$\mathbf{M}_1 = \{y: y_1 = Y_1(y_2)\} \quad \text{and} \quad \mathbf{M}_2 = \{y: y_2 = Y_2(y_1)\} \quad (6.32)$$

respectively. It follows that $S_1 = Y_1(S_2)$ and $S_2 = Y_2(S_1)$. For $\partial^2 G/\partial y_1\,\partial y_2 = 0$, the equation of $\mathbf{M}_1$ is $y_1 = S_1$ and that of $\mathbf{M}_2$ is $y_2 = S_2$. Note that the curve $\mathbf{M}_1$ (or $\mathbf{M}_2$) is the locus of all points y at which the function $G(y)$ achieves a minimum for a given y_2 (or y_1).

iv. *Optimal Ordering Strategies When $K_1 = K_2 = K$ and $L(\mathbf{y})$ Is Nonseparable*

For this case $\partial^2 G/\partial y_1\,\partial y_2 \gtrless 0$. Consider the three functions $\psi(\mathbf{y}) = 0$, $y_1 = \psi_1(y_2)$, and $y_2 = \psi_2(y_1)$ defined, respectively, by the following implicit relations:

$$\psi(\mathbf{y}) = G(\mathbf{y}) - K - G(\mathbf{S}) = 0 \tag{6.33}$$

$$G[\psi_1(y_2), y_2] = K + G[Y_1(y_2), y_2] \tag{6.34}$$

$$G[y_1, \psi_2(y_1)] = K + G[y_1, Y_2(y_1)] \tag{6.35}$$

In the space E^2 referred to the set of Cartesian coordinates (y_1, y_2), define the region $\Omega = \{\mathbf{y}: \mathbf{y} \le \mathbf{S}\}$ and the set of points (Figure 6.11)

$$\Gamma = \{\mathbf{y}: \mathbf{y} \in \Omega, \psi(\mathbf{y}) = 0\}$$

$$\Gamma_1 = \{\mathbf{y}: y_1 = \psi_1(y_2), y_2 > S_2\}$$

$$\Gamma_2 = \{\mathbf{y}: y_1 > S_1, y_2 = \psi_2(y_1)\}$$

Finally, construct the regions

$$o_1 = \{\mathbf{y}: y_1 \le \psi_1(y_2), y_2 > S_2\}$$

$$o_2 = \{\mathbf{y}: y_1 > S_1, y_2 \le \psi_2(y_1)\}$$

$$o_{12} = \{\mathbf{y}: \mathbf{y} \in \Omega, K + G(\mathbf{S}) < G(\mathbf{y})\}$$

$$\bar{o} = E^2 - (o_1 + o_2 + o_{12})$$

The optimal ordering strategy obtained by solving system (6.30) has the following configuration:

(1) If $\mathbf{x} \in o_1$, then order commodity 1 only to bring its inventory level up to $\mathbf{M}_1$.

(2) If $\mathbf{x} \in o_2$, then order commodity 2 only to bring its inventory level up to $\mathbf{M}_2$.

(3) If $\mathbf{x} \in o_{12}$, then order commodities 1 and 2 simultaneously to bring their inventory level up to $\mathbf{S}$.

(4) If $\mathbf{x} \in \bar{o}$, then do not order anything.

As an illustration, assume that initially $\mathbf{x} \in \Omega$, and let us determine the optimal strategies. One way of approaching the problem would be to find out under which conditions, say, strategy 3 is optimal, that is, preferred to strategies 1, 2, and 4. From system (6.30), we note that strategy 3 is preferred to strategy 1 iff (if and only if)

$$\min_{\mathbf{y} \geq \mathbf{x}} \{K - \mathbf{c}\mathbf{x}^T + G(\mathbf{y})\} < \min_{y_1 \geq x_1, y_2 = x_2} \{K - \mathbf{c}\mathbf{x}^T + G(\mathbf{y})\}$$

or
$$K - \mathbf{c}\mathbf{x}^T + G(\mathbf{S}) < K - \mathbf{c}\mathbf{x}^T + G[Y_1(x_2), x_2]$$

which is always satisfied, since $\mathbf{S}$ is the absolute minimum of the function $G(\mathbf{y})$. Similarly, it can be shown that strategy 3 is always preferred to strategy 2. Finally, strategy 3 will be preferred to strategy 4 iff

$$\min_{\mathbf{y} \geq \mathbf{x}} \{K - \mathbf{c}\mathbf{x}^T + G(\mathbf{y})\} < \min_{\mathbf{y} = \mathbf{x}} \{-\mathbf{c}\mathbf{x}^T + G(\mathbf{y})\}$$

or
$$K - \mathbf{c}\mathbf{x}^T + G(\mathbf{S}) < -\mathbf{c}\mathbf{x}^T + G(\mathbf{x})$$

or
$$K + G(\mathbf{S}) < G(\mathbf{x})$$

Hence, if initially $\mathbf{x} \in o_{12}$, it is optimal to use strategy 3 or order both commodities simultaneously; in addition, by bringing the joint stock level up to $\mathbf{S}$, that is, by ordering a quantity $\mathbf{S} - \mathbf{x}$, the total expected cost will be minimized. We are now left with the task of determining what is optimal if $\mathbf{x} \in \Omega - o_{12}$. Using again system (6.30), it may be verified that strategy 4 is preferred to all other strategies; it is then optimal not to order anything, in which case the minimal expected cost per period is $L(\mathbf{x})$.

Figure 6.11 illustrates the optimal ordering schemes for all possible initial inventory states; for example, if initially the stock level is A, it is optimal to order commodity 1 only and bring point A to point B.

v. *Optimal Ordering Strategies When $K_1 = K_2 = K$ and $L(\mathbf{y})$ Is Separable*

In this case $L(\mathbf{y})$ is expressible as (6.24) and $\partial^2 G/\partial y_1 \partial y_2 = 0$. The analysis is very similar to the previous one. The optimal ordering schemes are illustrated in Figure 6.12. $\mathbf{M}_1$ and Γ_1 are lines parallel to the y_2 axis, and $\mathbf{M}_2$ and Γ_2

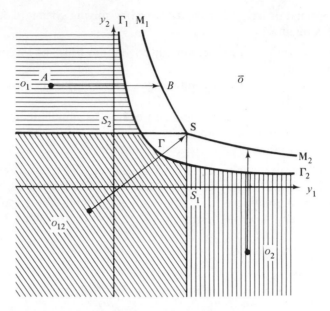

Figure 6.11. Regions of optimal strategies when $K_1 = K_2 = K$ and $L(y)$ is nonseparable and strictly convex.

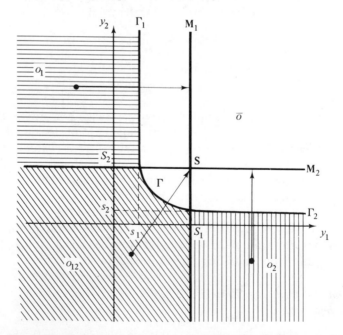

Figure 6.12. Regions of optimal strategies when $K_1 = K_2 = K$ and the function $L(y)$ is separable and strictly convex.

are lines parallel to the y_1 axis. For $i, j = 1, 2$ and $i \neq j$, let

$$G_i(y_i) = c_i y_i + L_i(y_i) \tag{6.36}$$

Then
$$G(\mathbf{y}) = G_1(y_1) + G_2(y_2) \tag{6.37}$$

and S_i is the solution to the equation

$$\frac{\partial G(\mathbf{y})}{\partial y_i} = \frac{dG_i(y_i)}{dy_i} = 0 \tag{6.38}$$

Also
$$\Gamma_i = \{\mathbf{y}: y_i = S_i, y_j > S_j\}$$

where s_i is the solution to the equation

$$G(s_i, y_j) = K + G(S_i, y_j)$$

or
$$G_i(s_i) - G_i(S_i) = K \tag{6.39}$$

The uniqueness of s_i is guaranteed, since $G_i(y_i)$ is a strictly convex function of y_i. We note in particular that the equations dictating the values of $\mathbf{S}$ and $\mathbf{s}$ are the same ones obtained when each commodity is treated independently of the other; in other words, the pair (s_i, S_i) obtained for the ith commodity is the same as if the ith commodity were studied separately, that is, under an (s, S) policy (see Chapter 5).

Using equation (6.33), we obtain for Γ

$$\Gamma = \left\{\mathbf{y}: \mathbf{y} \in \Omega, K + \sum_{i=1}^{i=2} [G_i(S_i) - G_i(y_i)] = 0\right\} \tag{6.40}$$

An Application: Assume $L_i(y_i)$ is of the form of expression (6.26); then for $i = 1, 2$,

$$G_i(y_i) = (h_i + p_i) \int_0^{y_i} \Phi_i(\xi_i) \, d\xi_i + p_i \bar{D}_i - (p_i - c_i)y_i \tag{6.41}$$

Now
$$\frac{dG_i(y_i)}{dy_i} = 0 = (h_i + p_i)\Phi_i(y_i) - (p_i - c_i) \tag{6.42}$$

Hence, if $p_i > c_i$, an $S_i > 0$ can be obtained from

$$\Phi_i(S_i) = \frac{p_i - c_i}{h_i + p_i} \tag{6.43}$$

An expression for $G_i(y_i) - G_i(S_i)$ is, $y_i < S_i$,

$$G_i(y_i) - G_i(S_i) = (p_i - c_i)(S_i - y_i) - (h_i + p_i) \int_0^{S_i - y_i} \Phi_i(S_i - \xi) \, d\xi \tag{6.44}$$

Since s_i is the solution to equation (6.39), we obtain, using relation (6.44) and setting $Q_i = S_i - s_i$, the following equation for Q_i:

$$(p_i - c_i)Q_i - (h_i + p_i) \int_0^{Q_i} \Phi_i(S_i - \xi) \, d\xi = K \tag{6.45}$$

The implicit form of $\psi(\mathbf{y}) = 0$ dictating the configuration of Γ is

$$K + \sum_{i=1}^{i=2} \left[(p_i - c_i)(S_i - y_i) - (h_i + p_i) \int_0^{S_i - y_i} \Phi_i(S_i - \xi) \, d\xi \right] = 0 \tag{6.46}$$

The analysis is next illustrated by two examples involving two different classes of demand distributions.

Example 6.3

Consider a situation when the demand for the commodities has marginal distribution of the exponential type. For $i = 1, 2$, let

$$\varphi_i(\xi_i) = \lambda_i e^{-\lambda_i \xi_i}, \tag{6.47}$$

Then
$$\Phi_i(\xi_i) = 1 - e^{-\lambda_i \xi_i}, \quad \lambda_i > 0, \ 0 \le \xi_i < \infty \tag{6.48}$$

Substituting (6.48) in (6.43) and (6.45), we obtain as in Example 5.4, for $i = 1, 2$,

$$S_i = \frac{1}{\lambda_i} \ln \frac{h_i + p_i}{h_i + c_i} \tag{6.49}$$

and
$$\frac{c_i + h_i}{\lambda_i K} (e^{\lambda_i Q_i} - 1 - \lambda_i Q_i) = 1 \tag{6.50}$$

The quantity $X_i = \lambda_i Q_i$ can be obtained as the positive intercept of the curves (see Figure 5.29)

$$Y = e^{X_i} \quad \text{and} \quad Y = \frac{\lambda_i K}{c_i + h_i} + 1 + X_i$$

The equation $\psi(\mathbf{y}) = 0$ is obtained from (6.46) as

$$\sum_{i=1}^{i=2} \frac{c_i + h_i}{\lambda_i K} (e^{\lambda_i Z_i} - 1 - \lambda_i Z_i) = 1 \tag{6.51}$$

where
$$Z_i = S_i - y_i, \quad i = 1, 2$$

For small values of $\lambda_i Q_i$, we have, using the power series expansion for $e^{\lambda_i Z_i}$,

$$e^{\lambda_i Z_i} = 1 + \lambda_i Z_i + \frac{1}{2!}(\lambda_i Z_i)^2 + \cdots \tag{6.52}$$

Substituting expression (6.52) in (6.50), we obtain

$$\frac{c_i + h_i}{\lambda_i K} \frac{1}{2!}(\lambda_i Q_i)^2 \approx 1$$

Hence
$$Q_i \approx \sqrt{\frac{2K}{\lambda_i(c_i + h_i)}} \tag{6.53}$$

Substituting expression (6.52) in equation (6.51), we obtain

$$\frac{1}{2}\frac{c_1 + h_1}{\lambda_1 K}(\lambda_1 Z_1)^2 + \frac{1}{2}\frac{c_2 + h_2}{\lambda_2 K}(\lambda_2 Z_2)^2 \approx 1$$

or
$$\frac{Z_1^2}{Q_1^2} + \frac{Z_2^2}{Q_2^2} \approx 1 \tag{6.54}$$

Hence, for small $\lambda_i Q_i$, $i = 1, 2$, Q_i is approximately given by the square root formula (6.53), and Γ is approximately an arc of an ellipse with center at **S** and semiaxes Q_1 and Q_2.

Example 6.4

Consider now the case when demand for each commodity has a uniform marginal distribution. For commodity i, $i = 1, 2$, let

$$\varphi_i(\xi_i) = \begin{cases} \dfrac{1}{a_i}, & 0 < \xi_i \le a_i \\ 0, & a_i < \xi_i < \infty \end{cases} \tag{6.55}$$

$$\Phi_i(\xi_i) = \begin{cases} \dfrac{\xi_i}{a_i}, & 0 < \xi_i \le a_i \\ 1, & a_i < \xi_i < \infty \end{cases} \tag{6.56}$$

Substituting (6.56) in equations (6.43) and (6.45), respectively, we obtain, for $i = 1, 2$,

$$S_i = \frac{a_i(p_i - c_i)}{h_i + p_i} \tag{6.57}$$

and
$$Q_i = \sqrt{\frac{2a_i K}{h_i + p_i}} \tag{6.58}$$

Note that for $p_i > c_i$, $0 < S_i < a_i$, the existence of S_i is guaranteed over the interval $(0, a)$ since $\varphi_i(\xi_i)$ is continuous and strictly positive over that interval. From equation (6.46), we obtain for the equation of Γ

$$\frac{Z_1^2}{2a_1 K/(h_1 + p_1)} + \frac{Z_2^2}{2a_2 K/(h_2 + p_2)} = 1$$

where $Z_i = S_i - y_i$, $i = 1, 2$. Hence Γ is an arc of an ellipse with center at $\mathbf{S}$ and semiaxes Q_1 and Q_2.

For this particular example, we can state the following procedure for optimal ordering strategies. Let x_i refer to the initial stock level of commodity i, $i = 1, 2$; then

(1) If $x_1 \leq S_1 - Q_1$ and $x_2 > S_2$, order $S_1 - x_1$ units of commodity 1 and do not order commodity 2.

(2) If $x_1 > S_1$ and $x_2 \leq S_2 - Q_2$, do not order commodity 1 and order $S_2 - x_2$ units of commodity 2.

(3) If $x_1 \leq S_1$, $x_2 \leq S_2$, and $[(S_1 - x_1)^2/Q_1^2] + [(S_2 - x_2)^2/Q_2^2] > 1$, order $S_1 - x_1$ units of commodity 1 and $S_2 - x_2$ units of commodity 2.

(4) Otherwise, do not order anything.

vi. Optimal Ordering Strategies When $K \neq K_1 \neq K_2$, $\max (K_1, K_2)$ $\leq K \leq K_1 + K_2$ and $L(y_1, y_2)$ Is Separable

We shall first formalize the optimal ordering strategies and illustrate later the methodology used to derive these strategies. Let again $\mathbf{S} = (S_1, S_2)$ be the absolute minimum of the function $G(\mathbf{y})$ as given in expression (6.37) and define Γ as in equation (6.40), where $\Omega = \{\mathbf{y} : \mathbf{y} \leq \mathbf{S}\}$. Let s_1', s_2', s_1'', and s_2'' be, respectively, solutions to the following system of equations:

$$(K - K_2) + G_1(S_1) = G_1(s_1')$$
$$(K - K_1) + G_2(S_2) = G_2(s_2') \tag{6.59}$$
$$K_1 + G_1(S_1) = G_1(s_1'') \tag{6.60}$$
$$K_2 + G_2(S_2) = G_2(s_2'')$$

Define the set of points (see Figure 6.13)

$$\Gamma_1 = \{\mathbf{y} : y_1 = s_1'', y_2 > s_2'\}$$
$$\Gamma_2 = \{\mathbf{y} : y_1 > s_1', y_2 = s_2''\}$$

Finally, construct the regions

$$o_1 = \{\mathbf{y} : y_1 \leq s_1'', y_2 > s_2'\}$$
$$o_2 = \{\mathbf{y} : y_1 > s_1', y_2 \leq s_2''\}$$
$$o_{12} = \{\mathbf{y} : \mathbf{y} \in \Omega, K + G(\mathbf{S}) < G(\mathbf{y}), y_1 \leq s_1', y_2 \leq s_2'\}$$
$$\bar{o} = E^2 - (o_1 + o_2 + o_{12})$$

The optimal ordering strategies have the following configuration:

(1) If $\mathbf{x} \in o_1$, then order commodity 1 only to bring its stock level up to $y_1 = S_1$.

(2) If $\mathbf{x} \in o_2$, then order commodity 2 only to bring its stock level up to $y_2 = S_2$.

(3) If $\mathbf{x} \in o_{12}$, then order both commodities simultaneously to bring their stock level up to $\mathbf{y} = \mathbf{S}$.

(4) If $\mathbf{x} \in \bar{o}$, then do not order anything.

The delineation of the optimal ordering strategies for this case is more difficult than in the case of invariant fixed setup cost. To illustrate the procedure, let us determine under which conditions strategy 1 is optimal if $\mathbf{x} \leq \mathbf{S}$. Referring to relation (6.30), strategy 1 is preferred to strategy 3 iff

$$\min_{y_1 \geq x_1, y_2 = x_2} \{K_1 - \mathbf{c}\mathbf{x}^T + G(\mathbf{y})\} < \min_{y \geq x} \{K - \mathbf{c}\mathbf{x}^T + G(\mathbf{y})\}$$

or

$$G(S_1, x_2) < (K - K_1) + G(S_1, S_2)$$

and, using expression (6.37),

$$G_2(x_2) < (K - K_1) + G_2(S_2) \tag{6.61}$$

Hence all points in the set $\{\mathbf{x}: \mathbf{x} \leq \mathbf{S}, x_2 > s_2'\}$ satisfy inequality (6.61).

Now strategy 1 is preferred to strategy 4 iff

$$\min_{y_1 \geq x_1, y_2 = x_2} \{K_1 - \mathbf{c}\mathbf{x}^T + G(\mathbf{y})\} < \min_{y = x} \{-\mathbf{c}\mathbf{x}^T + G(\mathbf{y})\}$$

or

$$K_1 + G(S_1, x_2) < G(x_1, x_2)$$

and, using expression (6.37), this reduces to

$$K_1 + G_1(S_1) < G_1(x_1) \tag{6.62}$$

Hence all points in the set $\{\mathbf{x}: \mathbf{x} \leq \mathbf{S}, x_1 < s_1''\}$ satisfy inequality (6.62). We next show that if

$$\mathbf{x} \in \{\mathbf{x}: \mathbf{x} \leq \mathbf{S}, x_1 < s_1'', x_2 > s_2'\}$$

then strategy 1 is preferred to strategy 2. This can be demonstrated by showing that strategy 3 is preferred to strategy 2, and this will be the case iff

$$\min_{y \geq x} \{K - \mathbf{c}\mathbf{x}^T + G(\mathbf{y})\} < \min_{y_1 = x_1, y_2 \geq x_2} \{K_2 - \mathbf{c}\mathbf{x}^T + G(\mathbf{y})\}$$

or

$$K + G(S_1, S_2) < K_2 + G(x_1, S_2)$$

or, using expression (6.37),

$$K - K_2 < G_1(x_1) - G_2(S_1) \tag{6.63}$$

Now, by assumption, $K \leq K_1 + K_2$, and for the set considered, relation (6.62) is satisfied; hence

$$K - K_2 \leq K_1 < G_1(x_1) - G_1(S_1)$$

and inequality (6.63) is always satisfied. From equation (6.40), we can write

$$\Gamma = \{y: y \le S, G_1(y_1) + G_2(y_2) = K + G_1(S_1) + G_2(S_2)\}$$

By adding equation (6.59) to equation (6.60), it can be seen that $(s_1'', s_2') \in \Gamma$. Using similar reasonings, the optimal strategies for other initial stock levels can be determined. Figure 6.13 illustrates the optimal ordering schemes for

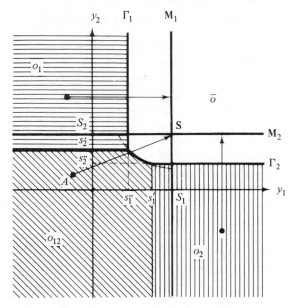

Figure 6.13. Regions of optimal strategies when Max (K_1, K_2) $\le K \le K_1 + K_2$ and the function $L(y)$ is separable and strictly convex.

all possible initial inventory states. For example, if initially the stock level of both commodities is defined by the coordinates of the point $A \in o_{12}$, then it is optimal to order both commodities simultaneously and bring their stock level up to $S = (S_1, S_2)$.

Using the information of Example 6.4, we obtain, for $i, j = 1, 2$ and $i \ne j$,

$$S_i = \frac{a_i(p_i - c_i)}{h_i + p_i}, \qquad Q_i = \sqrt{\frac{2a_iK}{h_i + p_i}}$$

$$s_i' = S_i - \sqrt{\frac{2a_i(K - K_j)}{h_i + p_i}}, \qquad s_i'' = S_i - \sqrt{\frac{2a_iK_i}{h_i + p_i}}$$

If then x_i refers to the initial stock level of commodity 1, the optimal ordering rules would be as follows:

(1) If $x_1 \leq s_1''$ and $x_2 \geq s_2'$, order $S_1 - x_1$ units of commodity 1 and do not order commodity 2.

(2) If $x_1 > s_1'$ and $x_2 \leq s_2''$, do not order commodity 1 and order $S_2 - x_2$ units of commodity 2.

(3) If $x_1 \leq s_1'$, $x_2 \leq s_2'$, and $[(S_1 - x_1)^2/Q_1^2] + [(S_2 - x_2)^2/Q_2^2] > 1$, order $S_1 - x_1$ units of commodity 1 and $S_2 - x_2$ units of commodity 2.

(4) Otherwise, do not order anything.

6.5. Single-Period Probabilistic Models— The Multicommodity Problem

In this section, the analysis of a two-commodity inventory system is partially extended to an m-commodity inventory system, $m = 1, 2, \ldots$.

Three cost elements affect the operation of the system: the procurement cost, holding cost, and penalty cost. The procurement cost will be assumed to have a fixed component incurred only if an order is placed, but not otherwise, and a variable component proportional to the quantity of each commodity ordered. The demand for the commodities in successive periods is described by a vector random variable with a known joint density function, independently distributed from period to period. Immediate delivery of orders will be assumed.

For commodity i, $i = 1, 2, \ldots, m$, let

x_i = stock on hand at the beginning of the period prior to making any decision

z_i = quantity ordered at the beginning of the period ($z_i \geq 0$)

y_i = stock on hand at the beginning of the period just after making a decision $y_i = z_i + x_i > 0$

c_i = variable cost in obtaining one unit of commodity i

D_i = random variable describing the demand ξ_i for commodity i during the period ($\xi_i > 0$)

Define the following vectors:

$$\mathbf{x} = (x_1, x_2, \ldots, x_m)$$
$$\mathbf{z} = (z_1, z_2, \ldots, z_m)$$
$$\mathbf{y} = (y_1, y_2, \ldots, y_m)$$
$$\mathbf{c} = (c_1, c_2, \ldots, c_m)$$
$$\mathbf{D} = (D_1, D_2, \ldots, D_m)$$
$$\xi = (\xi_1, \xi_2, \ldots, \xi_m)$$

Let

$k(\mathbf{z})$ = fixed order cost

$\varphi(\xi)$ = joint density function of demand

$L(\mathbf{y})$ = total expected holding and shortage cost measured at the end of the period

$C(\mathbf{x})$ = minimum total expected cost when following an optimal policy

In general, the possible values taken by $k(\mathbf{z})$ will be equal to the number 2^m of elements in the set of alternative ordering decisions, an order being placed in $2^m - 1$ cases. Let

$$G(\mathbf{y}) = \mathbf{c}\mathbf{y}^T + L(\mathbf{y})$$

Then
$$C(\mathbf{x}) = \min_{\mathbf{y} \geq \mathbf{x}} \{k(\mathbf{y} - \mathbf{x}) - \mathbf{c}\mathbf{x}^T + G(\mathbf{y})\}$$

Restricting our attention to the case when $L(\mathbf{y})$ is differentiable, strictly convex, and separable, then

$$L(y) = \sum_{i=1}^{i=m} L_i(y_i)$$

Let
$$G_i(y_i) = c_i y_i + L_i(y_i)$$

Then
$$G(\mathbf{y}) = \sum_{i=1}^{i=m} G_i(y_i)$$

Hence
$$C(\mathbf{x}) = \min_{\mathbf{y} \geq \mathbf{x}} \{k(\mathbf{y} - \mathbf{x}) - \mathbf{c}\mathbf{x}^T + G(\mathbf{y})\}$$

$$= \min_{\mathbf{y} \geq \mathbf{x}} \{k(\mathbf{y} - \mathbf{x}) - \sum_{i=1}^{i=m} c_i x_i + \sum_{i=1}^{i=m} G_i(y_i)\}$$

The function $G(\mathbf{y})$ achieves its unique absolute minimum at a point $\mathbf{S} \equiv (S_1, S_2, \ldots, S_m)$, where S_i is the unique solution to the equation $dG_i(y_i)/dy_i = 0$; we restrict again our attention to the case when $S_i > 0$ for all $i = 1, 2, \ldots, m$. Let $\mathbf{r} = (r_1, r_2, \ldots, r_k)$, $k = 1, 2, \ldots, m$, be those elements of $\mathbf{x}$ which are greater than or equal to their corresponding S_i values, and let $\mathbf{t} = (t_1, t_2, \ldots, t_{m-k})$ be the remaining elements. It is evident that it is optimal not to order those commodities corresponding to the elements of r. For those commodities corresponding to the elements of t, it is necessary to compare for the 2^{m-k} courses of action the minimum total expected costs and select that course of action corresponding to the minimum of these costs.

Assume, for example, that $m = 4$ and that $\mathbf{S} = (10, 21, 14, 18)$ and $\mathbf{x} = (12, 18, 15, 9)$. Then $x_1 > S_1$ and $x_3 > S_3$; thus no order will be placed for commodities 1 and 3, and it is necessary to compare the possible alternatives for commodities 2 and 4. The problem is thus reduced to a two-commodity problem with four possible courses of action.

Even with a possible reduction in dimensionality, this enumeration method for obtaining the optimal policy is not too advantageous. The separability condition was useful in computing the optimal value of $S = (S_1, S_2, \ldots, S_m)$; unfortunately, it does not help to overcome the enumeration procedure to define the optimal policy.

For invariant fixed ordering costs, a further reduction in dimensionality is possible. If for commodity i the quantity s_i is defined as the solution to the equation

$$G_i(s_i) = K + G_i(S_i), \quad i = 1, 2, \ldots, m$$

then if $x_i < s_i$ for at least one of the commodities, an order would necessarily be placed for all commodities for which $x_i < S_i$. This can be demonstrated by showing that if $L(\mathbf{y})$ is separable, then the set

$$\Gamma = \{\mathbf{y}: \mathbf{y} \in \Omega, G(\mathbf{y}) = K + G(\mathbf{S})\}$$

is convex, and hence that $\Gamma \subset \{\mathbf{y}: \mathbf{y} < \mathbf{S}, y \geq \mathbf{s}\}$, where $\mathbf{s} = (s_1, s_2, \ldots, s_m)$.

For the given example, let $\mathbf{s} = (5, 15, 6, 10)$; then since $s_2 < x_2 < S_2$ and $x_4 < s_4$, an order would be placed for commodities 2 and 4 to bring their stock level, respectively, up to 21 and 18. On the other hand, suppose that $\mathbf{s} = (5, 15, 6, 8)$; then $s_2 < x_2 < S_2$ and $s_4 < x_4 < S_4$; in this case, no optimal ordering rules can be ascertained on a knowledge of $\mathbf{s}$ and $\mathbf{S}$ alone: it is necessary to know in addition the configuration of Γ.

As an illustration, the derivations in Example 6.3 can easily be extended to m commodities to yield

$$S_i = \frac{1}{\lambda_i} \ln \frac{h_i + p_i}{h_i + c_i}, \quad p_i > c_i, i = 1, 2, \ldots, m$$

Also $Q_i = S_i - s_i$ is the unique solution to the equation

$$\frac{c_i + h_i}{\lambda_i K}(e^{\lambda_i Q_i} - 1 - \lambda_i Q_i) = 1, \quad i = 1, 2, \ldots, m$$

and

$$\Gamma = \left\{ \mathbf{y}: \mathbf{y} \in \Omega, \sum_{i=1}^{i=m} \left[\frac{c_i + h_i}{\lambda_i K}(e^{\lambda_i Z} - 1 - \lambda_i Z_i) \right] = 1 \right\}$$

where

$$Z_i = S_i - y_i$$

SELECTED REFERENCES

[1] CLARK, A. J., and H. SCARF, "Optimal Policies for a Multi-echelon Inventory Problem," *Management Science*, Vol. 6, No. 4, July 1960.

[2] FABRYCKY, W. J., and J. BANKS, *Procurement and Inventory Systems*, Reinhold Publishing Corporation, New York, 1967.

[3] Goswick, T. E., *Mixed Ordering Policies in Multi-Product, Multi-Installation Inventory Systems with Variant Set-up Costs*, Ph.D. Thesis, University of Florida, 1972.

[4] Gross, D., *An Investigation of Centralized Inventory Control in Multi-Location Supply Systems*, Ph.D. Thesis, Cornell University, 1962; University Microfilms, Inc., Ann Arbor.

[5] Hadley, G., and T. M. Whitin, *Analysis of Inventory System*, Prentice-Hall, Inc., Englewood Cliffs, New Jersey, 1963.

[6] Hanssmann, F., *Operations Research in Production and Inventory Control*, John Wiley & Sons, Inc., New York, 1961.

[7] Scarf, H. E., D. M. Gilford, and M. W. Shelly (Eds.), *Multistage Inventory Models and Techniques*, Stanford University Press, Stanford, California, 1963.

[8] Sivazlian, B. D., "A Multi-Commodity Inventory System with Set-up Costs," *Opsearch*, Vol. 7, No. 4, December 1970.

PROBLEMS

1. In the deterministic problem involving the joint ordering of m commodities with no shortages allowed, the expression for the total cost per unit time expressed as a function of the order quantities Q_i, $i = 1, 2, \ldots, m$, may be written as

$$(TC)_J = \sum_{i=1}^{i=m} \left(c_i D_i + K_i \frac{D_i}{Q_i} + h_i \frac{Q_i}{2} \right)$$

subject to

$$\frac{Q_1}{D_1} = \frac{Q_2}{D_2} = \cdots = \frac{Q_m}{D_m} = t$$

where t is the common time interval between orders.

   Using the techniques of Lagrange multipliers for handling the constraint, solve for the optimum value of the Q_i, $i = 1, 2, \ldots, m$, and recover the results of Section 6.3. Find the optimal value of the Lagrange multipliers, and give a qualitative interpretation for these multipliers.

2. A small printing company is using five different types of papers and inks, all purchased from a single supplier. Papers have to be ordered in units of 500 sheets (reams) and inks in units of gallons. It is estimated that the cost of placing orders is

   $10 if an order is placed for each item separately
   $15 if an order is placed for one category of items, papers or inks
   $25 if an order is placed for all 10 items

The unit purchase cost and the average yearly consumption for each item are as follows:

Item	Unit Cost, $	Unit Consumption per Year	
Paper 1	2.50 per ream	2000	reams
Paper 2	4.00	1500	
Paper 3	5.00	1000	
Paper 4	5.00	1000	
Paper 5	7.50	700	
Ink 1	18.00 per gallon	10	gallons
Ink 2	18.00	12	
Ink 3	12.00	15	
Ink 4	12.00	18	
Ink 5	10.00	20	

Assuming a 20% rate of return on investment, and disallowing shortages, determine an optimal ordering policy.

3. Consider the two-commodity deterministic problem where shortage in the form of complete backlog is allowed. Denote by p_i, $i = 1, 2$, the shortage cost per unit per unit time.
   (a) Determine the minimum total cost per unit time under an individual ordering policy.
   (b) Determine the minimum total cost per unit time under a joint ordering policy.
   (c) Under which condition is a joint ordering policy preferred to an individual ordering policy.
   (d) Show that under invariant fixed setup cost a joint ordering policy is always preferred to an individual ordering policy.
   (e) Show that under an optimal joint ordering policy the proportion of time commodity i is short is $h_i/(h_i + p_i)$, $i = 1, 2$.

4. Extend all the results of Problem 3 to an arbitrary number m of commodities.

5. In the single-period two-commodity probabilistic problem, consider the case when $K_1 = K_2 = K$, and assume that the density functions of the demand are respectively

$$\varphi_1(\xi_1) = \lambda_1 e^{-\lambda_1 \xi_1}, \qquad \lambda_1 > 0, 0 < \xi_1 < \infty$$
$$\varphi_2(\xi_2) = \lambda_2 \xi_2 e^{-\lambda_2 \xi_2}, \quad \lambda_2 > 0, 0 < \xi_2 < \infty$$

Determine the equations for S_1^*, S_2^* and Γ.

6. In the single-period two commodity problem, the joint distribution function of the demands (D_1, D_2) is given by

$$\varphi(\xi_1, \xi_2) = \begin{cases} \dfrac{2}{a^2}, & 0 \le \xi_1 < \xi_2 \le a, a > 0 \\ 0, & \text{otherwise} \end{cases}$$

Assume now that the holding and shortage costs for each product are proportional to the square of the amount at end of period. Determine an expression for $L(y_1, y_2)$.

7. Consider the following echelon structure. Assume that the demands at each of the lowest echelons are deterministic. The parameters associated with each echelon are as shown. No shortages are allowed. The Q_{ij}, $i = 1, 2, 3$, $j = 1, 2$ express the amount ordered from the next highest echelon. The fixed cost of a joint order is always K. Formulate mathematically an expression for the total cost of operating this echelon system per unit time.

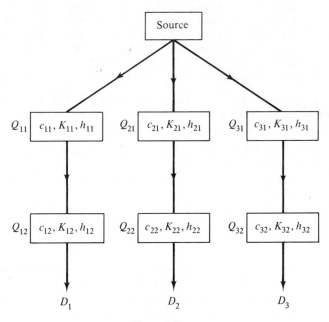

Figure P7.

8. A Christmas tree dealer is trying to determine how many trees of two kinds he may stock. He already has 100 large trees that cost him $6 each and that he can sell for $12 each. He is notified that a new shipment of trees has arrived and that no more shipments will be on their way. The new shipment includes large trees that would cost him again $6 apiece, as well as smaller trees that would cost him $4 apiece with a suggested sales price of $8 apiece. Any trees not sold during the season have to be disposed of at a complete loss. The dealer has to pay $200 for the rental of a large open lot to transact his business. It is estimated that demand for the large trees is normally distributed with mean 200 and a variance of 200, while the demand for the small trees is normally distributed with mean 300 and a variance of 200. Determine how many trees of each kind the dealer should buy to maximize his profit.

9. Establish an equivalence between a three-commodity single location inventory

system and a serial system consisting of three echelons. Develop in both cases expressions for the long run total cost per unit time consisting of procurement costs and holding costs. No shortages are allowed.

10. In the deterministic two-commodity single location system, let Q_1 and Q_2 be the order quantities for commodities 1 and 2 respectively. Assume that no shortage is allowed and that the storage capacity is limited so that

$$\alpha_1 Q_1 + \alpha_2 Q_2 \leq A$$

where $\alpha_1 > 0$, $\alpha_2 > 0$ and $A > 0$. Determine expressions for the optimal values of Q_1 and Q_2 which minimizes the long run total cost per unit time when using

(a) An individual ordering policy

(b) A joint ordering policy.

11. As in the single-commodity single installation inventory systems, it is possible to select for multi-commodity multi-installation inventory systems an objective function which minimizes the total discounted cost over an infinite horizon period. Formulate and solve the discounted problem for the deterministic m-commodity problem, $m = 1, 2, \ldots,$ when no shortages are allowed and under invariant setup cost.

REPLACEMENT THEORY

7.1. Introduction

The models developed in this chapter correspond to situations in which equipment deteriorates with age; that is, the longer the equipment is retained, the higher the cost of operating it. Thus, as an alternative, it may be profitable to acquire a new equipment that is more economical to operate. The fundamental problem that one is faced with is to make an appropriate balance between the cost of increased upkeep of the old equipment and the acquisition cost and reduced upkeep of a new equipment.

Equivalent models to replacement are generated by the set of maintenance-type problems, in which, following maintenance action, the equipment is restored to a state of as good as new. Thus, although in actual replacement problems the old equipment is physically substituted by another equipment, the process itself can be considered as one of regeneration whereby the equipment's age is brought back to zero, in a way similar to maintenance. Both replacement and maintenance theory involve the management of durable equipment.

In economic theory it is customary to divide the management of real property into the management of inventories and the management of durable equipment, the former giving rise to the theory of inventory of commodities

7

and the latter to the replacement and maintenance of equipment. Both theories are given a common economic interpretation through the theory of production, inventories supplying commodities to the production process, and durable equipment supplying services. This similarity goes beyond this conceptual level, for it is feasible to attach a common denominator to the mathematical theory underlying the abstract modeling of inventory and replacement; this matter will be discussed later in the chapter.

The reader is referred to Alchian [1], Dean [8], and Terborgh [17] for some of the early works in replacement theory. More recent works appear in the other references listed at the end of the chapter.

i. Classification of Equipment

Equipment may be classified as continuously operating equipment or intermittently operating equipment.

A *continuously operating* equipment is characterized by the fact that it is operated continuously in time. Surveillance radars and generators of power stations are typical examples. For this type of equipment, it is customary to measure aging along the time scale.

An *intermittently operating* equipment is characterized by the fact that its operation depends on the user's request; thus the equipment ages only when it provides service. A typical example of this equipment is a car.

Both continuously operating and intermittently operating equipment may be subject to breakdown and serviced for repair.

ii. *Input Parameters*

At this point, it is necessary to distinguish between two variables that play an important role in replacement problems; these are the equipment *service age* and the equipment *chronological age*.

The equipment *service age u* is defined as the total number of registered operational units by the equipment since acquisition. For example, in the case of a vehicle, the operational units would be miles, and the age would then be the number of miles registered by the vehicle odometer. The equipment *chronological age t* represents the clock time elapsed since acquisition date at zero service age. The average usage rate of the equipment since acquisition is u/t, while the instantaneous usage rate of an equipment in operation is du/dt.

For a continuously operating equipment not subject to breakdown and for which service aging is measured in time units, $du/dt = 1$; that is, service age is equal to chronological age.

Three basic cost parameters are generally incurred in replacement-type problems. These are

(1) The *nonrecurring* or *fixed cost K* in securing new equipment, expressed in terms of dollars. This usually includes the research and development costs, if any, the initial acquisition cost, the cost of equipment installation, and the cost of initial equipment break-in.

(2) The *recurring* or *operations cost* per unit of aging or usage of the existing equipment and the new equipment, expressed as a function of either service age or chronological age. This includes such costs as maintenance, repair, and actual running or operating cost. In general, for a large variety of equipment, this cost is an increasing function of equipment age; that is, the more the equipment is used, the higher the recurring cost. It is customary to express the recurring cost in the form of a function $C(u)$, which represents the instantaneous operations cost per unit of aging for an equipment of service age u. Thus, when the equipment ages from u to $u + du$, a total recurring cost in the amount $C(u)\,du$ is incurred. $C(u)$ will be termed the *operations* or *recurring cost rate function;* it will be assumed to be positive for all $u \geq 0$.

(3) The *salvage value* of discarded or replaced equipment. When a used equipment is replaced, it is often possible to salvage it and thus experience a dollar return. The salvage value of an equipment is a

nonincreasing function of its age. Rather than considering salvage as a separate entity, it is convenient to lump it with the nonrecurring cost of the new equipment to be acquired and redefine it as $K(X)$, where X is the replacement age of the old equipment. Thus $K(X)$ would be a nondecreasing function of X.

iii. *Replacement Policy—Decision Variables*

A replacement policy is a set of rules that dictates whether to keep or replace an existing equipment. In this chapter we shall discuss replacement policies based on service age, although similar-type policies may be formulated based on chronological age. Depending on whether the decision process to keep or replace an existing equipment is carried over continuous or discrete time, the theory analyzes continuous review replacement systems (Figure 7.1) or periodic review replacement systems (Figure 7.2).

For a continuous review replacement system, a typical replacement policy would dictate: "Replace the existing equipment when it reaches a specified age X by a new equipment; otherwise, keep the present equipment." The replacement age X is clearly a decision variable whose value may be determined by the selection of an appropriate objective function.

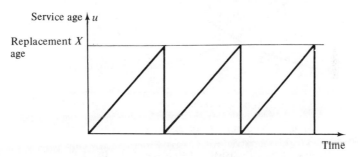

Figure 7.1. Continuous review replacement system.

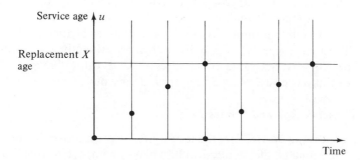

Figure 7.2. Periodic review replacement system.

iv. *Objective Functions*

A new equipment will usually lose value in a continuous fashion as it ages; simultaneously, the cost of repairs, operation, and maintenance will, in general, be an increasing function of the equipment age. The additive contribution of the loss in value and increased costs of upkeep could then be weighted with the higher value and lower operating costs associated with the acquisition of a new equipment.

In the simplest type of replacement situations, all pieces of equipment installed in the future are assumed to have identical cost and benefit characteristics as the existing equipment for which replacement is being contemplated. The problem then consists in determining the *optimum* or the *economic* replacement age, for replacing an existing equipment by a new equipment so as to minimize the total contribution of these three costs (Figure 7.3).

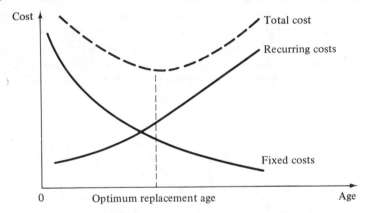

Figure 7.3. Optimum or economic replacement age.

To achieve optimal decisions, two types of objective functions may be selected:

(1) The total (expected) cost per unit time; equivalently, this is the total of all (expected) undiscounted costs of present and future investments and operations divided by the number of time units chosen.

(2) The total (expected) discounted cost; equivalently, this is the (expected) current value of all present and future investment and operations costs, using an appropriate discount factor.

v. *Economic Life and Service Life*

The concept of *service life*, (or useful life, or prospective life, or life expectancy, or estimated life, or physical life) plays an important role in an economic analysis. The service life of an equipment is the estimated number of years that equipment can physically be used in performing the function for which it was acquired.

The economic life of an equipment is determined on the basis of an economic analysis, whereas the service life is strictly a characteristic of the equipment. It is evident that the economic life cannot exceed the service life. Also, it is quite possible that an equipment may not possess an economic life. For example, equipment where the operations or recurring cost is expressed by a constant function can be shown to have an "infinite" time interval between acquisition and replacement; that is, it is most economical to keep the equipment until it reaches its service life prior to replacing it. Under these conditions, the service life should be used to perform an economic comparison between alternative courses of action. Such an analysis is justified only when an economic life cannot be determined. In what follows, we assume that the equipment has an economic life whose value is less than its service life.

vi. *Similarity Between Replacement Models and Inventory Models*

In a previous discussion we mentioned briefly that the modelings of re placement and inventory problems bear fundamental similarities. The reader will recall that in the basic inventory model for a single commodity discussed in Chapter 5, within the context of the formulation of a cost objective function, three input parameters, namely, order or setup cost, holding cost, and demand rate, interacted with a decision variable, the order quantity. The optimum order quantity was subsequently determined by balancing the contribution of the fixed charge with the holding cost. In a replacement problem the corresponding input parameters are, respectively, the fixed replacement cost, the operations cost, and the rate of service aging, while the decision variable is the replacement age. The optimum replacement age is determined by balancing the contribution of the fixed cost with the operations cost.

In general, whereas in an inventory problem the level of stock of a commodity defined a state forming the basis for a decision to either order or not, in a replacement problem the level of service age of the equipment item identifies the state upon which a decision to either replace or keep the equipment item is made. And whereas in an inventory problem stock level is reduced from demand requirement and increased through procurement, in a replacement problem service age is accumulated through equipment operation and terminated through equipment replacement.

This basic similarity between replacement models and inventory models will be pointed out throughout the present chapter. The reader may easily verify, for example, that if Figure 7.1 is rotated 180° about the abscissa, the familiar sawtooth diagram of the basic inventory model is obtained.

7.2. Continuous Review Replacement Systems: Deterministic Models

i. *Undiscounted Model*

a. AGED EQUIPMENT REPLACED BY EQUIPMENT OF THE SAME TYPE; NO SALVAGE VALUE

We begin our quantitative analysis with the development of a simple model, which we shall refer to as the *basic* model. In addition to playing a conceptual role, the solutions obtained from this model can often be shown to be good approximations to solutions of more complex replacement problems. The following assumptions underlie this model:

(1) All present and future equipment items operate continuously in time, and equipment aging is expressed as the amount of time the equipment has been in operation; thus, in a time interval θ, the equipment ages by a quantity θ.

(2) Equipment downtime associated with failure, repair, maintenance, or otherwise, is neglected.

(3) At time origin, the old equipment has just reached its replacement age.

(4) The planning horizon is infinite.

(5) All pieces of equipment installed in the future have identical characteristics.

(6) Replacement occurs according to the following policy: "Replace an equipment whenever it reaches age X by a new identical equipment having age zero."

(7) The two competing cost elements are the fixed cost in replacing an existing equipment and the recurring costs associated with the operations of an existing equipment (no salvage value).

(8) The objective function is selected as the long run total cost per unit time.

The problem consists then in determining the optimal value of the replacement age X so as to minimize the long run total cost per unit time. Let

$K =$ nonrecurring cost or initial fixed cost

$C(u) =$ operations or recurring cost rate function expressed in terms of equipment age u, $0 \leq u < \infty$

$X =$ replacement age of equipment

Consider one replacement cycle consisting of an interval between two succes-

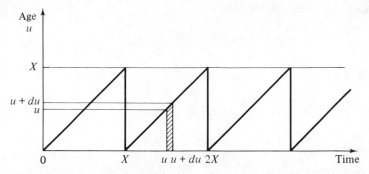

Figure 7.4. Replacement cycles of aging equipment in continuous review systems.

sive replacements. Referring to Figure 7.4, the total cost incurred over the cycle is

$$K + \int_0^X C(u)\,du$$

Therefore, the long-run total cost per unit time expressed as a function of X is

$$E(X) = \frac{K + \int_0^X C(u)\,du}{X} \tag{7.1}$$

We shall next investigate conditions under which $E(X)$ achieves a relative minimum. To obtain the optimal value of X, say X^*, differentiate $E(X)$ with respect to X, and set the result equal to zero; then

$$\frac{dE(X)}{dX} = 0 = \frac{XC(X) - \left[K + \int_0^X C(u)\,du\right]}{X^2}$$

or

$$XC(X) - \left[K + \int_0^X C(u)\,du\right] = 0$$

or

$$XC(X) - \int_0^X C(u)\,du = K \tag{7.2}$$

Integrating the quantity $\int_0^X C(u)\,du$ by parts, we obtain

$$\int_0^X C(u)\,du = uC(u)\Big|_0^X - \int_0^X u\,dC(u)$$

$$= XC(X) - \int_0^X u\,dC(u)$$

Substituting in equation (7.2) yields

$$XC(X) - \left[XC(X) - \int_0^X u \, dC(u) \right] = K$$

Thus if an $X^* > 0$ exists, it should be the solution to the equation

$$\int_0^{X^*} u \, dC(u) = K \tag{7.3}$$

Example 7.1

Let us assume that the operations cost rate function is linear; that is,

$$C(u) = au \qquad a > 0, 0 \leq u < \infty$$

Then equation (7.3) becomes

$$\int_0^X au \, du = K$$

Hence

$$\frac{aX^2}{2} = K$$

and the optimum replacement age is

$$X^* = \sqrt{\frac{2K}{a}}$$

This formula should be compared with the familiar Wilson's square root formula in inventory theory (see Chapter 5).

Example 7.2

Let $C(u)$ have the exponential form

$$C(u) = ae^{bu}, \quad a > 0, b > 0, 0 \leq u < \infty$$

Then equation (7.3) becomes

$$\int_0^X abue^{bu} \, du = K$$

Integrating by parts, we obtain

$$aXe^{bX} - \int_0^X ae^{bu} \, du = K$$

or

$$aXe^{bX} - \frac{a}{b}(e^{bX} - 1) = K$$

Thus the optimum replacement age is solution to the transcendental equation

$$bX - 1 = \left(\frac{Kb}{a} - 1\right)e^{-bX}$$

provided $Kb/a > 1$. Let $Y = bX$; then

$$Y - 1 = \left(\frac{Kb}{a} - 1\right)e^{-Y}$$

A graphical procedure to solve this equation for Y is illustrated in Figure 7.5.

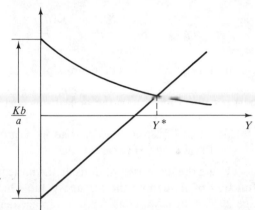

Figure 7.5. Solution of the equation $Y - 1 = \left(\frac{Kb}{a} - 1\right)e^{-Y}$.

Now we shall prove that optimality at $X^* > 0$ is guaranteed if $C(u)$ is strictly increasing in u. A condition for the existence of a relative minimum for $E(X)$ is for the second derivative of $E(X)$ with respect to X evaluated at X^* to be positive. Now from (7.1) we obtain after simplification

$$\frac{d^2 E(X)}{dX^2} = \frac{X^2 C'(X) - 2XC(X) + 2K + 2\int_0^X C(u)\,du}{X^3}$$

$$= \frac{C'(X)}{X} - \frac{2}{X^3}\left[XC(X) - \int_0^X C(u)\,du - K\right] \tag{7.4}$$

Using equation (7.2) in expression (7.4), we obtain

$$\left.\frac{d^2 E(X)}{dX^2}\right|_{X=X^*} = \frac{C'(X^*)}{X^*}$$

We thus note that if $C(u)$ is a strictly increasing function of u, $0 \leq u < \infty$, then $C'(u) > 0$ for all values of u; hence, provided $X^* > 0$, $[d^2 E(X)/dX^2]$ > 0, and an optimal solution will always exist that will minimize $E(X)$.

The reader should verify that if $C(u) = a$ (a constant) for $0 \leq u < \infty$, then $E(X)$ is a strictly decreasing function of X; thus it is optimal to keep the equipment for as long as possible. Situations in which the equipment upkeep is not changing with the length of equipment operation are quite common in practice, a light bulb or a hydraulic pipe being typical examples. For such cases, it is most economical to replace the equipment when it reaches its useful life, a rather routine and intuitive practice.

An interesting economic interpretation may be attached to the optimal solution to the replacement problem, if such a solution exists. At $X = X^*$, relation (7.2) may be written as

$$\frac{K + \int_0^{X^*} C(u)\, du}{X^*} = C(X^*)$$

But from (7.1), the left side expression is $E(X^*)$. Thus, at optimality $E(X^*)$ $= C(X^*)$. Thus, the equipment should be replaced at age X^* when the total cost per unit time equals the operations cost rate at age X^*.

b. Aged Equipment Replaced by Equipment of the Same Type; Effect of Salvage Value

Using the same assumptions as the basic model, let $S(X)$, a nonincreasing function of X, denote the salvage value when the equipment is replaced at age X. Again, if we consider one replacement cycle, the expression for the total cost per unit time is

$$E(X) = \frac{K + \int_0^X C(u)\, du - S(X)}{X}$$

For convenience, and without loss in generality, we define the function

$$K(X) = K - S(X)$$

This function defines the net cost of acquistion when replacing the old equipment age X with a new equipment. In the special case when $K(X) = K$ for all values of X, then salvage value equals zero. Since $S(X)$ is a decreasing function of X, then $K(X)$ will be a nondecreasing function of X, (see Figure 7.6). The expression for the long run total cost per unit time becomes

$$E(X) = \frac{K(X) + \int_0^X C(u)\, du}{X}$$

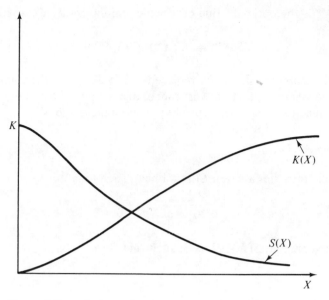

Figure 7.6 Variations of the salvage function $S(X)$ and $K(X)$ $= K - S(X)$.

Differentiating $E(X)$ with respect to X and setting the result equal to zero, we obtain

$$\frac{dE(X)}{dX} = 0 = \frac{X[K'(X) + C(X)] - \left[K(X) + \int_0^X C(u)\, du\right]}{X^2}$$

or

$$X[K'(X) + C(X)] - \left[K(X) + \int_0^X C(u)\, du\right] = 0 \qquad (7.5)$$

Integrating $\int_0^X C(u)\, du$ by parts yields

$$XK'(X) + XC(X) - K(X) - XC(X) + \int_0^X u\, dC(u) = 0$$

Thus if an optimum value of $X > 0$ exists, it is a solution to the equation

$$\int_0^X u\, dC(u) = K(X) - XK'(X)$$

We also note that expression (7.5) may be written as

$$\frac{K(X) + \int_0^X C(u)\, du}{X} = K'(X) + C(X)$$

Thus, if X^* is the optimal value of X which minimizes $E(X)$, we must have

$$E(X^*) = K'(X^*) + C(X^*)$$

Hence, the equipment should be replaced at age X^* when the total cost per unit time equals the operations cost rate at age X^* plus the rate of change of the net acquisition cost when replacing an equipment age X^*.

Example 7.3

Assume that the function $C(u)$ is linear, such that

$$C(u) = au, \quad a > 0$$

We also assume that $S(X)$ is linear, such that for $\alpha > 0$,

$$S(X) = \begin{cases} K - \alpha X, & 0 \leq X < \dfrac{K}{\alpha} \\[2mm] 0, & \dfrac{K}{\alpha} \leq X < \infty \end{cases}$$

Then, the function $K(X) = K - S(X)$ is

$$K(X) = \begin{cases} \alpha X, & 0 \leq X < \dfrac{K}{\alpha} \\[2mm] K, & \dfrac{K}{\alpha} \leq X < \infty \end{cases}$$

We consider the two cases when $0 \leq X < K/\alpha$ and $K/\alpha \leq X < \infty$ (see Figure 7.7).

Case 1: $0 \leq X < \dfrac{K}{\alpha}$

The expression for the long run total cost per unit time is

$$E(X) = \frac{\alpha X + \displaystyle\int_0^X au \, du}{X}$$

$$= \alpha + \frac{aX}{2}$$

which is an increasing function of X.

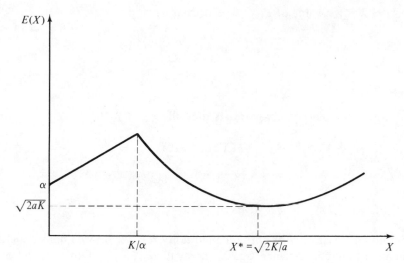

Figure 7.7. The function $E(X) = \begin{cases} \alpha = (aX/2), & 0 \leq X < K/\alpha \\ (K/X) + (aX/2), & K/\alpha \leq X < \infty \end{cases}$

Case 2: $\dfrac{K}{\alpha} \leq X < \infty$

The expression for the long run total cost per unit time is

$$E(X) = \frac{K + \int_0^X au\,du}{X}$$

$$= \frac{K + \frac{1}{2}aX^2}{X}$$

$$= \frac{K}{X} + \frac{aX}{2}$$

which is a convex function of X. A minimum at an interior point of the interval $K/\alpha \leq X < \infty$ will exist at

$$X^* = \sqrt{\frac{2K}{a}}$$

iff

$$X^* > \frac{K}{\alpha}$$

or

$$\sqrt{\frac{2K}{a}} > \frac{K}{\alpha}$$

or

$$K < \frac{2\alpha^2}{a}$$

At $X \doteq X^*$, the value of $E(X)$ would then be

$$E(X^*) = K\sqrt{\frac{a}{2K}} + \frac{a}{2}\sqrt{\frac{2X}{a}}$$
$$= \sqrt{2aK}$$

An optimum replacement age exists then iff

$$E(X^*) < E(0)$$

or

$$\sqrt{2aK} < \alpha$$

or

$$K < \frac{\alpha^2}{2a}$$

Combining this inequality with the inequality $K < 2\alpha^2/a$, we finally conclude that an optimum replacement age X^* exists iff $K < \alpha^2/2a$. Otherwise an optimum or an economic replacement age does not exist.

ii. *Discounted Model*

a. AGED EQUIPMENT REPLACED BY EQUIPMENT OF THE SAME TYPE

In dealing with an objective function that measures the total discounted cost, the present worth of all future investments and operations cost are evaluated using appropriate discounting parameters. We shall first consider a situation identical to the basic model, where the equipment has age X at time origin and is to be replaced immediately; later the existing equipment age will be assumed to be arbitrary. We shall use the same symbols and notations as in the basic model, and, again, follow a replacement policy whereby the equip-

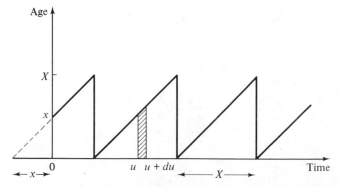

Figure 7.8. Discounting cost over a given replacement cycle.

ment is replaced once it reaches a given age X. Let

$i =$ interest rate in dollars/[(dollar)(unit time)]

$e^{-i} =$ discount factor over continuous time

(1) Discounted Replacement Model; No Salvage Value

Over one replacement cycle of length X, the discounted cost at the beginning of the cycle is (Figure 7.8)

$$K + \int_0^X e^{-iu}C(u)\,du$$

The total discounted cost starting with an equipment of age X is

$$\bar{E}(X) = \left[K + \int_0^X e^{-iu}C(u)\,du\right] + e^{iX}\left[K + \int_0^X e^{-iu}C(u)\,du\right]$$

$$+ e^{-2iX}\left[K + \int_0^X e^{-iu}C(u)\,du\right] + \cdots$$

$$= \frac{K + \int_0^X e^{-iu}C(u)\,du}{1 - e^{-iX}}$$

This expression bears similarity with expression (7.1). We shall now show that irrespective of the initial equipment age x, $0 \le x \le X$, the optimization problem with respect to X is not affected.

The total discounted cost starting with an equipment of age 0 is

$$\bar{E}_1(X) = \int_0^X e^{-iu}C(u)\,du + e^{-iX}\left[K + \int_0^X e^{-iu}C(u)\,du\right] + \cdots$$

$$= \frac{Ke^{-iX} + \int_0^X e^{-iu}C(u)\,du}{1 - e^{-iX}} = \frac{K + \int_0^X e^{-iu}C(u)\,du}{1 - e^{-iX}} - K$$

The total discounted cost starting with an equipment of age x, $(0 \le x < X)$ is

$$\bar{E}_2(X) = \int_x^X e^{-iu}C(u)\,du + e^{-iX}\left[K + \int_0^X e^{-iu}C(u)\,du\right] + \cdots$$

Now

$$\int_x^X e^{-iu}C(u)\,du = \int_0^X e^{-iu}C(u)\,du - \int_0^x e^{-iu}C(u)\,du$$

Hence

$$\bar{E}_2(X) = -\int_0^x e^{-iu}C(u)\,du + \frac{Ke^{-iX} + \int_0^X e^{-iu}C(u)\,du}{1 - e^{-iX}}$$

$$= \frac{K + \int_0^X e^{-iu}C(u)\,du}{1 - e^{-iX}} - \left(K + \int_0^x e^{-iu}C(u)\,du\right)$$

It is evident that in all cases the minimization of the total discounted costs $\bar{E}(X)$, $\bar{E}_1(X)$, and $\bar{E}_2(X)$ with respect to X will yield the same optimal value for X.

Prior to solving this optimization problem, we provide an alternative formulation to the discounted problem and simultaneously consider the more general problem where K is a function of the replacement age X, thus accounting for salvage value. We shall derive a functional equation to arrive at an explicit expression for the total discounted cost.

(2) Use of the Functional Equation in the Discounted Replacement Model; Effect of Salvage Value

Define

$C(u) =$ operating cost rate function for the equipment aged u;
$\quad i =$ interest rate;
$\quad X =$ replacement age of equipment;
$K(X) =$ nonrecurring costs (fixed cost plus salvage value) expressed as a function of the replacement age X;
$g_0(x) =$ minimum total discounted cost when the initial equipment age is x, $x \geq 0$, and no replacement is initiated, that is, the equipment is kept;
$g_1(x) =$ minimum total discounted cost when the initial equipment age is x, $x \geq 0$, and a replacement is initiated.

We consider each of the functions $g_0(x)$ and $g_1(x)$ separately.

We note that if a replacement is initiated at age x, an immediate cost of $K(x)$ is incurred; further, the minimum total discounted cost associated with the new equipment age zero, replacing the old one is $g_0(0)$. Thus

$$g_1(x) = K(x) + g_0(0), \quad 0 \leq x < \infty$$

If now initially, $t = 0$, no replacement takes place, and the equipment is allowed to operate, then the corresponding total discounted cost is $g_0(x)$. To structure the function $g_0(x)$, we consider the state of the equipment when it ages by an amount dx from time $t = 0$ to time $t = dt$ and reaches age $x + dx$. During this interval dt, an operations cost in the amount $C(x)\,dx$ will be experienced. When the equipment reaches age $x + dx$ at time dt, the total discounted cost from time dt onwards is $g_0(x + dx)$, and this last quantity, discounted to time $t = 0$, is $(1 - i\,dt)g_0(x + dx)$. We then have

$$g_0(x) = C(x)\,dx + (1 - i\,dt)g_0(x + dx), \quad 0 \leq x < \infty$$

The solution to this functional equation will yield the explicit form of the function $g_0(x)$, whose structure may be noted to depend on the function $C(x)$.

Figure 7.9 displays the functions $g_1(x)$ and $g_0(x)$ for $0 \leq x < \infty$. It is

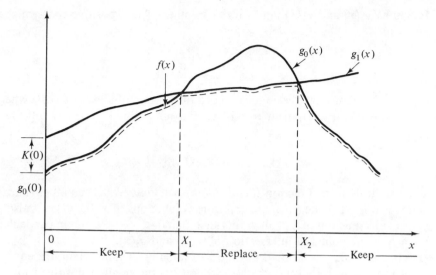

Figure 7.9. Variation of $f(x) = \text{Min}\{g_0(x), g_1(x)\}$, the minimum total discounted cost function when starting with an equipment age x.

clear that if $K(x)$ is continuous and a nondecreasing function of x, then $g_1(x)$ is also continuous and a nondecreasing function of x. Further, since

$$g_1(0) = K(0) + g_0(0)$$

it follows that

$$g_1(0) > g_0(0)$$

For the time being, we shall assume $g_0(x)$ to be a continuous function of x. Its variation will depend on the function $C(x)$, and may look, for example, as in Figure 7.9.

It is evident that the *optimal* policy to be opted, will depend on the values of the functions $g_1(x)$ and $g_0(x)$. Thus, whenever

$$g_0(x) < g_1(x)$$

a no replacement or keep policy will be followed, while if

$$g_0(x) > g_1(x)$$

then a replacement will be initiated. For the particular functions $g_0(x)$ and $g_1(x)$ exhibited in Figure 7.9, the equipment will be kept if its initial age x is such that $0 \le x < X_1$. It will be replaced, if $X_1 \le x < X_2$. Finally, it will be kept if $X_2 \le x < \infty$. The age intervals over which the equipment is kept or replaced, following an optimal policy, will depend on the variations of the functions $g_0(x)$ and $g_1(x)$, and on the number of real positive roots of the equation

$$g_0(x) = g_1(x)$$

If now we let

$f(x) =$ minimum total discounted cost to time origin of all future costs when the initial equipment age is x, and an optimal policy is followed

then

$$f(x) = \min \{g_0(x), g_1(x)\}$$

The variation of the function $f(x)$ is as shown in Figure 7.9. We note, in particular, that if $g_0(x)$ and $g_1(x)$ are continuous functions of x, then $f(x)$ is also a continuous function of x at all points such that $0 \le x < \infty$, and in particular at each of the points of intersection of $g_0(x)$ and $g_1(x)$.

In what follows, we shall assume that initially, at time origin, the equipment age x is such that $0 \le x \le X$, where X is the smallest positive root of the equation $g_0(x) = g_1(x)$. Thus X is the replacement age when following an optimal policy. It is also clear that at no future time can the equipment age exceed the value of X. Within the framework of this assumption, we may then write

$$f(x) = \min \begin{cases} g_1(x), & x = X \\ g_0(x), & 0 \le x < X \end{cases}$$

and the continuity condition at $x = X$, namely

$$g_1(X) = g_0(X)$$

Equivalently, $f(x)$ may be written as:

$$f(x) = \min \begin{cases} K(x) + f(0), & x = X \\ C(x)\, dx + (1 - i\, dt)f(x + dx), & 0 \le x < X \end{cases}$$

In this functional equation the first alternative represents the cost of replacing the existing equipment once it reaches age X by a new equipment, thus incurring an immediate replacement cost $K(X)$ plus the discounted cost associated with all future programs when starting with an equipment age 0. In the second alternative, no replacement occurs while the equipment ages from x to $x + dx$ in an interval of time dt; this results in the incurrence of a recurring cost in the amount of $C(x)\, dx$ plus the discounted cost to time origin of the discounted cost associated with all future programs when the equipment has age $x + dx$.

For $0 \leq x < X$, when no replacement occurs, $f(x)$ satisfies the relation

$$f(x) = C(x)\, dx + (1 - i\, dt)\, f(x + dx)$$
$$= C(x)\, dx + \left(1 - i\frac{dt}{dx}\, dx\right) f(x + dx)$$

We assume that when the equipment ages continuously in time, such aging is measured in time unit and is independent of the age of the equipment and time, so that $(dx/dt) = 1$. We then have for $f(x)$ the relation

$$f(x) = C(x)\, dx + (1 - i\, dx)\, f(x + dx)$$
$$= C(x)\, dx + f(x + dx) - if(x + dx)\, dx$$

Equivalently, this may be written

$$\frac{f(x + dx) - f(x)}{dx} - if(x + dx) = C(x)$$

Taking limits on both sides of this relation as $dx \to 0$, we obtain

$$\lim_{dx \to 0} \left\{ \frac{f(x + dx) - f(x)}{dx} - if(x + dx) \right\} = C(x)$$

Assuming $f(x)$ is differentiable in the interval $0 \leq x < X$, we can write

$$\frac{df(x)}{dx} - if(x) = C(x), \quad 0 \leq x < X$$

The solution of this linear differential equation may be obtained by multiplying both sides of the equation by the integrating factor e^{-ix}, to yield the differential equation

$$\frac{d}{dx}[e^{-ix}f(x)] = e^{-ix}C(x)$$

Integration of this equation yields for solution

$$f(x) = e^{ix}f(0) - e^{ix} \int_0^x e^{-iu}C(u)\, du, \quad 0 \leq x < X \qquad (7.6)$$

where we have selected as a convenient arbitrary constant the value of $f(0)$. The economic interpretation of this relation is as follows: the total discounted cost to time origin starting with an equipment of age x is equal to the total discounted cost to time $t = -x$, that is,

$$f(0) - \int_0^x e^{-iu}C(u)\, du$$

compounded back to time origin, resulting in the product of this last expression by e^{ix}.

At $x = X$, a replacement occurs, and $f(x)$ satisfies the relation

$$f(X) = K(X) + f(0) \tag{7.7}$$

From expressions (7.6) and (7.7), the continuity of $f(x)$ at $x = X$ would imply

$$K(X) + f(0) = e^{iX}f(0) - e^{iX} \int_0^X e^{-iu}C(u) \, du$$

Hence, solving for $f(0)$, we obtain

$$f(0) = \frac{K(X) + e^{iX} \int_0^X e^{-iu}C(u) \, du}{e^{iX} - 1}$$

$$= \frac{K(X)e^{-iX} + \int_0^X e^{-iu}C(u) \, du}{1 - e^{-iX}}$$

This last expression measures the total discounted cost when starting with an equipment age zero. The replacement age X is to be selected to satisfy the optimality condition, that is to minimize $f(0)$. Also, from relation (7.6), it is evident that, given $x \min_x f(x)$ is equivalent to $\min_x f(0)$.

(3) Optimization Problem When $K(X) = K = $ Constant

We note then that $f(0) = \bar{E}_1(X)$. Since the minimization of the three functions $\bar{E}(X)$, $\bar{E}_1(X)$, and $\bar{E}_2(X)$ will yield the same optimum, consider the function

$$\bar{E}(X) = \frac{K + \int_0^X e^{-iu}C(u) \, du}{1 - e^{-iX}} \tag{7.8}$$

Differentiating $\bar{E}(X)$ with respect to X, and setting the result to zero, yields

$$(1 - e^{-iX})e^{-iX}C(X) - ie^{-iX}\left[K + \int_0^X e^{-iu}C(u) \, du\right] = 0 \tag{7.9}$$

The optimum value of X, say X^*, is given as the solution to the equation

$$\frac{K + \int_0^{X^*} e^{-iu}C(u) \, du}{1 - e^{-iX^*}} = \frac{C(X^*)}{i}$$

or

$$\bar{E}(X^*) = \frac{C(X^*)}{i}$$

Thus the optimal policy is to replace the equipment at age X^* when the total discounted cost starting with an equipment of age X^* equals $C(X^*)/i$. Relation (7.9) can be written after dividing both sides by e^{-iX}

$$C(X) - e^{-iX}C(X) - iK - i\int_0^X e^{-iu}C(u)\, du = 0 \qquad (7.10)$$

Integrating by parts the integral expression, we obtain

$$i\int_0^X e^{-iu}C(u)\, du = \left| -e^{-iu}C(u) \right|_0^X + \int_0^X e^{-iu}\, dC(u)$$
$$= -e^{-iX}C(X) + C(0) + \int_0^X e^{-iu}\, dC(u) \qquad (7.11)$$

Substituting expression (7.11) in equation (7.10) yields

$$C(X) - iK - C(0) = \int_0^X e^{-iu}\, dC(u)$$

or

$$\int_0^X dC(u) - iK = \int_0^X e^{-iu}\, dC(u)$$

and finally

$$\int_0^X (1 - e^{-iu})\, dC(u) = iK \qquad (7.12)$$

The solution to this algebraic equation will yield the optimal value of X, X^*. In general, the equation will have a transcendental form, and may be solved either numerically or graphically. The reader should verify that if the function $C(u)$ is a strictly increasing function of u to infinity a minimum solution always exists.

Example 7.4

Let
$$C(u) = au + b, \quad a > 0, b > 0$$
Then
$$dC(u) = a\, du$$

and the optimal value of X is given from (7.12) by the equation

$$a\int_0^X (1 - e^{-iu})\, du = iK$$

or

$$\left| u + \frac{1}{i}e^{-iu} \right|_0^X = \frac{Ki}{a}$$

or

$$X + \frac{1}{i}e^{-iX} - \frac{1}{i} = \frac{Ki}{a}$$

or

$$e^{-iX} = 1 - iX + \frac{Ki^2}{a}$$

The optimal value of $iX = Y$ is a solution to the equation

$$e^{-Y} = \left(1 + \frac{Ki^2}{a}\right) - Y$$

A graphical solution to this equation may be obtained by plotting the two functions e^{-Y} and $[1 + (Ki^2/a)] - Y$ and determining the abscissa of their intercept. This is depicted graphically in Figure 7.10. For small values of Y, e^{-Y} may be expanded as a power series of Y to yield

$$1 - Y + \frac{Y^2}{2!} - \cdots \approx 1 + \frac{Ki^2}{a} - Y$$

Thus

$$Y \approx \sqrt{\frac{2Ki^2}{a}}$$

Hence

$$X^* \approx \sqrt{\frac{2K}{a}}$$

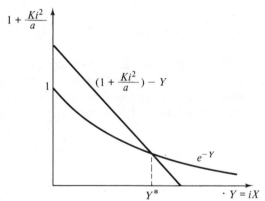

Figure 7.10. On the solution of the equation $e^{-Y} = \left(1 + \frac{Ki^2}{a}\right) - Y.$

b. Aged Equipment Replaced by Equipment of a Different Type

Suppose that at the time of replacing the existing equipment (equipment 1), the available new equipment (equipment 2) has different cost and benefit characteristics. Furthermore, assume that at the time of replacing equipment 2 the new piece of equipment available (equipment 3) is different from equipment 2, and so on, until the replacement of equipment N. Following this last replacement, all other equipment to be installed in the future are assumed to have identical characteristics to equipment N. For equipment $j, j = 1, 2, \ldots, N$, let

$X_j =$ replacement age

$K_j(X_j)$ = nonrecurring cost of replacing equipment j at age X_j by new equipment $j + 1$

$C_j(u)$ = recurring cost rate function for equipment j

$f_j(x)$ = minimum total discounted cost when starting with equipment j age x and following an optimal policy

Then, we obtain the following system of functional equations for the functions $f_j(x)$

$$f_N(x) = \min \begin{cases} K_N(x) + f_N(0), & x = X_N \\ C_N(x)\, dx + (1 - i\, dt)f_N(x + dx), & 0 \le x < X_N \end{cases} \quad (7.13)$$

and, for $j = 1, 2, \ldots, N - 1$,

$$f_j(x) = \min \begin{cases} K_j(x) + f_{j+1}(0), & x = X_j \\ C_j(x)\, dx + (1 - i\, dt)f_j(x + dx), & 0 \le x < X_j \end{cases} \quad (7.14)$$

The procedure for solving these equations is to work backward, starting from $j = N$, and recursively until an expression for $f_1(x)$ is obtained. As defined, $f_1(x)$ is the minimum total discounted cost starting with the existing equipment age x and following an optimal policy at all future times. The solution for $f_N(x)$ and the determination of X_N are computed in a way similar to $f(x)$ and X previously obtained in Section 7.2-ii-a-(2).

To illustrate the procedure, consider the case when $N = 2$; then, for $0 \le x < X_2$, we have, from equation (7.6), as solution to (7.13)

$$f_2(x) = e^{ix}f_2(0) - e^{ix} \int_0^x e^{-iu}C_2(u)\, du$$

When $x = X_2$, replacement occurs, and we have

$$f_2(X_2) = K_2(X_2) + f_2(0)$$

Continuity at $x = X_2$ gives the condition

$$K(X_2) + f_2(0) = e^{iX_2}f_2(0) - e^{iX_2} \int_0^{X_2} e^{-iu}C_2(u)\, du$$

Hence
$$f_2(0) = -K_2(X_2) + \frac{K_2(X_2) + \int_0^{X_2} e^{-iu}C_2(u)\, du}{1 - e^{-iX_2}} \quad (7.14)$$

Let X_2^* be the value of X_2 that minimizes $f_2(0)$. Now for $j = 1$ we have, from (7.14), when $0 \le x < X_1$,

$$f_1(x) = e^{ix}f_1(0) - e^{ix} \int_0^x e^{-iu}C_1(u)\, du$$

and for $x = X_1$

$$f_1(X_1) = K_1(X_1) + f_2(0)$$

Again, continuity at $x = X_1$ yields the condition

$$K_1(X_1) + f_2(0) = e^{iX_1}f_1(0) - e^{iX_1} \int_0^{X_1} e^{-iu} C_1(u) \, du$$

Hence, solving for $f_1(0)$, we get

$$f_1(0) = \int_0^{X_1} e^{-iu} C_1(u) \, du + e^{-X_1}[K_1(X_1) + f_2(0)]$$

Since $\min_{X_1} f_1(x)$ and $\min_{X_1} f_1(0)$ will yield the same optimum value of X_1, say X_1^*, the latter can then be obtained directly from this last expression knowing the explicit form of $f_2(0)$ as given by (7.15).

In the special case when $K_1(x) = K_2(x) = K = $ constant, we obtain by minimizing $f_2(0)$ with respect to X_2

$$\frac{C_2(X_2^*)}{i} = K + f_2(0)$$

It is then easily verified that the minimization of $f_1(0)$ with respect to X_1 yields

$$\frac{C_1(X_1^*)}{i} = K + f_2(0) = \frac{C_2(X_2^*)}{i}$$

Knowing the value of X_2^*, the value of X_1^* may be obtained from the relation $C_1(X_1^*) = C_2(X_2^*)$.

As would have been expected, the determination of the optimal value X_2^* in no way depends on the replacement age X_1^* of the old equipment; however, the optimal value X_1^* can only be determined once the value of X_2^* is known. This may be accomplished by a simple graphical procedure. The two functions $C_1(u)$ and $C_2(u)$ are plotted on the same graph as functions of u (Figure 7.11). A vertical line drawn from $u = X_2^*$ intersects $C_2(u)$ at a point D. A horizontal line drawn from D intersects $C_1(u)$ at a point B. Finally, X_1^* is determined as the abscissa of the point B.

Extending the analysis to the case of arbitrary N, $N = 1, 2, \ldots$, it can be verified that if the fixed acquisition cost K is the same for all replacements, the optimum values of $X_1, X_2, \ldots, X_n$ are solutions to the system of equations:

$$i[K + f_N(0)] = C_N(X_N^*) = C_{N-1}(X_{N-1}^*) = \cdots = C_2(X_2^*) = C_1(X_1^*)$$

An optimal solution will always exist if the functions $C_j(u)$, $j = 1, 2, \ldots, N$, $0 \leq u < \infty$, are strictly increasing functions of u to infinity.

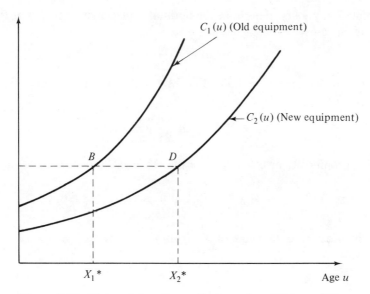

Figure 7.11. Determining the optimum value X_1^* from X_2^*, when old equipment is replaced by equipment of different type.

Example 7.5

A piece of equipment is presently 12 years old, and replacement is being contemplated by a new piece of equipment. Salvage value is negligible, and it is assumed that all future replacements will be made with equipment having identical cost characteristics as the new equipment. Assume that the fixed cost of acquisition is \$120,000 and the operations cost rate functions for the present equipment and the new equipment are respectively

$$C_1(u) = 500 + 2{,}400u \text{ dollars per year}$$

and $\qquad C_2(u) = 600 + 3{,}600u$ dollars per year

If interest is computed at the rate of .100 \$/[(\$)(year)], determine whether the present equipment should be replaced.

The parameters of interest are:

$$x = 12 \text{ years}$$
$$K = \$120{,}000$$
$$i = .10 \text{ \$/[(\$)(year)]}$$

Since the operations cost rate functions are linear, we will use the results of

Example 7.4. We first determine the optimal replacement age X_2^* of the new equipment. Now

$$C_2(u) = b_2 + a_2 u$$

where $b_2 = 600$ and $a_2 = 3,600$. Let $i X_2^* = Y_2$. Then Y_2 is the solution to the equation

$$e^{-Y_2} = \left(1 + \frac{Ki^2}{a_2}\right) - Y_2$$

Solving graphically or otherwise, we find $Y_2 = .945$, hence

$$X_2^* = \frac{.945}{.10} = 9.45 \text{ years}$$

We now determine the value of X_1^* from the equation

$$C_1(X_1^*) = C_2(X_2^*)$$

or

$$500 + 2,400 X_1^* = 600 + 3,600 X_2^*$$
$$= 600 + (3,600)(9.45)$$

Solving for X_1^* we obtain

$$X_1^* = \frac{600 + (3,600)(9.45) - 500}{2,400}$$

$$= \frac{34,100}{2,400} = 14.25 \text{ years}$$

Hence, the present equipment should be kept for 2.25 additional years before being replaced by the new equipment.

iii. *Replacement of Multiple Equipment Items*

Many practical situations do arise in which a decision maker is faced with the problem of managing various equipment items performing identical tasks. The replacement policy may be on an individual basis; that is, each and every equipment item is managed separately and independently from any other equipment item. An alternative replacement policy would be to consider the aggregate set of equipment items and replace all of them by a new set of equipment items. Since, in general, the existing equipment items are of different age, a group replacement policy will necessitate the determination of the proper replacement timing. In practice, several factors may weigh favorably toward the use of a group replacement policy. For example, a quantity discount may be available in the acquisition of new equipment items or it may be possible to substitute the existing equipment items with a set of new equipment items not necessarily in the same number, but performing identical functions.

The selection of an individual replacement policy or a group replacement policy will ultimately be dictated by economic considerations. Thus the minimum cost for each of the two policies should be computed separately, and the policy yielding the least cost should be implemented.

In what follows we develop a discounted cost model for the replacement of multiple equipment items. We assume that all future replacements are made with equipment items identical to the new equipment items. Further, we assume that the acquisition cost has a fixed component and a component which is proportional to the number of equipment units acquired.

For the replacement of N existing equipment items, let

$\quad C_j(u) =$ operations cost rate function of equipment item j,
$\qquad j = 1, 2, \ldots, N$
$\quad x_j =$ age of equipment item j at time $t = 0$
$\quad \theta =$ time at which the first joint replacement is to occur
$\quad X =$ replacement age in future generations
$\quad I =$ interest rate
$\quad C(u) =$ operations cost rate function of new equipment items
$\quad \tilde{K} =$ fixed cost associated with an acquisition
$\quad A =$ unit cost (variable) associated with the acquisition of a single new equipment item
$\quad \tilde{N} =$ number of new equipment items

The following is obtained for the discounted cost for equipment item $j, j = 1, 2, \ldots, N$, just prior to its first replacement:

$$e^{ix_j}\left[\int_0^{x_j+\theta} e^{-iu}C_j(u)\,du - \int_0^{x_j} e^{-iu}C_j(u)\,du\right] = e^{ix_j}\int_{x_j}^{x_j+\theta} e^{-iu}C_j(u)\,du$$

The total discounted cost of existing items and all future acquisitions is:

$$F(X, \theta) = \sum_{j=1}^{j=N} e^{ix_j}\int_{x_j}^{x_j+\theta} e^{-iu}C_j(u)\,du + e^{-i\theta}\frac{(\tilde{K} + A\tilde{N}) + \tilde{N}\int_0^X e^{-iu}C(u)\,du}{1 - e^{iX}}$$

The optimal value of X and θ will be determined. The equation

$$\frac{\partial F(X, \theta)}{\partial X} = 0$$

yields

$$\frac{(\tilde{K} + A\tilde{N}) + \tilde{N}\int_0^X e^{-iu}C(u)\,du}{1 - e^{-iX}} = \frac{\tilde{N}C(X)}{i}$$

from which the optimal value of X, X^*, can be obtained.

This equation and the one obtained for the single equipment discounted model are identical, with K replaced by $(\tilde{K} + A\tilde{N})$ and $C(u)$ replaced by $\tilde{N}C(u)$. Thus from (7.12) the optimal value of X is a solution to the equation

$$\tilde{N}\int_0^X (1 - e^{-iu})\, dC(u) = i(\tilde{K} + A\tilde{N})$$

The optimum value θ^* of θ is obtained by differentiating $F(X, \theta)$ with respect to θ and setting the result to zero, yielding

$$\frac{\partial F(X, \theta)}{\partial \theta} = 0 = \sum_{j=1}^{j=N} e^{ix_j}e^{-i(x_j+\theta)}C_j(x_j + \theta) - ie^{-i\theta}\tilde{F}(X^*)$$

where $$\tilde{F}(X^*) = \frac{(\tilde{K} + A\tilde{N}) + \tilde{N}\int_0^{X^*} e^{-iu}C(u)\, du}{1 - e^{-iX^*}} = \frac{\tilde{N}C(X^*)}{i}$$

Thus if a θ^* exists, it is a solution to the equation

$$\sum_{j=1}^{j=N} C_j(x_i + \theta^*) = \tilde{N}C(X^*)$$

Example 7.6

Let

$$C_j(u) = au, \quad \text{for all } j, \quad j = 1, 2, \ldots, N; a > 0$$
$$C(u) = bu, \quad b > 0$$

The optimal value X^* is the solution to the equation

$$b\tilde{N}\int_0^{X^*} (1 - e^{-iu})\, du = i(\tilde{K} + A\tilde{N})$$

The optimal value θ^* is the solution to the equation

$$a\sum_{j=1}^{j=N}(x_j + \theta^*) = \tilde{N}bX^*$$

or $$\sum_{j=1}^{j=N} x_j + N\theta^* = \frac{\tilde{N}bX^*}{a}$$

Thus $$\theta^* = \frac{1}{N}\left[\frac{\tilde{N}bX^*}{a} - \sum_{j=1}^{j=N} x_j\right]$$

Note that a solution is valid only for $\theta^* > 0$.

7.3. Continuous Review Replacement Systems: Stochastic Models

So far we have assumed that the effect of equipment downtime associated with repairs was negligible. When incorporating such downtimes, it is necessary to segregate between the two distinct phases in the equipment states, the operating phase and the downtime phase. We assume that downtime is caused by equipment breakdowns, and the length of a given downtime is the time necessary to repair the equipment and set it back into operation. In general, an uncertainty element is present both in the frequency of occurrence of downtimes and the length of downtime. We shall assume that the equipment service age is not affected by a downtime. The replacement policy in effect is to replace the equipment by a new identical equipment when service age X is reached. A sample function of the equipment aging and replacement process is shown in Figure 7.12.

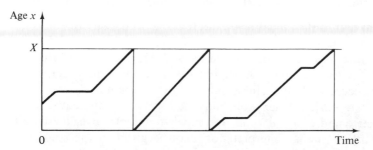

Figure 7.12. A sample function for a continuously operating equipment subject to failure, repair, and replacement.

In what follows, we shall make the following assumptions:

(1) The probability that the equipment age is u and in an operating state will fail as it ages from u to $u + du$ is $[\lambda(u)du + o(du)]$, and is independent of the time at which such failure occurs.

(2) The length of a downtime or repair time is a random variable having a negative exponential distribution with parameter μ. Hence, the probability that the equipment in a state of breakdown at time t will be repaired in the time interval $(t, t + dt]$ is $[\mu \, dt + o(dt)]$.

(3) The repair time is independent of the failure time.

We shall also assume that at time $t = 0$, the equipment age is x, where $0 \leq x < X$. As we shall point out later, it will be reasonable to attach a cost while the equipment is operating and a cost while the equipment is in a state of repair. Thus, the operations cost will be broken down into two basic components, namely an operating cost and a repair cost. We also note that, while in an operating state, the amount of aging equals the elapsed time; thus $du = dt$.

i. *Undiscounted Model*

Let

$P_1(t, u)\, du$ = probability that at time t, the equipment is in an operating condition and its age is between u and $u + du$;

$P_0(t, u)\, du$ = probability that at time t, the equipment is in a repair state and its age is between u and $u + du$;

We then have the following relations

$$P_1(t + dt, u + du) = P_1(t, u)[1 - \lambda(u)du + o(du)] + P_0(t, u + du)$$
$$\times [\mu\, dt + o(dt)]$$
$$P_0(t + dt, u) \qquad = P_0(t, u)[1 - \mu dt + o(dt)] + P_1(t, u)\,[\lambda(u)\, du + o(du)]$$

The first of these relations states that the probability that at time $t + dt$ the equipment will be of age $u + du$ and operating, is equal to the sum of the probabilities of two mutually exclusive and exhaustive events, that is; (a) the probability that at time t the equipment is of age u and operating and no failure occurs while the equipment ages by an amount du over the time interval $(t, t + dt]$; (b) the probability that at time t, the equipment is in a failed state and of age $u + du$, and repair terminates over the time interval $(t, t + dt]$. A similar interpretation may be given to the second relation. These two relations may be written alternatively as

$$P_1(t + dt, u + du) - P_1(t, u)$$
$$= -\lambda(u)\, P_1(t, u)\, du + \mu P_0(t, u + du)\, dt + o(du) + o(dt)$$
$$P_0(t + dt, u) - P_0(t, u)$$
$$= -\mu P_0(t, u)\, dt + \lambda(u)\, P_1(t, u)\, du + o(du) + o(dt)$$

Dividing both sides of these expressions by dt, and noting that $du = (du/dt)\, dt$, we can write respectively

$$\frac{P_1(t + dt, u + du) - P_1(t, u + du)}{dt} + \frac{P_1(t, u + du) - P_1(t, u)}{du}\frac{du}{dt}$$

$$= -\lambda(u)P_1(t, u)\frac{du}{dt} + \mu P_0(t, u + du) + \frac{o(du)}{du}\frac{du}{dt} + \frac{o(dt)}{dt}$$

$$\frac{P_0(t + dt, u) - P_0(t, u)}{dt} = -\mu P_0(t, u) + \lambda(u)P_1(t, u)\frac{du}{dt} + \frac{o(du)}{du}\frac{du}{dt} + \frac{o(dt)}{dt}$$

Taking the limit on both sides of each expression as $dt \rightarrow 0$, (and hence as $du \rightarrow 0$), yields respectively for $t \geq 0$ and $0 \leq u < X$

$$\frac{\partial P_1(t, u)}{\partial t} + \frac{\partial P_1(t, u)}{\partial u} = -\lambda(u)P_1(t, u) + \mu P_0(t, u)$$

$$\frac{\partial P_0(t, u)}{\partial t} = -\mu P_0(t, u) + \lambda(u)P_1(t, u)$$

(7.16)

At this stage, it must be noted that the decision to keep or replace the equipment has not played a role in setting up the system of equations (7.16). This system of equations would still have been valid in a situation where no replacement occurs and the equipment is repaired everytime it fails to recover its age at failure. The distinctive point to note however is that if replacement occurs at age X, then, for all time $t \geq 0$, the equipment age cannot exceed X, since at time $t = 0$, the initial equipment age is x, $0 \leq x < X$. Further, since replacement occurs as soon as age X is reached, then for all $t \geq 0$, we must have

$$P_1(t, X^-) = P_1(t, 0)$$

We now assume steady state conditions and let

$$\varphi_1(u) = \lim_{t \to \infty} P_1(t, u)$$

$$\varphi_0(u) = \lim_{t \to \infty} P_0(t, u)$$

where $0 \leq u < X$. The system of equations (7.16) reduces then to

$$\frac{d\varphi_1(u)}{du} = -\lambda(u)\varphi_1(u) + \mu\varphi_0(u) \tag{7.17}$$

$$= -\mu\varphi_0(u) + \lambda\varphi_1(u) \tag{7.18}$$

with the condition

$$\varphi_1(X^-) = \varphi_1(0) \tag{7.19}$$

To solve the system of equations (7.17), (7.18) and (7.19), we note from equations (7.17) and (7.18) that

$$\frac{d\varphi_1(u)}{du} = 0$$

for all $0 \leq u < X$. It thus follows that

$$\varphi_1(u) = A \tag{7.20}$$

where A is an arbitrary constant to be determined. In addition, it should be noted that solution (7.20) satisfies condition (7.19). In order to obtain $\varphi_1(u)$, we use relation (7.20) in equation (7.18) to yield

$$\varphi_0(u) = A\frac{\lambda(u)}{\mu}, \quad 0 \leq u < X \tag{7.21}$$

Let now

U = random variable denoting the steady state age of the equipment
$\varphi(u)\,du$ = probability that in the steady state, the equipment age will be between u and $u + du$, that is

$$\varphi(u)\,du = P\{u < U \leq u + du\}, \quad 0 \leq u < X$$

Since the equipment can only be either in an operating state or a failed state, it follows then from relations (7.20) and (7.21) that

$$\varphi(u) = \varphi_1(u) + \varphi_0(u)$$
$$= A + A\frac{\lambda(u)}{\mu}$$
$$= A\left[1 + \frac{\lambda(u)}{\mu}\right], \quad 0 \leq u < X \tag{7.22}$$

Because $\varphi(u)$ is a probability density function, the normalizing condition

$$\int_0^X \varphi(u)\,du = 1$$

must prevail. Thus, using expression (7.22) we have

$$\int_0^X A\left[1 + \frac{\lambda(u)}{\mu}\right] du = 1 \tag{7.23}$$

Let now

$$\Lambda(X) = \int_0^X \lambda(u)\,du \tag{7.24}$$

Then, solving for A in equation (7.23) and using (7.24) we obtain

$$A = \frac{1}{X + \dfrac{\Lambda(X)}{\mu}} \tag{7.25}$$

Using (7.25) in expressions (7.20) and (7.21) yields respectively

$$\varphi_1(u) = \frac{1}{X + \dfrac{\Lambda(X)}{\mu}}, \quad 0 \leq u < X \tag{7.26}$$

$$\varphi_0(u) = \frac{\lambda(u)}{\mu\left[X + \dfrac{\Lambda(X)}{\mu}\right]}, \quad 0 \leq u < X \tag{7.27}$$

Let

> $T =$ random variable denoting the time interval between two successive replacements in the steady state.

We may obtain an expression for $E[T]$ by noting that the probability the equipment is operating is

$$\int_0^X \varphi_1(u)\, du = \frac{X}{X + \dfrac{\varLambda(X)}{\mu}}$$

This is the same as the frequency of time the equipment is operating over a cycle T, and that is $X/(E[T])$. Hence

$$E[T] = X + \frac{\varLambda(X)}{\mu} \tag{7.28}$$

We are now in a position to write an expression for the long run total expected cost per unit time $E(X)$, as a function of the replacement age X.

In setting up the cost expressions it is necessary to distinguish between the cost associated while the equipment is in an operating state and the cost associated while the equipment is in an idle state. Let

> $\tilde{C}(u)\, du =$ cost of equipment operation between ages u and $u + du$, given that the equipment is in an operating state
>
> $r(u) =$ cost associated with one unit time of downtime for an equipment aged u; $r(u)$ is sometimes referred to as the repair cost per unit time

As before, K will be used to denote the nonrecurring or fixed acquisition cost of a new equipment. We then have using expression (7.28) for the expected acquisition cost per unit time

$$\frac{K}{E[T]} = \frac{K}{X + \dfrac{\varLambda(X)}{\mu}} \tag{7.29}$$

Also, the expected operation cost per unit time is using expression (7.26)

$$\int_0^X \tilde{C}(u)\varphi_1(u)\, du = \frac{\displaystyle\int_0^X \tilde{C}(u)\, du}{X + \dfrac{\varLambda(X)}{\mu}} \tag{7.30}$$

And the expected repair cost per unit time is using expression (7.27)

$$\int_0^X r(u)\varphi_0(u)\, du = \frac{\dfrac{1}{\mu}\displaystyle\int_0^X r(u)\lambda(u)\, du}{X + \dfrac{\varLambda(X)}{\mu}} \tag{7.31}$$

Hence the total expected cost per unit time which is the sum of (7.29), (7.30) and (7.31) is

$$E(X) = \frac{K + \int_0^x \left[\tilde{C}(u) + \frac{r(u)}{\mu} \lambda(u) \right] du}{X + \frac{\Lambda(X)}{\mu}}$$

The optimum value of X may be obtained by differentiating $E(X)$ with respect to X and setting the result equal to zero.

Special Case: $\lambda(u) = \lambda$; $r(u) = r$. Then $\Lambda(X) = \lambda X$. The expression for $E(X)$ can be written as

$$E(X) = \frac{K + \int_0^x \tilde{C}(u)\, du + \frac{r\lambda}{\mu} X}{X + \frac{\lambda}{\mu} X} = \frac{1}{1 + \frac{\lambda}{\mu}} \frac{K + \int_0^x \left[\frac{r\lambda}{\mu} + \tilde{C}(u) \right] du}{X}$$

yielding an expression similar to the basic model. Note that the equivalent operations cost rate function is

$$\frac{r\lambda}{\mu} + \tilde{C}(u) = C(u)$$

The optimum value of X is then from (7.3) a solution to the equation

$$\int_0^x d\left[\frac{r\lambda}{\mu} + \tilde{C}(u) \right] = K$$

or

$$\int_0^x d\tilde{C}(u) = K$$

This expression is independent of λ, μ, and r; hence the only relevant cost element affecting X^* is the operating cost function $\tilde{C}(u)$.

ii. *Discounted Model*

The situation is similar to the previous problem except that cost is discounted over infinite time. Let for $0 \leq x < X$

$f_1(x) =$ minimum total expected discounted cost when the equipment is initially of age x and is operating

$f_0(x) =$ minimum total expected discounted cost when the equipment is initially of age x and is not operating (i.e., in a state of repair)

$i =$ interest rate

The same set of assumptions are assumed to prevail as in the undiscounted model, and used equipment aged X is replaced by an identical new equipment. The following functional equation is obtained as in Section 7.2-ii-a-(2):

$$f_1(x) = \min \begin{cases} K + f_1(0) & x = X \\ \tilde{C}(x)\, dx + (1 - i\, dt)[1 - \lambda(x)\, dx] f_1(x + dx) \\ \quad + (1 - i\, dt)\lambda(x)\, dx\, f_0(x) & 0 \le x < X \end{cases}$$

The first alternative is the cost of replacing the existing equipment by a new equipment in an operating state. The second alternative represents the cost of no replacement involving the cost of equipment operations and the sum of the discounted costs associated with the equipment in an operating state and age $x + dx$, and the equipment which has just failed and age x. For $0 \le x < X$, no replacement occurs and $f_1(x)$ satisfies the relation

$$f_1(x) = \tilde{C}(x)\, dx + (1 - i\, dt)[1 - \lambda(x)\, dx] f_1(x + dx)$$
$$+ (1 - i\, dt)\lambda(x)\, dx\, f_0(x)$$

or

$$0 = \tilde{C}(x)\, dx + [f_1(x + dx) - f_1(x)] - [i\, dt + \lambda(x)\, dx] f_1(x + dx)$$
$$+ \lambda(x) f_0(x)\, dx + o(dx)$$

Dividing by dx, and letting $dx \to 0$, we obtain for $dx/dt = 1$

$$0 = \tilde{C}(x) + \frac{df_1(x)}{dx} - [i + \lambda(x)] f_1(x) + \lambda(x) f_0(x), \quad 0 \le x < X \quad (7.32)$$

to obtain an expression for $f_0(x)$, we note that for an equipment which has just failed, given that the repair time is v, the total expected discounted cost is (Figure 7.13)

$$\int_0^v r(x) e^{-i\theta}\, d\theta + f_1(x) e^{-iv}$$

Since repair time has the negative exponential distribution with parameter μ, it follows that

$$f_0(x) = \int_0^\infty \left[\int_0^v r(x) e^{-i\theta}\, d\theta + f_1(x) e^{-iv} \right] \mu e^{-\mu v}\, dv$$
$$= \int_0^\infty \left[\frac{r(x)}{i} (1 - e^{-iv}) + f_1(x) e^{-iv} \right] \mu e^{-i\mu}\, dv \quad (7.33)$$
$$= \frac{1}{i + \mu} [\mu f_1(x) + r(x)]$$

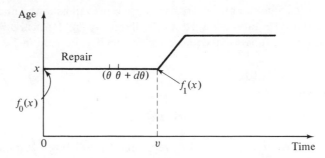

Figure 7.13. $f_0(x) = \int_0^\infty \left[\int_0^v r(x)e^{-i\theta} \, d\theta + f_1(x)e^{-iv} \right] \mu e^{-\mu v} \, dv.$

Substituting $f_0(x)$ in (7.33) in equation (7.32), we obtain the following first-order linear differential equation for $f_1(x)$, $0 \le x < X$:

$$0 = \tilde{C}(x) + \frac{df_1(x)}{dx} - [i + \lambda(x)]f_1(x) + \frac{\lambda(x)}{i + \mu}[\mu f_1(x) + r(x)]$$

or
$$\frac{df_1(x)}{dx} - \frac{i^2 + i\mu + i\lambda(x)}{i + \mu}f_1(x) = -\tilde{C}(x) - \frac{\lambda(x)r(x)}{i + \mu}$$

After integration, the expression obtained for $f_1(x)$ is

$$f_1(x) = \exp\left[ix + \frac{i}{i + \mu} \int_0^x \lambda(\tau) \, d\tau \right] f_1(0)$$
$$+ \exp\left[ix + \frac{i}{i + \mu} \int_0^x \lambda(\tau) \, d\tau \right] \tag{7.34}$$
$$\cdot \int_0^x \exp\left\{ -\left[iu + \frac{i}{i + \mu} \int_0^u \lambda(\tau) \, d\tau \right] \right\}\left[-\tilde{C}(u) - \frac{\lambda(u)r(u)}{i + \mu} \right] du$$

It is evident from relation (7.34) that minimizing $f_1(x)$ with respect to X is equivalent to minimizing $f_1(0)$ with respect to X; the next objective is to determine $f_1(0)$.

Since the policy is to replace the equipment at age X (and the equipment must be in an operating state at age X), it follows that for $x = X$

$$f_1(X) = K + f_1(0) \tag{7.35}$$

Let
$$\int_0^x \lambda(\tau) \, d\tau = \Lambda(x)$$

From relations (7.34) and (7.35), the continuity of $f(x)$ at $x = X$ would imply that

$$K + f_1(0) = \exp\left[iX + \frac{i}{i+\mu}\Lambda(X)\right]f_1(0)$$
$$- \exp\left[iX + \frac{i}{i+\mu}\Lambda(X)\right]\int_0^X \exp\left\{-\left[iu + \frac{i}{i+\mu}\Lambda(u)\right]\right\}$$
$$\times \left[\tilde{C}(u) + \frac{\lambda(u)r(u)}{i+\mu}\right]du$$

Solving for $f_1(0)$, we obtain

$$f_1(0) = -K + \frac{K + \int_0^X \exp\left\{-\left[iu + \frac{i}{i+\mu}\Lambda(u)\right]\right\}\left[\tilde{C}(u) + \frac{\lambda(u)r(u)}{i+\mu}\right]du}{1 - \exp\left\{-\left[iX + \frac{i}{i+\mu}\Lambda(X)\right]\right\}}$$

(7.36)

The optimum value of X is then obtained by differentiating $f_1(0)$ with respect to X and setting the result to zero.

Special Case: $\lambda(u) = \lambda$; $r(u) = r$. Relation (7.36) can be written

$$f_1(0) = -K + \frac{K + \int_0^X \exp\left[-i\left(1 + \frac{\lambda}{i+\mu}\right)u\right]\tilde{C}(u)\,du}{1 - \exp\left[-i\left(1 + \frac{\lambda}{i+\mu}\right)X\right]}$$

$$+ \frac{\int_0^X \frac{\lambda r}{i+\mu}\exp\left[-i\left(1 + \frac{\lambda}{i+\mu}\right)u\right]du}{1 - \exp\left[-i\left(1 + \frac{\lambda}{i+\mu}\right)X\right]}$$

$$= -K + \frac{K + \int_0^X \exp\left[-i\left(1 + \frac{\lambda}{i+\mu}\right)u\right]\tilde{C}(u)\,du}{1 - \exp\left[-i\left(1 + \frac{\lambda}{i+\mu}\right)X\right]}$$

(7.37)

$$+ \frac{\frac{\lambda r}{i+\mu}\cdot\frac{i+\mu}{i(i+\mu+\lambda)}\left|\exp\left[-i\left(1 + \frac{\lambda}{i+\mu}\right)u\right]\right|_0^X}{1 - \exp\left[-i\left(1 + \frac{\lambda}{i+\mu}\right)X\right]}$$

$$= -K + \frac{K + \int_0^X \exp\left[-i\left(1 + \frac{\lambda}{i+\mu}\right)u\right]\tilde{C}(u)\,du}{1 - \exp\left[-i\left(1 + \frac{\lambda}{i+\mu}\right)X\right]} + \frac{\lambda r}{i(i+\mu+\lambda)}$$

We note that the minimization of $f_1(0)$ given in relation (7.37) is similar to the minimization of expression (7.8), where i in (7.8) is replaced with

$i[1 + \lambda/(i + \mu)]$. Thus the optimum value of X is given by a modified form of equation (7.12), or

$$\int_0^X \left[1 - \exp\left[-i\left(1 + \frac{\lambda}{i + \mu}\right)u\right]\right] dC(u) = i\left(1 + \frac{\lambda}{i + \mu}\right)K$$

7.4. Periodic Review Replacement Systems: Deterministic Models

In a periodic review system, a decision to keep or replace an existing equipment is made at the beginning of equal time intervals. The set of models generated under these conditions is like the set of models previously discussed, except that time is measured over a discrete scale. Because of this similarity, we shall restrict our attention to two models, one illustrating the undiscounted case and the other the discounted case.

i. *Undiscounted Model*

The assumptions in this model are similar to the assumptions outlined in the development of the basic model with salvage value. The review period is selected as the unit time and the operating equipment is assumed to age by

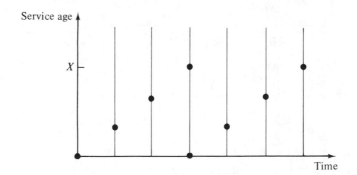

Figure 7.14. Replacement cycles of aging equipment. Periodic review system.

a unit time quantity over each period of time. Equipment is replaced once it reaches age X, $X = 1, 2, \ldots$, (Figure 7.14). We shall use the same symbols as in the basic model and let

$L(x)$ = operations or recurring cost during a given period when at the beginning of the period the equipment is of age x

It is evident that

$$L(x) = \int_x^{x+1} C(u) \, du$$

From Figure 7.14, the total cost incurred over one replacement cycle is

$$K(X) + \sum_{x=0}^{x=X-1} L(x)$$

Therefore, the long-run total cost per unit time is

$$E(X) = \frac{K(X) + \sum_{x=0}^{x=X-1} L(x)}{X}$$

The optimum value of X, $X = 1, 2, \ldots$, is obtained by minimizing $E(X)$ with respect to X (see Problem 7.7).

It is easy to verify that if $C(x)$ is introduced in this expression relation (7.1) is obtained. In many practical situations, however, the recurring cost of the equipment is often given in terms of $L(x)$, in which cases the last expression for $E(X)$ would be useful.

ii. *Discounted Model*

We shall formulate the discounted problem using the functional equation approach. Define α as the discount factor for one period. Let $f(x)$ be the minimum total discounted cost over an infinite horizon period when starting with an equipment age x and when following an optimal policy. Then

$$f(x) = \min \begin{cases} K(x) + f(0), & \text{(replace)} \\ L(x) + \alpha f(x+1), & \text{(keep)} \end{cases}$$

This functional equation may be solved recursively. Let X be the replacement age, and assume $x \leq X$. Then, when $x = 0, 1, 2, \ldots, X - 1$, the decision is to keep the equipment; hence

$$f(0) = L(0) + \alpha f(1)$$
$$f(1) = L(1) + \alpha f(2)$$
$$\ldots$$
$$f(X - 1) = L(X - 1) + \alpha f(X)$$

and, by successive substitution, we obtain

$$f(0) = L(0) + \alpha[L(1) + \alpha f(2)]$$
$$= L(0) + \alpha L(1) + \alpha^2[L(2) + \alpha f(3)] \qquad (7.38)$$
$$\ldots$$
$$= L(0) + \alpha L(1) + \alpha^2 L(2) + \cdots + \alpha^{X-1} L(X - 1) + \alpha^X f(X)$$

When $x = X$, the decision is to replace the equipment; hence

$$f(X) = K(X) + f(0) \tag{7.39}$$

Eliminating $f(X)$ from equations (7.38) and (7.39) yields

$$f(0) = \sum_{x=0}^{x=X-1} \alpha^x L(x) + \alpha^X [K(X) + f(0)]$$

and, solving for $f(0)$, we obtain

$$f(0) = \frac{\alpha^X K(X) + \displaystyle\sum_{x=0}^{x=X-1} \alpha^x L(x)}{1 - \alpha^X} \tag{7.40}$$

The optimum value of X, $X = 1, 2, \ldots$, is obtained by minimizing the function $f(0)$ with respect to X. Expression (7.40) should be compared to expression (7.8) when $\alpha = e^{-i}$ and $L(x) = \int_x^{x+1} C(u)\, du$. In Problem 7.8, the reader is asked to derive the necessary and sufficient conditions for determining the optimal value of X when $K(X) = K$.

7.5. Periodic Review Replacement Systems: Stochastic Models

i. *Undiscounted Model*

We consider a periodic review replacement system operating under the following policy: At the end of each period, the equipment age is examined and a decision is made to keep or replace the equipment. If the equipment age is less than a given level $X > 0$, the equipment is kept; if on the other hand, the equipment age equals to or exceeds the level X, the equipment is replaced by a new identical equipment. (See Figure 7.15.) Let for $t = 1, 2, \ldots$

$D(t) =$ amount of equipment aging during period t

$U(t) =$ equipment age at the end of period t before a decision

$Y(t) =$ equipment age at the beginning of period t following a decision

Under the given replacement policy, we have

$$U(t+1) = \begin{cases} D(t+1) & \text{if } X \le U(t) < \infty \\ U(t) + D(t+1) & \text{if } 0 \le U(t) < X \end{cases}$$

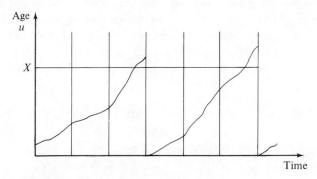

Figure 7.15 Sample function for an equipment subject to random aging and periodic replacement.

a. Stationary Solution

We assume that $\{D(t)\}$ form a sequence of independently and identically distributed nonnegative random variables from period to period, with distribution function $\Phi(\xi)$, $0 < \xi < \infty$, and PDF $\varphi(\xi) > 0$. The stochastic process $\{U(t), t = 1, 2, \ldots\}$ is Markovian since the state at time $t + 1$ depends only on the state of the immediately preceding time t. We shall assume that steady state exists and has been reached. Each of $U(t)$ and $Y(t)$ will be independent of the initial conditions and of the parameter t. We shall denote them respectively by U and Y. For the resulting stationary problem, we consider the state of the system at the beginning of a period following a decision and let

$\Psi(y)$ = distribution function of the equipment age Y, $-\infty < y < \infty$

M = probability of replacing the old equipment by a new equipment, that is, $M = P\{Y = 0\} = P\{X \le U < \infty\}$, $y = 0$

$\psi(y)dy$ = probability that the equipment age Y lies between y and $y + dy$, that is, $\psi(y)dy = P\{y < Y \le y + dy\} = P\{y < U \le y + dy\}$, $0 < y < X$

We may simply write

$$d\Psi(y) = \begin{cases} M, & y = 0 \\ \psi(y)\, dy, & 0 < y < X \\ 0, & \text{otherwise} \end{cases} \qquad (7.41)$$

The quantity M and the function $\psi(y)$ are unknowns to be determined as in Section 5.3-ii-b.

If we consider the transitions between two successive periods, we note that the equipment age is y, $0 < y < X$, at the beginning of a period if the previous period starts with a new equipment of age 0 and there is aging in the amount $\xi = y$; or if the previous period starts with an equipment age u, $0 < u < X$, and there is aging of the equipment in the amount $\xi = y - u$. Hence

$$\psi(y)\,dy = M\varphi(y)\,dy + \left\{\int_0^y \psi(u)\varphi(y-u)\,du\right\}dy$$

or $\qquad \psi(y) = M\varphi(y) + \int_0^y \psi(u)\varphi(y-u)\,du, \quad 0 < y < X \qquad (7.42)$

Equation (7.42) is similar to the integral equation (2.109) for the renewal density. Its solution is given by

$$\psi(y) = M\sum_{n=1}^{\infty}\varphi*^{(n)}(y), \quad 0 < y < X \qquad (7.43)$$

where $\varphi*^{(n)}(y)$ is the nth-fold convolution of $\varphi(y)$ with itself. Denoting

$$\tau(y) = \sum_{n=1}^{\infty}\varphi*^{(n)}(y), \quad 0 < y < X$$

we rewrite expression (7.43) as

$$\psi(y) = M\tau(y) \qquad (7.44)$$

From the definition of $\Psi(y)$, we have

$$\int_{-\infty}^{\infty} d\Psi(y) = 1.$$

Using (7.41), we get

$$M + \int_0^X \psi(y)\,dy = 1$$

or $\qquad M + \int_0^X M\tau(y)\,dy = 1$

Therefore, solving for M, yields

$$M = \frac{1}{1 + \int_0^X \tau(y)\,dy} \qquad (7.45)$$

b. OPTIMIZATION PROBLEM

Define

$K =$ fixed cost of recurring a new equipment

$C(u)du =$ operations cost when the equipment ages from u to $u + du$. We assume $C(u)$ to be strictly increasing in u.

$L(x) =$ expected recurring cost per period given the period starts with an equipment age x. Thus

$$L(x) = E\left[\int_x^{x+\xi} C(u)du\right] = \int_0^\infty \int_x^{x+\xi} C(u)\,\varphi(\xi)\,du\,d\xi \qquad (7.46)$$

Now, let $F(X)$ be the steady state total expected cost per period. $F(X)$ is the sum of the expected replacement cost and the expected operations cost:

$$\begin{aligned} F(X) &= KM + \int_{-\infty}^\infty L(u)\,d\Psi(u) \\ &= KM + ML(0) + \int_0^X L(u)\psi(u)\,du \end{aligned} \qquad (7.47)$$

Using expressions (7.44) and (7.45), we get

$$F(X) = \frac{K + L(0) + \int_0^X L(u)\tau(u)\,du}{1 + \int_0^X \tau(u)\,du} \qquad (7.48)$$

The value of X that minimizes $F(X)$ is obtained by differentiating (7.48) with respect to X and setting the result to zero. This yields

$$\frac{\left(1 + \int_0^X \tau(u)\,du\right)L(X)\tau(X) - \left[K + L(0) + \int_0^X L(u)\tau(u)\,du\right]\tau(X)}{\left[1 + \int_0^X \tau(u)\,du\right]^2} = 0$$

or $\qquad\qquad L(X) - L(0) + \int_0^X [L(X) - L(u)]\tau(u)\,du = K \qquad (7.49)$

or in terms of the derivative of $L(x)$:

$$\int_0^X L'(v)\,dv + \int_0^X \int_u^X L'(v)\tau(u)\,dv\,du = K$$

Interchanging the order of integration in the double integral term results in

$$\int_0^X L'(v)\,dv + \int_0^X \int_0^v L'(v)\tau(u)\,du\,dv = K$$

or $\qquad\qquad \int_0^X L'(v)\left\{1 + \int_0^v \tau(u)\,du\right\}dv = K$

Let

$$T(x) = \int_0^x \tau(u)\, du$$

Then the optimum value X^* of X is the solution to

$$\int_0^X L'(v)\{1 + T(v)\}\, dv = K \tag{7.50}$$

We now show that if $C(u)$ is strictly increasing in u, then the solution X^* to equation (7.50), if it exists, is unique. Now, from expression (7.41)

$$L'(x) = \int_0^\infty [C(x + \xi) - C(x)]\varphi(\xi)\, d\xi$$

Hence, if $C(u)$ is strictly increasing, the quantity

$$C(x + \xi) - C(x) > 0, \quad \xi > 0$$

for all $x > 0$; thus $L'(x) > 0$ for all $x > 0$. Next, since $\varphi(x) > 0$ for $x > 0$

$$\tau(x) = \sum_{n=1}^\infty \varphi*^{(n)}(x) > 0, \quad \text{for all } x > 0$$

Hence
$$T(x) = \int_0^x \tau(u)\, du > 0, \quad \text{for all } x > 0$$

and is a strictly increasing function of x. Thus, in relation (7.50), the integrand term in the left side expression is a positive function of v. Therefore, the quantity

$$\int_0^X L'(v)\{1 + T(v)\}\, dv$$

is a strictly increasing function of X. Thus if equation (7.50) has a solution, it is unique.

We have not as yet shown that the solution X^* to equation (7.50) minimizes the function $F(X)$. To do so, we evaluate the second derivative of $F(X)$ with respect to X at $X = X^*$. Now in expression (7.48), let

$$U = K + L(0) + \int_0^X L(u)\tau(u)\, du$$

and
$$V = 1 + \int_0^X \tau(u)\, du$$

Then
$$F(X) = \frac{U}{V}$$

$$F'(X) = \frac{dF(X)}{dX} = \frac{VU' - UV'}{V^2}$$

$$F''(X) = \frac{d^2 F(X)}{dX^2} = \frac{V^2(VU'' - UV'') - 2VV'(VU' - UV')}{V^4}$$

At $X = X^*$ we have

$$F'(X^*) = 0$$

Hence
$$VU' - UV'|_{X=X^*} = 0$$

Thus

$$F''(X^*) = \frac{VU'' - UV''}{V^2}$$

$$= \frac{1}{\left[1 + \int_0^{X^*} \tau(u)\, du \right]^2}$$

$$\times \left\{ \left[1 + \int_0^{X^*} \tau(u)\, du \right][L'(X^*)\tau(X^*) + L(X^*)\tau'(X^*)] \right.$$

$$\left. - \tau'(X^*)\left[K + L(0) + \int_0^{X^*} L(u)\tau(u)\, du \right] \right\}$$

$$= \frac{1}{\left[1 + \int_0^{X^*} \tau(u)\, du \right]^2}$$

$$\times \left\{ \left[1 + \int_0^{X^*} \tau(u)\, du \right]L'(X^*)\tau(X^*) + \tau'(X^*) \right.$$

$$\left. \times \left[L(X^*) + L(X^*)\int_0^{X^*} \tau(u)\, du - K - L(0) - \int_0^{X^*} L(u)\tau(u)\, du \right] \right\}$$

And using equation (7.49), we finally obtain

$$F''(X^*) = \frac{L'(X^*)\tau(X^*)}{1 + \int_0^{X^*} \tau(u)\, du} \tag{7.51}$$

Since $L'(x) > 0$ and $\tau(x) > 0$ for all $x > 0$, it follows that $F''(X^*) > 0$. Hence the function $F(X)$ is minimized at $X = X^*$.

Example 7.7

Let

$$\varphi(\xi) = \lambda e^{-\lambda \xi}, \quad 0 < \xi < \infty, \quad \lambda > 0$$

$$C(u) = au + b, \quad a > 0, \quad b > 0$$

From expression (7.46),

$$L(x) = \int_0^\infty \int_x^{x+\xi} C(u)\varphi(\xi)\, du\, d\xi$$

$$L'(x) = \int_0^\infty [C(x+\xi) - C(x)]\varphi(\xi)\, d\xi$$

$$= \int_0^\infty [a(x+\xi) - ax]\lambda e^{-\lambda\xi}\, d\xi$$

$$= a\lambda \int_0^\infty \xi e^{-\lambda\xi}\, d\xi$$

$$= \frac{a}{\lambda}$$

Now

$$\tau(u) = \sum_{n=1}^\infty \varphi *^{(n)}(u)$$

$$= \sum_{n=1}^\infty \lambda e^{-\lambda u} \frac{(\lambda u)^{n-1}}{\Gamma(n)} = \lambda$$

Hence

$$T(u) = \int_0^u \tau(v)\, dv = \lambda u$$

Finally, we use equation (7.50) to obtain X^*:

$$\int_0^{X^*} \frac{a}{\lambda}(1 + \lambda u)\, du = K$$

or

$$\frac{aX^*}{\lambda} + \frac{aX^{*2}}{2} = K$$

or

$$\lambda^2 X^{*2} + 2\lambda X^* - \frac{2\lambda^2 K}{a} = 0$$

or

$$\lambda X^* = -1 + \sqrt{1 + \frac{2\lambda^2 K}{a}}$$

If $\lambda^2 K/a$ is much larger than unity, we have approximately

$$X^* \approx \sqrt{\frac{2K}{a}}$$

It is interesting to compare this with the result obtained in Example 7.1 and to note that the optimum value of X^* will not depend on λ for large $\lambda^2 K/a$.

c. THE GENERAL CASE WHEN $C(u)$ IS LINEAR

Consider now the case when

$$C(u) = au + b, \quad a > 0, \quad b > 0$$

Then

$$L'(u) = a\bar{D}$$

where
$$\bar{D} = \int_0^\infty \varphi(\xi)\, d\xi$$

is the expected aging over one period. Then equation (7.50) becomes

$$a\bar{D} \int_0^{X^*} \left[1 + \int_0^u \tau(v)\, dv \right] du = K$$

Hence
$$X^* + \int_0^{X^*} \int_0^u \tau(v)\, dv\, du = \frac{K}{a\bar{D}}$$

Interchanging the order of integration, we get

$$X^* + \int_0^{X^*} \int_v^{X^*} \tau(v)\, du\, dv = \frac{K}{a\bar{D}}$$

or, finally,

$$X^* + \int_0^{X^*} (X^* - v)\tau(v)\, dv = \frac{K}{a\bar{D}}$$

Define the function $m(x)$ such that

$$m(x) = x + \int_0^x (x - v)\tau(v)\, dv, \quad x > 0 \tag{7.52}$$

Then X^* is solution to the equation

$$m(X^*) = \frac{K}{a\bar{D}} \tag{7.53}$$

Define the Laplace transform of a function $f(x)$ as (see Appendix)

$$\bar{f}(s) = \int_0^\infty e^{-sx} f(x)\, dx$$

The Laplace transformation of relation (7.52) yields immediately

$$\bar{m}(s) = \frac{1}{s^2} + \frac{1}{s^2}\bar{\tau}(s) \tag{7.54}$$

Using the results of Section 2.8-iii on renewal theory, we have

$$\bar{\tau}(s) = \frac{\bar{\varphi}(s)}{1 - \bar{\varphi}(s)}$$

We can write expression (7.54) as

$$\bar{m}(s) = \frac{1}{s^2}\left[1 + \frac{\bar{\varphi}(s)}{1 - \bar{\varphi}(s)} \right]$$

$$= \frac{1}{s^2}\left[\frac{1}{1 - \bar{\varphi}(s)} \right] \tag{7.55}$$

The optimum value X^* is then obtained by inverting (7.55) first to obtain $m(x)$, and then solving for X^* using equation (7.53).

Example 7.8

Suppose that

$$\varphi(\xi) = \lambda e^{-\lambda \xi}, \quad 0 < \xi < \infty, \quad \lambda > 0$$

Then

$$\bar{\varphi}(s) = \frac{\lambda}{\lambda + s}$$

and expression (7.55) yields

$$\bar{m}(s) = \frac{1}{s^2 \left[1 - \dfrac{\lambda}{\lambda + s} \right]} \qquad (7.56)$$

$$= \frac{\lambda}{s^3} + \frac{1}{s^2}$$

Inverting (7.56), we find that

$$m(x) = \frac{\lambda}{2} x^2 + x \qquad (7.57)$$

From (7.53) and (7.57), we obtain

$$\frac{\lambda}{2} X^{*2} + X^* = \frac{K\lambda}{a}$$

from which X^* is obtained as in Example 7.7.

Example 7.9

$$\varphi(\xi) = \lambda^2 \xi e^{-\lambda \xi}, \quad 0 < \xi < \infty, \quad \lambda > 0$$

Then

$$\bar{\varphi}(s) = \left(\frac{\lambda}{\lambda + s} \right)^2 \qquad (7.58)$$

and expression (7.55) becomes

$$\bar{m}(s) = \frac{1}{2 \left[1 - \dfrac{\lambda^2}{(\lambda + s)^2} \right]} \qquad (7.59)$$

$$= \frac{(\lambda + s)^2}{s^3 (2\lambda + s)}$$

Expansion of $\bar{m}(s)$ in partial fractions yields

$$\bar{m}(s) = -\frac{1}{8\lambda(2\lambda + s)} + \frac{\lambda}{2s^3} + \frac{3}{4s^2} + \frac{1}{8\lambda s}$$

which may be inverted to obtain

$$m(x) = -\frac{1}{8\lambda}e^{-2\lambda x} + \frac{1}{\lambda}x^2 + \frac{3}{4}x + \frac{1}{8\lambda}$$

Thus, using equation (7.53), we find that X^* satisfies the following transcendental equation, which may be solved by graphical or numerical techniques:

$$8X^{*2} + 6X^* + 1 - e^{-2\lambda X^*} = \frac{4\lambda^2 K}{a}$$

d. Asymptotic Analysis for Large X^*

We wish to consider approximate solution of equation (7.53) for large values of X^*. Expansion of $\bar{\varphi}(s)$ about $s = 0$ yields (Section 1.4-iii)

$$\bar{\varphi}(s) = 1 - \bar{D}s + \tfrac{1}{2}(\bar{D}^2 + \sigma^2)s^2 - \cdots$$

where $\sigma^2 = \text{Var}[D]$

Then for small s, a first approximation to expression (7.55) is

$$\bar{m}(s) = \frac{1}{s^2\{1 - [1 - \bar{D}s + \cdots]\}}$$

$$\approx \frac{1}{s^3 \bar{D}} \tag{7.60}$$

The inverse of expression (7.60) yields as a first approximation to $m(x)$ for large x

$$m(x) \approx \frac{x^2}{2\bar{D}}$$

Hence

$$\frac{X^{*2}}{2\bar{D}} \approx \frac{K}{a\bar{D}}$$

or

$$X^* \approx \sqrt{\frac{2K}{a}}$$

For a better approximate expression to X^*, the reader is referred to Problem 11.

SELECTED REFERENCES

[1] ALCHIAN, A., "Economic Replacement Policy," RAND Report R-224, April 1952.

[2] BARLOW, R. E., and F. PROSCHAN, *Mathematical Theory of Reliability*, John Wiley & Sons, Inc., New York, 1967.

[3] BECKMAN, M. J., *Dynamic Programming of Economic Decisions*, Springer-Verlag, Berlin, 1968.

[4] BELLMAN, R., "Notes in the Theory of Dynamic Programming-III: Equipment Replacement Policy," RAND Report P-632, Jan. 1955.

[5] BELLMAN, R., and S. DREYFUS, *Applied Dynamic Programming*, Princeton University Press, Princeton, N. J., 1962.

[6] CHURCHMAN, C. W., R. L. ACKOFF, and E. L. ARNOFF, *Introduction to Operations Research*, John Wiley & Sons, Inc., New York, 1957.

[7] COX, R. D., *Renewal Theory*, John Wiley & Sons, Inc., New York, 1962.

[8] DEAN, J., *Capital Budgeting*, Columbia University Press, New York, 1951.

[9] FABRYCKY, W. J., P. M. GHARE, and P. E. TORGERSEN, *Industrial Operations Research*, Prentice-Hall, Inc., Englewood Cliffs, N. J., 1972.

[10] HOWARD, R., *Dynamic Programming and Markov Processes*, The MIT Press, Cambridge, Mass., 1960.

[11] JORGENSEN, D. W., J. J. McCALL, and R. RADNER, *Optimal Replacement Policy*, North-Holland Publishing Company, Amsterdam, 1967.

[12] KAUFMANN, A., *Methods and Models of Operations Research*, Prentice-Hall, Inc., Englewood Cliffs, N. J., 1963.

[13] MASSÉ, P., *Optimal Investment Decisions*, Prentice-Hall, Inc., Englewood Cliffs, N. J., 1962.

[14] MORSE, P., *Queues, Inventories and Maintenance*, John Wiley & Sons, Inc., New York, 1958.

[15] SAVAGE, R. L., "Cycling," *Naval Research Logistics Quarterly*, Vol. 3, pp. 163–175, 1956.

[16] SIVAZLIAN, B. D., "On a Discounted Replacement Problem with Arbitrary Repair Time Distribution", *Management Science*, Vol. 19, No. 11, pp. 1301–1309, 1973.

[17] TERBORGH, B., *Dynamic Equipment Policy*, McGraw-Hill Book Company, New York, 1949.

PROBLEMS

1. A machine costs $50,000. Annual operating costs are $2000 for each of the first 2 years, $4000 for each of the second 2 years, $6000 for each of the third 2 years, and so on. What is the best age to replace the machine if the current interest rate is 10% and the used machine has no salvage value?

2. In the basic model, determine the optimum replacement age for the case when the operations cost rate function $C(u)$ takes the following forms:
   (a) $C(u) = a\sqrt{u}$, $a > 0$.
   (b) $C(u) = ae^{bu}$, $a > 0$, $b > 0$.

(c) $C(u) = a(1 - e^{-bu})$, $a > 0$, $b > 0$.
(d) $C(u) = a + bu^2$, $a > 0$, $b > 0$.

3. A mining firm expects that the deposits in a recently discovered mine will last for about 50 years. Among other resources, the operation of the mine requires a special equipment with an acquisition price of $100,000. Being a special-purpose machine, the equipment has negligible salvage value, and it is expected that available new equipment in the future will not change appreciably in price or characteristics. The equipment is to operate continuously, and its recurring cost rate function is expected to be $C(u) = 5000 + 10,000u$ dollars per year when equipment has aged u years. Determine the time interval between equipment replacement.

4. A truck owner has three trucks, two of which are 3 years old and the third one is 1 year old. The three trucks are of the same make and type, each with a purchase price of $10,000 and operations cost per annum as follows:

Age of truck (year)	1	2	3	4	5	6	7	8
Operations cost (dollars/year)	1500	1700	1900	2100	2500	3000	3500	4000
Resale price (dollars)	7000	5000	3000	1000	500	500	500	500

The truck owner is contemplating the replacement of these three trucks by two new trucks, each with 50% more capacity than one of the old ones. The price of a new truck is $12,000 and its operations cost per annum is as follows:

Age of truck (year)	1	2	3	4	5	6	7	8
Operations cost (dollars/year)	2000	2300	2600	3000	4000	5000	6000	7000
Resale price (dollars)	9000	8000	6000	4000	2000	1000	500	500

Assume that all future replacements, if any, will be made using two of the new type of trucks and neglect the time value of money. Determine the time at which all three trucks should be simultaneously replaced.

5. In the basic model it was assumed that in a time interval θ the equipment ages by a quantity θ. Suppose that the equipment ages at an average rate $\bar{D}$, so that if x is the equipment age and t is time, $du/dt = \bar{D}$ (rather than $du/dt = 1$). Formulate the expression for the long-run total cost per unit time, and determine the conditions for the existence of an optimum equipment replacement age.

6. Consider a system consisting of a large number N of identical components that are increasingly prone to failure with age (e.g., light bulbs). Let $f(\cdot)$ denote the probability density function for the time to failure of each component. The policy for replacing components may take one of the following different forms:
   (1) Individual replacement policy: Under this policy, each component is replaced by a new component as soon as it fails, and the cost of replacing a particular component is $\$C_2$ per unit.

(2) Group replacement policy: Under this policy, each component is replaced by a new component as soon as it fails; simultaneously, all components are replaced at fixed intervals T by new components. When group replacement occurs, even those components which have not failed are replaced. Let the cost of replacing a particular component when group replacement occurs be $\$C_1$ per unit ($C_1 < C_2$).

(a) Let $U(t)\, dt$ be the proportion of component failures in the time interval $(t, t + dt]$ when following an individual replacement policy. Show that $U(t)$ is a solution to the renewal equation

$$U(t) = f(t) + \int_0^t f(t - x)U(x)\, dx$$

(b) Show that

$$\lim_{t \to \infty} U(t) = \frac{1}{\mu}$$

where

$$\mu = \int_0^\infty tf(t)\, dt$$

(c) Show that the average cost per unit time under a group replacement policy is

$$C_G(T) = \frac{C_2 \int_0^T U(t)\, dt + C_1 N}{T}$$

Determine T^*, the optimum value of T.

(d) The average cost per unit time under an individual replacement policy is $C_I = C_2/\mu$. Thus a group replacement policy is preferred to an individual replacement policy if and only if $C_G(T^*) < C_I$. Show that if the time-to-failure distribution of each component is negative exponential it is always best to use an individual replacement policy.

7. In the periodic review deterministic undiscounted model, we have seen that the expression for the long run total cost per unit time is when $K(x) = K$

$$E(X) = \frac{K + \sum_{x=0}^{x=X-1} L(x)}{X}$$

The optimum value X^* of X which minimizes $E(X)$ is then dictated by the inequalities

$$E(X^*) \le E(X^* - 1)$$
$$E(X^*) \le E(X^* + 1)$$

Determine the conditions for optimality when
(a) $L(x) = a = $ constant, $a > 0$.
(b) $L(x) = ax$, $a > 0$.
(c) $L(x) = ax^r$, $a > 0$, $r > 0$.

8. Solve Problem 7 in the case of the discounted model.

9. The total number of hours that an intermittently operating piece of equipment is used per month is approximately normally distributed with mean equal to 200 hours and standard deviation equal to 50 hours. The operations cost rate function is linear in form and is given by

$$C(u) = .001u + 50 \text{ dollars/operating hour}$$

If the initial acquisition cost of the piece of equipment is $5,000 and salvage is negligible, determine when replacement should occur.

10. In a periodic review replacement system with stochastic aging, let

$L(x) =$ cost of operations per period when equipment age is x at the beginning of the period
$K =$ fixed acquisition cost
$\varphi(\cdot) =$ probability density function of the amount of equipment aging per period
$\alpha =$ discount factor
$X =$ replacement age

If $f(x)$ is the minimum total expected discounted cost over an infinite horizon period when following an optimal policy, show that for $0 \leq x < X$, $f(x)$ satisfies the functional equation

$$f(x) = L(x) + \alpha \int_x^X f(y)\varphi(y - x)\, dy + \alpha[K + f(0)] \int_X^\infty \varphi(y - x)\, dy$$

11. Show that a better approximation to expression (7.60) is

$$\bar{m}(s) \approx \frac{1}{\bar{D}} \frac{1}{s^3} + \frac{1}{2}\left(1 + \frac{\sigma^2}{\bar{D}^2}\right)\frac{1}{s^2}$$

Hence, show that the approximate optimal value of X is

$$X^* \approx \frac{1}{2}\left[-\bar{D}\left(1 + \frac{\sigma^2}{\bar{D}^2}\right) + \sqrt{\frac{8K}{a} + \bar{D}^2\left(1 + \frac{\sigma^2}{\bar{D}^2}\right)^2}\right]$$

Show that this last result is exact when D has a negative exponential distribution.

INFORMATION THEORY

8.1. Introduction

In the present chapter we introduce the student to the study of information theory. While we do not intend to imply that information theory belongs to operations research, we include the present chapter for several reasons. First, information theory can be used to solve certain decision problems, some examples of which will be exhibited later, which might otherwise be less easily solved. Moreover, information has become a topic of much interest today. One hears a great deal about the mass of information being generated daily, and, similarly, there is much written about systems for collecting, storing, and making rapidly available for use information in a variety of settings; for example, there is much interest in information systems to support management decision making in business and industry; there are automated systems that retrieve documents or references to documents from libraries; artificial satellites collect and transmit to earth information for use in predicting weather or in monitoring the activities of unfriendly nations; information-processing systems have been used to diagnose illnesses after accepting as inputs the particular symptoms.

Given the present-day emphasis on information, it was felt that an in-

8

troduction to a theory of information would be both of interest and possible future use to the reader.

First, let us consider the problem of measuring information.

8.2. A Measure of Information

i. *Intuitive Considerations*

"Information" is a common word in everyday language—any person could render a reasonable definition. On the other hand, however, if a person is the recipient of a piece of information and is then asked "how much" information he has received, we are less certain of his response, because this is a much less familiar concept.

One might arrive at crudely quantitative assessments as to the quantity of information contained in a piece of information. For example, suppose that we put ourselves in the position of one about to receive some information from another party, whose truthfulness is not questionable; that is, whatever he tells us is true.

Suppose, as we stand in a downpour, he approaches, stating, "It's raining." Now, have we received much information? We would like the reader to think not; that is, we claim here that it is reasonable to conclude that if presented a piece of information which was *already* known, then, in fact, *no* information has been received, for we have not been informed. Under the same damp circumstances, suppose that the informant states, "The sun will be out all day tomorrow." How much information has been received? We say some, without being specific, because we have been informed as to something about which we didn't *know*, although we would not be surprised by the event that was described.

As a final example, still in the rain, let us suppose that the information is, "The rain will continue steadily for 40 days and 40 nights." In this case, how much information is received? We say we have received information and, indeed, a great deal *more* than in the second example, because in the third example a statement is made whose truth we would *never* have surmised, whereas the second statement is not at all so surprising.

What we have done is to relate, in very imprecise terms, the quantity of information with the a priori *likelihood of validity of the statement made;* in fact, the relationship has been that the quantity is *inversely proportional* to that likelihood.

Now, there are no physical or natural peculiarities of information which dictate that this relationship is one which truly describes a property of the quantity of information in a statement or an event (here we say "or an event," since we could as well have eliminated the informant, dealt with the events themselves, and spoken of the amounts of information given by the events).

It is suggested, however, that the preceding is a *reasonable* approach, and we pursue it.

Presuming that however information is to be measured, our *reasonable* relationship is to be preserved, there are many measures, clearly, which satisfy it.

ii. *Basic Assumptions*

To better illustrate this fact, let us be specific in our assumptions. We assume the following.

There is a finite† set S of events $\{s_1, s_2, \ldots, s_n\}$. The probability of occurrence of s_i is p_i, $i = 1, \ldots, n$, and we have

$$\sum_{i=1}^{i=n} p_i = 1$$

---

†Infinite sets are quite acceptable, but we restrict the cardinality of S for specificity only.

Suppose that we observe which event occurs, and find it to be s_k. We have now reduced our concept of likelihood to that of probability, so previous remarks imply that the amount of information we obtain from the occurrence of s_k will be somehow inversely proportional to p_k; that is, denoting the amount of information obtained from the occurrence of s_k by $I(s_k)$, we have $I(s_k) > I(s_r)$ if $p_k < p_r$.

As mentioned, $I(s_k)$ could be defined in an infinite number of ways so that our inverse relationship is satisfied. For example, define $I(s_k) = \alpha/p_k$, $\alpha > 0$, and we have an infinity of such measures, one for each positive α. Another possibility is $I(s_k) = 1 - p_k$, and so on.

But there are other properties, intuition tells us, that should be satisfied by a measure of information, and one of these was hinted at in our first example situation: if one event of S, say s_k, has unit probability of occurrence, then we should have $I(s_k) = 0$, for no information is obtained if we knew beforehand which event would occur.

Whatever the form chosen for $I(s_k)$, one can then write the *expected value* of the information obtained by the occurrence of an event in S as

$$\sum_{i=1}^{i=n} p_i I(s_i)$$

or, since we are only interested in the probabilities of the events and not their actual natures, we may write

$$\sum_{i=1}^{i=n} p_i I(p_i)$$

iii. *Further Intuitional Considerations*

Now, to change our point of view, if the occurrence of s_k provides $I(p_k)$ units of information, then if we could obtain those particular $I(p_k)$ units of information, we could deduce that s_k occurred. This dual interpretation for $I(p_k)$ is quite important and should be fully understood. From this second viewpoint, then, we can interpret

$$\sum_{i=1}^{i=n} p_i I(p_i)$$

as the expected amount of information needed to determine which event of S has occurred. Thus

$$\sum_{i=1}^{i=n} p_i I(p_i)$$

is a measure of our *uncertainty* about which event of S has occurred or will occur.

When would our uncertainty be greatest? It seems evident that uncertainty is maximum when the events are equally probable; that is, when $p_1 = p_2 = \cdots = p_n$, and this is another requirement that one might place upon a measure of information.

At any rate, one can write down a number of properties that intuition seems to require to be possessed by a measure of information. To obtain an exact expression, then, for $I(s_k)$ or $I(p_k)$ it would seem proper to attempt to derive all such functions which satisfy the list of properties (and we do not intend to imply that our list matches that of concern to early workers in the field), if indeed there do exist such functions. (Hopefully, of course, there would be precisely one such function.)

iv. *Choice of Measure*

The early workers of whom we speak, however, were not completely rigorous and precise, and a measure was chosen which seemed to satisfy all the intuitive requirements (see, e.g., Reference [5]). It was

$$I(p_k) = -\log_2 p_k \qquad (8.1)$$

Later work yielded the rigorous result that, given the list of desired properties, not all of which have been mentioned here, (8.1) was a measure of the desired kind and was, in fact, except for a multiplicative constant, *unique* [3]. Now, the multiplicative constant would serve only to change the base of the logarithm; but in this chapter we shall use, except as otherwise described, base 2, though the base will not always be written.

Base 2 was, of course, a natural choice since information theory was initially concerned with communication, and binary transmission is most common.

Selection of expression (8.1) as a measure of information then yields

$$-\sum_{i=1}^{i=n} p_i \log_2 p_i = H(p_1, \ldots, p_n) \qquad (8.2)$$

as our expected information or expected uncertainty. The function H is known as the *entropy function*.

v. *Unit of Information*

Having a measure of information quantity, what units should be attached to the resulting numbers? It was decided to define the unit as follows.

Let $S = \{s_1, s_2\}$ and let $p_1 = p_2 = \frac{1}{2}$. Then

$$H(\tfrac{1}{2}, \tfrac{1}{2}) = -[\tfrac{1}{2} \log_2 \tfrac{1}{2} + \tfrac{1}{2} \log_2 \tfrac{1}{2}]$$
$$= -[\tfrac{1}{2}(-1) \cdot 2]$$
$$= 1 \text{ unit of information}$$

That is, the unit of information is taken to be that expected quantity of information yielded by the occurrence of one of two equally probable events or, equivalently, the quantity necessary to uniquely specify which of two equally probable events has occurred. The name given to this unit is the *bit*, which is a name commonly used in other contexts, but which should be distinguished from these other usages. For example, "bit" is also a contraction of binary digit, and one finds binary digits being referenced as bits. For others, a bit is a hardware component capable of representing either of the two binary digits 0, 1.

Example 8.1

Let S be a set of events $\{s_1, s_2, s_3, s_4, s_5, s_6\}$ with respective probabilities $\frac{1}{2}, \frac{1}{4}, \frac{1}{16}, \frac{1}{16}, \frac{1}{16}, \frac{1}{16}$. One of these events has occurred, and we would like to discover which one it was.

To do this, we pick a subset T of events and ask some omniscient observer, "Was it an element of T?" His reply, "yes" or "no," will convey one bit of information *if that response results in dichotomizing the events into two equally probable subsets*. Now this is entirely consistent with our definition of a bit; in the case of two equally probable events, one such reply, in *every* case, suffices to determine the identity of the occurring events whether the reply be yes or no, and regardless of which of s_1, s_2 we choose to ask about. The one difference is that each event is actually a *set* of elementary events now.

Thus our first inquiry would be whether it was s_1 that occurred, and this is the *only* way the above dichotomy could be obtained. Also, if one were interested in isolating the event that occurred with as few inquiries as possible, would he not query a most probable event first? Furthermore, considering the remaining events at any time as comprising two subsets, one containing the event of interest, the other not, since $H(\frac{1}{2}, \frac{1}{2}) \geq H(\alpha, 1 - \alpha)$ for any $0 \leq \alpha \leq 1$, we maximize the information obtained from one response by arranging the two subsets such that their probabilities are equal.

With probability $\frac{1}{2}$, we discover the event with the acquisition of 1 bit of information; otherwise, we ask next whether the event was s_2. If it was, then we have found the event with two bits of information, and this happens with probability $\frac{1}{4}$. Otherwise, we know the event to be one of s_3, s_4, s_5, s_6, so we choose any two of these equally probable events, which gives two equally probable subsets, and ask if it was one of that subset. If so, we find the event itself on the *next* inquiry. If not, we also find the event on the *next* inquiry. So with probability $4(\frac{1}{16}) = \frac{1}{4}$, we require 4 bits to isolate the event.

The expected amount of information necessary to isolate the event is then

$$\tfrac{1}{2}\cdot 1 + \tfrac{1}{4}\cdot 2 + \tfrac{1}{4}\cdot 4 = 2$$

which is precisely $H(\frac{1}{2}, \frac{1}{4}, \frac{1}{16}, \frac{1}{16}, \frac{1}{16}, \frac{1}{16})$.

vi. *Maximizing H: Concavity*

Now, under the same circumstance regarding submitting queries and obtaining responses, suppose the probabilities were such that the creation of two equally probable subsets was impossible. What should one do? The answer is given by the fact that $H(x, 1 - x)$, $0 \leq x \leq 1$, is a *strictly concave* function. In fact, $H(p_1, p_2, \ldots, p_n)$ is a strictly concave function of $n - 1$ variables, the nth being determined by $\sum_{i=1}^{i=n} p_i = 1$ after $n - 1$ of them are specified.

$H(x, 1 - x)$ may be graphed as in Figure 8.1.

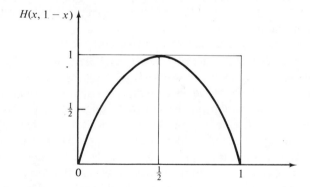

Figure 8.1. The entropy function $H(x, 1 - x)$.

Notice that we have $H(0, 1) = H(1, 0) = 0$. The resulting expression in both cases is $1 \log_2 1 + 0 \log_2 0$, the second term of which is not defined. However, the reader will verify that

$$\lim_{x \to 0} x \log_2 x = 0$$

so that taking $0 \log_2 0 = 0$ is acceptable.

At any rate, owing to the strict concavity and symmetry of $H(x, 1 - x)$ on $0 \leq x \leq 1$, to maximize the information obtained from a single response, one should choose subsets so that they are as *nearly* equal as possible in probability. In our present situation, with $H(x, 1 - x)$ this means that if x cannot be $\frac{1}{2}$ it should be as close as possible to $\frac{1}{2}$.

The reader may ask about the case where the inquiry and response can be more general; for example, can we partition the set of events into $k \leq n$ subsets and obtain a response which identifies one of those k subsets as that which contains the event in question. The proper strategy of interrogation can then be most easily seen by considering H to involve logarithms to base k, if we prefer to have one response convey at most 1 unit of information; for

$$\max \left\{ -\left(\sum_{i=1}^{i=k} Z_i \log_k Z_i \right) \right\}$$

subject to
$$\sum_{i=1}^{i=k} Z_i = 1, \quad Z_i \geq 0$$

occurs for $Z_1 = Z_2 = \cdots = Z_k = 1/k$ and is $-\log_k(1/k) = 1$ unit, but the unit is not a "bit" unless $k = 2$. Thus we should choose our subsets so that they are as nearly equal in probability as possible, which means in this case that

$$\sum_{i=1}^{i=k} \left(Z_i - \frac{1}{k} \right)^2$$

where Z_i is the probability of subset i, should be as small as possible.

Clearly, however, the strategy is the same regardless of the base of the logarithm, it is just that changing the base makes the maximum information received equal to 1 unit. With k equally probable subsets the maximum *bits* exceed 1, which they should, since knowing one of $k > 2$ exhaustive subsets is more informative than knowing one of two exhaustive subsets, if we choose subsets in each case as wisely as possible.

vii. *Remarks*

Several remarks are appropriate at this point. It should be clear now that in assessing "how much information" we shall require a probability distribution. Thus in many kinds of problems where there is no such distribution, but where the question "how much information" might seem interesting and appropriate, the question must be unanswered in our present terms.

The solution of any problem requires the utilization of the given information and the subsequent drawing of the proper inferences. Thus it might be hoped that a theory of information would have broad and important applications in at least showing *how best* to go about solving given problems, as it did in our simple search problem above. Although much more can be done with the notion of information and uncertainty than will be evident from our brief chapter, it is safe to say that the impact of information theory has not been so profound on other areas of study.

The various concepts of information theory carry over quite nicely to the case of continuous probability distributions, but we restrict ourselves here to the discrete variety.

8.3. Measures of Other Information Quantities

i. *Assumptions and Calculations of Probabilities*

Thus far we have mentioned only the amount of information yielded by the occurrence of a simple event and, equivalently, the amount of information necessary to decide that such an event has occurred. Now we wish to consider, for example, the amount of information specified by the occurrence of an event *given that a perhaps related event has occurred*, or the amount specified by the occurrence of one event *about the occurrence of some other event*.

Example 8.2

As an example of a familiar instance in which the occurrence of one event gives some information about the occurrence of another, consider evaluating a function f unimodal on $[a, b]$ at two points $a \leq x_1 < x_2 \leq b$. The events described by the relative magnitudes $f(x_1)$, $f(x_2)$ give various pieces of information about other events, for example, about the position of the global extremum of f on $[a, b]$.

We wish, then, to assume the existence of two sets X, Y, each finite; say X has m elements, the events $x_1, x_2, \ldots, x_m$ and Y has n elements, the events $y_1, y_2, \ldots, y_n$. On the set $X \cdot Y$, there is defined the joint probability distribution

$$P\{x_i, y_j\} = P\{x_i \in X \text{ occurs and } y_j \in Y \text{ occurs}\}$$

From this joint distribution may be computed the marginal distributions

$$P\{x_i\} = P\{x_i \in X \text{ occurs}\} = \sum_{j=1}^{j=n} P\{x_i, y_j\}, \quad \text{for any } i$$

and
$$P\{y_j\} = P\{y_j \in Y \text{ occurs}\} = \sum_{i=1}^{i=m} P\{x_i, y_j\}, \quad \text{for any } j$$

Then, it is also possible to calculate the conditional probabilities

$$P\{x_i \mid y_j\} = \frac{P\{x_i, y_j\}}{P\{y_j\}}, \quad i = 1, 2, \ldots, m; j = 1, 2, \ldots, n$$

$$P\{y_j \,|\, x_i\} = \frac{P\{x_i, y_j\}}{P\{x_i\}}, \quad i = 1, 2, \ldots, m; \, j = 1, 2, \ldots, n$$

ii. Defining Other Quantities

Now the product space $X \cdot Y$, being finite and possessing a distribution $P\{x_i, y_j\}$, may be considered a single set of simple events, so it is natural to define

$$I(x_i, y_j) = -\log P\{x_i, y_j\} \tag{8.3}$$

as the *amount of information provided by the joint occurrence of the pair* (x_i, y_j) *from the event space* $X \cdot Y$.

To return to the initial question, the *amount of information provided about the occurrence of the event* x_i *by the occurrence of* y_j is defined to be

$$\log \frac{P\{x_i \,|\, y_j\}}{P\{x_i\}} = I(x_i \longleftarrow y_j) \tag{8.4}$$

where we have attempted to use a notation which conveys the fact that y_j is implying something about the occurrence of x_i.

To better understand the reason for this definition, let us ask how we might *reasonably* define the amount of information provided by the occurrence of x_i, *given that* y_j *has already occurred*, a quantity which we shall denote $I(x_i \,|\, y_j)$.

The m probabilities $P\{x_1 \,|\, y_j\}, P\{x_2 \,|\, y_j\}, \ldots, P\{x_m \,|\, y_j\}$ sum to unity for any j and are the probabilities of occurrence of the respective events of X, given the occurrence of y_j.

Thus the amount of information provided by x_i, given that y_j is known to have occurred, is defined to be

$$I(x_i \,|\, y_j) = -\log P\{x_i \,|\, y_j\} = -\log \frac{P\{x_i, y_j\}}{P\{y_j\}} \tag{8.5}$$

Returning to the definition (8.4), we have

$$I(x_i \longleftarrow y_j) = \log P\{x_i \,|\, y_j\} - \log P\{x_i\} = I(x_i) - I(x_i \,|\, y_j)$$

In words, this is the amount of information provided by the occurrence of x_i alone, minus the amount of information provided by x_i when it is known that y_j occurred. Examining the word definition, we see that this difference is *the amount of information which* y_j *provides about* x_i. We begin with the amount provided by x_i alone and subtract that which x_i provides once y_j is known—the difference must have been provided about x_i's occurrence by y_j's occurrence.

Example 8.3

To illustrate some of these concepts, let us construct sets X and Y and the appropriate probabilities as follows: Let

$$X = \{x_1, x_2, x_3, x_4, x_5\}$$
$$Y = \{y_1, y_2, y_3, y_4\}$$

$P\{x_i, y_j\}$	y_1	y_2	y_3	y_4	$P\{x_i\}$
x_1	$\frac{1}{4}$	$\frac{1}{8}$	$\frac{1}{16}$	$\frac{1}{16}$	$\frac{1}{2}$
x_2	$\frac{1}{8}$	$\frac{1}{16}$	$\frac{1}{32}$	$\frac{1}{32}$	$\frac{1}{4}$
x_3	$\frac{1}{16}$	$\frac{1}{32}$	$\frac{1}{64}$	$\frac{1}{64}$	$\frac{1}{8}$
x_4	$\frac{1}{32}$	$\frac{1}{64}$	$\frac{1}{128}$	$\frac{1}{128}$	$\frac{1}{16}$
x_5	$\frac{1}{32}$	$\frac{1}{64}$	$\frac{1}{128}$	$\frac{1}{128}$	$\frac{1}{16}$
$P\{y_j\}$	$\frac{1}{2}$	$\frac{1}{4}$	$\frac{1}{8}$	$\frac{1}{8}$	

The conditional probabilities $P\{x_i | y_j\}$ may then be calculated as

$P\{x_i|y_j\}$	y_1	y_2	y_3	y_4
x_1	$\frac{1}{2}$	$\frac{1}{2}$	$\frac{1}{2}$	$\frac{1}{2}$
x_2	$\frac{1}{4}$	$\frac{1}{4}$	$\frac{1}{4}$	$\frac{1}{4}$
x_3	$\frac{1}{8}$	$\frac{1}{8}$	$\frac{1}{8}$	$\frac{1}{8}$
x_4	$\frac{1}{16}$	$\frac{1}{16}$	$\frac{1}{16}$	$\frac{1}{16}$
x_5	$\frac{1}{16}$	$\frac{1}{16}$	$\frac{1}{16}$	$\frac{1}{16}$

Given that y_j has occurred, for any j, the probabilities of the events x_1, x_2, x_3, x_4 and x_5 are, in order, $\frac{1}{2}, \frac{1}{4}, \frac{1}{8}, \frac{1}{16}$, and $\frac{1}{16}$. Thus, for example

$$I(x_1 | y_4) = -\log \tfrac{1}{2} = 1 \text{ bit}$$
$$I(x_4 | y_1) = -\log \tfrac{1}{16} = 4 \text{ bits}$$
$$I(x_3 | y_3) = -\log \tfrac{1}{8} = 3 \text{ bits}$$

iii. Averages and Asymmetry

As might be suspected then, the *average* amount of information provided by *the occurrence* of an event of X, given that y_j has occurred, is defined to be

$$I(X|y_j) = \sum_{i=1}^{i=m} P\{x_i|y_j\} \log P\{x_i|y_j\} \tag{8.6}$$

Using equation (8.6), it follows that the average amount of information provided by an X event, given the occurrence of a Y event, may be expressed as

$$\sum_{j=1}^{j=n} P\{y_j\}I(X|y_j) = -\sum_{j=1}^{j=n} P\{y_j\}\left[\sum_{i=1}^{i=m} P\{x_i|y_j\} \log P\{x_i|y_j\}\right]$$

$$= -\sum_{i=1}^{i=m}\sum_{j=1}^{j=n} P\{x_i|y_j\}P\{y_j\} \log P\{x_i|y_j\} \qquad (8.7)$$

$$= -\sum_{i=1}^{i=m}\sum_{j=1}^{j=n} P\{x_i, y_j\} \log P\{x_i|y_j\} = I(X|Y)$$

and clearly $I(X|Y) \geq 0$.

We might also ask about quantities of information provided by Y events, *given* the occurrence of X events. That is, what is $I(y_j|x_i)$? By definition, we would write

$$I(y_j|x_i) = -\log P\{y_j|x_i\},$$

and since

$$P\{y_j|x_i\} = \frac{P\{x_i, y_j\}}{P\{x_i\}}$$

$$I(y_j|x_i) = -\log \frac{P\{x_i, y_j\}}{P\{x_i\}}$$

Thus, in general, $I(y_j|x_i) \neq I(x_i|y_j)$. For example, from the present example probabilities

$$I(x_4|y_1) = -\log \frac{1/32}{1/2} = 4 \text{ bits}$$

whereas
$$I(y_1|x_4) = -\log \frac{1/32}{1/16} = 1 \text{ bit}$$

From the probabilities, we see that when y_1 occurs, x_4 is not too likely to have occurred simultaneously, whereas knowing that x_4 has occurred, it is not too surprising that y_1 has occurred.

The average amount of information provided by the occurrence of a Y, given the occurrence of an X event, is defined, entirely similar to the reverse situation (8.7), by

$$I(Y|X) = -\sum_{i=1}^{i=m}\sum_{j=1}^{j=n} P\{y_j, x_i\} \log \frac{P\{y_j, x_i\}}{P\{x_i\}}$$

$$= -\sum_{i=1}^{i=m}\sum_{j=1}^{j=n} P\{x_i, y_j\} \log \frac{P\{x_i, y_j\}}{P\{x_i\}} \geq 0$$

iv. *Further Sample Calculations and General Results*

a. IMPLICATION OF INDEPENDENCE

Returning to our sample probabilities of Example 8.3, let us evaluate several mutual information quantities:

$$I(x_2 \longleftarrow y_4) = \log \frac{P\{x_2 | y_4\}}{P\{x_2\}} = \log \frac{1/4}{1/4} = \log 1 = 0$$

$$I(y_4 \longleftarrow x_2) = \log \frac{P\{y_4 | x_2\}}{P\{y_4\}} = \log \frac{P\{x_2, y_4\}}{P\{x_2\}P\{y_4\}} = \log 1 = 0$$

$$I(x_3 \longleftarrow y_2) = \log \frac{1/8}{1/8} = \log 1 = 0$$

Thus in each of the three previous calculations we find that the one event provides *no* information about the other. But this is exactly as it should be, for if one considers the table of joint and marginal distributions, he finds, for all *i* and *j*,

$$P\{x_i, y_j\} = P\{x_i\}P\{y_j\}$$

That is, the events x_i and y_j are *independent*, in which case one event *cannot* give information about the other.

b. SIGN RESULTS AND INTERPRETATIONS

From the defining equation we see that $I(x_i \longleftarrow y_j)$ is not obviously non-negative; if, in fact, $P\{x_i\} > P\{x_i | y_j\}$, then $I(x_i \longleftarrow y_j) < 0$. Can this happen? To decide, consider the following joint distribution:

$P\{x_i, y_j\}$	y_1	y_2	y_3	$P\{x_i\}$
x_1	$\frac{45}{100}$	$\frac{45}{100}$	$\frac{1}{100}$	$\frac{91}{100}$
x_2	$\frac{2}{100}$	$\frac{2}{100}$	$\frac{1}{100}$	$\frac{5}{100}$
x_3	$\frac{1}{100}$	$\frac{2}{100}$	$\frac{1}{100}$	$\frac{4}{100}$
$P\{y_j\}$	$\frac{48}{100}$	$\frac{49}{100}$	$\frac{3}{100}$	

Then

$$I(x_1 \longleftarrow y_3) = \log \frac{P\{x_1 | y_3\}}{P\{x_1\}} = \log \frac{1/3}{45/100}$$

$$= \log \frac{100}{135} < 0$$

Here $P\{x_1\} > P\{x_1 | y_3\}$.

Thus the definition implies that y_3 provides a negative amount of information about the occurrence of x_1. How should this be interpreted? Expanding $I(x_i \longleftarrow y_j)$ into

$$I(x_i \longleftarrow y_j) = I(x_i) - I(x_i | y_j)$$

we see that if $I(x_i \leftarrow y_j)$ is negative, more information is obtained from the occurrence of x_i, given that y_j has occurred, than from the occurrence of x_i alone. Ordinarily, one would imagine that having observed something about the system, he would know more, and thus obtain less information from the occurrence of a related event. Here the opposite is true; thus we might say that we have been *misled* by the occurrence of y_j, so that if x_i is then discovered to have occurred more information is obtained.

The average amount of information about x_i that is provided by the occurrence of a Y event and the average amount of information about y_j that is provided by an X event are, respectively, then

$$\sum_{j=1}^{j=n} P\{x_i|y_j\}I(x_i \leftarrow y_j) = \sum_{j=1}^{j=n} P\{x_i|y_j\} \log \frac{P\{x_i|y_j\}}{P\{x_i\}} \tag{8.8}$$

$$= I(x_i \leftarrow Y)$$

and

$$\sum_{i=1}^{i=m} P\{y_j|x_i\}I(y_j \leftarrow x_i) = \sum_{i=1}^{i=m} P\{y_j|x_i\} \log \frac{P\{y_j|x_i\}}{P\{y_j\}} \tag{8.9}$$

$$= I(y_j \leftarrow X)$$

Relation (8.9) expresses the conditional probabilities as $P\{y_j|x_i\}$, whereas we originally assumed the probabilities to be given as $P\{x_i|y_j\}$. But if given in one of these forms, they may always be expressed in the other, since

$$P\{y_j|x_i\} = \frac{P\{x_i, y_j\}}{P\{x_i\}} = \frac{P\{x_i|y_j\}P\{y_j\}}{P\{x_i\}} \tag{8.10}$$

Relation (8.10), then, may be used to obtain the one form from the other.

Also, expressions may be written for $I(X \leftarrow y_j)$ and $I(Y \leftarrow x_i)$. These are

$$I(X \leftarrow y_j) = \sum_{i=1}^{i=m} P\{x_i|y_j\}I(x_i \leftarrow y_j) = \sum_{i=1}^{i=m} P\{x_i|y_j\} \log \frac{P\{x_i|y_j\}}{P\{x_i\}} \tag{8.11}$$

and

$$I(Y \leftarrow x_i) = \sum_{j=1}^{j=n} P\{y_j|x_i\}I(y_j \leftarrow x_i) = \sum_{j=1}^{j=n} P\{y_j|x_i\} \log \frac{P\{y_j|x_i\}}{P\{y_j\}} \tag{8.12}$$

We have seen that the mutual information expressions can be negative, purely as a result of the magnitudes of the probabilities involved.

Let us consider one of the expected mutual information expressions, $I(X \leftarrow y_j)$, for example, and inquire into its possible signs. We already know that $I(X \leftarrow y_j)$ can be zero; this will happen if $x_1, x_2, \ldots, x_m$ are all independent of y_j.

We wish to make use of an elementary inequality in what follows. The inequality is most useful in proving theorems in information theory.

The inequality, which we ask the reader to prove, is that

$$\log_e x \leq x - 1, \quad \text{for all } x \geq 0 \tag{8.13}$$

Using (8.13), we have

$$I(X \longleftarrow y_j) = \sum_{i=1}^{i=m} P\{x_i | y_j\} \log \frac{P\{x_i | y_j\}}{P\{x_i\}} \leq \log e \sum_{i=1}^{i=m} P\{x_i | y_j\} \left[\frac{P\{x_i | y_j\}}{P\{x_i\}} - 1 \right]$$

Thus

$$-I(X \longleftarrow y_j) \leq \log e \sum_{i=1}^{i=m} P\{x_i | y_j\} \left[\frac{P\{x_i\}}{P\{x_i | y_j\}} - 1 \right]$$

$$= \log e \sum_{i=1}^{i=m} [P\{x_i\} - P\{x_i | y_j\}] = \log e[1 - 1] = 0$$

Thus we have $I(X \longleftarrow y_j) \geq 0$, for any j. Similarly, one shows that any other expected mutual information is also always nonnegative.

The implication of this fact, that is, that the *average* mutual information is always nonnegative, is that, although one might be misled on some occasions, on the *average* he will *not* be. In other words, on the average the mutual information is not misleading; knowledge of the occurrence of an event in one set will on the average contribute to one's knowledge about the occurrence of an event in the other set.

Examining the functions $\log_e x$ and $x - 1$, one sees that they are equal at $x = 1$ and *only* there. Thus

$$\frac{P\{x_i\}}{P\{x_i | y_j\}} - 1 = \log \frac{P\{x_i\}}{P\{x_i | y_j\}}$$

if and only if

$$\frac{P\{x_i\}}{P\{x_i | y_j\}} = 1$$

that is, if and only if $P\{x_i\} = P\{x_i | y_j\}$, which is equivalent to saying that x_i, y_j are independent.

Thus our expected mutual information measures will be zero when and only when the events involved in the respective expressions are independent, that is, when x_i, Y are independent or when y_j, X are independent.

v. Relationships Among Information Quantities

There are numerous relationships existing among the quantities defined thus far. Let us obtain some of these and comment on their interpretations.

First, we have

$$I(X \longleftarrow Y) = \sum_{i=1}^{i=m} \sum_{j=1}^{j=n} P\{x_i, y_j\} \log \frac{P\{x_i|y_j\}}{P\{x_i\}}$$

$$= \sum_{i=1}^{i=m} \sum_{j=1}^{j=n} P\{x_i, y_j\} \log \left(\frac{\frac{P\{y_j|x_i\}P\{x_i\}}{P\{y_j\}}}{P\{x_i\}} \right) \qquad (8.14)$$

$$= \sum_{i=1}^{i=m} \sum_{j=1}^{j=n} P\{x_i, y_j\} \log \frac{P\{y_j|x_i\}}{P\{y_j\}}$$

$$= I(Y \longleftarrow X)$$

Thus the average amount of information that an observed Y event provides about the occurrence of an X event is the *same* as the average amount of information that an observed X event provides about the occurrence of a Y event. This result is intuitively satisfying, since we are given just a joint distribution, a probabilistic description of *pairs* of events.

Next we have

$$I(X \longleftarrow Y) = \sum_{i=1}^{i=m} \sum_{j=1}^{j=n} P\{x_i, y_j\} \log P\{x_i|y_j\} + H(X) \qquad (8.15)$$

The first term on the right side of (8.15) is clearly nonpositive; hence it is clear that

$$I(X \longleftarrow Y) \leq H(X) \qquad (8.16)$$

Using (8.14), it is clear that

$$I(X \longleftarrow Y) \leq H(Y) \qquad (8.17)$$

Inequality (8.16) tells us that the average amount of information which the occurrence of a Y event provides about the occurrence of an X event is never greater than the average amount of information provided by the occurrence of an X event itself. Again, one would expect (8.16), owing to the uncertainties in inferring which X event has occurred from knowledge of the Y event that occurred.

The reader can formulate the verbal expression for (8.17) as well as for the inequalities that follow from (8.16) and (8.17) by virtue of (8.14); that is,

$$I(Y \longleftarrow X) \leq H(Y)$$
$$I(Y \longleftarrow X) \leq H(X)$$

Returning to (8.15), it is seen that that equation may be written

$$I(X \longleftarrow Y) = H(X) - I(X|Y) \qquad (8.18)$$

or $\qquad I(X \longleftarrow Y) = H(Y) - I(Y|X) = I(Y \longleftarrow X) \qquad (8.19)$

A verbal equivalent of (8.18) and a justification can be easily extrapolated from our comments following the definition of $I(x_i \leftarrow y_j)$. Equation (8.18) simply represents the average of $I(x_i \leftarrow y_j)$ over the set $X \cdot Y$. Equation (8.19) may be likewise described.

vi. *A Word on More Complicated Distributions*

One may consider cases where there are more than two sets of events and define various conditional and mutual information quantities in similar fashion to that of the preceding section.

For example, with sets X, Y, and Z having m, n, and p elements, respectively, and a probability distribution $P\{x_i, y_j, z_k\}$, one defines, for example, $I(x_i \leftarrow y_j \mid z_k)$ as the amount of information provided about the occurrence of x_i by the occurrence of y_j given the occurrence of z_k, and $I(y_j \mid x_i z_k)$ as the amount of information provided by the occurrence of y_j given the *joint* occurrence of x_i and z_k.

8.4. Application 1—Heavy-Coin Problem

We wish to apply some of our basic results to the solution of a particular problem, which is a generalization of a problem with which the reader is likely familiar.

The problem is a coin-weighing problem. A particular puzzle-type version goes as follows:

There are given 12 coins, one of which, though identical in appearance to all the others, is heavier. It is desired to locate this heavier coin, and to accomplish this, we have available an equal arm balance. By a "weighing" is meant the putting of a subset of the coins on each of the balance pans and then observing the result. The problem is to isolate the heavy coin in the smallest number of weighings.

For 12 coins, three weighings will always suffice, and no less than three will guarantee finding the heavy coin.

The flow charts given in Figures 8.2 and 8.3 illustrate an optimal strategy. The heavy coin is known to be within a subset of four coins after the weighing of Figure 8.2.

In the event of either of the first two possibilities in Figure 8.3, the heavy coin is found in *two* weighings; in the third event, one more weighing suffices, so the foregoing strategy isolates the counterfeit in (at most) three weighings.

The generalization we would like to treat, then, is the case where there are n coins and where we ask for the minimum number of weighings that is guaranteed to isolate the heavy one, or, given a number of weighings, the

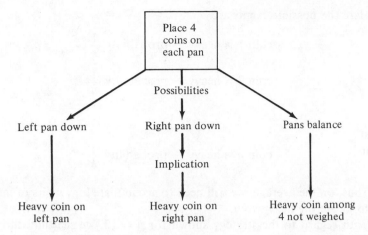

Figure 8.2. Optimal first weighing for the heavy coin problem.

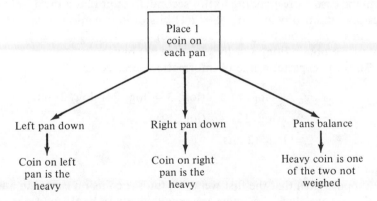

Figure 8.3. Optimal second weighing for the heavy coin problem.

maximum number of coins that can be accommodated. In the process we shall discover the optimal weighing strategy.

First, it is clear that, knowing nothing about the degree of heaviness of the odd coin, observing the results of a weighing with unequal numbers of coins on the pans will not always provide information. (Can you construct examples where information is provided with unequal numbers of coins?) It is clear that every weighing must have the same number of coins on each pan in order that the measurement be certain to provide information.

Next the expected amount of information necessary to isolate the heavy coin is

$$H\left(\frac{1}{n}, \frac{1}{n}, \cdots, \frac{1}{n}\right) = -\log_2 \frac{1}{n} = \log_2 n \text{ bits}$$

Here the possible events are

$$\text{coin 1 is heavy} - \text{probability } \frac{1}{n}$$

$$\text{coin 2 is heavy} - \text{probability } \frac{1}{n}$$

$$\vdots \qquad\qquad \vdots$$

$$\text{coin } n \text{ is heavy} - \text{probability } \frac{1}{n}$$

Thus, on the average, we will need to accumulate $\log_2 n$ bits of information to locate the heavy coin.

With regard to the strategy shown for $n = 12$, we see that during the first-stage weighings we get $I(\frac{1}{3}) = \log_2 3$ bits of information, regardless of the outcome. At stage two we get either $I(\frac{1}{4}) = \log_2 4$ or $I(\frac{1}{2}) = \log_2 2$, and it is in the case corresponding to this second amount that a third weighing is necessary, and it provides $\log_2 2 = I(\frac{1}{2})$ bits of information.

Thus with probability $\frac{1}{2}$ we obtain $(\log_2 3 + \log_2 4)$ bits, and with probability $\frac{1}{2}$ we obtain $(\log_2 3 + \log_2 2 + \log_2 2)$ bits.

Thus the expected value of information received is

$$\tfrac{1}{2}(\log_2 3 + \log_2 4) + \tfrac{1}{2}(\log_2 3 + \log_2 2 + \log_2 2) \text{ bits}$$
$$= \tfrac{1}{2}(\log_2 12) + \tfrac{1}{2}(\log_2 12)$$
$$= \log_2 12 \text{ bits}$$

as anticipated.

Suppose then that the first weighing finds x coins on each pan and $n - 2x$ coins not weighed. The coins are equally likely to be the odd one. Therefore,

$$P\{\text{left pan down}\} = \frac{x}{n}$$

$$P\{\text{right pan down}\} = \frac{x}{n}$$

$$P\{\text{pans balance}\} = \frac{n - 2x}{n}$$

The expected amount of information provided by the outcome of the weighing then is $H[x/n, x/n, (n - 2x)/n]$. Clearly, we want to maximize this information; thus, the ideal strategy is to have $x/n = (n - 2x)/n = \frac{1}{3}$. This, however, may not be possible, owing to the divisors of n. As we know, then, the best policy is to separate the original n coins into three subsets that are as nearly equal as possible, and we also require that two of the subsets contain identical numbers.

If n is divisible by 3, $x = n/3$, and

$$H(\tfrac{1}{3}, \tfrac{1}{3}, \tfrac{1}{3}) = -\log_2 \tfrac{1}{3} = \log_2 3$$

If we are able to obtain this maximum $\log_2 3$ *bits of information during each weighing*, then k weighings provide $k \log_2 3$ bits of information.

To have isolated the heavy coin after these k weighings, we know we must have $k \log_2 3 \geq \log_2 n$ or $\log_2 3^k \geq \log_2 n$, or $3^k \geq n$.

8.5. Application 2—Odd-Coin Problem

A variation of the preceding problem is the case where the type of deviation of the distinguished coin is unknown; that is, it may be lighter or heavier than the others. The problem is to locate the odd coin and establish whether it is lighter or heavier, which, as a matter of fact will be known when the odd coin is isolated.

Given n coins, there are $2n$ equally probable possibilities for the odd coin since each of the n coins may be the odd one, and the odd coin may be lighter or heavier than the others. Thus we must accumulate $\log_2 2n$ bits of information to isolate the odd coin.

Again, there must be equal numbers of coins on the weighing pans in order that a weighing actually provide any information.

The complicating factor here is that if we observe one pan to go down there may be a heavy coin on *that* pan *or* a light coin on the other pan; it is impossible to distinguish which in the absence of additional information.

If one is hasty in drawing conclusions, he might decide that the best strategy is to have, if possible, the same number of coins off the pans as on, since the result of the weighing will, in this case, leave one with either the set of coins on both pans or the set not on the pans. But this reasoning is not correct, since finding the pans out of balance does not leave one with the coins on the pans and *no* additional information.

If we have x coins on each pan, in fact, the following conditions and deductions are possible.

1. Left pan down. We have either a heavy on the left or a light on the right. Thus there are $2x$ possibilities.
2. Right pan down. There is either a heavy on the right or a light on the left. Again, there are $2x$ possibilities.
3. Pans balance. There is either a heavy one out or a light one out. So there are $2(n - 2x)$ possibilities.

To maximize expected information received, we need to have as nearly as possible $x = n - 2x$.

The maximum expected information obtained from a single weighing is then

$$\log_2 3 \text{ bits}$$

From k weighings, we obtain a maximum of $\log_2 3^k = k \log_2 3$ bits of information, whereas $\log_2 2n$ bits is the average amount needed.

Suppose that we obtain the maximum possible expected information with each weighing (impossible, in general) and that k weighings provide exactly $\log_2 2n$ bits of information. Then

$$\log_2 3^k = \log_2 2n$$

or $$3^k = 2n$$

Thus $$n = \frac{3^k}{2} \tag{8.20}$$

is the maximum number of coins we could treat with k weighings. But n as defined in (8.20) is not an integer. However, $3^k - 1$ is even for any k, so

$$n = \frac{3^k - 1}{2} \tag{8.21}$$

is the maximum number of coins we could handle with the k weighings.

As k varies, $(k = 1, 2, \ldots)$, n as defined by (8.21) assumes the values $n = 1, 4, 13, 40, \ldots$

In general the values of interest for n are always one greater than the product of 3 and some positive integer. (There is no problem for $n = 1$.) That is, n has the general form $n = 3m + 1$ for some positive integer m.

To see this fact, it is easy to show that for a fixed positive integer k the equation

$$3m + 1 = \frac{3^k - 1}{2} \tag{8.22}$$

has a unique nonnegative integer solution m, for by (8.22)

$$3m = \frac{3^k - 1}{2} - 1 = \frac{3^k - 3}{2} = \frac{3(3^{k-1} - 1)}{2}$$

Now by (8.21)

$$\frac{3^{k-1} - 1}{2}$$

is a nonnegative integer, and so there is a nonnegative integer m satisfying (8.22).

The best strategy is then to place m coins on the left, m on the right, and to keep $m + 1$ coins off the pans.

Suppose that they balance. Then the problem becomes to isolate the coin and determine its type from among $2(m + 1)$ possibilities in $k - 1$ weighings. That is, we must have

$$\log_2 3^{k-1} \geq \log_2 2(m + 1)$$

or
$$3^{k-1} \geq 2m + 2 \tag{8.23}$$

But
$$m = \frac{n - 1}{3}$$

so (8.23) requires

$$3^{k-1} \geq \frac{2(n - 1)}{3} + 2$$

or
$$3^k \geq 2(n + 2)$$

But, by (8.21), we cannot accomplish this subtask in $k - 1$ weighings.

From (8.21), then, the next smaller number of coins for which k weighings could suffice is

$$n = \frac{3k - 3}{2} \tag{8.24}$$

Thus (8.24), given k, provides an upper bound on the number of coins that can be accommodated, although we shall not show that n in (8.24) is sufficient in general.

At this point it would be well to consider the strategy for $n = 12$. We know now we first have four coins on the left pan, four on the right, and four not weighed.

Case 1

Suppose the pans balance. Then the odd coin is one of the four not weighed.

Several observations allow the design of a best second weighing. First, if *only* coins from the distinguished set of four are involved in the second weighing, three weighings will not always suffice. But the only other alternative is to use some of the eight coins known to be true, and, fortunately, these may be used to very good advantage. In that regard, however, it is necessary to make a further assumption; if some coins from the set known to contain the odd coin are placed on the same pan as some known to be true coins, then these sets remain distinguishable *after* the weighing. Without this assumption, which may not always be justifiable, our efforts will fail.

The second weighing then has, we suppose,

t_1 true coins and x_1 of the 4 on the left pan

t_2 true coins and x_2 of the 4 on the right pan

$4 - x_1 - x_2$ of those 4 not on either pan

and, of course,

$$t_1 + x_1 = t_2 + x_2$$
$$P\{\text{left pan down}\} = (x_1 + x_2)/8$$
$$P\{\text{right pan down}\} = (x_1 + x_2)/8$$
$$P\{\text{balance}\} = 2(4 - x_1 - x_2)/8$$

Ideally, we would like

$$x_1 + x_2 = 2(4 - x_1 - x_2)$$

or $$3x_1 + 3x_2 = 8 \tag{8.25}$$

in order to maximize expected information from the second weighing; but (8.25) has no solution for nonnegative integers x_1, x_2.

We come as close as possible to satisfying (8.25), and hence to achieving equal probabilities, by taking one of the x equal to 2 and the other to 1.

Thus set $x_1 = 2$, $x_2 = 1$. There are then a number of ways to select t_1, t_2, and we may as well set $t_1 = 0$, $t_2 = 1$. There is one coin of the questionable four not weighed.

Depending upon the outcome of the second weighing, a particular third weighing will always solve the problem, and we leave the finding of these strategies for the reader. Is there another best second weighing?

Case 2

Suppose that the left pan went down during weighing one.

Let A be those which *were* on the left pan; B, those on the right; C, those not weighed during weighing one, and which we know, therefore, to be true coins.

For the second weighing, we use the following notation:

x_{LA} = number of coins from A put on left pan

x_{RB} = number of coins from B put on right pan

$x_{LB}, x_{LC}, x_{RA}, x_{RC}$ are defined in obvious fashion

Coins not weighed during weighing two are denoted x_A, x_B according to their location during weighing one.

The number of ways the left pan can go down, then, for weighing two is $x_{LA} + x_{RB}$; $x_{RA} + x_{LB}$ is the number of ways for the right pan to be down; and they will balance in $x_A + x_B$ ways.

We must have

$$x_{LA} + x_{LB} + x_{LC} = x_{RA} + x_{RB} + x_{RC}$$

$$x_A + x_{LA} + x_{RA} = 4$$

$$x_B + x_{LB} + x_{RB} = 4$$

and $\qquad x_{LA} + x_{RB} + x_{RA} + x_{LB} + x_A + x_B = 8 \qquad$ (8.26)

And it is desirable that

$$x_{LA} + x_{RB} = x_{RA} + x_{LB} = x_A + x_B \qquad (8.27)$$

Again we find that (8.27) cannot be satisfied, but we may satisfy system (8.26) and come as close as possible to satisfying (8.27) by setting

$$x_{LA} = 1, x_{RA} = 1, x_A = 2, x_{LB} = 2,$$

$$x_{RB} = 1, x_B = 1, x_{LC} = 0, x_{RC} = 1.$$

(Are there other solutions?)

We leave as an exercise the description of all third weighings and the inferences to be drawn, depending upon the results of the second weighing described.

8.6. Application 3—Simple Coding Problems

There is no question that insofar as applications are concerned, information theory has had its greatest influence, as might be suspected, in the areas of information transmission and coding theory. Although we do not wish to expend much space on the topic of coding, this is a most important and interesting area, and we include an application to it.

Suppose that we have a set S of n words (composed of strings of symbols from some alphabet). Messages, strings of these source words, occur, and it is desired to communicate these over some communication channel. There is a coding alphabet E composed of r symbols, and only these symbols may be transmitted across the given channel.

To accomplish this we assign a different string of code symbols to each given word, and when a message arrives to be sent, we transmit the corresponding string of code symbols, with no spaces between coded words.

For example, let

$$S = \{a, b, c, d, e\}$$

Let $E = \{0, 1\}$

Suppose that we associate words with source words as follows:

$$
\begin{aligned}
a &\sim 10 \\
b &\sim 01 \\
c &\sim 00 \\
d &\sim 11 \\
e &\sim \ 1
\end{aligned}
\tag{8.28}
$$

Let an arriving message be *baacd;* we transmit

$$0110100011 \tag{8.29}$$

Decoding is accomplished sequentially; that is, *incoming symbols are scanned sequentially and original words decoded as they are discovered.* Following that rule, we have a problem with the code as described by (8.28), for the string (8.29) will be decoded

$$bebcbe\dagger$$

which is quite different from what was transmitted. Our code (8.28) is ambiguous; that is, there are messages with more than a single interpretation, which is clearly undesirable for the transmission of information. Thus we want our codes to be unambiguous, or *uniquely decipherable*, that is, such that every legitimate message is decodable in one and only one way.

For the same set S the following code is uniquely decipherable:

$$
\begin{aligned}
a &\sim 01 \\
b &\sim 10 \\
c &\sim 00 \\
d &\sim 110 \\
e &\sim 111
\end{aligned}
\tag{8.30}
$$

For an arbitrary set S and an arbitrary alphabet E, there will be, in general, a large number of ways to construct a uniquely decipherable code for S. In fact, if the length in symbols of the longest code word is allowed to

---

†We assume no transmission errors.

grow arbitrarily large, so also will the number of uniquely decipherable codes for S.

We assume the existence of n numbers p_i with the properties

$$\sum_{i=1}^{i=n} p_i = 1, \quad p_i \geq 0, \text{ all } i = 1, \ldots, n$$

and such that

$$P\{\text{next word arriving to be sent is the } i\text{th}\}$$
$$= P\{\text{next code word transmitted is the } i\text{th code word}\} = p_i$$

Thus we have a stationary probability distribution for arriving words, or, equivalently, for transmitted code words. At any time, the next word is independent of the time of occurrence and of all previous words. This simplifying assumption is violated by many familiar kinds of messages, to be sure. Natural language messages provide a perfect example.

Let us suppose that the code symbols are equally costly to transmit, so that the *number of symbols in a code word is a measure of its cost*.

Let l_i, $i = 1, \ldots, n$, be the *length* of code word i, that is, the number of symbols in code word i, or the *cost* of code word i.

At some point in time, denote the length of the next code word transmitted by the random variable $\mathcal{L}$.

Then the expected value of $\mathcal{L}$ is

$$E[\mathcal{L}] = \sum_{i=1}^{i=n} p_i l_i$$

A desirable objective then is to choose a set of uniquely decipherable code words such that we minimize $E[\mathcal{L}]$, or

$$\min \left\{ E[\mathcal{L}] = \sum_{i=1}^{i=n} p_i l_i \right\} \tag{8.31}$$

The expected value of the amount of information necessary to determine which word has occurred is $H(p_1, p_2, \ldots, p_n) = H(S)$. This information must be communicated by the code chosen.

Suppose that we can locate a code in which the average amount of information communicated by *one code symbol* is maximized. This will be possible we know *if and only if the code symbols have equal probabilities of occurring at a given time* and that maximum amount of information is $\log r$.

With such an ideal code the expected value of the information provided by one code word is

$$E[\mathcal{L}] \log r$$

and if we are to deduce the correct word from the given message we must have

$$E[\mathcal{L}] \log r \geq H(S)$$

which implies

$$E[\mathcal{L}] \geq \frac{H(S)}{\log r} \tag{8.32}$$

The inequality (8.32) specifies a lower bound on the expected word length or word cost of a uniquely decipherable code for the given distribution and alphabet. Inequality (8.32) says that the best one can hope to do is $H(S)/\log r$, *although that lower bound may not be achievable.*

Now our lower bound says nothing about when (8.32) might be satisfied as an equality or how one goes about finding an optimal code.

We shall not pursue here the finding of optimal codes, but there are very simple algorithms available which accomplish that task under the assumptions made. The algorithms are quite different from classical optimization procedures and require no familiarity with mathematical programming. It is true, however, that the problem of finding an optimal code has a mathematical statement. To write it, we must first present without proof an important, related result.

Lemma 8.1

In order that the positive integers $l_1, l_2, \ldots, l_n$ be the lengths of the code words in a uniquely decipherable n-word code over some alphabet of r symbols, it is necessary and sufficient that

$$\sum_{i=1}^{i=n} r^{-l_i} \leq 1 \tag{8.33}$$

It is important to notice that the Lemma poses *no* conditions on the composition of code words—only on their lengths. Thus, although we have

$$\sum_{i=1}^{i=n} r^{-l_i} \leq 1$$

it will generally be possible to construct a set of code words with lengths $l_1, \ldots, l_n$ that do *not* comprise a uniquely decipherable code. For example, with $n = 5, r = 3$, let $l_1 = 1, l_2 = 1, l_3 = l_4 = l_5 = 2$. Then

$$\sum_{i=1}^{i=n} r^{-l_i} = 3^{-1} + 3^{-1} + 3 \cdot 3^{-2} = 1$$

But $\{a, b, ab, aa, ac\}$ is obviously not uniquely decipherable.

At any rate, a mathematical statement of the problem of finding an optimal code is

$$\min \sum_{i=1}^{i=n} p_i l_i$$

subject to

$$\sum_{i=1}^{i=n} r^{-l_i} \leq 1 \tag{8.34}$$

$$l_i \text{ an integer, } i = 1, \ldots, n$$

It should be emphasized that there are methods for obtaining an optimal code which are much more efficient than could be expected from applying any method for solving nonlinear integer programming problems to (8.34).

Theorem 8.1 answers the question about attaining the lower bound.

Theorem 8.1

A uniquely decipherable code with lengths $l_1, l_2, \ldots, l_n$ satisfies

$$E[\mathfrak{L}] = \frac{H(S)}{\log r}$$

if and only if $l_1, \ldots, l_n$ satisfy

$$\frac{I(p_i)}{\log r} = l_i, \quad i = 1, \ldots, n \tag{8.35}$$

Proof: If we have

$$l_i = \frac{I(p_i)}{\log r}$$

then

$$\sum_{i=1}^{i=n} p_i l_i = \sum_{i=1}^{i=n} \frac{p_i I(p_i)}{\log r} = \frac{1}{\log r} H(S)$$

Furthermore, since

$$l_i = \frac{I(p_i)}{\log r}$$

$$l_i \log r = I(p_i) = -\log p_i$$

or

$$r^{-l_i} = p_i$$

Thus

$$\sum_{i=1}^{i=n} r^{-l_i} = \sum_{i=1}^{i=n} p_i = 1$$

so, by Lemma 8.1, such a set $\{l_i\}$ does correspond to a uniquely decipherable code.

On the other hand, we have always

$$E[\mathcal{L}] \geq \frac{H(S)}{\log r}$$

or

$$H(S) - E[\mathcal{L}] \log r \leq 0$$

or

$$-\sum_{i=1}^{i=n} p_i \log p_i - \log r \sum_{i=1}^{i=n} p_i l_i \leq 0$$

That is,

$$\sum_{i=1}^{i=n} p_i \log \frac{r^{-l_i}}{p_i} \leq 0$$

But, using the inequality (8.13),

$$\sum_{i=1}^{i=n} p_i \log\left(\frac{r^{-l_i}}{p_i}\right) \leq \log e \sum_{i=1}^{i=n} p_i \left(\frac{r^{-l_i}}{p_i} - 1\right)$$

with equality if and only if

$$\frac{r^{-l_i}}{p_i} = 1, \quad \text{for all } i$$

Thus we have

$$H(S) - E[\mathcal{L}] \log r = 0$$

if and only if

$$r^{-l_i} = p_i, \quad i = 1, \ldots, n$$

or

$$l_i = \frac{I(p_i)}{\log r}, \quad i = 1, \ldots, n$$

Of course, whatever the p_i and whatever the value of r, one can define a set of *real numbers* $\{l_i\}$ that satisfy (8.35) and hence achieve the lower bound; but the $\{l_i\}$ must be integers for the expression $E[\mathcal{L}]$ to make sense.

Summarizing, then, given a set of probabilities $\{p_i\}$ and a value of r, if we find that

$$\frac{I(p_i)}{\log r}$$

is a positive integer for all i the lower bound

$$\frac{H(S)}{\log r}$$

is achievable; otherwise, it is not.

If the numbers involved are cooperative, numbers of bits and numbers of binary digits are identical.

Let the probability distribution be

$$P = \{\tfrac{1}{2}, \tfrac{1}{4}, \tfrac{1}{8}, \tfrac{1}{16}, \tfrac{1}{16}\}$$

for which the code

```
1
0  0
0  1  0
0  1  1  0
0  1  1  1
```

is optimal and does, in fact, achieve the lower bound.

Here the self-information of any event (code word) is identical to its length in binary digits, which is exactly what the equation (8.35) says.

SELECTED REFERENCES

[1] BRILLOUIN, L., *Science and Information Theory*, Academic Press, Inc., New York, 1956.

[2] FANO, R., *Transmission of Information*, The MIT Press, Cambridge, Mass.; John Wiley & Sons, Inc., New York, 1961.

[3] KHINCHIN, A., *Mathematical Foundations of Information Theory*, Dover Publications, Inc., 1957.

[4] RÉNYI, A., *Probability Theory*, North Holland Publishing Company, Amsterdam, 1970.

[5] SHANNON, C., "A Mathematical Theory of Communication," *Bell System Technical Journal*, Vol. 27, p. 623, 1948.

[6] WIENER, N., *Cybernetics*, Technology Press, MIT, Cambridge, Mass.; John Wiley & Sons, Inc., New York, 1948.

PROBLEMS

1. How many bits of information are communicated by each of the following statements (known to be true).
   (a) A man at the crap table rolled craps six consecutive rolls.
   (b) In a game of five-card poker, with four participants, one man was dealt three queens in his first five cards.
   (c) In the same hand, a man was dealt a pair of eights.

(d) An n vector of 0's and 1's was chosen at random from the set of all such n vectors and it had no adjacent 1's.

(e) Assuming a batting average is a precise measure of one's probability of getting a hit during any time at bat that day, consider
 (1) A .250 hitter got three hits in four at bat.
 (2) A .300 hitter got one hit in four at bat.

(f) An object was hidden in one of n locations and the probability that it was in location i was $2i/n(n + 1)$. A person selected a location entirely at random, looked in that location, and there was the object.

2. For each of the following circumstances, state how much information is communicated by the event described. Two boys and two girls are throwing darts. For any one throw, the boys, B_1, B_2 have probabilities $\frac{1}{4}$, $\frac{1}{8}$, respectively, of hitting the bull's-eye. Girls G_1, G_2 have comparable probabilities $\frac{1}{2}$, $\frac{1}{32}$. Each throws four darts. The events of interest are
 (a) There was a total of two bull's-eyes.
 (b) There was a total of two bull's-eyes and B_2 got one of them.
 (c) There were three bull's-eyes and the girls got more than the boys.
 (d) There were three total bull's-eyes, and G_2 had none of them.

3. With the same setting as in Problem 2, consider each of the following pairs of events, and for each, tell how much information the occurrence of A provides about the occurrence of B.
 (a) A: The girls had two bull's-eyes.
  B: The total was less than five.
 (b) A: The total was less than five.
  B: The girls had two bull's-eyes.
 (c) A: The total was four.
  B: G_2 had four.
 (d) A: B_1 had more bull's-eyes than G_2.
  B: G_1 had more bull's-eyes than B_2.
 (e) A: Some three of the players had a total of seven.
  B: B_1 had three.

4. For each of the following pairs of events, A, B, compute $I(A \mid B)$:
 (a) A: The total was less than five.
  B: The girls had two.
 (b) A: The total was four.
  B: G_2 had four.
 (c) A: For one round, the teams tied.
  B: B_1, G_1 each had zero.

5. A particular young lady enjoys appearing capricious and unpredictable. Thus she devises a bizarre scheme for changing her hair color. She begins each week as a blonde and the following daily transitions are possible:

 The transitions that she makes depend upon the outcome of an experiment each evening—she flips a fair coin and rolls a fair die. Denoting these random variables by X and Y respectively, the transitions take place according to the following table.

	Blonde	Red	Black	Brown
Blonde	$X = H, Y \leq 4$	$X = H, Y > 4$	$X = T$	
Red	$X = T, Y \neq 4$	$X = T, Y = 4$		$X = H$
Black		$X = T, Y = 6$	$X = H, Y \geq 2$	Anything else
Brown		$X = H, 1 \leq Y \leq 4$		Anything else

How much information is contained then, in each of the following:
(a) She was blonde every day one week.
(b) She used only two colors one week.
(c) How much information is provided by the fact that she was a redhead on Wednesday about the event that she will be blonde on Friday?
(d) Given that she had black hair on Wednesday, how much information is provided by her hair being brown on Friday.
(e) How much information is provided by the pair of events, she was a blonde on Tuesday and a redhead on Friday?

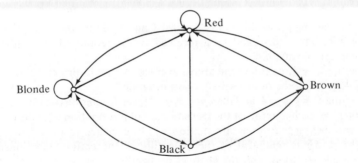

Figure P5.

6. Jane, Linda, and Charme have different methods of voting in a particular five-office, two-party election. Linda flips a fair coin and votes a straight ticket for party 1 or party 2, as the coin is a head or a tail. Charme flips a coin with $P\{\text{head}\} = \frac{3}{4}$ and votes a straight ticket, party 1 or party 2, as her coin was a tail or a head, respectively. Jane, being more interested in candidates than parties, flips a fair coin for *each* office and votes party 2 or party 1 according to a fixed rule for associating party with outcome, but she refuses to divulge that rule. Answer the following:
(a) How much information is provided by the existence of a ballot in favor of all party 2 candidates about the event, "that ballot is Jane's"?
(b) Same as part (a), but for party 1.
(c) Same as part (a), but ... "is Charme's."
(d) Same as part (a), but ... "is Linda's."

(e) How much information is given in the event that all three ballots are identical?

(f) How much information is given by the existence of exactly two identical ballots about the event that Jane associates heads with party 1?

(g) Same as part (f) but with party 2.

7. Suppose that there is a communication channel over which the symbols 0 and 1 are transmitted. Transmission errors occur and the probabilities of received symbols are given by the following diagram.

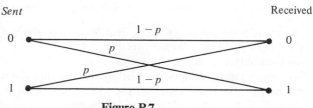

Figure P 7.

For this "binary symmetric channel," the probability that the symbol 0(1) is received, given that 0(1) was sent, is $1 - p$, while the probability that 0(1) is received, given that 1(0) is sent is p, $0 < p < 1$. (In practice, most likely $0 < p \ll 1$.)

(a) Find sending probabilities q, $1 - q$ for 0 and 1 so that the average amount of information provided by a received symbol about which symbol was sent is a maximum.

(b) Letting $q = \frac{1}{2}$, calculate the above average mutual information for $p = 0$, $\frac{1}{4}, \frac{1}{2}, \frac{3}{4}, 1$. Do the results satisfy one's intuition?

(c) Letting $p = \frac{1}{4}, q = \frac{1}{2}$, and then $p = \frac{1}{2}, q = \frac{1}{2}$, verify relations (8.18) and (8.19). Interpret each in terms of the transmission of information over the channel, remembering that some information will be expected to be lost in noise.

(d) The reader is reminded of the generality of the entities mentioned, for example, symbols sent, the channel probability of error, p, the communications channel itself. A situation that could be described in the same terms might involve the sending out of real or decoy articles by one party, and the observation of these and possible error in identification by another. Being precise about the various entities involved, describe several other situations in such terms.

(e) With p arbitrary, find q, $1 - q$ so that the average self-information of the received symbol is a maximum. Then find values to minimize the same quantity.

8. Suppose that f is defined on $0, 1, 2, \ldots, n$. Suppose that we select i, j, and compute $f(i)$, $f(j)$, $0 \leq i < j \leq n$, and find $f(i) > f(j)$. If, at the outset, the maximum value of f was considered equally probable to occur at $0, 1, \ldots,$ or n, what is the average amount of information provided about the location of the max by the event $f(i) > f(j)$?

Answer the same question if f is known to be unimodal.

9. Using concepts and results from information theory, the quantities (8.35) were found as optimal for minimizing expected word length. Obtain the same result, using the methods of mathematical optimization.

10. Carry out, as far as possible, along the lines of that for the heavy-coin problem, the analysis of the case where it is known that *exactly* two of the n coins are heavy.

11. Consider the heavy-coin problem, one heavy in 12 coins, but with the balance given to errors. If the balance has probability p of being in error, what is the average information provided by the first weighing? Will two more weighings suffice then?

12. Is the code $\{000, 001, 010, 011, 100, 101, 110, 111\}$ optimal for the uniform distribution of eight elements? At any time, what is the probability that the next symbol received is 0? 1? Does this imply that the code is optimal? Explain.

13. Complete the finding of optimal weighing strategies for all possible situations in the odd-coin problem.

14. For the odd-coin problem with n coins, describe the first two optimal weighing strategies.

15. Show that the bracketed expression of (8.31) is the average length of a word in an infinitely long message.

16. Supply a proof of inequality (8.13).

LOCATION THEORY

9.1. Introduction

Location theory deals essentially with the problem of optimum geographic location of one or more sources to meet the requirements of a given number of destinations. This is a logistics-type problem of interest to military, industrial, and urban management.

The problem first posed early in the 17th century by the noted mathematician Fermat, and solved in 1640 by Toricelli, considered the location of a point on a plane so as to minimize the sum of the distances from that point to three other given points. It was not until 1837 that Steiner gave the conditions to be satisfied for a point to be located in such a way that the sum of its distances to n given destinations, $n = 3, 4, \ldots$, is a minimum.

The study took a new dimension at the beginning of this century when economists such as A. Weber [6], and A. Losch [11] became interested in this class of problem and incorporated, in addition to distance, socioeconomic factors. With the advent of new techniques in operations research, location theory is being given increasing attention by interested investigators.

To motivate our discussion, consider the following problem: Given $n > 1$ towns within a province with specific geographic locations, where should a

9

market be located for the population of these towns to transact business, if the total road distances from this market to all the towns is to be a minimum? Here, depending on circumstances, the site of the potential market might be *unrestricted*, in which case any geographic location would be a candidate site, or such site could be *restricted*, in which case only a finite number of candidate sites would be considered.

Going back to our example, it seems that there is no reason why one cannot talk about setting more than one market within the province. This makes sense if one approaches the problem from an economic point of view and incorporates such factors as cost of road construction and cost of setting and building a specific market. If all possible alternatives were available, a proper economic balance could be achieved between the cost of opening markets and the total cost of road construction. Thus one could specify in an optimum fashion the total number of markets, the location of each of these markets, and the towns which these markets would serve.

Needless to say, this added degree of freedom increases considerably the complexity of the problem, for one is not only confronted with a location problem (number and sites of sources), but, simultaneously, with an allocation problem (defining which source supplies which destination).

Location problems are of interest to the military who are constantly faced with the problem of optimum strategic deployment of forces. Both the military and industry at one time or another have to decide on where to locate a depot center or a repair and manufacturing site to provide economically services and goods to a given number of destinations.

The present chapter attempts to provide some insights in the theory of location. However, the analysis is far from exhaustive, and, indeed, much remains to be done in formalizing a methodology to tackle the general location problem.

9.2. Unrestricted Single-Source Location Problem

Let us consider the following problem: n, $n > 2$, destinations are located at specific geographic sites over some region. The coordinates of destination i, $i = 1, 2, \ldots, n$, are (x_i, y_i) and the requirement at this destination is m_i. Let (x, y) be the coordinates of the source point P. The Euclidean distance from the source point to destination i is (Figure 9.1)

$$d_i = \sqrt{(x - x_i)^2 + (y - y_i)^2} \tag{9.1}$$

We shall assume that the source point has adequate resources to meet the requirements at each destination and that the cost of supplying from the source to destination i is a function of the requirements at i, m_i and the

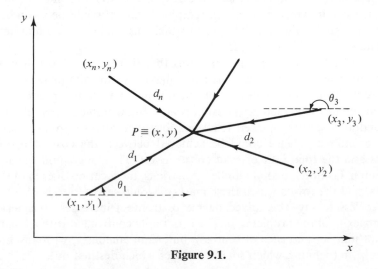

Figure 9.1.

distance d_i from i to P, say $g_i(m_i, d_i)$. The total cost expression is thus

$$S(x, y) = \sum_{i=1}^{i=n} g_i(m_i, d_i) = \sum_{i=1}^{i=n} g_i[m_i, \sqrt{(x - x_i)^2 + (y - y_i)^2}] \quad (9.2)$$

The problem then consists of determining the value of (x, y) that will minimize expression (9.2). This is a problem in differential calculus, and the necessary conditions for a minimum are

$$\frac{\partial S}{\partial x} = 0 = \frac{\partial S}{\partial y} \quad (9.3)$$

thus yielding two equations in the two unknowns x and y. Of particular interest is the case when for c_i, a constant,

$$g_i[m_i, \sqrt{(x - x_i)^2 + (y - y_i)^2}] = c_i m_i \sqrt{(x - x_i)^2 + (y - y_i)^2}, \\ i = 1, 2, \ldots, n \quad (9.4)$$

that is, when the cost of supplying a total quantity m_i from the source P to destination i, $i = 1, 2, \ldots, n$, is proportional to the quantity supplied and the shipping distance. In such a case, the objective function (9.2) becomes

$$S(x, y) = \sum_{i=1}^{i=n} c_i m_i \sqrt{(x - x_i)^2 + (y - y_i)^2} \quad (9.5)$$

and conditions (9.3) reduce to the following, provided $x \neq x_i$ and $y \neq y_i$ for all i:

$$\frac{\partial S}{\partial x} = 0 = \sum_{i=1}^{i=n} \frac{c_i m_i (x - x_i)}{\sqrt{(x - x_i)^2 + (y - y_i)^2}} \quad (9.6)$$

$$\frac{\partial S}{\partial y} = 0 = \sum_{i=1}^{i=n} \frac{c_i m_i (y - y_i)}{\sqrt{(x - x_i)^2 + (y - y_i)^2}} \quad (9.7)$$

That the function (9.5) possesses an absolute minimum can be established by showing that it is a convex function of the variables x and y possessing a relative minimum. (See e.g., Chapter 2 of *Optimization Techniques in Operations Research*, also by B.D. Sivazlian and L.E. Stanfel, Prentice-Hall, Inc., Englewood Cliffs, N. J., 1975.) From the properties of convex functions, it is sufficient to show that the function

$$Z(x, y) = \sqrt{x^2 + y^2}$$

is convex; then the function

$$c_i m_i \sqrt{(x - x_i)^2 + (y - y_i)^2}, \quad i = 1, 2, \ldots, n$$

will also be convex. Finally, the sum of convex functions being convex,

$$\sum_{i=1}^{i=n} c_i m_i \sqrt{(x - x_i)^2 + (y - y_i)^2}$$

will necessarily be convex.

Now the function $Z(x, y)$ represents the surface of an inverted cone whose vertex coincides with the origin and whose axis is the Z axis (Figure 9.2). $Z(x, y)$ is a convex function since any line segment joining two distinct points on the surface lie on or above the surface.

Let θ_i be the angle between the x axis and the vector joining destination i to source P (Figure 9.1). We shall always measure this angle in a counter-clockwise direction. We shall also assume that the source point does not coincide with any of the destinations. Then

$$\cos \theta_i = \frac{x - x_i}{\sqrt{(x - x_i)^2 + (y - y_i)^2}}, \quad i = 1, 2, \ldots, n \qquad (9.8)$$

and

$$\sin \theta_i = \frac{y - y_i}{\sqrt{(x - x_i)^2 + (y - y_i)^2}}, \quad i = 1, 2, \ldots, n \qquad (9.9)$$

Using relations (9.8) and (9.9) in equations (9.6) and (9.7), respectively, we obtain

$$\sum_{i=1}^{i=n} c_i m_i \cos \theta_i = 0 \qquad (9.10)$$

and

$$\sum_{i=1}^{i=n} c_i m_i \sin \theta_i = 0 \qquad (9.11)$$

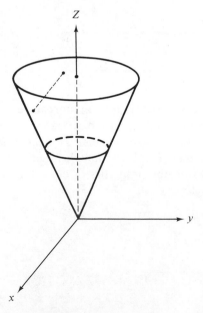

Figure 9.2. $Z(x, y) = \sqrt{x^2 + y^2}$ is a convex function.

The system of reference axes over which the coordinates of the destinations and the source are defined can be arbitrarily selected, and, indeed, equations (9.10) and (9.11) will hold irrespective of the position of these axes. We can thus state that the point P is an optimum source point if each of the weighted sums of the cosines and the sines of the angles between the lines joining the set of destinations and the source point, and an arbitrary line in the plane is equal to zero, the weight associated with destination i being $c_i m_i$.

For the case when $c_i m_i = 1$, for all $i = 1, 2, \ldots, n$, the conditions are for the sum of the distances from the source to all the destinations to be a minimum. This particular case was first treated in its general context by Tedenat (1810), although Steiner (1837) was the first to establish the required conditions.

One might think at first that equations (9.10) and (9.11) are not sufficient to determine all the values of θ_i, $i = 1, 2, \ldots, n$, except possibly for the case of two destinations. This however is not true if one makes the following two observations. First, to determine the location of P, it suffices to determine the values of any two of the θ_i angles associated with two different destinations: the intersection of the sides of these angles will uniquely determine the location of P. Second, the geographic location of the destinations delineates a specific geometric structure, a polygon whose vertices are the destinations, and there exist certain intrinsic relationships between the geometric features of such polygons. Thus the two conditions (9.10) and (9.11) could be adequate to locate the coordinates of P.

Furthermore, despite the attractive simplicity of equations (9.10) and (9.11), their solution is not so obvious. In fact, until now there is no general *analytic* or *geometric* procedure to solve this system of equations if the number of destinations exceeds three. There exists however an interesting and rather elegant analogue procedure based on a problem in mechanics, which we shall next discuss.

i. *Mechanical Analogue Solution*

Consider a massless particle P in equilibrium under the effect of n coplanar forces $F_1, F_2, \ldots, F_n$ acting simultaneously on P. Let x, y be a system of Cartesian reference axes, and let $\theta_1, \theta_2, \ldots, \theta_n$ be the angles between the x axis and the vectors $F_1, F_2, \ldots, F_n$, respectively. Since the particle P is in equilibrium,

$$\sum_{i=1}^{i=n} F_i \cos \theta_i = 0 \qquad (9.12)$$

and

$$\sum_{i=1}^{i=n} F_i \sin \theta_i = 0 \qquad (9.13)$$

and the vectors $F_1, F_2, \ldots, F_n$ form the sides of the well-known polygon of forces (Figure 9.3).

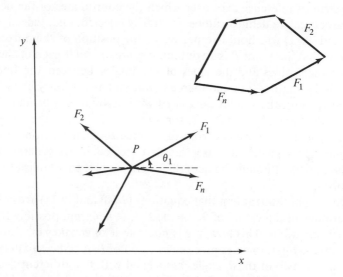

Figure 9.3. Particle P in equilibrium under the effect of coplanar forces F_i ($i = 1, 2, \ldots, n$). The polygon of forces is also pictured.

The similarity between these last two equations and equations (9.10) and (9.11) is quite evident. In fact, the source location problem is analogous to the following problem in mechanics:

A massless particle P, free to move over the xy plane experiences coplanar forces $c_1 m_1, c_2 m_2, \ldots, c_n m_n$ passing, respectively, through n given points whose coordinates are $(x_1, y_1), (x_2, y_2), \ldots, (x_n, y_n)$. Determine the position of equilibrium of the particle P.

Particle P acts as the original source point P, and if it is to be in equilibrium, equations (9.10) and (9.11) have to be satisfied. This analogue reformulation of the problem becomes quite useful in building a contrivance to solve the source location problem. The idea of such a contrivance was first adopted by George Pick, who developed an analog model from the Varignon frame (invented by Varignon to demonstrate the parallelogram of forces).

The location of each of the n destinations is translated to a scale on a vertical frame, and a small pulley is fixed at each of these points. A thread is made to run over each pulley, and the inner ends of the n threads are joined together. Weights proportional to the quantities $c_i m_i$, $i = 1, 2, \ldots, n$, at the corresponding ith destination are attached to the other ends of the threads. The junction point of the thread will by itself move into position and locate the optimum source point P (Figure 9.4).

The construction of this frame is simple and quite inexpensive. Inaccuracies in the reading of the coordinates of P can be introduced if excessive

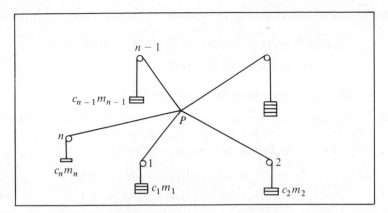

Figure 9.4. Mechanical analogue for the optimum location of a single source point.

friction is present in the rollers. A polygon of forces can easily be drawn to check the accuracy of the results.

Remark. So far we have assumed that an equilibrium point exists that will balance all the forces. It is quite possible that this might not be the case. Thus, for example, if any of the weights in Figure 9.4 at any corner exceeds the sum of the weights of all others, the minimum point P will move to that particular corner. It can be shown that this corresponds to the case when the optimum source point coincides with the destination at that corner. The advantage of the mechanical analog model is thus that one can obtain the minimum point for any situation involving a finite number of destinations. The reader is referred to Kuhn [10] for an analysis of the general problem using mathematical programming.

ii. *Geometric Solution When the Number of Destinations Equals 3, $n = 3$*

Using relations (9.10) and (9.11), it is possible to arrive at a geometric solution to the single source location problem when the number of destinations equals 3. Let $D_i = c_i m_i$, $i = 1, 2, \ldots, n$; then, for $n = 3$, equations (9.10) and (.911) become

$$D_1 \cos \theta_1 + D_2 \cos \theta_2 + D_3 \cos \theta_3 = 0 \qquad (9.14)$$

and
$$D_1 \sin \theta_1 + D_2 \sin \theta_2 + D_3 \sin \theta_3 = 0 \qquad (9.15)$$

Thus
$$D_1 \cos \theta_1 = -D_2 \cos \theta_2 - D_3 \cos \theta_3 \qquad (9.16)$$

and
$$D_1 \sin \theta_1 = -D_2 \sin \theta_2 - D_3 \sin \theta_3 \qquad (9.17)$$

Squaring and adding equations (9.16) and (9.17), we obtain

$$D_1^2 = D_2^2 + D_3^2 + 2D_2D_3(\cos\theta_2 \cos\theta_3 + \sin\theta_2 \sin\theta_3)$$
$$= D_2^2 + D_3^2 + 2D_2D_3 \cos(\theta_3 - \theta_2) \tag{9.18}$$

Similarly, the following expressions can be formed:

Since D_1, D_2, and D_3 are known, if a solution exists, equations (9.18), (9.19), and (9.20) can be solved for $\theta_2 - \theta_1$, $\theta_3 - \theta_2$, and $\theta_1 - \theta_3$. Let α_1, α_2 and α_3 be, respectively, the values of these quantities, that is

$$D_2^2 = D_1^2 + D_3^2 + 2D_1D_3 \cos(\theta_1 - \theta_3) \tag{9.19}$$
$$D_3^2 = D_1^2 + D_2^2 + 2D_1D_2 \cos(\theta_2 - \theta_1) \tag{9.20}$$

$$-\theta_1 + \theta_2 = \alpha_1$$
$$-\theta_2 + \theta_3 = \alpha_2 \tag{9.21}$$
$$\theta_1 - \theta_3 = \alpha_3$$

A set of independent solutions is $\theta_2 - \theta_1 = \alpha_1$ and $\theta_3 - \theta_2 = \alpha_2$.

We note that α_1, α_2, and α_3 can be obtained geometrically as follows: Equations (9.18), (9.19), and (9.20) can also be written as

$$D_1^2 = D_2^2 + D_3^2 - 2D_2D_3 \cos[\pi - (\theta_3 - \theta_2)] \tag{9.22}$$
$$D_2^2 = D_1^2 + D_3^2 - 2D_1D_3 \cos[\pi - (\theta_1 - \theta_3)] \tag{9.23}$$
$$D_3^2 = D_2^2 + D_1^2 - 2D_1D_2 \cos[\pi - (\theta_2 - \theta_1)] \tag{9.24}$$

In this format, the relations can be recognized as those giving the length of a side of a triangle in terms of the two other sides and the cosine of the angle in between them. If the conditions

$$|D_3 - D_2| < D_1 < D_3 + D_2$$
$$|D_3 - D_1| < D_2 < D_3 + D_1 \tag{9.25}$$
$$|D_2 - D_1| < D_3 < D_2 + D_1$$

are satisfied, we could construct a triangle with sides D_1, D_2, and D_3, and the three angles of this triangle would be $|\pi - (\theta_2 - \theta_1)|$, $|\pi - (\theta_3 - \theta_2)|$, and $|\pi - (\theta_1 - \theta_3)|$. The supplement of these angles, that is, the exterior angles of the triangle, then determines the absolute values of the angles $\theta_2 - \theta_1$, $\theta_3 - \theta_2$, and $\theta_1 - \theta_3$. Indeed, conditions (9.25) are those for the existence of the quantities α_1, α_2, and α_3, and consequently of the angles θ_1, θ_2, and θ_3. More specifically, they constitute the set of relations to be satisfied by the quantities $D_i = c_i m_i$, $i = 1, 2, 3$, if an equilibrium point P is to exist. If any of the six inequalities in expressions (9.25) is not satisfied, the source point P coincides with one of the three destinations.

Having thus determined $\theta_2 - \theta_1$ and $\theta_3 - \theta_2$, we now come to our original problem and attempt to locate geometrically the source point P. There is no loss in generality if, for the purpose of discussion, $\theta_2 - \theta_1$ and $\theta_3 - \theta_2$ are assumed to be positive quantities. Let us construct the triangle ABC, where A, B, and C are, respectively, the locations of destinations 1, 2, and 3, and select the x axis to be parallel to the side AB of this triangle (Figure 9.5). If P is the position of the minimum source location, then

$$\angle APB = \theta_2 - \theta_1$$

and

$$\angle CPB = \theta_3 - \theta_2$$

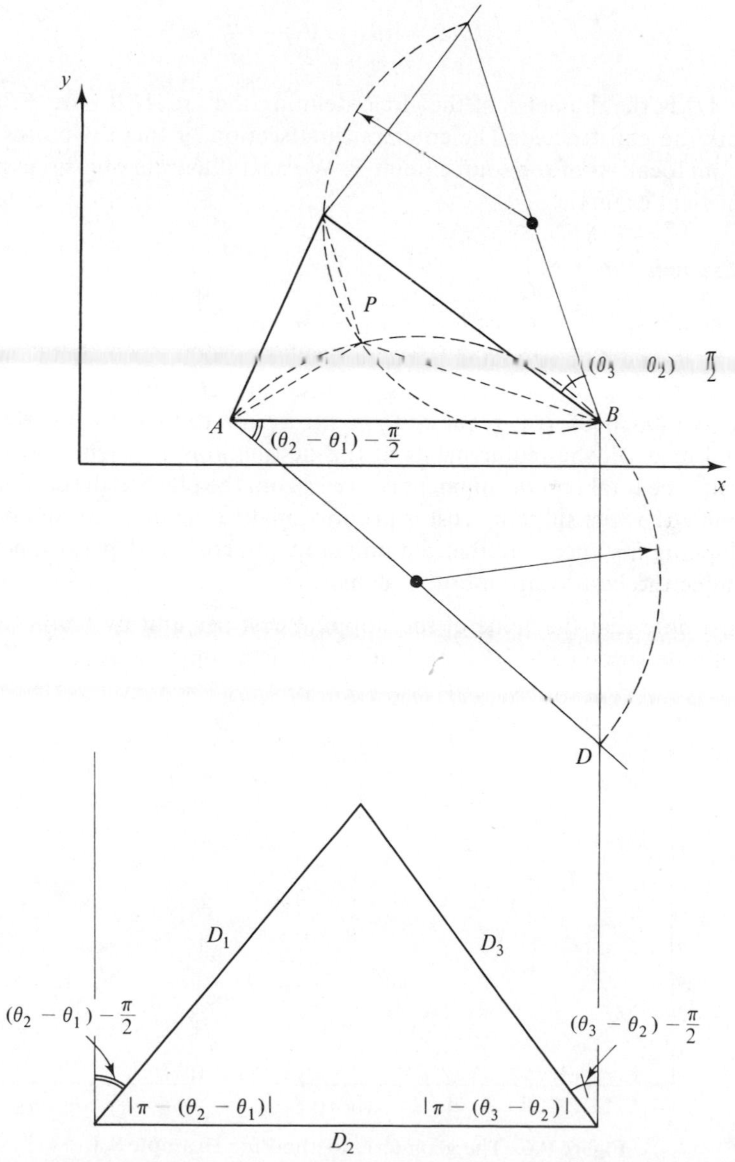

Figure 9.5. Geometric solution when $n = 3$. $\angle APB = \theta_2 - \theta_1$, $\angle BPC = \theta_3 - \theta_2$.

P therefore lies at the intersection of two loci. The first is the locus of all angles whose value is $\theta_2 - \theta_1$ such that the sides of the angles pass through the fixed points *A* and *B*. The second is the locus of all angles whose value is $\theta_3 - \theta_2$ such that the sides of the angles pass through the fixed points *B* and *C*. From basic Euclidean geometry, it is known that these two loci are arcs of circles. Let us construct one of the arcs, say *APB*. Draw the perpendicular *BD* to the side *AB* and draw the side *AD* of the angle

$$\angle BAD = (\theta_2 - \theta_1) - \frac{\pi}{2}$$

Then *AD* is the diameter of the circle defining the arc $\overset{\frown}{APB}$. Arc $\overset{\frown}{CPB}$ can similarly be constructed. The common intersection of these two arcs is the optimum location of the source point *P*. We next illustrate our discussion by a numerical example.

Example 9.1

The average annual demand for a given product at three different locations *A*, *B*, and *C* is estimated to be, respectively, 5000, 6000, and 8000 tons. The coordinates of these locations in miles are $A = (0, 0)$, $B = (600, 0)$, and $C = (400, 600)$. It is required to locate a plant to manufacture the product and meet all the requirements at the destinations in such a way as to minimize the total cost of shipment per year from the plant to the destinations. Assuming that the shipping cost is proportional to the quantity shipped and the shipping distance, and that the unit shipping cost is \$1 per ton per mile, determine the best location of the plant.

We first note that the value of the shipping cost per unit over unit distance need not be specified for this particular problem since it does not affect the

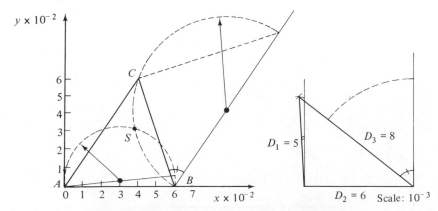

Figure 9.6. The geometric method for Example 9.1.

optimum solution. Also, the sites of the destinations need not be specified by their Cartesian coordinates, for it is sufficient to know the relative position of the destinations. The geometric construction is illustrated in Figure 9.6. The optimum location of the source point can be determined solely with the use of a ruler and compass. From the diagram, P should be located approximately at the point whose coordinate is (385,300).

The geometric method of solution was first discovered by W. Launhardt in 1882. It was rediscovered in 1909 by A. Weber.

iii. *Is the Location of the Optimum Source Point the Center of Gravity?*

Some people advocate as the optimal location of the source point the center of gravity of a system consisting of masses $D_i = c_i m_i$ located at destination i, $i = 1, 2, \ldots, n$. This would be true if the shipping cost were proportional to the square of the distance from the source to a given destination. For in this case, the objective function (9.2) becomes

$$S(x, y) = \sum_{i=1}^{i=n} c_i m_i [(x - x_i)^2 + (y - y_i)^2] \tag{9.26}$$

and the optimum value of (x, y) is then given by the two equations

$$\frac{\partial S}{\partial x} = 0 = \sum_{i=1}^{i=n} 2 c_i m_i (x - x_i) \tag{9.27}$$

and
$$\frac{\partial S}{\partial y} = 0 = \sum_{i=1}^{i=n} 2 c_i m_i (y - y_i) \tag{9.28}$$

These equations when solved yield

$$x = \frac{\sum_{i=1}^{i=n} c_i m_i x_i}{\sum_{i=1}^{i=n} c_i m_i} \quad \text{and} \quad y = \frac{\sum_{i=1}^{i=n} c_i m_i y_i}{\sum_{i=1}^{i=n} c_i m_i} \tag{9.29}$$

Those familiar with basic mechanics will remember that the center of gravity of a system is the point about which the system has the least moment of inertia. It is easy to verify that relation (9.26) represents in fact the moment of inertia of a system consisting of masses $D_i = c_i m_i$ located at points i, $i = 1, 2, \ldots, n$, about the point (x, y).

iv. *Other Methods To Determine the Optimum Source Location*

We next describe three other methods that can be used to determine the optimum source location for any given finite number of destinations.

a. Method of Isocost Contours

Suppose that the cost expression $S(x, y)$ as defined in expression (9.5) is computed for various coordinate points (x, y). It is evident that to each of these points there corresponds a unique value of $S(x, y)$. If we connect by a smooth curve all points corresponding to a given cost value, the resultant will be a set of contour curves known as *isocosts*. In Figure 9.7 these isocosts have been plotted for the three-destination case of Example 9.1, for values of $S(x, y) \times 10^{-5}$ equal to 71, 72, 73, 75, and 80. It is sufficient to compute $S(x, y)$ for points lying within and close to the triangular area ABC and which are the nodes of an arbitrarily selected rectangular networks whose grid lines are parallel to the coordinate axes. Since the function $S(x, y)$ is convex, the location of the optimum source point lies somewhere inside the region bounded by isocost curve 71. A grid with finer meshes can then be drawn within this last contour to obtain additional isocost curves and thus determine the optimum point with greater accuracy. For our example, refering to Figure 9.7, the optimum source point is approximately located at (390, 300) with $S(390, 300) = 68.8 \times 10^5$.

b. Iterative Solution of the Extremal Equations

Using the notation $d_i = \sqrt{(x - x_i)^2 + (y - y_i)^2}$, $i = 1, 2, \ldots, n$, the extremal equations (9.6) and (9.7) can be written, respectively, as

$$\sum_{i=1}^{i=n} \frac{c_i m_i (x - x_i)}{d_i} = 0, \qquad \sum_{i=1}^{i=n} \frac{c_i m_i (y - y_i)}{d_i} = 0$$

or

$$x \sum_{i=1}^{i=n} \frac{c_i m_i}{d_i} = \sum_{i=1}^{i=n} \frac{c_i m_i x_i}{d_i}$$

and

$$y \sum_{i=1}^{i=n} \frac{c_i m_i}{d_i} = \sum_{i=1}^{i=n} \frac{c_i m_i y_i}{d_i}$$

Solving for x and y, we obtain

$$x = \frac{\displaystyle\sum_{i=1}^{i=n} \frac{c_i m_i x_i}{d_i}}{\displaystyle\sum_{i=1}^{i=n} \frac{c_i m_i}{d_i}}, \qquad y = \frac{\displaystyle\sum_{i=1}^{i=n} \frac{c_i m_i y_i}{d_i}}{\displaystyle\sum_{i=1}^{i=n} \frac{c_i m_i}{d_i}}$$

Since d_i is a function of x and y, the following iterative method is suggested to compute x and y [2]. Let the superscript k, $k = 1, 2, 3, \ldots$, indicate the kth iteration. Starting with the values of x and y given in equations (9.29),

$$x^0 = \frac{\displaystyle\sum_{i=1}^{i=n} c_i m_i x_i}{\displaystyle\sum_{i=1}^{i=n} c_i m_i}, \qquad y^0 = \frac{\displaystyle\sum_{i=1}^{i=n} c_i m_i y_i}{\displaystyle\sum_{i=1}^{i=n} c_i m_i}$$

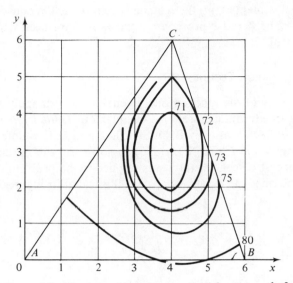

Figure 9.7. The isocost contour method for Example 9.1.

we obtain for the first iterated values of x and y

$$x^1 = \frac{\sum_{i=1}^{i=n} \frac{c_i m_i x_i}{d_i^0}}{\sum_{i=1}^{i=n} \frac{c_i m_i}{d_i^0}}, \qquad y^1 = \frac{\sum_{i=1}^{i=n} \frac{c_i m_i y_i}{d_i^0}}{\sum_{i=1}^{i=n} \frac{c_i m_i}{d_i^0}}$$

Continuing in this fashion, the expression for the kth iterated value of x and y will be

$$x^k = \frac{\sum_{i=1}^{i=n} \frac{c_i m_i x_i}{d_i^{k-1}}}{\sum_{i=1}^{i=n} \frac{c_i m_i}{d_i^{k-1}}}, \qquad y^k = \frac{\sum_{i=1}^{i=n} \frac{c_i m_i y_i}{d_i^{k-1}}}{\sum_{i=1}^{i=n} \frac{c_i m_i}{d_i^{k-1}}}$$

where $d_i^k = \sqrt{(x^k - x_i)^2 + (y^k - y_i)^2}, \quad k = 0, 1, 2, \ldots$

The iteration ends when two successive pair of values of (x, y) become equal within a specified accuracy level. Applying this method to the problem in Example 9.1, we obtain the following iterated values of x and y:

Iteration k	0	1	2	3	...	14	15
$x^k \times 10^{-2}$	3.58	3.78	3.83	3.84	...	3.84	3.84
$y^k \times 10^{-2}$	2.53	2.67	2.76	2.83	...	3.02	3.02

With the initial values (x^0, y^0) the scheme is seen to yield a convergent result fairly quickly. The iterative procedure appears to have been first suggested by Weiszfeld [13].

c. Use of Search Techniques

The convexity of the function $S(x, y)$ enables one to apply successfully a number of search techniques (see e.g., Chapter 7 of *Optimization Techniques in Operations Research* also by B. D. Sivazlian and L. E. Stanfel, Prentice-Hall, Inc., Englewood Cliffs, N.J., 1975) to solve for the optimum source point. The steepest-descent method was applied to the problem of Example 9.1 and the intermediate iterative steps are illustrated in Figure 9.8.

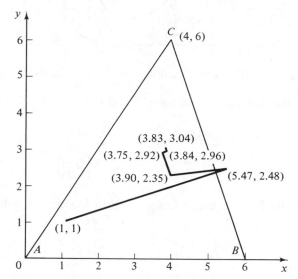

Figure 9.8. The method of steepest ascent for Example 9.1.

v. *Special Extreme Cases*

So far, we have restricted our attention to the case of a denumerably finite number of destinations whose sites are not collinear. These restrictions can be removed and some important classes of problems can be generated. We consider first the case when all the destinations lie along a single straight line. Next, we consider the rectilinear distance location problem. Finally, we analyze the case when the destination points are continuously distributed over a given geographic plane area.

a. CASE OF COLLINEAR DESTINATIONS

Francis [5] treats the problem of locating a source point when all the destinations lie along a single straight line. To illustrate this case, suppose that a new machine is to be located in a plant along an aisle where there are n existing machines all located along this aisle. The new machine "interacts" with each of the old ones, and the degree of interaction with machine i, $i = 1, 2, \ldots, n$, is a measurable quantity m_i. For example, the new machine could be a conveyor belt to transfer the outputs of the existing machines. The output from machine i would then constitute the quantity m_i to be moved to the location of the conveyor belt prior to its transfer. Let x_i be the distance of destination i from an origin 0. The position of this origin and the labeling of the destinations are such that $0 \leq x_1 < x_2 < \cdots < x_n$; that is, the first destination is the closest to the origin, the second destination is the next closest, and so on, the last destination being the farthest away from the origin (Figure 9.9).

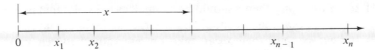

Figure 9.9. The collinear destination problem.

It is evident that the source point should be positioned on the same straight line as the destinations and should lie somewhere on the closed line segment joining the first and last destination. Let x be the distance from the source to the origin, $x \geq x_1$, and let c_i be the cost of moving one unit from destination i to the source point. The expression for the total cost function for this case reduces to

$$S(x) = \sum_{i=1}^{i=n} c_i m_i \sqrt{(x - x_i)^2} \tag{9.30}$$

The reader can easily verify that the techniques of differential calculus cannot be used to determine the optimum value of x. In fact, expression (9.30) is equivalent to an expression involving the sum of the absolute values of $c_i m_i (x - x_i)$, $i = 1, 2, \ldots, n$. Thus the problem reduces to

$$\min \left\{ S(x) = \sum_{i=1}^{i=n} c_i m_i |x - x_i| \right\} \tag{9.31}$$

subject to

$$0 \leq x_1 < x_2 < \cdots < x_n \tag{9.32}$$

$$x_1 \leq x \leq x_n \tag{9.33}$$

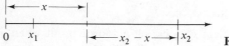

Figure 9.10.

We shall show that an optimum solution to the problem as posed by expressions (9.31), (9.32), and (9.33) exists. To motivate the reader, we first consider the case of two destinations, $n = 2$ (Figure 9.10). The objective function can be written

$$S(x) = c_1 m_1 (x - x_1) + c_2 m_2 (x_2 - x)$$
$$= (c_1 m_1 - c_2 m_2)x + (c_2 m_2 x_2 - c_1 m_1 x_1) \tag{9.34}$$

The location of the source should be selected such that (9.34) is minimized and the inequality $x_1 \leq x \leq x_2$ is satisfied. Since (9.34) is a linear function of x and the quantity $(c_2 m_2 x_2 - c_1 m_1 x_1)$ is a constant, the optimal solution for x will solely be dictated by the sign of $c_1 m_1 - c_2 m_2$. Thus

(1) If $c_1 m_1 > c_2 m_2$, then x should be as small as possible; that is, $x = x_1$.
(2) If $c_1 m_1 < c_2 m_2$, then x should be as large as possible; that is, $x = x_2$.
(3) If $c_1 m_1 = c_2 m_2$, then x can take on *any* value between x_1 and x_2 *including* the end points x_1 and x_2.

It is interesting to note that, for this simple case, the optimum location of the source is not dictated by the magnitude of the distance between the two destinations. Furthermore, the optimum solution need not be unique; however, *an* optimum site for the source is always a destination point.

Consider now the case of n destinations $n \geq 2$, and let j be a positive integer $1 \leq j \leq n$. First, we show that if an optimum solution exists some destination point is an optimum solution. Assume that the optimum source location is at a distance $x = x_j + a$ from the origin where $0 \leq a \leq (x_{j+1} - x_j)$, that is, the source is somewhere between the jth and $(j + 1)$st destination (Figure 9.11). Expression (9.31) for the cost function can be written as

$$S(x_j + a) = \sum_{i=1}^{i=n} c_i m_i |x_j + a - x_i|$$

$$= \sum_{i=1}^{i=n} c_i m_i (x_j + a - x_i) + \sum_{i=j+1}^{i=n} c_i m_i (x_i - x_j - a) \tag{9.35}$$

$$= \left(\sum_{i=1}^{i=j} c_i m_i - \sum_{i=j+1}^{i=n} c_i m_i \right)(x_j + a) - \left(\sum_{i=1}^{i=j} c_i m_i x_i - \sum_{i=j+1}^{i=n} c_i m_i x_i \right) \tag{9.36}$$

Let
$$A_j = \sum_{i=1}^{i=j} c_i m_i - \sum_{i=j+1}^{i=n} c_i m_i \tag{9.37}$$

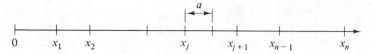

Figure 9.11.

and
$$B_j = \sum_{i=1}^{i=j} c_i m_i x_i - \sum_{i=j+1}^{i=n} c_i m_i x_i \qquad (9.38)$$

Then
$$S(x_j + a) = A_j(x_j + a) - B_j \qquad (9.39)$$

The quantities A_j and B_j are constants; hence, following the same reasoning as for the case of two destinations,

(1) If $A_j > 0$, then $x_j + a$ should be as small as possible; that is, $a = 0$ and the source point should be located at destination j.

(2) If $A_j < 0$, then $x_j + a$ should be as large as possible; that is, $a = x_{j+1} - x_j$ and the source point should be located at destination $j + 1$.

(3) If $A_j = 0$, then the quantity a can take on any value in its range of definition, including the end values $a = 0$ and $a = x_{j+1} - x_j$. Thus, although the source point has an infinite number of optimum locations that can be anywhere between destinations j and $j + 1$, an optimum location is still destination j or destination $j + 1$.

From these considerations, it follows that a search for an optimal solution can be effected by considering only the finite set of points consisting of the locations of the destinations and ignoring all points on the open line segments joining any two successive destinations.

Consider now the three contiguous destinations $j - 1$, j, and $j + 1$. Destination j is an optimum solution if and only if

$$S(x_j) - S(x_{j+1}) \le 0, \quad j = 1, 2, \ldots, n - 1 \qquad (9.40)$$
and
$$S(x_j) - S(x_{j-1}) \le 0, \quad j = 2, 3, \ldots, n \qquad (9.41)$$

From expression (9.35),

$$S(x_j) = \sum_{i=1}^{i=j} c_i m_i(x_j - x_i) + \sum_{i=j+1}^{i=n} c_i m_i(x_i - x_j) \qquad (9.42)$$

$$S(x_{j+1}) = \sum_{i=1}^{i=j+1} c_i m_i(x_{j+1} - x_i) + \sum_{i=j+2}^{i=n} c_i m_i(x_i - x_{j+1}) \qquad (9.43)$$

$$S(x_{j-1}) = \sum_{i=1}^{i=j-1} c_i m_i(x_{j-1} - x_i) + \sum_{i=j}^{i=n} c_i m_i(x_i - x_{j-1}) \qquad (9.44)$$

Thus

$$
\begin{aligned}
S(x_j) - S(x_{j+1}) = {} & \left\{ \sum_{i=1}^{i=j} c_i m_i (x_j - x_i) + \sum_{i=j+1}^{i=n} c_i m_i (x_i - x_j) \right\} \\
& - \left\{ \sum_{i=1}^{i=j} c_i m_i (x_{j+1} - x_i) + c_{j+1} m_{j+1} (x_{j+1} - x_{j+1}) \right. \\
& \left. + \sum_{i=j+1}^{i=n} c_i m_i (x_i - x_{j+1}) - c_{j+1} m_{j+1} (x_{j+1} - x_{j+1}) \right\} \\
= {} & \sum_{i=1}^{i=j} c_i m_i [(x_j - x_i) - (x_{j+1} - x_i)] \qquad (9.45) \\
& + \sum_{i=j+1}^{i=n} c_i m_i [(x_i - x_j) - (x_i - x_{j+1})] \\
= {} & \sum_{i=1}^{i=j} c_i m_i (x_j - x_{j+1}) + \sum_{i=j+1}^{i=n} c_i m_i (x_{j+1} - x_j) \\
= {} & (x_{j+1} - x_j) \left[\sum_{i=j+1}^{i=n} c_i m_i - \sum_{i=1}^{i=j} c_i m_i \right]
\end{aligned}
$$

Since by definition $x_{j+1} > x_j$, then inequality (9.40) will be satisfied iff

$$
\sum_{i=1}^{i=j} c_i m_i \geq \sum_{i=j+1}^{i=n} c_i m_i, \quad j = 1, 2, \ldots, n-1 \qquad (9.46)
$$

Using a similar approach, we find

$$
\begin{aligned}
S(x_j) - S(x_{j-1}) = {} & \left\{ \sum_{i=1}^{i=j-1} c_i m_i (x_j - x_i) + c_j m_j (x_j - x_j) \right. \\
& \left. + \sum_{i=j}^{i=n} c_i m_i (x_i - x_j) - c_j m_j (x_j - x_j) \right\} \\
& - \left\{ \sum_{i=1}^{i=j-1} c_i m_i (x_{j-1} - x_i) + \sum_{i=j}^{i=n} c_i m_i (x_i - x_{j-1}) \right\} \\
= {} & \sum_{i=1}^{i=j-1} c_i m_i (x_j - x_i - x_{j-1} + x_i) \qquad (9.47) \\
& + \sum_{i=j}^{i=n} c_i m_i (x_i - x_j - x_i + x_{j-1}) \\
= {} & (x_j - x_{j-1}) \left[\sum_{i=1}^{i=j-1} c_i m_i - \sum_{i=j}^{i=n} c_i m_i \right]
\end{aligned}
$$

Again, since by definition $x_j > x_{j-1}$, inequality (9.41) will be satisfied iff

$$
\sum_{i=1}^{i=j-1} c_i m_i \leq \sum_{i=j}^{i=n} c_i m_i, \quad j = 2, 3, \ldots, n \qquad (9.48)
$$

For $j = 1$ to be an optimum solution, it is necessary to satisfy condition (9.46), and for $j = n$ to be an optimum solution, it is necessary to satisfy condition (9.48). For any of $j = 2, 3, \ldots, n - 1$ to be an optimum solution, conditions (9.46) and (9.48) must be simultaneously satisfied.

In addition to forming a set of necessary conditions, inequalities (9.40) and (9.41) constitute a set of sufficient conditions. Since these inequalities are satisfied by relations (9.46) and (9.48), respectively, it is possible to devise a rather simple procedure to generate the optimum solution(s). Define the quantity

$$M(\alpha, \beta) = \sum_{i=\alpha}^{i=\beta} c_i m_i, \quad 1 \leq \alpha \leq \beta \leq n \tag{9.49}$$

Thus the quantity $M(1, j)$, $j \geq 1$, represents the forward partial sum of the quantities $c_i m_i$, starting with $c_1 m_1$ and adding to it successively $c_2 m_2$, $c_3 m_3$, $\ldots, c_j m_j$. The quantity $M(j, n)$, $j \leq n$, represents the backward partial sum of the quantities $c_i m_i$, starting with $c_n m_n$ and adding to it successively $c_{n-1} m_{n-1}$, $c_{n-2} m_{n-2}, \ldots, c_j m_j$.

Table 9.1

j	$c_j m_j$	Col. 1 $M(1,j)$	Col. 2 $M(j+1, n)$	Col. 3 $M(1, j-1)$	Col. 4 $M(j, n)$
1	$c_1 m_1$	$M(1,1)$	$M(2, n)$	—	$M(1, n)$
2	$c_2 m_2$	$M(1,2)$	$M(3, n)$	$M(1, 1)$	$M(2, n)$
3	$c_3 m_3$	$M(1,3)$	$M(4, n)$	$M(1, 2)$	$M(3, n)$
$\vdots$	$\vdots$	$\vdots$	$\vdots$	$\vdots$	$\vdots$
$r-1$	$c_{r-1} m_{r-1}$	$M(1, r-1)$	$M(r, n)$	$M(1, r-2)$	$M(r-1, n)$
$r = s$	$c_r m_r$	$M(1, r)$	$M(r+1, n)$	$M(1, r-1)$	$M(r, n)$
$r+1$	$c_{r+1} m_{r+1}$	$M(1, r+1)$	$M(r+2, n)$	$M(1, r)$	$M(r+1, n)$
$\vdots$	$\vdots$	$\vdots$	$\vdots$	$\vdots$	$\vdots$
$n-2$	$c_{n-2} m_{n-2}$	$M(1, n-2)$	$M(n-1, n)$	$M(1, n-3)$	$M(n-2, n)$
$n-1$	$c_{n-1} m_{n-1}$	$M(1, n-1)$	$M(n, n)$	$M(1, n-2)$	$M(n-1, n)$
n	$c_n m_n$	$M(1, n)$	—	$M(1, n-1)$	$M(n, n)$

One can then easily form Table 9.1. Note that column 3 is column 1 shifted downward by one row, and that column 4 is column 2 shifted downward by one row. For some values of j, say $j = r, r + 1, \ldots, n - 1, 1 \leq r \leq n - 1$, inequality (9.46) will be satisfied. This is represented by the unhatched portion of columns 1 and 2. Also, for some values of j, say $j = 2, 3, \ldots, s, 2 \leq s \leq n$, inequality (9.48) will be satisfied. This is represented by the unhatched portion of columns 3 and 4. Since an optimum solution always exists, we must have

$r \leq s$. The set of values of j for which all of columns 1, 2, 3, and 4 remain unhatched define the set j of optimum solution(s). If this set consists of a single value of j, the optimum solution is unique; then $s = r$ (case of Table 9.1) and the location of the source should be at a distance $x = x_r$ from the origin. If the set of optimum solutions consists of more than one value of j, then an infinite number of optimum solutions exist. These solutions are the set of points located at a distance x from the origin 0 such that $x_r \leq x \leq x_s$.

It is interesting to note that as long as the ordering of the destinations along the line is the same and the quantity $c_i m_i$ is associated with destination i the formal structure of the optimum solution will be the same and will be independent of the relative distances between two successive destinations.

Example 9.2

Our example illustrates the preceding discussions. Let us assume that the number of destinations is $n = 7$ and that $c_i m_i = i$, $i = 1, 2, \ldots, 7$. The computation of the optimum source location is given in Table 9.2. The source should be sited at $j = 5$ and the associated minimum cost is 40.

Table 9.2

j	$c_j m_j$	$M(1, j)$	$M(j+1, n)$	$M(1, j-1)$	$M(j, n)$	$S(j)$
1	1	1	27	—	28	113
2	2	3	25	1	27	86
3	3	6	22	3	25	64
4	4	10	18	6	22	48
5	5	15	13	10	18	40
6	6	21	7	15	13	42
7	7	28	—	21	7	56

A. The Median Condition. By adding the quantity $\sum_{i=1}^{i=j} c_i m_i$ on both sides of inequality (9.46), one obtains

$$2 \sum_{i=1}^{i=j} c_i m_i \geq \sum_{i=1}^{i=n} c_i m_i, \quad j = 1, 2, \ldots, n-1$$

or

$$\sum_{i=1}^{i=j} c_i m_i \geq \tfrac{1}{2} \sum_{i=1}^{i=n} c_i m_i$$

Similarly, adding the quantity $\sum_{i=1}^{i=j-1} c_i m_i$ on both sides of inequality (9.48), one obtains

$$2 \sum_{i=1}^{i=j-1} c_i m_i \leq \sum_{i=1}^{i=n} c_i m_i, \quad j = 2, 3, \ldots, n$$

or

$$\sum_{i=1}^{i=j-1} c_i m_i \leq \tfrac{1}{2} \sum_{i=1}^{i=n} c_i m_i$$

These two inequalities when combined yield

$$\sum_{i=1}^{i=j-1} c_i m_i \leq \tfrac{1}{2} \sum_{i=1}^{i=n} c_i m_i \leq \sum_{i=1}^{i=j} c_i m_i$$

The resultant condition should be compared to the condition for locating the median of a discretely distributed random variable (see Chapter 1, Problem 3).

B. A Linear Programming Formulation

The optimization problem

$$S(x) = \sum_{i=1}^{i=n} c_i m_i |x - x_i|$$

may be formulated as a linear programming problem (see e.g., Chapter 3 of *Optimization Techniques in Operations Research* also by B. D. Sivazlian and L.E. Stanfel, Prentice-Hall, Inc., Englewood Cliffs, N.J., 1975) as follows: For $i = 1, 2, \ldots, n$, define the variables u_i and v_i where

$$u_i = \begin{cases} x - x_i, & x \geq x_i \\ 0, & x < x_i \end{cases}$$

and

$$v_i = \begin{cases} 0, & x \geq x_i \\ x_i - x, & x < x_i \end{cases}$$

Then, the optimization problem may be equivalently stated as

$$\min \sum_{i=1}^{i=n} c_i m_i (u_i + v_i)$$

subject to

$$x - u_i + v_i = x_i, \quad i = 1, 2, \ldots, n$$

$$\begin{aligned} u_i &\geq 0, \\ v_i &\geq 0, \end{aligned} \quad i = 1, 2, \ldots, n.$$

The constraints $x - u_i + v_i = x_i$, $i = 1, 2, \ldots, n$ assure that if for a given i, u_i takes on a positive value, v_i must be equal to zero; on the other hand, if v_i takes on a positive value, then u_i must be equal to zero.

b. THE RECTILINEAR DISTANCE PROBLEM

The collinear problem may be easily extended to two and three dimensional cases involving rectangular movement. For example, in the two dimensional case, referring to a set of Cartesian coordinates, the movements would be restricted to paths parallel to the coordinate axes only; the corresponding location problem is often referred to as the *rectilinear distance problem*. A typical problem may be the location of new machines in an existing plant layout where only movement along aisles parallel to the walls is permitted. If again we denote by (x_i, y_i), $i = 1, 2, \ldots, n$, the coordinates of the n given destinations, and by (x, y) the unknown coordinate of the source point, then the objective function corresponding to expression (9.5) but using rectilinear distances is:

$$S(x, y) = \sum_{i=1}^{i=n} c_i m_i [|x - x_i| + |y - y_i|]$$

$$= \sum_{i=1}^{i=n} c_i m_i |x - x_i| + \sum_{i=1}^{i=n} c_i m_i |y - y_i|$$

Since the minimum value of $S(x, y)$ is the minimum value of the first sum plus the minimum value of the second sum, the optimum location of (x, y) may be found as in the one-dimensional collinear case by minimizing each sum separately.

c. CASE OF CONTINUOUSLY DISTRIBUTED DESTINATIONS

We now consider the case when the destination points are continuously distributed over a given plane domain R. This model constitutes a reasonable approximation to a number of practical problems. Thus a city planner might be interested in the optimum location of a sewage process plant or of a fresh-water surge tank or of an electric substation to meet the requirements of the city dwellers. It is implicitly assumed that the urban population is sufficiently dense that the demand requirements for a service can be viewed as continuously distributed over the geographic area of the city. This demand can be expressed mathematically by a density that expresses the demand per unit area at a specific point of the domain R as a function of the coordinates of the point.

We shall refer the system to the set of Cartesian coordinates u and v (Figure 9.12). Let $m(u, v)$ be the density function expressing the demand at the point (u, v); thus the demand requirement over the area element $du\, dv$ is

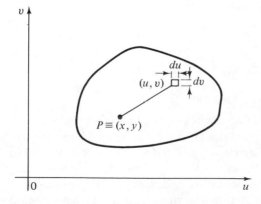

Figure 9.12. Continuously distributed destinations.

$m(u, v)\, du\, dv$. Let (x, y) be the location of the source point P, and assume that the cost of meeting the requirement at a given point is proportional to the demand density at this point and its distance from the source P. The total cost function is then

$$S(x, y) = c \iint\limits_{R} m(u, v)\sqrt{(u - x)^2 + (v - y)^2}\, du\, dv \qquad (9.50)$$

where c is a constant of proportionality. Other cost functions can be formulated depending on the operating characteristics of the system. However, we shall restrict ourselves to the functional form (9.50) to be consistent with the linearity assumption we have so far adhered to.

Analytical arguments similar to the ones used in solving the discrete case can also be presented here. Thus the function $S(x, y)$ in equation (9.50) can be shown to be convex, and its first partial derivatives can be set equal to zero to determine the necessary conditions for the optimum values of x and y. These discussions being of a general character, we shall consequently illustrate the analysis by a specific example.

Example 9.3

Assume the domain R to be a semicircle (Figure 9.13) whose center is the origin of the u, v axes and whose base diameter is made to coincide arbitrarily with the v axis. Let a be the radius of the circle and let $m(x, y) = m =$ constant be the demand density function. By reason of symmetry, the source point P should be located on the u axis, and this specifies the value of $y = 0$.

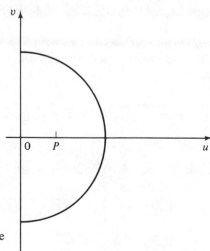

Figure 9.13. Case when the domain R is a semicircle.

Since every point on the semicircle satisfies the equation

$$u^2 + v^2 = a^2 \tag{9.51}$$

expression (9.50) for the cost function can be written as

$$S(x, 0) = 2cm \int_0^a \int_0^{\sqrt{a^2-v^2}} \sqrt{(u - x)^2 + v^2}\, du\, dv \tag{9.52}$$

The problem now consists in determining the optimum value of x that will minimize $S(x, 0)$. Differentiating $S(x, 0)$ with respect to x, we obtain

$$
\begin{aligned}
\frac{dS(x, 0)}{dx} &= -2cm \int_0^a \int_0^{\sqrt{a^2-v^2}} \frac{2(u - x)}{2\sqrt{(u - x)^2 + v^2}}\, du\, dv \\
&= -2cm \int_0^a \int_0^{\sqrt{a^2-v^2}} d(\sqrt{(u - x)^2 + v^2})\, dv \\
&= -2cm \int_0^a [\sqrt{(\sqrt{a^2 - v^2} - x)^2 + v^2} - \sqrt{x^2 + v^2}]\, dv
\end{aligned}
\tag{9.53}
$$

The optimum value of x is obtained by setting expression (9.53) equal to zero, and this yields the following equation in x:

$$\int_0^a \sqrt{(\sqrt{a^2 - v^2} - x)^2 + v^2}\, dv = \int_0^a \sqrt{x^2 + v^2}\, dv \tag{9.54}$$

Let $e = x/a$, $0 \le e \le 1$, and let $v = at$. Substituting in equation (9.54), we get

$$\int_0^1 \sqrt{(\sqrt{1 - t^2} - e)^2 + t^2}\, dt = \int_0^1 \sqrt{e^2 + t^2}\, dt \tag{9.55}$$

The new unknown variable in this equation is the quantity e. If we let $t = \cos \theta$ in the left integral and if we evaluate explicitly the right-hand integral, we obtain

$$\int_0^{\pi/2} \sqrt{1 + e^2 - 2e \sin \theta}\, \sin \theta\, d\theta = \frac{1}{2}\left(\sqrt{1 + e^2} + e^2 \ln \frac{1 + \sqrt{1 + e^2}}{e}\right) \tag{9.56}$$

This equation can be solved graphically or numerically. Define the functions

$$f_1(e) = \int_0^{\pi/2} \sqrt{1 + e^2 - 2e \sin \theta}\, \sin \theta\, d\theta \tag{9.57}$$

and $\qquad f_2(e) = \frac{1}{2}\left(\sqrt{1 + e^2} + e^2 \ln \frac{1 + \sqrt{1 + e^2}}{e}\right) \tag{9.58}$

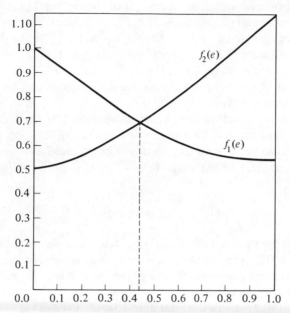

Figure 9.14. Plot of the functions $f_1(e)$ and $f_2(e)$.

The plot of these two functions is shown in Figure 9.14. The integral defining the function $f_1(e)$ can be evaluated explicitly for $e = 0$ and $e = 1$ to yield, respectively, $f_1(0) = 1$ and $f_1(1) = \frac{4}{3}(\sqrt{2} - 1)$. For other values of e, the integral should be computed numerically. It is easy to verify that $f_2(0) = \frac{1}{2}$ and $f_2(1) = 1.15$. The abscissa of the point of intersection of the two curves defines the optimum value of e. This is found to be approximately .441. Thus the optimum location of the source point P is at a distance of .441 a from the origin on the u axis.

9.3. Unrestricted Multisource Problem

We now examine the problem of a multiplicity of sources to be located simultaneously [2]. We shall assume that each source has an unlimited supply capacity and that the unit shipping cost from a given source to a given destination is independent of the source output. Under these assumptions, it is clear that a set of optimum solutions will be such that a given destination will be supplied by not more than one source. Hence the search for an optimum solution will be restricted to this set.

To understand the characteristics of the multisource problem and the extent by which it differs from the single-source problem, it is necessary to analyze the cost structure of the operation of the system. Let l, $l \geq 1$, be the number of sources and (x_{sj}, y_{sj}), $j = 1, 2, \ldots, l$, be the Cartesian coordinates of these sources. We shall let again (x_i, y_i), $i = 1, 2, \ldots, n$, be the Cartesian coordinates of the set of n given destinations, $n \geq l$. The unknown variables in this problem are l and (x_{sj}, y_{sj}), $j = 1, 2, \ldots, l$.

In addition to the distribution cost, we shall assume that associated with each source there are capital depreciation and operating costs which can be expressed by some function of l, $g_1(l)$, independent of the location of the sources. This function will be a monotone increasing function of l; that is, higher depreciation and operating costs will be incurred as more sources are added. For example, one particular form of $g_1(l)$ might be a linear function of l:

$$g_1(l) = a + bl, \quad a \geq 0, b > 0 \tag{9.59}$$

where a and b are constants that could be determined empirically.

Let now $g_2(l)$ represent the *minimum* cost of supplying the set of n destinations from these l sources. Suppose that, somehow, the values of $g_2(l)$ have been obtained and tabulated for various values of l. On the one extreme, when $l = 1$, $g_1(1)$ represents the minimum cost of distribution associated with the location of a single source. At the other extreme, when $l = n$, the minimum cost results in $g_2(n) = 0$, since the optimum is simply obtained by locating all n sources at the site of each of the given n destinations. Intuitively, the function $g_2(l)$ should be a monotone decreasing function of l since the distribution costs can be decreased by increasing the number of sources.

The two plots of $g_1(l)$ and $g_2(l)$ might look as shown in Figure 9.15. The total cost of supply operation is given by

$$E(l) = g_1(l) + g_2(l) \tag{9.60}$$

The qualitative plot of $E(l)$ is also shown in Figure 9.15. This function will in general achieve an absolute minimum for some value(s) of $l = l^*$, $1 \leq l^* \leq n$, and this determines the optimum number of sources to be sited.

In this analysis, the crucial problem is the determination of the function $g_2(l)$. This necessitates that for all possible values of l, $l = 1, 2, \ldots, n$, the optimum solution be generated and the corresponding minimum distribution cost be evaluated. The solution to this problem is not an easy one for, in addition to not knowing the location of the l sources, we do not know which source is to supply which destinations except for $l = 1$ and $l = n$. To overcome partially this difficulty in formulating the objective function, Cooper [2] defines a multiplier δ_{ij} such that $\delta_{ij} = 1$ if the ith destination is supplied by the jth source and $\delta_{ij} = 0$ otherwise. Equivalently, any allocation of n destinations to l sources can be expressed by the matrix $[\delta_{ij}][4]$, which will

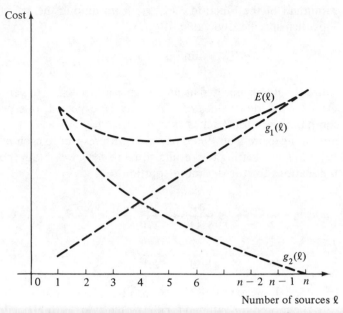

Figure 9.15. The function $E(l) = g_1(l) + g_2(l)$.

be termed the *allocation matrix*. The total number of allocation matrices is given by

$$\sigma(n, l) = \frac{1}{l!} \sum_{k=0}^{k=l} \binom{l}{k} (-1)^k (l - k)^n \tag{9.61}$$

This is equivalent to the combinatorial problem of placing n different objects in l indistinguishable cells with no cell empty.

Let m_i, $i = 1, 2, \ldots, n$, denote the requirement at the ith destination, and c_i, $i = 1, 2, \ldots, n$, be the unit cost of shipping one unit over a unit distance from a given source to destination i. Define the vectors

$$\mathbf{x}_s = (x_{s1}, x_{s2}, \ldots, x_{sl})$$
$$\mathbf{y}_s = (y_{s1}, y_{s2}, \ldots, y_{sl}) \tag{9.62}$$

Assuming that cost is proportional to distance, then for *any given allocation matrix* the total cost of distribution is

$$S_l(\mathbf{x}_s, \mathbf{y}_s) = \sum_{j=1}^{j=l} \sum_{i=1}^{i=n} \delta_{ij} c_i m_i \sqrt{(x_{sj} - x_i)^2 + (y_{sj} - y_i)^2} \tag{9.63}$$

This function is to be minimized for all of the $\sigma(n, l)$ different allocation matrices. The optimum allocation matrix can be determined by selecting the absolute minimum of the function $S_i(\mathbf{x}_s, \mathbf{y}_s)$ from among the set of $\sigma(n, l)$ minima. This then specifies the value of $g_2(l)$.

$$g_2(l) = \min_{\sigma(n, l)} \{\min S_i(\mathbf{x}_s, \mathbf{y}_s)\} \qquad (9.64)$$

The procedure should be repeated as many times as necessary to generate the function $g_2(l)$ for different values of l. As may be expected, this procedure presents computational drawbacks.

Differentiating expression (9.63) partially with respect to each of x_{sj} and y_{sj}, $j = 1, 2, \ldots, l$, and setting the results equal to zero, we obtain the following system equations for the optimum solution:

$$\sum_{i=1}^{i=n} \delta_{ij} c_i m_i \frac{x_{sj} - x_i}{\sqrt{(x_{sj} - x_i)^2 + (y_{sj} - y_i)^2}} = 0, \quad j = 1, 2, \ldots, l \qquad (9.65)$$

$$\sum_{i=1}^{i=n} \delta_{ij} c_i m_i \frac{y_{sj} - y_i}{\sqrt{(x_{sj} - x_i)^2 + (y_{sj} - y_i)^2}} = 0, \quad j = 1, 2, \ldots, l \qquad (9.66)$$

The system of equations (9.65) and (9.66) can be solved iteratively in a manner similar to the procedure outlined for the single-source problem. Let

$$d_{ij} = \sqrt{(x_{sj} - x_i)^2 + (y_{sj} - y_i)^2} \qquad (9.67)$$

Then the extremal equations (9.65) and (9.66) become

$$\sum_{i=1}^{i=n} \frac{\delta_{ij} c_i m_i (x_{sj} - x_i)}{d_{ij}} = 0, \quad j = 1, 2, \ldots, l \qquad (9.68)$$

$$\sum_{i=1}^{i=n} \frac{\delta_{ij} c_i m_i (y_{sj} - y_i)}{d_{ij}} = 0, \quad j = 1, 2, \ldots, l \qquad (9.69)$$

Again, if the superscript k denotes the kth iteration, $k = 1, 2, \ldots$, then for $j = 1, 2, \ldots, l$, the iteration equations are

$$x_{sj}^k = \frac{\sum_{i=1}^{i=n} \frac{\delta_{ij} c_i m_i x_i}{d_{ij}^{k-1}}}{\sum_{i=1}^{i=n} \frac{\delta_{ij} c_i}{d_{ij}^{k-1}}}, \qquad y_{sj}^k = \frac{\sum_{i=1}^{i=n} \frac{\delta_{ij} c_i m_i y_i}{d_{ij}^{k-1}}}{\sum_{i=1}^{i=n} \frac{\delta_{ij} c_i}{d_{ij}^{k-1}}} \qquad (9.70)$$

A set of convenient initial values is

$$x_{sj}^0 = \frac{\sum_{i=1}^{i=n} \delta_{ij} c_i m_i x_i}{\sum_{i=1}^{i=n} \delta_{ij} c_i m_i}, \qquad y_{sj}^0 = \frac{\sum_{i=1}^{i=n} \delta_{ij} c_i m_i y_i}{\sum_{i=1}^{i=n} \delta_{ij} c_i m_i} \qquad (9.71)$$

Example 9.4

Our example is taken from Cooper [2]. Consider the problem of seven destinations and two sources. The destination locations are

i	1	2	3	4	5	6	7
(x_i, y_i)	(15, 15)	(5, 10)	(10, 27)	(16, 8)	(25, 14)	(31, 23)	(22, 29)

Also, $c_i m_i = 1$, $i = 1, 2, \ldots, 7$. Here $\sigma(7, 2) = 63$; thus associated with $l = 2$ there are 63 possible allocation matrices. For each of these allocation matrices, the optimal source locations are determined and the associated minimum cost measured. The minimum from among these minima determines the optimal allocation matrix and the optimum coordinates of the sources. The results are illustrated in Figure 9.16. The first source is located at (15.42, 12.05) and supplies destinations 1, 2, 4, and 5; the second source is located at (22.00, 29.00) and supplies destinations 3, 6, and 7. It so happens that in this particular example the second source coincides with destination 7.

The preceding method is exact and can be used to solve small problems on a digital computer. For moderately large values of l and n, the quantity $\sigma(n, l)$ can be very large. Thus $\sigma(10, 4) = 723,680$. This combinatorial diffi-

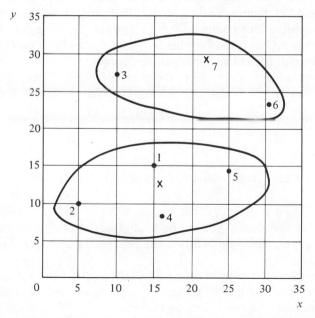

Figure 9.16. Cooper's example of 2 sources and 7 destinations.

culty makes the amount of computation involved prohibitive for many practical problems.

Cooper [2] suggests a number of heuristic procedures to overcome the computational difficulty; however, none of these procedures guarantees the optimality of the solution. One heuristic approach consists of assuming that the location of the l sources will coincide with the location of l of the destinations. The minimum cost is then computed for the total of $\sum_{l=1}^{l=n} \binom{n}{l}$ combinations. The minimum from among these minima is then selected as a "good" solution. As was pointed out before, for $l = n$ the optimum location of the sources is indeed the location of the n destinations. Also, for $l = n - 1$, it is optimal to locate the sources at the destinations. This immediately follows from the fact that there will be exactly one source which will supply two destinations; in such a case, it was shown that it is optimal to locate the source at a destination point. The specific $n - 1$ destinations constituting the sites of the sources can be investigated by considering all $\binom{n}{n-1} = n$ possibilities. In general, however, when $1 \leq l \leq n - 2$, the optimum location of the sources need not coincide with the location of the destinations.

Another procedure suggested from the previous discussions would be the following: Since it is possible to determine rather easily the values of the function $g_2(l)$ at the three values of $l = 1$, $n - 1$, n, and since the function $g_2(l)$ is monotone decreasing, one could determine the value of the function $g_2(l)$ at some other point, say at $l = w$, and then make a rough plot of $g_2(l)$ passing through the four points $g_2(l)$, $g_2(w)$, $g_2(n - 1)$, and $g_2(n)$. Assuming that the function $g_1(l)$ can be determined empirically, the function $E(l) = g_1(l) + g_1(l)$ can be plotted, and an approximate optimum value of l can be found. A suggested value of w is $w = l/2$ to the nearest integer. Having determined the number of sources to site, the problem left to face is then to locate optimally these sources. Again, this being a heuristic approach, the optimality of l is not guaranteed.

9.4. Restricted Single-Source Location Problem

Quite often, it so happens that, on the basis of certain considerations, the selection of the source(s) has to be made from a set of finite number of candidate sites; we shall say then that the problem is a *restricted* source location problem.

Let n be the total number of destinations and λ the total number of candidate source sites. Let c_{ij} be the cost of shipping one unit of product from source i to destination j, $i = 1, 2, \ldots, \lambda$ and $j = 1, 2, \ldots, n$; it will be useful to consider subsequently a matrix whose element in the ith row and jth column is c_{ij}. Let m_j be the total requirement at destination j and assume that all sources have an infinite capacity. We consider first the single-source problem.

The single-source problem can simply be solved through a straight-forward enumeration procedure. Let us form the matrix $[c_{ij}m_j]$, which we shall call the *distribution cost matrix*. If source i is the selected site, then the total distribution cost is $\sum_{j=1}^{j=n} c_{ij}m_j$. That source for which the condition $\min_i \{\sum_{j=1}^{j=n} c_{ij}m_j\}$ is satisfied is clearly the optimum selection. Example 9.5 illustrates the method.

Example 9.5

We are given seven destinations, $E_1, \ldots, E_7$, and four candidate source sites S_1, S_2, S_3, S_4. Table 9.3 gives the matrix $\lfloor c_{ij} \rfloor$ and the requirements at each of the seven destinations. The distribution cost matrix is shown in Table 9.4. The optimum source site should clearly be S_1 for which the distribution cost is minimum and is equal to 312.

Table 9.3

m_j	5	12	6	10	8	10	13
E_j S_i	E_1	E_2	E_3	E_4	E_5	E_6	E_7
S_1	6	8	3	4	9	3	2
S_2	5	4	4	10	7	6	6
S_3	8	4	9	7	2	5	3
S_4	4	8	11	6	1	3	3

Table 9.4

E_j S_i	E_1	E_2	E_3	E_4	E_5	E_6	E_7	$\sum_{j=1}^{j=7} c_{ij}m_j$
S_1	30	96	18	40	72	30	26	<u>312</u>
S_2	25	48	24	100	56	60	78	391
S_3	40	48	54	70	16	50	39	317
S_4	20	96	66	60	32	30	39	333

Example 9.6

To illustrate an extension of the single-source problem, consider a single factory located at F_1 with a maximum weekly production capacity of 1200 units of a given commodity, which supplies four warehouses A, B, C, D, whose weekly requirements are, respectively, 300, 400, 300, and 200 units. It is expected that over the next coming years the weekly requirement at warehouses C and D will increase by 500 and 400 units, respectively, while the requirements at warehouses A and B will not change. Management is contemplating the building of a second factory to absorb the excess in the future requirements. Three sites $F_2, F_3,$ and F_4 are considered for possible locations of the second factory. The following table gives the cost of producing and shipping one unit of the commodity from the various possible plants to the warehouses:

	A	B	C	D
F_1	1.8	2.2	1.5	2.3
F_2	1.9	1.4	1.8	1.6
F_3	1.6	2.9	1.9	2.4
F_4	2.0	2.3	1.4	1.8

We wish to select a site for the second plant, and determine the weekly allocation of each plant to each warehouse, such that the new plant is to absorb the excess weekly demand of 900 units. The problem can be solved as follows: each of factory sites $F_2, F_3,$ and F_4 is paired, simultaneously, with factory F_1, and the cost associated with the best allocation is determined in each case using the transportation algorithm (see e.g., Chapter 6 of *Optimization Techniques in Operations Research* also by B. D. Sivazlian and L. E. Stanfel, Prentice-Hall, Inc. Englewood Cliffs, N.J., 1975). A simple cost comparison would then indicate the optimum site. The optimal allocation tableau for the three cases are as shown:

Case 1

From \ To	A(300)	B(400)	C(800)	D(600)
F_1 (1200)	300	100	800	
F_2 (900)		300	0	600

$$\text{cost} = (300 \times 1.8) + (100 \times 2.2) + (800 \times 1.5) + (600 \times 1.6)$$
$$= \$3340$$

Case 2

From \ To	A(300)	B(400)	C(800)	D(600)
F_1 (1200)	0	400	800	
F_3 (900)	300		0	600

$$\text{cost} = (400 \times 2.2) + (800 \times 1.5) + (300 \times 1.6) + (600 \times 2.4)$$
$$= \$4000$$

Case 3

From \ To	A(300)	B(400)	C(800)	D(600)
F_1 (1200)	300	400	500	
F_4 (900)			300	600

$$\text{cost} = (300 \times 1.8) + (400 \times 2.2) + (500 \times 1.5)$$
$$+ (300 \times 1.4) + (600 \times 1.8)$$
$$= \$3670$$

Hence F_2 is selected as the optimal site.

9.5. Restricted Multisource Location Problem

i. *Method of Complete Enumeration*

The composite cost factors considered in the study of the unrestricted multisource problem are also relevant for the restricted multisource problem. Thus the chosen number of sources l and the allocation of the destinations to the sources will be dictated by the functions $g_1(l)$ and $g_2(l)$ as previously defined. Again, the critical problem is the characterization of the function $g_2(l)$. At the outset, it should be noted that there is no loss in generality if we assume that the *total* number of candidate sources, m, is less than or equal to the total number of destinations, n, $m \le n$. If $m > n$, then it is possible to form a one-to-one minimum cost allocation of the n destinations to exactly n out of the m sources, and thus eliminate the $m - n$ remaining sources not entering in this allocation.

The function $g_2(l)$, $l = 1, \ldots, m$, will be a monotone nonincreasing function of l. The complete enumeration procedure can again be used here to characterize $g_2(l)$. Thus, for each source combination taken l at a time, $1 \leq l \leq m$, the minimum cost $g_2(l)$ is determined for a given l; the process is then repeated for the $\binom{m}{l}$ cases. Clearly, the total number of allocations is $\sum_{l=1}^{l=m} \binom{m}{l} = 2^m - 1$. For small or moderately large numbers of sources, the problem can be easily handled on a digital computer. The total number of possible allocations for $m = 10$ and $m = 20$ is, respectively, 1023 and 1,048,575. Although the last figure is commendable, the basic operations involved are extremely simple; the procedure is illustrated using the data of Example 9.5.

Example 9.7

Tables 9.4, 9.5, 9.6, and 9.7 illustrate the allocations when the sources are taken 1, 2, 3, and 4 at a time, respectively. The results are summarized in Table 9.8. The distribution cost matrix could also have been used in setting up Tables 9.5, 9.6, and 9.7. For example, in Table 9.5, when the selected sources are S_1 and S_2, it is optimal to allocate destinations E_3, E_4, E_6, and E_7 to S_1, and destinations E_1, E_2, and E_5 to S_2.

Table 9.5

m_j	5	12	6	10	8	10	13	
S $\diagdown$ E	E_1	E_2	E_3	E_4	E_5	E_6	E_7	Total Cost
S_1	6	8	3	4	9	3	2	243
S_2	5	4	4	10	7	6	6	
S_1	6	8	3	4	9	3	2	208
S_3	8	4	9	7	2	5	3	
S_1	6	8	3	4	9	3	2	262
S_4	4	8	11	6	4	3	3	
S_2	5	4	4	10	7	6	6	272
S_3	8	4	9	7	2	5	3	
S_2	5	4	4	10	7	6	6	253
S_4	4	8	11	6	4	3	3	
S_3	8	4	9	7	2	5	3	267
S_4	4	8	11	6	4	3	3	

Table 9.6

m_j	5	12	6	10	8	10	13	
S $\;\;\;E$	E_1	E_2	E_3	E_4	E_5	E_6	E_7	Total Cost
S_1	6	8	3	4	9	3	2	
S_2	5	4	4	10	7	6	6	203
S_3	8	4	9	7	2	5	3	
S_1	6	8	3	4	9	3	2	
S_2	5	4	4	10	7	6	6	214
S_4	4	8	11	6	4	3	3	
S_1	6	8	3	4	9	3	2	
S_3	8	4	9	7	2	5	3	198
S_4	4	8	11	6	4	3	3	
S_2	5	4	4	10	7	6	6	
S_3	8	4	9	7	2	5	3	237
S_4	4	8	11	6	4	3	3	

Table 9.7

	5	12	6	10	8	10	13	
S $\;\;\;E$	E_1	E_2	E_3	E_4	E_5	E_6	E_7	Total Cost
S_1	6	8	3	4	9	3	2	
S_2	5	4	4	10	7	6	6	198
S_3	8	4	9	7	2	5	3	
S_4	4	8	11	6	4	3	3	

Table 9.8

l	1	2	3	4
Source combination	S_1	S_1, S_3	S_1, S_3, S_4	S_1, S_2, S_3, S_4
$g_2(l)$	312	208	198	198

9.6. Conclusion

The discussion of this chapter has been restricted to analyzing quantitatively certain classes of location problems. Another important class of location problem is the "warehouse location" problem first treated by Baumol and Wolfe [1]. The problem consists of selecting the optimum location pattern for the warehouses of a manufacturing concern from among a given set of candidate warehouses. There is a distribution cost in shipping products from the factories to the warehouses and from the warehouses to the customers. In addition, there is a warehouse operating cost. The method the authors use involves solving a sequence of transportation problems that converge toward a *local* optimum. The reader is also referred to [9] for an alternative algorithmic approach

No attempt has been made to go through the various location theories put forth by economists. A work of particular interest is the classic theory of A. Weber [6]. A similar work is presented by Isard [8] who restates the Weberian theory, and brings out its shortcomings.

SELECTED REFERENCES

[1] BAUMOL, W. Y., and P. WOLFE, "A Warehouse Location Problem," *Operations Research*, Vol. 6, March–April 1958.

[2] COOPER, L., "Location-Allocation Problems," *Operations Research*, Vol. 11, No. 3, May–June 1963.

[3] Courant, R., and H. ROBBINS, *What Is Mathematics*, Oxford University Press, New York, 1941.

[4] ELMAGHRABY, S. E., *The Design of Production Systems*, Van Nostrand Reinhold Company, New York, 1966.

[5] FRANCIS, R. L., "A Note on the Optimum Location of New Machines in Existing Plant Layout," *The Journal of Industrial Engineering*, Vol. 14, No. 1, Jan.–Feb. 1963.

[6] FRIEDRICK, C. H., *Alfred Weber's Theory of the Location of Industries*, University of Chicago Press, Chicago, 1929.

[7] HALEY, K. B., "The Siting of Depots," *The International Journal of Production Research*, Vol. 2, pp. 41–45, 1963.

[8] ISARD, W., *Location and Space-Economy*, Technology Press, MIT, Cambridge, Mass., 1956.

[9] KHUMAWALA, B. M., "An Efficient Branch and Bound Algorithm for the Warehouse Location Problem," *Management Science*, Vol. 18, pp. B-718–B-731, 1972.

[10] KUHN, H. W., "One Pair of Dual Non Linear Problems," *Nonlinear Programming*, Chapter 3 (J. Abadie, ed.), J. Wiley and Sons, Inc., New York, 1967.

[11] LOSCH, A., *The Economics of Location*, John Wiley & Sons, Inc., New York, 1967.

[12] POLYA, G., *Induction and Analogy in Mathematics*, Princeton University Press, Princeton, N.J., 1954.

[13] WEISZFELF, E., Sur le Point pour lequel la Somme des Distances de n Points Donnés est Minimum, *Tôhoku Math. J.*, Vol. 43, pp. 355–386 (1936).

PROBLEMS

1. (a) Illustrate by a realistic example a location-allocation problem
    (1) With n destinations, $n > 2$, and one source.
    (2) With n destinations, $n > 2$, and m sources, $m > 1$.
    (3) With destinations distributed continuously over a geographic area and m sources, $m > 1$.
    (b) By carefully defining your symbols and notations and by selecting a suitable objective function, formulate the mathematical problem of optimization for the examples of part (a). (Do not solve.)

2. Consider an isosceles triangle ABC where $AB = AC$.
    (a) Show that if the angle A is less than 120° then, irrespective of the location of the vertex A on the axis of symmetry, the point S that minimizes the sum $SA + SB + SC$ is the vertex of an isoceles triangle, where $SB = SC$ such that the angle BSC measures 120°.
    (b) Show that the point G which minimizes the sum $SA^2 + SB^2 + SC^2$ depends on the location of the vertex A.
    (c) If the angle A measures more than 120°, prove that the points S and A coincide.

3. Attempt to develop a method (other than the ones given in the text) to solve the m destinations, $m > 3$, and one-source location-allocation problem. Illustrate your method by a numerical example.

4. Consider the unrestricted single-source 3 destinations problem and refer to Figure 9.5. Show that when the requirements at all destinations are equal and the shipping cost is proportional to the quantity shipped and the shipping distance, $\angle APB = \angle BPC = \angle CPA = 120°$, provided that none of the interior angles of the triangle ABC exceeds 120°. Show also that if any of the angles, say $\angle CAB$, exceeds 120° the location of the optimum source is A; that is, the source point coincides with that destination where the angle exceeds 120° (see [13] for an optical interpretation of this problem).

5. The mission of an aircraft carrier is to destroy simultaneously three coastal targets A, B, and C. Targets A and C are located north of the M sea, while target B is located south of the M sea. The coordinates in miles of the targets

are $A = (500, 800)$, $B = (1100, 400)$, and $C = (1400, 1200)$. For each target a squadron of four aircrafts is assigned to form the task force. Assume that

(1) All aircrafts have exactly the same characteristics.
(2) Takeoff time, bombing time, and landing time are negligible.
(3) Fuel consumption by the aircrafts is proportional to the time in flight.
(4) All 12 aircrafts are expected to return safely to base.

(a) Suppose that the time in flight is proportional to the distance traveled. Determine the coordinates of the carrier for the launching and the landing of the aircrafts so as to minimize the total fuel consumption of the airplanes.
(b) Suppose that the time in flight is proportional to the square of the distance from the carrier. This could be the case if the aircrafts encounter a resistance that increases in intensity as they approach the targets and decreases in intensity as they recede from the target. Determine the coordinates of the carrier for the same objective as part (a).
(c) Suppose that the objective is to minimize the mission time, that is, the total time necessary to complete *all* missions. Assume that all aircrafts take off simultaneously and that the time in flight is proportional to the distance traveled. What would be the coordinates of the carrier? (The mission time is not the sum of the time in flight for each aircraft.)

6. Show that if the destinations in a plane are located at the vertices of a convex quadrilateral and the requirements at all destinations are unity the intersection of the diagonals is the point of minimum source location.

7. Given a tetrahedron with vertices at the points A, B, C, and D, assume that there is a point P inside the tetrahedron such that the sum of its distances from the four vertices

$$AP + BP = CP + DP$$

is a minimum. Show that the angles $\angle APB$ and $\angle CPD$ are equal and are bisected by the same straight line. Consider also the extreme case in which the points A, B, C, and D, all in the same plane, are the four vertices of a convex quadrilateral $ABCD$ (see [12]).

8. By using the method of isocost contours, determine the optimum location of the source point for the five destinations with the following coordinates and requirements:

Destination i	1	2	3	4	5
Coordinate (x_i, y_i)	(0, 6)	(2, 16)	(10, 19)	(14, 14)	(14, 12)
Requirements m_i	200	300	100	250	250

9. At equally spaced intervals along an aisle, 10 automatic machines are operating. The machines are labeled with a number equal to the natural sequence in which they appear along the aisle. A single control board provides an operator

with the characteristics of all the machines simultaneously. If a particular machine needs attention, the operator will immediately leave his post at the control board to adjust the machine. The expected frequency P_j that machine j will need adjustment over a week period is given by the following table:

M/c j	1	2	3	4	5	6	7	8	9	10
P_j	10	15	10	20	20	15	10	25	20	20

(a) Assuming that the call on the operator for adjustment by two or more machines is negligible, determine the optimum location of the control board if the objective is to minimize the expected distance traveled by the operator over one week.

(b) Suppose that the machines can be reordered along the aisle; what is the best ordering sequence of the machines if the objective is the same as for part (a)?

10. For the case of continuously distributed destinations, assume the domain R to be an equilateral triangle, and a constant demand density function. Determine the optimum source location which minimizes expression (9.50).

A.1. Definition

Let $f(t)$ be a function of t, $0 \leq t < \infty$. The *Laplace transform* of $f(t)$, denoted by $\mathcal{L}\{f(t)\}$ or $\bar{f}(s)$, if it exists, is defined by

$$\mathcal{L}\{f(t)\} = \bar{f}(s) = \int_0^\infty e^{-st} f(t)\, dt$$

where s is either a real or complex quantity.

Example A.1

Find the Laplace transform of the following functions ($0 \leq t < \infty$);
(a) $f(t) = t$
(b) $f(t) = e^{-at}$
(a) We have

$$\mathcal{L}\{t\} = \bar{f}(s) = \int_0^\infty e^{-st} t\, dt$$

Integrating by parts we obtain

$$\bar{f}(s) = \left| -\frac{1}{s} e^{-st} t \right|_0^\infty + \frac{1}{s} \int_0^\infty e^{-st}\, dt$$

$$= -\frac{1}{s} \lim_{t \to \infty} t e^{-st} + \frac{1}{s} \int_0^\infty e^{-st}\, dt$$

THE LAPLACE TRANSFORM

If now s is real, then, for $s > 0$

$$\lim_{t \to \infty} t\, e^{-st} = 0$$

and

$$\int_0^\infty e^{-st}\, dt = \frac{1}{s}$$

Hence, for $s > 0$

$$\mathcal{L}\{t\} = \frac{1}{s^2}$$

(b) Here we have

$$\mathcal{L}\{e^{-at}\} = f(s) = \int_0^\infty e^{-st}\, e^{-at}\, dt$$

$$= \int_0^\infty e^{-(s+a)t}\, dt$$

If $s > -a$, the integral exists, and we have

$$\mathcal{L}\{e^{-at}\} = \frac{1}{s+a}$$

In Table A.1, a selected list of functions and their Laplace transforms is presented. The reader is encouraged to obtain the results of this table directly from the definition and to determine the values of s for the existence of each of the integrals.

Table A.1. SELECTED LIST OF LAPLACE TRANSFORMS

$f(t)$	$\bar{f}(s)$	$f(t)$	$\bar{f}(s)$
1	$\dfrac{1}{s}$	$J_0(at)$	$\dfrac{1}{\sqrt{s^2 + a^2}}$
t	$\dfrac{1}{s^2}$	$I_0(at)$	$\dfrac{1}{\sqrt{s^2 - a^2}}$
$t^n \quad (n > -1)$	$\dfrac{\Gamma(n+1)}{s^{n+1}}$	$\dfrac{t^n e^{at}}{\Gamma(n+1)} \quad (n > -1)$	$\dfrac{1}{(s-a)^{n+1}}$
e^{at}	$\dfrac{1}{s-a}$	$\begin{cases} a & t=0 \\ 0 & t>0 \end{cases}$	a
$\sin at$	$\dfrac{a}{s^2 + a^2}$	$e^{bt} \sin at$	$\dfrac{a}{(s-b)^2 + a^2}$
$\cos at$	$\dfrac{s}{s^2 + a^2}$	$e^{bt} \cos at$	$\dfrac{s-b}{(s-b)^2 + a^2}$
$\sinh at$	$\dfrac{a}{s^2 - a^2}$	$e^{bt} \sinh at$	$\dfrac{a}{(s-b)^2 - a^2}$
$\cosh at$	$\dfrac{s}{s^2 - a^2}$	$e^{bt} \cosh at$	$\dfrac{s-b}{(s-b)^2 - a^2}$

Example A.2

Find the Laplace transform of the function

$$f(t) = \begin{cases} 1 & t = 0 \\ t & 0 < t < 2 \\ 2 & t = 2 \\ e^{-t} & 2 < t < \infty \end{cases}$$

By definition

$$\bar{f}(s) = \int_0^\infty e^{-st} f(t)\, dt$$

$$= e^{-s \cdot 0}(1) + \int_0^2 e^{-st} t\, dt + e^{-2s} + \int_2^\infty e^{-st} e^t\, dt$$

$$= 1 + \left| t\left(-\frac{1}{s} e^{-st}\right) \right|_0^2 - \frac{1}{s} \int_0^2 e^{-st}\, dt$$

$$+ e^{-2s} - \frac{1}{(s+1)} \left| e^{-(s+1)t} \right|_2^\infty, \quad s > 0$$

$$= 1 - \frac{2}{s} e^{-2s} - \frac{1}{s^2}(1 - e^{-2s})$$

$$+ e^{-2s} - \frac{1}{(s+1)^2} e^{-2(s+1)}, \quad s > 0$$

A.2. Some Properties of the Laplace Transforms

a. THE LINEARITY PROPERTY

Let c_1 and c_2 be constants. If $\tilde{f}_1(s)$ and $\tilde{f}_2(s)$ are the Laplace transforms of the functions $f_1(t)$ and $f_2(t)$ respectively, then

$$\mathcal{L}\{c_1 f_1(t) + c_2 f_2(t)\} = c_1 \mathcal{L}\{f_1(t)\} + c_2 \mathcal{L}\{f_2(t)\}$$
$$= c_1 \tilde{f}_1(s) + c_2 \tilde{f}_2(s)$$

The proof of this property is left as an exercise.

b. THE CONVOLUTION PROPERTY

Let the convolution of the two functions $f_1(t)$ and $f_2(t)$ be defined by

$$f_1(t) * f_2(t) = \int_0^t f_1(u) f_2(t - u)\, du$$

Then

$$\mathcal{L}[f_1(t) * f_2(t)] = \mathcal{L}\{f_1(t)\} \cdot \mathcal{L}\{f_2(t)\}$$
$$= \tilde{f}_1(s) \tilde{f}_2(s)$$

Proof
By definition

$$\mathcal{L}\{f_1(t) * f_2(t)\} = \int_0^\infty e^{-st} \left[\int_0^t f_1(u) f_2(t - u)\, du \right] dt$$
$$= \int_0^\infty \int_0^t e^{-st} f_1(u) f_2(t - u)\, du\, dt$$

Interchanging the order of integration we obtain

$$\mathcal{L}\{f_1(t) * f_2(t)\} = \int_0^\infty \int_u^\infty e^{-st} f_1(u) f_2(t - u)\, dt\, du$$
$$= \int_0^\infty f_1(u) \left[\int_u^\infty e^{-st} f_2(t - u)\, dt \right] du$$

If we make the change in variable $v = t - u$ in the integral expression in brackets, we obtain

$$\mathcal{L}\{f_1(t) * f_2(t)\} = \int_0^\infty f_1(u) \left[\int_0^\infty e^{-s(u+v)} f_2(v)\, dv \right] du$$
$$= \left[\int_0^\infty e^{-su} f_1(u)\, du \right] \left[\int_0^\infty e^{-sv} f_2(v)\, dv \right]$$
$$= \mathcal{L}\{f_1(t)\} \mathcal{L}\{f_2(t)\}$$

It is easy to verify that given the three functions of t, $f_1(t)$, $f_2(t)$ and $f_3(t)$

$$f_1(t)*f_2(t) = f_2(t)*f_1(t)$$
$$f_1(t)*[f_2(t)*f_3(t)] = [f_1(t)*f_2(t)]*f_3(t)$$
$$= f_1(t)*f_2(t)*f_3(t)$$
$$f_1(t)*[f_2(t) + f_3(t)] = f_1(t)* f_2(t) + f_1(t)*f_3(t)$$

c. THE DIFFERENTIATION THEOREM

If $f'(t)$ is the derivative of the continuous function $f(t)$, then, provided each of the Laplace transforms of $f(t)$ and $f'(t)$ exist,

$$\mathcal{L}\{f'(t)\} = s\bar{f}(s) - f(0)$$

Proof

$$\mathcal{L}\{f'(t)\} = \int_0^\infty e^{-st} f'(t)\, dt$$

Integrating by parts we obtain

$$\mathcal{L}\{f'(t)\} = \left| e^{-st} f(t) \right|_0^\infty + s \int_0^\infty e^{-st} f(t)\, dt$$
$$= \lim_{t\to\infty} e^{-st} f(t) - f(0) + s \int_0^\infty e^{-st} f(t)\, dt$$

Provided the limit term tends to zero, we have

$$\mathcal{L}\{f'(t)\} = s\bar{f}(s) - f(0)$$

d. THE INTEGRATION THEOREM

Let $F(t) = \int_0^t f(u)\, du$ and $F(0) = 0$, then

$$\bar{F}(s) = \mathcal{L}\left\{ \int_0^t f(u)\, du \right\} = \frac{1}{s}\bar{f}(s)$$

The proof of this theorem follows from the differentiation theorem

e. THE FINAL VALUE THEOREM

If the Laplace transform of $f(t)$ and $f'(t)$ exist, then

$$\lim_{t\to\infty} f(t) = \lim_{s\to 0} s\bar{f}(s)$$

Proof
We have

$$\lim_{s\to 0} \mathcal{L}\{f'(t)\} = \lim_{s\to 0} s\bar{f}(s) - f(0)$$

or

$$\lim_{s\to 0} s\bar{f}(s) = \lim_{s\to 0} \mathcal{L}\{f'(t)\} + f(0)$$

Now
$$\lim_{s \to 0} \mathcal{L}\{f'(t)\} = \lim_{s \to 0} \int_0^\infty e^{-st} f'(t) \, dt$$

$$= \int_0^\infty f'(t) \, dt$$

$$= \lim_{t \to \infty} f(t) - f(0)$$

Hence
$$\lim_{s \to 0} s\bar{f}(s) = \lim_{t \to \infty} f(t) - f(0) + f(0)$$

$$= \lim_{t \to \infty} f(t)$$

A.3. The Inverse Laplace Transform

If $\bar{f}(s)$ is the Laplace transform of $f(t)$, then $f(t)$ is called the *inverse Laplace transform* of $\bar{f}(s)$, and we write

$$f(t) = \mathcal{L}^{-1}\{\bar{f}(s)\}$$

Clearly, given $\bar{f}(s)$, $f(t)$ is solution to the integral equation

$$\bar{f}(s) = \int_0^\infty e^{-st} f(t) \, dt$$

The inverse Laplace transform possesses the linearity property, so that if $\bar{f}_1(s)$ and $\bar{f}_2(s)$ are respectively the Laplace transforms of $f_1(t)$ and $f_2(t)$, and c_1 and c_2 are constants, then

$$\mathcal{L}^{-1}\{c_1\bar{f}_1(s) + c_2\bar{f}_2(s)\} = c_1\mathcal{L}^{-1}\{\bar{f}_1(s)\} + c_2\mathcal{L}^{-1}\{\bar{f}_2(s)\}$$
$$= c_1 f_1(t) + c_2 f_2(t)$$

Often, the Laplace transform of a function will be known, and it will be required to invert the given Laplace transform to obtain the original function. The problem of inverting Laplace transforms may be solved in a number of ways. For example, if the Laplace transform is a rational function of the form

$$\bar{f}(s) = \frac{a_0 + a_1 s + \cdots + a_m s^m}{b_0 + b_1 s + \cdots + b_n s^n}$$

where the degree of the numerator m is less than the degree of the denominator n (m, n being integers), and the denominator has distinct roots then, it is possible to apply the method of partial fraction and decompose the rational function to the form

$$\bar{f}(s) = \frac{A_1}{s - \alpha_1} + \frac{A_2}{s - \alpha_2} + \cdots + \frac{A_n}{s - \alpha_n}$$

where A_i and α_i, $i = 1, 2, \ldots, n$, are in general complex quantities whose values may be determined using the well known theorems related to partial fraction expansion.

Using the linearity property, one then may invert $\bar{f}(s)$ and obtain

$$\mathcal{L}^{-1}\{\bar{f}(s)\} = f(t)$$
$$= A_1 \mathcal{L}^{-1}\left\{\frac{1}{s - \alpha_1}\right\} + A_2 \mathcal{L}^{-1}\left\{\frac{1}{s - \alpha_2}\right\} + \cdots + A_n \mathcal{L}^{-1}\left\{\frac{1}{s - \alpha_n}\right\}$$
$$= A_1 e^{\alpha_1 t} + A_2 e^{\alpha_2 t} + \cdots + A_n e^{\alpha_n t}$$

We illustrate the method of inversion by two examples

Example A.3

Let
$$\bar{f}(s) = \frac{1}{s(s^2 + 1)}$$

Then
$$\bar{f}(s) = \frac{1}{s(s + i)(s - i)} \quad (i = \sqrt{-1})$$
$$= \frac{A_1}{s} + \frac{A_2}{s + i} + \frac{A_3}{s - i}$$
$$= \frac{1}{s} + \frac{-\frac{1}{2}}{s + i} + \frac{-\frac{1}{2}}{s - i}$$

Term by term inversion yields

$$f(t) = 1 - \tfrac{1}{2}e^{-it} - \tfrac{1}{2}e^{it}$$
$$= 1 - \tfrac{1}{2}(e^{-it} + e^{it})$$
$$= 1 - \cos t$$

Example A.4

Let
$$\bar{f}(s) = \frac{1}{s^3 + 1}$$

Then
$$\bar{f}(s) = \frac{1}{(s + 1)(s^2 - s + 1)}$$
$$= \frac{1}{(s + 1)(s - \alpha_1)(s - \alpha_2)}$$

where
$$\alpha_1 = \frac{1 + \sqrt{1 - 4}}{2} = \frac{1}{2} + i\frac{\sqrt{3}}{2}$$
$$\alpha_2 = \frac{1 - \sqrt{1 - 4}}{2} = \frac{1}{2} - i\frac{\sqrt{3}}{2}$$

We may write
$$\bar{f}(s) = \frac{A_1}{s+1} + \frac{A_2}{s-\alpha_1} + \frac{A_3}{s-\alpha_2}$$

The constants A_1, A_2, and A_3 are computed as

$$A_1 = \frac{1}{3}, \quad A_2 = \frac{2}{-3 + i3\sqrt{3}}, \quad A_3 = \frac{2}{-3 - i3\sqrt{3}}$$

Thus, the inverse Laplace transform is

$$f(t) = A_1 e^{-t} + A_2 e^{\alpha_1 t} + A_3 e^{\alpha_2 t}$$

$$= \frac{1}{3} e^{-t} + \frac{2}{-3 + i3\sqrt{3}} e^{(1/2 + i\sqrt{3}/2)t} + \frac{2}{-3 - i3\sqrt{3}} e^{(1/2 - i\sqrt{3}/2)t}$$

$$= \frac{1}{3} e^{-t} + \frac{2}{3} e^{t/2} \left[\frac{1}{-1 + i\sqrt{3}} e^{i(\sqrt{3}/2)t} + \frac{1}{-1 - i\sqrt{3}} e^{-i(\sqrt{3}/2)t} \right]$$

$$= \frac{1}{3} e^{-t} + \frac{2}{3} e^{t/2} \left[-\frac{1}{2} \cos \frac{\sqrt{3}}{2} t + \frac{\sqrt{3}}{2} \sin \frac{\sqrt{3}}{2} t \right]$$

$$= \frac{1}{3} \left\{ e^{-t} - 2e^{t/2} \left[\sin \frac{\pi}{6} \cos \frac{\sqrt{3}}{2} t - \cos \frac{\pi}{6} \sin \frac{\sqrt{3}}{2} t \right] \right\}$$

$$= \frac{1}{3} \left\{ e^{-t} - 2e^{t/2} \sin \left(\frac{\pi}{6} - \frac{\sqrt{3}}{2} t \right) \right\}$$

A.4. Solving Differential Equations Using Laplace Transforms

The technique of Laplace transforms may readily be used to solve a system of ordinary linear differential equations with constant coefficients, by changing such system into one of linear algebraic equations. We illustrate the method by the following example.

Example A.5

Solve the simultaneous system of differential equations

$$\frac{dy_1(t)}{dt} = 2y_1(t) - 3y_2(t)$$

$$\frac{dy_2(t)}{dt} = -2y_1(t) + y_2(t)$$

given the initial conditions $y_1(0) = 8$; $y_2(0) = 3$.

Let $\bar{y}_1(s)$ and $\bar{y}_2(s)$ be respectively the Laplace transforms of the unknown functions $y_1(t)$ and $y_2(t)$. Then, taking the Laplace transform on both sides of each equation, we obtain

$$s\bar{y}_1(s) - y_1(0) = \quad 2\bar{y}_1(s) - 3\bar{y}_2(s)$$
$$s\bar{y}_2(s) - y_2(0) = -2\bar{y}_1(s) + \quad \bar{y}_2(s)$$

Using the initial conditions, we obtain the following system of linear algebraic equations in the unknowns $\bar{y}_1(s)$ and $\bar{y}_2(s)$:

$$(s - 2)\bar{y}_1(s) + 3\bar{y}_2(s) = 8$$
$$2\bar{y}_1(s) + (s - 1)\bar{y}_2(s) = 3$$

Using the method of determinants, we obtain as the solution of this system of equations

$$\frac{\bar{y}_1(s)}{\begin{vmatrix} 8 & 3 \\ 3 & s-1 \end{vmatrix}} = \frac{\bar{y}_2(s)}{\begin{vmatrix} s-2 & 8 \\ 2 & 3 \end{vmatrix}} = \frac{1}{\begin{vmatrix} s-2 & 3 \\ 2 & s-1 \end{vmatrix}}$$

Hence

$$\bar{y}_1(s) = \frac{8s - 17}{(s + 1)(s - 4)} = \frac{5}{s + 1} + \frac{3}{s - 4}$$

and

$$\bar{y}_2(s) = \frac{3s - 22}{(s + 1)(s - 4)} = \frac{5}{s + 1} - \frac{2}{s - 4}$$

Term by term inversion of $\bar{y}_1(s)$ and $\bar{y}_2(s)$ yields

$$y_1(t) = 5e^{-t} + 3e^{4t}$$
$$y_2(t) = 5e^{-t} + 2e^{4t}$$

For a more comprehensive treatment of Laplace transforms where other forms of inversion are treated, the reader is referred to *Laplace Transforms*, by M. R. SPIEGEL, Schaum Publishing Co, New York, 1965.

INDEX

Allocation matrix, 507
Average amount of information, 456–57

Bernoulli distribution, 15
Bernoulli process:
  independent, 72–92
    generalization of, 84–85
    limit form of, 86–92
  Markov-dependent, 95–104
Binomial distribution, 15–16, 18–19
Birth and death processes, 125–29
Birth processes, as continuous-parameter Markov
    chain, 114–20, 125–29

Central limit theorem, 59–60
Chapman-Kolmogorov equations, 106–8
Coding problems, simple, 469–75
Collinear destinations, case of, 495–500
Complementary cumulative function, 11–13
Complete enumeration, method of, 513–15
Compound distributions, 36–39
Conditional probability, 5
  density functions, 49
  mass functions, 20–21
Continuous distribution, 39
Continuous-parameter processes:
  defined, 68
  independent, 86–92
  Markov chains, 111–29
  queuing theory, example from, 68–70

Continuous random variables, 39–59
  expectation, 40–41, 48–49
  gamma distribution, 45–46, 150–51
  jointly distributed, 48–55
  Laplace transforms, 41–43, 55–59
  moment generating functions, 41–43
  negative exponential distribution, 44–45
  normal or Gaussian distribution, 46–48
  probability density function, 39–40, 55–59
  rectangular distribution, 43–44
  variance, 40–41
Continuous review inventory systems, 266–67
  deterministic models, 275–302
  stochastic models, 309–26
Continuous review replacement systems, 398–430
  deterministic models, 398–420
  stochastic models, 421–30
Continuously distributed destinations, 502–5
Counting process, 129
Cumulative distribution function, 6, 11–13

Death process, as continuous-parameter Markov chain,
    114, 120–29
Decay, deterministic inventory models with, 290–93
Decision variables:
  in inventory policies, 271–72
  in queuing system, 196–97
Density function, 39–40, 55–59
  failure, 144, 147–52
  joint probability, 48–49
  probability, 55–59

Deterministic inventory models, 275–308
  multicommodity multiinstallation systems, 353–70
    multicommodity problem, 368–70
    two-commodity problem, 353–68
  single-commodity single-installation systems, 275–308
    continuous review system, 275–302
    periodic review system, 302–8
Deterministic replacement models:
  continuous review systems, 398–420
  periodic review systems, 430–32
Difference equations, 97–102, 202–3
Differential equations, Laplace transforms and, 527–29
Differentiation theorem, 524
Discipline, queue, 193–94
Discounting, deterministic inventory models and, 281–84
Discrete parameter processes:
  defined, 68
  independent, 72–85
  Markov chains, 94–111
  queuing theory, example from, 68–70
Discrete random variables, 7–39
  Bernoulli distribution, 15
  binomial distribution, 15–16
  complementary cumulative function, 11–13
  compound distributions, 36–39
  expectation, 7, 8, 19–20
  generating function, 9–11
  geometric distribution, 17–18
  jointly distributed, 19–32
  negative binomial distribution, 18–19
  Poisson distribution, 16–17
  probability generating function, 9–11, 32–39
  probability mass function, 7–8
  uniform distribution, 13–15
  variance, 8–9
Distribution function, 6, 7
Distributions:
  continuous random variables, 39, 43–48
  discrete random variables, 13–19, 36–39
Dynamic programming formulation, deterministic inventory model, 300–302

Economic order quantity (EOQ) formula, 263, 275–80
Endogenous-type failures, stochastic models for, 152–62
Entropy function, 450
Enumeration, method of complete, 513–15
Equipment (*see* Replacement theory)
Equivalence problems, 352–53
Exhaustive events, 5
Exogenous-type failures, 141–42, 154–57
Expectation:
  continuous random variables, 40–41, 48–49
  discrete random variables, 7, 19–20
Expected number, queuing systems, 223–37
Exponential distribution, 44–45, 147
Extremal equations, iterative solution of, 492–94
Extreme value distribution, 149–50

Failures (*see* Reliability theory)
Final value theorem, 524–25
Finite production rate, 288–90
Finite sample space, 4
Fixed cost, 270
Flow-diagram approach, 128–29
  $M/M/r$ queuing system, 217–19

Gamma distribution, 45–46, 150–51
Gaussian distribution, 46–48

Generating function, 9–11, 32–39, 41–43
Geometric distribution, 17–18

Hazard function, 145–52
Hazard rate, 142
Heavy-coin problem, 462–65
Holding cost, 271
Homogeneous Poisson process, 88, 115–18
Horizon period, 272–73

Imbedded Markov chains, 225–28
Independent events, 5–6
Independent processes, 71–92
Independent random variables, 21–22, 32–36, 49–50, 55–58
Independent serial system, 164–65
Independent stochastic processes:
  continuous parameter, 86–92
  definition, 71–72
  discrete parameter, 72–85
  independent Bernoulli process, 72–92
Information theory, 446–79
  coding problems, simple, 469–75
  heavy-coin problem, 462–65
  measure of information, 447–54
    basic assumptions, 448–49
    choice of measure, 450
    concavity, 452–53
    intuitive considerations, 447–48, 449–50
    unit of information, 450–52
  measures of other information quantities, 454–62
    assumption and calculations of probabilities, 454–55
    asymmetry, 456–57
    averages, 456–57
    defining other quantities, 455–56
    implication of independence, 458
    relationships among quantities, 460–62
    sign results and interpretations, 458–60
  odd-coin problem, 465–69
Initial conditions, 71
Initial failure, 142
Input process, queuing systems, 192
Integration theorem, 524
Interarrival times, 90–91, 130
Intermittently operating equipment, 393–94
Inventory theory (*see* Multicommodity multiinstallation systems; Single-commodity single-installation inventory systems)
Inverse Laplace transform, 525–27
Isocost contours, method of, 492

Kendall's queue notation, 195–96

Laplace transforms, 41–43, 55–59, 520–29
  convolution property, 523–24
  defined, 520–22
  differentiation theorem, 524
  final value theorem, 524–25
  integration theorem, 524
  inverse, 525–27
  linearity property, 523
  solving differential equations using, 527–29
Lead time:
  deterministic inventory models and, 280–81
  stochastic inventory models and, 322–26
Life characteristic, 142
Life distribution, 144

Linear programming formulation, single-source location problem, 501–8
Little's formula, 197, 209
Location theory, 480–519
  multisource problem:
    restricted, 513–15
    unrestricted, 505–10
  single-source problem:
    restricted, 510–13
    unrestricted, 482–505
Lognormal distribution, 151–52

Marginal probability:
  density functions, 49
  mass functions, 20–21
Markov processes, 92–129
  Bernoulli process, Markov-dependent, 95–104
  birth and death processes, 114–29
  chains, 93–94, 225–28
  classification, 93–94
  continuous parameter, 111–29
  definition, 92–93
  discrete parameter, 94–111
  transition intensity, 113–14
Matrices, 102–6, 507
Mean time to failure, 144–45
Mechanical analogue solution, single-source location problem, 485–87
$M/G/1$ queuing system, modified, 224–42
Mixed distributions, 58
Mixed-type failures, 162–63
$M/M/\infty$ queuing system, 242–51
$M/G/1$ queuing system, 197–210
$M/M/r$ queuing system, 211–24
Moment generating functions, 41–43
Multicommodity multiinstallation inventory systems, 346–91
  classification of, 348–52
  deterministic models, 353–70
    multicommodity problem, 368–70
    two-commodity problem, 353–68
  equivalence problems, 352–53
  single-period probabilistic models, 370–87
    multicommodity problem, 385–87
    two-commodity problem, 370–85
Multisource location problem:
  restricted, 513–15
  unrestricted, 505–10
Mutually exclusive events, 4

No wear outs, 154–57
Number of renewals, 131–36

Odd-coin problem, 465–69
One-period model, 330–36
One-step transition probability matrix, 102–3
Optimal decision rules, stochastic inventory model, 314–16, 321–22, 339–42
Ordering policies, 350–52, 354–68
Ordering strategies, 371, 374–85

Parallel redundancy, 167–75
Penalty cost, 271
Periodic review inventory systems, 267
  deterministic models, 302–8
  stochastic models, 326–42
Periodic review replacement systems, 430–41
  deterministic models, 430–32
  stochastic models, 432–41

Planning period, 272
Poisson arrivals, 199–200, 202–10
Poisson distribution, 16–17
Poisson process, 87–89, 115–19, 243–44
Policies, inventory, 271–72
Position inventory, 269
Probabilistic inventory models, single-period, 370–87
  multicommodity problem, 385–87
  two-commodity problem, 370–85
Probability density function, 39–40, 55–59
Probability generating function, 9–11, 32–39
Probability mass function, 7–8, 19
Probability theory, 2–65
  central limit theorem, 59–60
  conditional probability, 5
  continuous random variables, 39–59
    expectation, 40–41, 48–49
    gamma distribution, 45–46, 150–51
    jointly distributed, 48–55
    Laplace transforms, 41–43, 55–59
    moment generating functions, 41–43
    negative exponential distribution, 44–45
    normal or Gaussian distribution, 46–48
    probability density function, 39–40, 55–59
    rectangular distribution, 43–44
    variance, 40–41
  discrete random variables, 7–39
    Bernoulli distribution, 15
    binomial distribution, 15–16
    complementary cumulative function, 11–13
    compound distrubutions, 36–39
    expectation, 7, 8, 19–20
    generating function, 9–11
    geometric distribution, 17–18
    jointly distributed, 19–32
    negative binomial distribution, 18–19
    Poisson distribution, 16–17
    probability generating function, 9–11, 32–39
    probability mass function, 7–8
    uniform distribution, 13–15
    variance, 8–9
  independent events, 5–6
  probability, 5–6
  random variables, 6–7
  sample space, 3–6
Procurement cost, 270

Queuing theory, 190–261
  control of single server queuing system, 252–58
  $M/G/1$ system, modified, 224–42
  $M/G/\infty$ system, 242–51
  $M/M/1$ system, 197–210
  $M/M/r$ system, 211–24
  problem, characteristics of, 196–97
  stochastic process example from, 68–70
  systems, characteristics of, 191–96

Random variables, 6–7
(*See also* Continuous random variables; Discrete random variables)
Range of definition, 8
Rectangular distribution, 43–44
Rectilinear distance problem, 501–2
Recurrence relations, 72–84
Redundancy, failure and, 166–78
Reliability function, 144
Reliability theory, 140–89
  failure phenomenon, 141–43
    stochastic processes underlying, 152–63

Reliability theory (*cont.*)
  improving reliability of system, methods for, 165–86
    design considerations, 184–86
    economic considerations, 184–86
    introduction of redundancy, 166–78
    repair or replacement, 178–84
    technological considerations, 166
  serial systems, 163–65
  statistical characteristics of system subject to failure, 143–52
Renewal theory, 129–36
  asymptotic result, 134
  counting process, 129
  definitions, 129–31
  expected number of renewals, 131–34
  number of renewals, 135–36
Repair:
  in redundant systems, 178–84
  in replacement systems, 421–30
Replacement theory, 392–445
  classification of equipment, 393–94
  continuous review replacement system, 398–430
    deterministic models, 398–420
    stochastic models, 421–30
  decision variables, 395
  economic life, 396–97
  input parameters, 394–95
  objective functions, 396
  periodic review replacement systems, 430–41
    deterministic models, 430–32
    stochastic models, 432–41
  replacement models, inventory models and, 397
  replacement policy, 395
  service life, 396–97

Sample space, 3–6
Search techniques, 494
Serial systems, 163–65
Service life, concept of, 396–97
Service mechanism, queuing systems, 192–93
Shortage, deterministic inventory models and, 284–88, 288–90
Single-commodity single-installation inventory systems, 262–345
  analysis, 266–71
  decision variables, 271–72
  deterministic models, 275–308
    continuous review system, 275–302
    periodic review system, 302–8
  examples, 264–65
  historical background, 263–65
  inventory policies, 271–72
  motives for carrying inventory, 265–66
  objective functions, 272–75
  stochastic models, 308–42
    continuous review system, 309–26
    periodic review system, 326–42
Single-period probabilistic inventory models:
  multicommodity problem, 385–87
  two-commodity problem, 370–85
Single server queuing system, control of, 252–58
Single-source location problem:
  restricted, 510–13
  unrestricted, 482–505
Standard deviation, 9, 41
Standby redundancy, 167, 175–78
Stationary transition probability function, 105
Steady-state component, 99

Steady-state distribution, 203–4, 206–9, 211–19, 221–24, 237–42, 246–51
Steady-state period review model, 336–42
Steady-state probabilities, 113
Steady-State Probability Law, 109–11
Stochastic inventory models, 308–42
  continuous review system, 309–26
  periodic review system, 326–42
Stochastic processes, 66–139
  classification, 68
  definition, 67–68
  independent processes, 71–92
    continuous parameter, 86–92
    definition, 71–72
    discrete parameter, 72–85
    independent Bernoulli process, 72–92
  Markov processes, 92–129
    Bernoulli process, Markov-dependent, 95–104
    birth and death processes, 114–29
    classification, 93–94
    continuous parameter, 111–29
    definition, 92–93
    discrete parameter, 94–111
    transition intensity, 113–14
  probability law of, 70–71
  queuing theory, example from, 68–70
  renewal theory, 129–36
    asymptotic result, 134
    counting process, 129
    definitions, 129–31
    expected number of renewals, 131–34
    number of renewals, 135–36
  underlying the failure phenomenon, 152–53
Stochastic replacement models, continuous review systems, 421–30
Stress generating environment, 152–53

Transient component, 99
Transition intensity, 113–14
Transition probability:
  distributions, 93, 94
  functions, 105
  matrices, 103–4, 105–6
Two-commodity problem:
  deterministic inventory models, 353–68
  single-period probabilistic inventory models, 370–85
Two-echelon single commodity systems, 365–68
Two-parameter Weibull distribution, 147–49

Uniform distribution, 13–15

Variable cost, 270–71
Variance:
  continuous random variables, 40–41
  discrete random variables, 7–8

Wear-out failure, 142
Wear-out process, 152–53, 158–62
Weibull distribution, two-parameter, 147–49
Wilson formula, 263, 275–80, 354–58

Yule-Furry birth process, 119–20